Student Solutions Manual

for

Dwyer and Gruenwald's

Precalculus

Ignacio Alarcón
Santa Barbara City College

Australia • Canada • Mexico • Singapore • Spain • United Kingdom • United States

Printed in the United States of America
1 2 3 4 5 6 7 08 07 06 05 04

Printer: Thomson/West

ISBN: 0-534-35289-8

For more information about our products, contact us at:
Thomson Learning Academic Resource Center
1-800-423-0563

For permission to use material from this text or product, submit a request online at
http://www.thomsonrights.com.
Any additional questions about permissions can be submitted by email to **thomsonrights@thomson.com.**

Thomson Brooks/Cole
10 Davis Drive
Belmont, CA 94002-3098
USA

Asia
Thomson Learning
5 Shenton Way #01-01
UIC Building
Singapore 068808

Australia/New Zealand
Thomson Learning
102 Dodds Street
Southbank, Victoria 3006
Australia

Canada
Nelson
1120 Birchmount Road
Toronto, Ontario M1K 5G4
Canada

Europe/Middle East/South Africa
Thomson Learning
High Holborn House
50/51 Bedford Row
London WC1R 4LR
United Kingdom

Latin America
Thomson Learning
Seneca, 53
Colonia Polanco
11560 Mexico D.F.
Mexico

Spain/Portugal
Paraninfo
Calle/Magallanes, 25
28015 Madrid, Spain

Contents

Chapter 5 Trigonometric Functions

Chapter 6 Trigonometric Identities and Equations

Chapter 7 Applications of Trigonometry

Chapter 8 Relations and Conic Sections

Chapter 9 Systems of Equations and Inequalities

Chapter 10 Integer Functions and Probability

For Deanna and George Gregg,

and for Susana, María Elisa, Mariana,

Ana Elena, and Rocío.

Preface

This manual consists of the solutions to the odd numbered exercises in the First Edition of "*Precalculus*", by David Dwyer and Mark Gruenwald. The Chapter Reviews and Chapter Tests Exercises are included as well.

In the preparation of this supplement, I have used *Microsoft Word 2000*, *Math Type V*, and *Derive 5 for Windows.*

I deeply thank Professors Dwyer and Gruenwald for the privilege of working on this supplement to their excellent book. One of the major strengths of their text is its wealth of exercises, both in number and variety.

My sincere thanks to my colleague at Santa Barbara City College, Jason Miner, for agreeing to be my accuracy checker. This has surely been an onerous task.

I would like to acknowledge the staff at Thomson-Brooks/Cole for their continued support and encouragement, in particular Paul Durantini, Jennifer Huber, John-Paul Ramin, Lisa Chow, and Rachel Sturgeon. I truly enjoy working with such a dedicated and energetic group of professionals.

Ignacio Alarcón
Santa Barbara City College
alarcon@sbcc.edu

December, 2003

Chapter 1. FOUNDATIONS AND FUNDAMENTALS

Section 1.1 The Real Number System

1. $\frac{6}{3}$ is natural, whole, integer, rational, real, and complex.

3. $\sqrt{7}$ is irrational, real, and complex.

5. $3-i$ is complex.

7. 0.8888888… is rational, real, and complex.

9. $\sqrt{-4}$ is equal to $2i$, so it is complex.

11. $-\sqrt{25}$ is equal to -5, so it is integer, rational, real, and complex.

13. The height in miles of a 1 foot-tall rooster is $\frac{1}{5280}$, which is rational, real, and complex.

15. $3>0$

17. $-\frac{6}{5}<-\frac{4}{5}$ **18.** $\frac{5}{6}>\frac{2}{3}$

In Exercises 19-23, each list of numbers is ordered from smallest to largest.

19. $-23, -22.9, 22.9, 23$

21. $3+\frac{1}{10}+\frac{4}{100}, \pi, \frac{22}{7}, 3.15$

23. $\{x | -2<x\leq 4\}$ corresponds to the interval $(-2,\ 4]$

25. $\{x | -7<x<-\frac{1}{2}\}$ corresponds to the interval $(-7,\ -\frac{1}{2})$

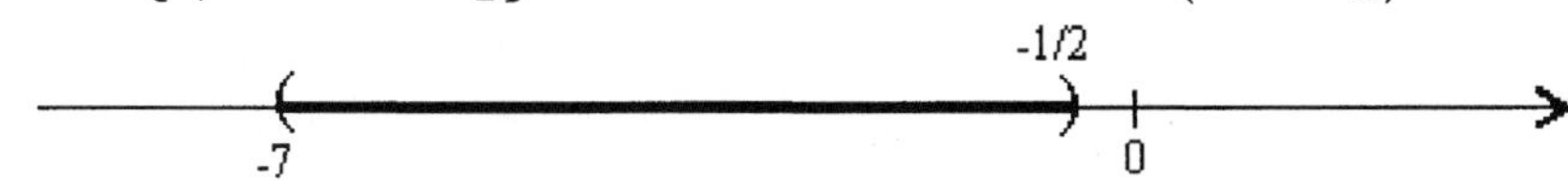

27. $\{x | x<-3\}$ corresponds to the interval $(-\infty,\ -3)$

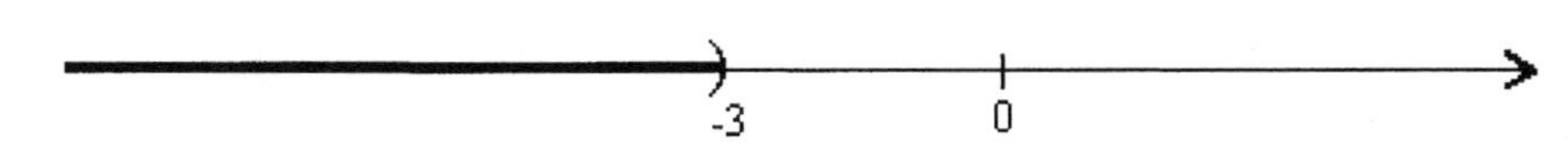

29. $\{x | x\geq\sqrt{2}\}$ corresponds to the interval $[\sqrt{2},\ \infty)$

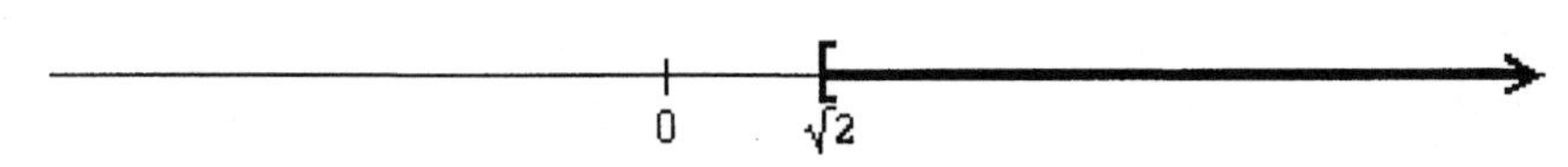

31. (3, 5) corresponds to $\{x \mid 3<x<5\}$

33. $(0,\ \infty)$ corresponds to $\{x \mid x>0\}$

35. $(100, \infty)$ corresponds to $\{x \mid x > 100\}$

37. $5 - 3\times 4 = -7$

39. $1.2 - 2.5 - 3.4 - 1.8 = -6.5$

41. $4\times[3-(4-5)] = 16$

43. $3\times 4 - 2 + 5\times 2 - 1 = 19$

45. $|-6| = 6$

47. $|3-1| = 2$

49. $\dfrac{|-3.7|}{-3.7} = -1$

51. $|x-1| = x-1$ if $x > 1$

53. $|y+2| = -y-2$ if $y < -2$

55. An expression for the distance between -3 and 7 is $|-3-7|$, and this distance is equal to 10.

57. An expression for the distance between $3x$ and 2 is $|3x-2|$.

59. An expression for the distance between $x+\frac{2}{3}$ and $-\frac{2}{3}$ is $\left|\left(x+\frac{2}{3}\right)-\left(-\frac{2}{3}\right)\right|$, or $\left|x+\frac{4}{3}\right|$.

In Exercises 61 – 70, answers may vary depending on the assumptions underlying the required estimates. Following are some specific estimates, and the underlying assumptions.

61. The estimate of the amount of money that a 3-pack-a-day smoker will spend in 30 years is $(3\times 365\times 30)p = 32850p$, where p is the price per pack. So, if p = \$3, the estimate of cigarette expenses for this individual is \$98,550. (*Note*: this is a gross underestimate, since it assumes no change in the price of a pack of cigarettes.)

63. The estimated number of families that Bill Gates's \$59.3 billion dollars could feed for one year, assuming that it takes \$30 a day to feed a family of four, or \$10,950 per year, is $\frac{59{,}300{,}000{,}000}{10{,}950} \approx 5{,}415{,}525$ or about 5.4 million families.

65. To estimate the side length of a cube that can contain \$1 billion in \$100 bills, we need the sizes of currency notes. According to the U.S. Department of the Treasury website http://www.ustreas.gov/education/faq/currency/, the dimensions of U.S. currency notes are: $2.6in\times 6.14in\times 0.0043in$. The number of \$100 bills that amount to \$1 billion is 10,000,000. If layers of 34×14 bills are laid out, with a separation of 0.1 inches between bills, we will have a base that measures $(34\times 2.6)+35\times 0.1 = 88.4+3.5 = 91.9$ *in* on one side, and $(14\times 6.14)+15\times 0.1 = 85.96+1.5 = 87.46$ *in* on the other. For a square base we allow for side length of 91.9 *in*. This allows 34×14=476 piles of bills. Each pile will hold $\frac{91.9}{0.0043}$ bills, or approximately 21,300 bills. This cube can contain $21{,}300\times 476 = 10{,}138{,}800$ bills, a tad over the ten million \$100 bills = \$1 billion. A cube of side length 91.9 *in* ≈ 7.7 *ft* will suffice.

67. Assuming an average number of 300 pages per book, with 35 lines per page, and 12 words per line, we have 126,000 words per book, or $725\times 126{,}000$ =91,350,000 words for the entire production of 725 books. To average per day over 75 years: $\frac{91{,}350{,}000}{75\times 365} \approx 3{,}340$ words/day.

69. **(a)** The discounted total bill for a \$3.50 and a \$2.80 meal is \$6.30 minus 10% of this, or \$6.30 - \$0.63 = \$5.67.

(b) The discounted bill for the \$3.50 meal is: \$3.50 – 10% of \$3.50, or \$3.15, and the discounted bill for the \$2.80 meal is: \$2.80 - \$0.28 = \$2.52. So the total for both checks is \$3.15 + \$2.52 = \$5.67.

(c) The discount of 10% is applied as: 10% of (\$3.50 + \$2.80) = 0.1(3.50 + 2.80) = 0.1(3.50) + 0.1(2.80) (distributive law). Of course, both schemes yield the same result.

Concepts and Critical Thinking

71. The statement: "All rational numbers have a finite number of decimal places" is false. (Example: 1/3 has a decimal representation with an infinite number of decimal places.)

73. The statement: "0 is the multiplicative identity" is false. 0 is actually the additive identity.

75. All negative integers, and 0, are examples of integers that are not natural.

77. The commutative property holds for addition and multiplication, but not for subtraction and division. For example: $5-4 \neq 4-5$, and $10 \div 5 \neq 5 \div 10$.

79. **(a)** Division is not commutative: $4 \div 2 \neq 2 \div 4$

(b) Multiplication is commutative

(c) Marriage is commutative

(d) Love is not commutative. Person *A* may love Person *B*, and Person *B* not love Person *A*.

(e) Taking the maximum of two numbers is a commutative operation.

(f) Addressing is not commutative. "Jamal speaks to Ann" does not imply "Ann speaks to Jamal."

(g) Implication is not commutative. "If John lives in San Francisco, then John lives in California" is a true statement. However, the statement "If John lives in California, then John lives in San Francisco" does not hold.

81. Rotations and flips commute with one another.

83. $1{,}000{,}003 \times 90 = (1{,}000{,}000 + 3) \times 90$, or 90 million plus 270 = 90,000,270.

85. **(a)** $3 \cdot 5 + 6 \div 2 = 15 + 6 \div 2 = 21 \div 2 = 10.5$, if we just did operations from left to right.

(b) $3 \cdot 5 + 6 \div 2 = 3.5 + 3 = 3 \cdot 8 = 24$, if we did operations from right to left.

(c) $3 \cdot 5 + 6 \div 2 = 3 \cdot 11 \div 2 = 3(5.5) = 16.5$, in the order: addition, division, multiplication.

(d) $3 \cdot 5 + 6 \div 2 = 3 \cdot 11 \div 2 = 33 \div 2 = 66$, in the order: addition, multiplication, division.

(e) $3 \cdot 5 + 6 \div 2 = 15 + 6 \div 2 = 21 \div 2 = 11.5$, in the order: multiplication, addition, division.

(f) $3 \cdot 5 + 6 \div 2 = 15 + 6 \div 2 = 15 \div 3 = 18$, in the order: multiplication, division, addition.

(g) $3 \cdot 5 + 6 \div 2 = 3.5 + 3 = 3 \cdot 8 = 24$, in the order: division, addition, multiplication.

(h) $3 \cdot 5 + 6 \div 2 = 3.5 + 3 = 15 + 3 = 18$, in the order: division, multiplication, addition.

(i) $3 \cdot 5 + 6 \div 2 = 15 + 6 \div 2 = 15 + 3 = 18$, using the correct order of operations.

There are six possible results of the computation $3 \cdot 5 + 6 \div 2$: 10.5, 11.5, 16.5, 18, 24, and 66 if we don't have any pre-assigned order of operations.

Chapter 1. Section 1.2 Exponents and Radicals

1. $\left(\frac{2}{5}\right)^3=\frac{8}{125}$

3. $\left(-\frac{1}{10}\right)^3=-\frac{1}{1000}$

5. $4^{-3}=\frac{1}{64}$

7. $(-1)^{-4}=1$

9. $-3^{-2}=-\frac{1}{9}$

11. $\left(4^{-2}\right)^3=\frac{1}{4096}$

13. $a^3a^6=a^9$

15. $\frac{r^5}{r^7}=\frac{1}{r^2}$

17. $\left(a^3b^{-2}\right)\left(a^{-2}b^{-4}\right)=\frac{a}{b^6}$

19. $\frac{p^{10}q^6}{p^4q^2}=p^6q^4$

21. $\frac{3^{-1}x^4y^{-2}}{3^{-2}x^{-2}y}=\frac{3x^6}{y^3}$

23. $\left(y^{-1}z^2\right)\left(yz^{-1}\right)^{-1}=\frac{z^3}{y^2}$

25. $i^{11}=-i$

27. $i^{701}i^3=1$

29. $3.04\times10^{-3}=0.00304$

31. $2.7\times10^{-1}=0.27$

33. $4{,}000{,}000=4\times10^6$

35. $0.0943=9.43\times10^{-2}$

37. $\left(2.5\times10^{-45}\right)\left(4\times10^{52}\right)=10^8$

39. $13{,}176{,}000{,}000{,}000{,}000{,}000{,}000{,}000=1.3176\times10^{25}$

41. $\sqrt[3]{27}=3$

43. $\sqrt[4]{\frac{1}{16}}=\frac{1}{2}$

45. $\sqrt{18}=3\sqrt{2}$

47. $\sqrt{5^8}=625$

49. $\sqrt{x^3y^7}=xy^3\sqrt{xy}$

51. $\sqrt{50a^{30}b^{20}c^{11}}=5a^{15}b^{10}c^5\sqrt{2c}$

53. $\sqrt{x^7(y+z)^4}=x^3(y+z)^2\sqrt{x}$

55. $\sqrt[3]{54s^6t^7}=3s^2t^2\sqrt[3]{2t}$

57. $\frac{3}{\sqrt{7}} = \frac{3\sqrt{7}}{7}$

59. $\sqrt[3]{\sqrt{x}} = \sqrt[6]{x}$

61. $\sqrt{-25} = 5i$

63. $\sqrt{3}\sqrt{15} = 3\sqrt{5}$

65. $\sqrt[3]{2xyz} \cdot \sqrt[3]{20x^2y^4z} = 2xy\sqrt[3]{5y^2z^2}$

67. $\frac{\sqrt{10}}{\sqrt{2}} = \sqrt{5}$

69. $\sqrt{6} + \sqrt{24} = 3\sqrt{6}$

71. $\sqrt[3]{x} = x^{1/3}$

73. $\frac{1}{\sqrt{x}} = x^{-1/2}$

75. $2^{3/2} = 2\sqrt{2}$

77. $b^{2/3} = \sqrt[3]{b^2}$

79. $16^{-1/2} = \frac{1}{4}$

81. $32^{2/5} = 4$

83. $x^{1/3}x^{2/3} = x$

85. $\left(81a^4b^{12}\right)^{1/4} = 3ab^3$

87. $\frac{z^2}{z^{3/2}} = z^{1/2}$

Applications

89. Receiving \$300,000 a day for 31 days adds up to \$9,300,000. On the other hand, receiving 1¢ the first day, 2¢ the second, 4¢ the third, and so on doubling the pay each successive day of the adds up to $S = .01 + 2(.01) + 2^2(.01) + 2^3(.01) + \ldots + 2^{30}(.01)$. Here, just the pay for the last day of May is $2^{30}(.01) = \$10,737,418.24$, by itself bigger than the total pay under the previous scheme.

91. **(a)** After the first bounce, the ball bounces to $\frac{2}{3}(1454) = 969\frac{1}{3}$ *ft*.

(b) After two bounces, the ball bounces to $\left(\frac{2}{3}\right)^2(1454) = 646\frac{2}{9}$ *ft*.

(c) After n bounces, the ball bounces to $\left(\frac{2}{3}\right)^n(1454)$ *ft*.

(d) After 24 bounces, the ball bounces to $\left(\frac{2}{3}\right)^{24}(1454) \approx .08637$ *ft*, or approximately 1.03 *in*.

93. At the constant rate of inflation of 4% per year, the estimated cost of a CD that sells for \$18 in 2003 will sell for $18(1.04)^{30} \approx \$58.38$. This assumes that the 4% rate applies to CDs.

95. If gravity is the only force present, the given formula for the velocity gives $v = 8\sqrt{33{,}000}$ $\approx 1460.5\ ft/\sec$. The fact that the individual fell in a portion of the fuselage implies that there was the initial force of the explosion acting on this individual, hence the much higher velocity at the time of impact with the ground.

97. If $v_0 = \sqrt{64h}$, and $h = 4t^2$, then: **(a)** As a function of *t*, $v_0 = 16t$ **(b)** If the hang time *t* is 1.036 seconds, the initial velocity is $v_0 \approx 16.576\ ft/\sec$ **(c)** If the hang time *t* is 1.036 seconds, the height of the jump is about 4.29 *ft*. This, with the height that Wilson would initially reach with his arms stretched upwards, would add up to $\approx 12\ ft$.

99. According to the time dilation formula,
$T = \frac{T_0}{\sqrt{1-v^2}} = \frac{50 \text{ years}}{\sqrt{1-\left(3.28554\times10^{-6}\right)^2}} \approx 50.0000000003 \text{ years}$, so the traveler will be younger 0.0000000003 years or .00946 seconds, almost 1/100 of one second.

Concepts and Critical Thinking

101. The statement: "For all real numbers *x* and integers *a* and *b*, $x^a x^b = x^{a+b}$" is true.

103. The statement: "11.72×10^{13} is not in scientific notation" is true. The corresponding scientific notation version for this number is actually 1.172×10^{14}.

105. The statement "$(\pm6)^2 = 36$" is true.

107. Answers may vary. An example of a number greater than 100 in scientific notation is 1.5×10^6.

109. An example of a number *x* for which $\sqrt{x^2} \neq x$ is $x = -3$, since $\sqrt{(-3)^2} = 3$. Any negative number *x* is suitable as an example in which $\sqrt{x^2} = -x$

111. "$\sqrt{25} = \pm5$" is not a correct statement.
The symbol $\sqrt{25}$ represents only the principal square root of 25, which is 5.

Chapter 1. Section 1.3 Algebraic Expressions

1. The expression $2a^2-5a+6$ is a polynomial, more specifically a trinomial of degree 2.

3. The expression $\frac{7}{2}k^4$ is a polynomial, more specifically a monomial of degree 4.

5. The expression $2u^5-u^{1/3}+4u$ is not a polynomial.

7. The expression $2x^3y^2+x^4-y^4$ is a polynomial, specifically a trinomial of degree 5.

9. $\left(3x^2+5x-9\right)+\left(x^2-2x+1\right)=4x^2+3x-8$

11. $x\left(\frac{1}{2}x-1\right)+3x\left(x+\frac{5}{3}\right)=\frac{7}{2}x^2+4x$

13. $(2a+1)(3a-2)=6a^2-a-2$

15. $\left(k-\frac{2}{3}\right)\left(k+\frac{2}{3}\right)=k^2-\frac{4}{9}$

17. $\left(2x^2+1\right)\left(2x^2-1\right)=4x^4-1$

19. $(3a+2b)^2=9a^2+12ab+4b^2$

21. $(2x+1)\left(x^2-6x+2\right)=2x^3-11x^2-2x+2$

23. $(x-y+2)(x+y-2)=x^2-y^2+4y-4$

25. $(a-b)^3=a^3-3a^2b+3ab^2-b^3$

27. $w(w+2)+5(w+2)=(w+5)(w+2)$

29. $2t^2-8=2(t-2)(t+2)$

31. $6n^2-19n+10=(3n-2)(2n-5)$

33. $z^4-2z^2+1=(z^2-1)^2$

35. $x^2+x-42=(x+7)(x-6)$

37. $16a^2+20a+6=2(2a+1)(4a+3)$

39. $t^2-7t+12=(t-4)(t-3)$

41. $4x^2-12xy+9y^2=(2x-3y)^2$

43. $6x^3y^3+48=6(xy+2)(x^2y^2-2xy+4)$

In Exercises 45-72, the equality holds only for values of the intervening variables in the domain of the expressions involved.

45. $\dfrac{4t^6-10t^4}{2t^2}=2t^4-5t^2$

47. $\dfrac{x^2-4}{x+2}=x-2$

49. $\dfrac{r^2+10r+25}{2r^2+4r-30}=\dfrac{r+5}{2r-6}$

51. $\dfrac{y^2-3y}{y-3}=y$

53. $\dfrac{15x^2}{16y^3}\cdot\dfrac{12y^3}{20x^3}=\dfrac{9}{16x}$

55. $\dfrac{1}{a+1}+\dfrac{1}{a-1}=\dfrac{2a}{a^2-1}$

57. $\dfrac{2q+1}{q^2+q}-\dfrac{q+2}{q+1}=\dfrac{1-q}{q}$

59. $\dfrac{x+3}{x+4}\div\dfrac{x^3+27}{x^2-16}=\dfrac{x-4}{x^2-3x+9}$

61. $\left(2a^2-8b^2\right)\cdot\dfrac{b-a}{a^2-4ab+4b^2}=\dfrac{4b^2-2ab-2a^2}{a-2b}$

63. $\dfrac{4x+20}{x^2+8x+16}\cdot\dfrac{x^2+4x}{x^2+7x+10}=\dfrac{4x}{x^2+6x+8}$

65. $\dfrac{t+2}{t^3+t^2}-\dfrac{1}{t+1}=\dfrac{2-t}{t^2}$

67. $\dfrac{x+5}{x^3+125}-\dfrac{1}{x+5}+\dfrac{x-5}{x^2-25}=\dfrac{1}{x^2-5x+25}$

69. $\dfrac{\frac{1}{x}+\frac{1}{y}}{\frac{1}{xy}}=x+y$

71. $\dfrac{\frac{1}{x}-\frac{1}{c}}{x-c}=-\dfrac{1}{xc}$

73. $\dfrac{2}{3-\sqrt{5}}=\dfrac{3+\sqrt{5}}{2}$

75. $(6+2i)-(3+4i)=3-2i$

77. $(2+3i)(3+4i)=-6+17i$

79. $\left(\sqrt{2}-4i\right)\left(\sqrt{2}+4i\right)=18$

81. $\dfrac{5}{2+i}=2-i$

83. $\dfrac{-18+13i}{5-2i}\cdot\dfrac{5+2i}{5+2i}=-4+i$

84. $\dfrac{3-2i}{5i}=-\dfrac{2}{5}-\dfrac{3}{5}i$

Applications

85. $d=\dfrac{\pi}{4}CB^2S\approx 455.03\ in^3$

87. If $c=60\ in$, $0.0015c^3\approx 324$ board feet. If the diameter $d=60\ in$, the circumference of the pine is $c=\pi d\approx 188.50\ in$, and $0.0015c^3\approx 10{,}046.74$ board feet.

89. If $t=1$, $h(1)=64\ ft$. If $t=2$, $h(2)=96\ ft$. The ball hits the ground when $t=5$ seconds.

91. The number of calories expended by the triathlete is, approximately, 7,680 calories.

93. If $t=90^\circ$, $r=60\%$, then $T_A\approx 99.65^\circ$. If $t=100^\circ$, $r=80\%$, then $T_A\approx 158.12^\circ$.

95. **(a)**

	Average time between arrivals (c)	Average time for processing order (s)	Average waiting time approximated by $1/(1/s-1/c)$
(i)	10	5	10
(ii)	10	6	15
(iii)	10	7	$23\frac{1}{3}$
(iv)	10	8	40
(v)	10	9	90

As the average time between arrivals and average processing time get close, the average waiting time increases, as we would expect.

(b) If the average time between arrivals is less than the average time for processing an order, the given formula for the average waiting time gives a negative answer, which doesn't have an adequate interpretation. The model ought to be used only if $c>s$.

Concepts and Critical Thinking

97. The statement: "The degree of the product of two polynomials is the sum of the degrees of the polynomials" is true.

99. The statement: "$(a+b)^2 = a^2 + b^2$" is false. For example, $(3+1)^2 = 16$, while $3^2 + 1^2 = 10$.

101. Answers may vary. An example: $5x^4 - 2$ is a fourth-degree binomial.

103. Answers may vary. An example: $x^2 - 9$ is a polynomial of degree 2 for which $x-3$ is a factor, since $x^2 - 9 = (x-3)(x+3)$.

105. The degree of the sum or difference of two polynomials is smaller than or equal to the largest of the degrees of each polynomial. The product of two polynomials is a polynomial. The degree of the product of two polynomials is equal to the sum of the degrees of the polynomials that are being multiplied.

107. If *x* The sum of two perfect cubes is never a prime number, since $a^3 + b^3 = (a+b)(a^2 - ab + b^2)$.

Chapter 1. Section 1.4 Graphs of Equations

1. The midpoint of the line segment that connects (1, 2) and (5, 4) is the point (3, 3). The distance between the endpoints is $2\sqrt{5}$.

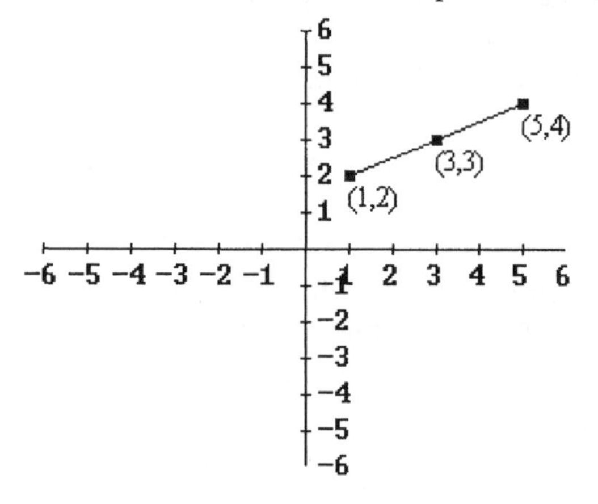

3. The midpoint of the line segment that connects $(-\frac{4}{3},2)$ and $(\frac{8}{3},1)$ is $(\frac{2}{3},\frac{3}{2})$. The distance between endpoints is $\sqrt{17}$.

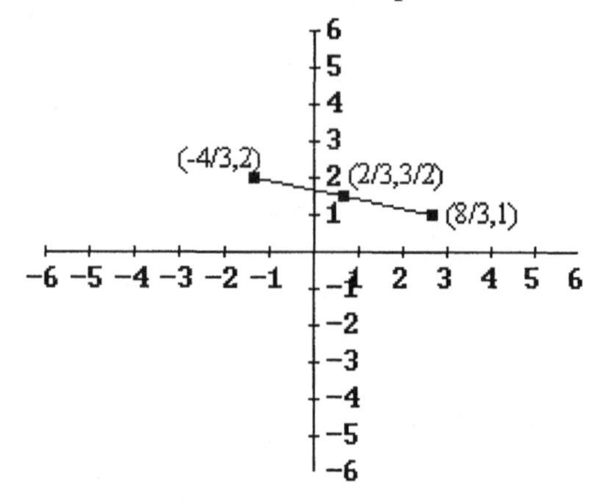

5. The opposite sides of this quadrilateral have equal lengths, so it is a parallelogram.

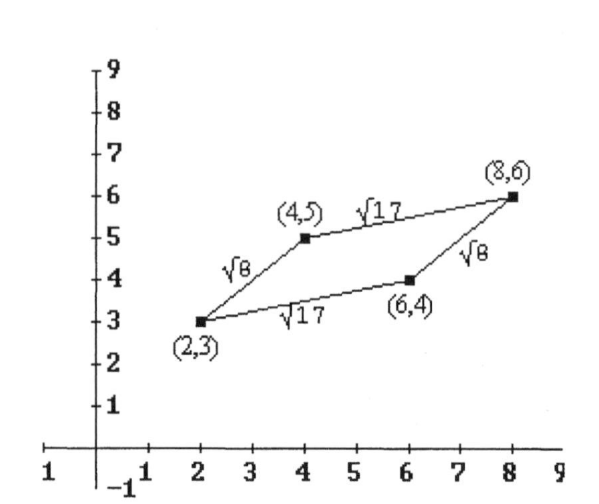

7. The sides of the triangle formed by the points (1,0), (2,3) and (4,−1) satisfy the Pythagorean Theorem. Therefore, this is a right triangle.

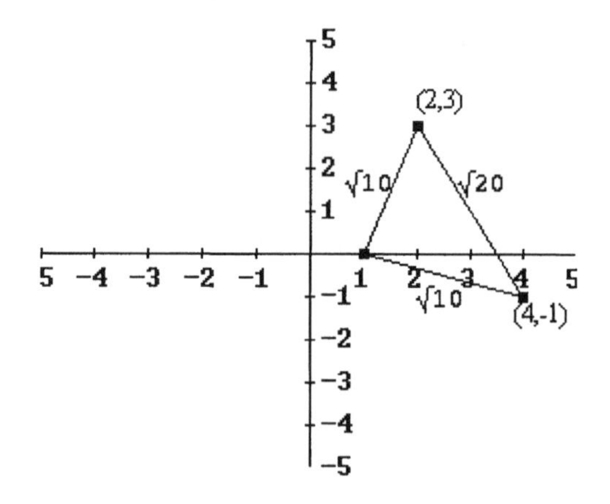

9. $8^2+6^2=100,\ 0^2+(-10)^2=100,$ and $\left(5\sqrt{2}\right)^2+\left(5\sqrt{2}\right)^2=100$. Then, all three points $(8,6)$, $(0,-10)$, and $\left(5\sqrt{2},5\sqrt{2}\right)$ are on the graph of the equation $x^2+y^2=100$.

11. $2=\dfrac{-2}{(-2)+1}$, $4=\dfrac{-\frac{4}{3}}{(-\frac{4}{3})+1}$, and $2-\sqrt{2}=\dfrac{\sqrt{2}}{\sqrt{2}+1}$ are all true. Therefore, all three points $(-2,2),(-\frac{4}{3},4)$, and $\left(\sqrt{2},2-\sqrt{2}\right)$ are on the graph of the equation $y=\frac{x}{x+1}$.

13. Figure A below shows the portion of the graph of $y=1000(x^3+1)$ corresponding to the viewing window Xmin$=-0.3$, Xmax=0.5, Ymin$=-300$, Ymax=1700. The points $(-0.1,999),(0,1000),(0.1,1001)$, $(0.2,1008)$, and $(0.3,1027)$ indicated on it.

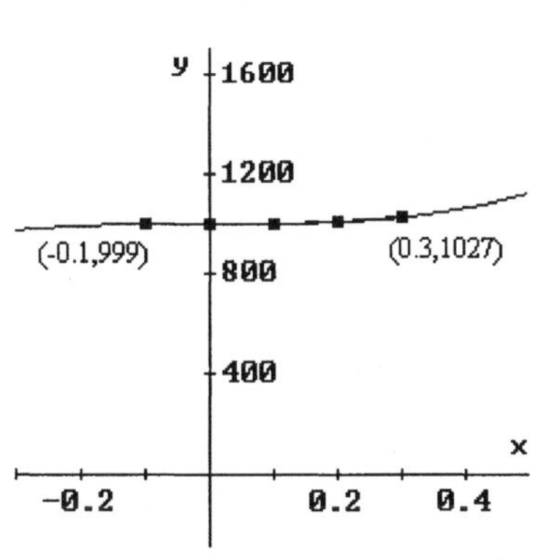

Figure A

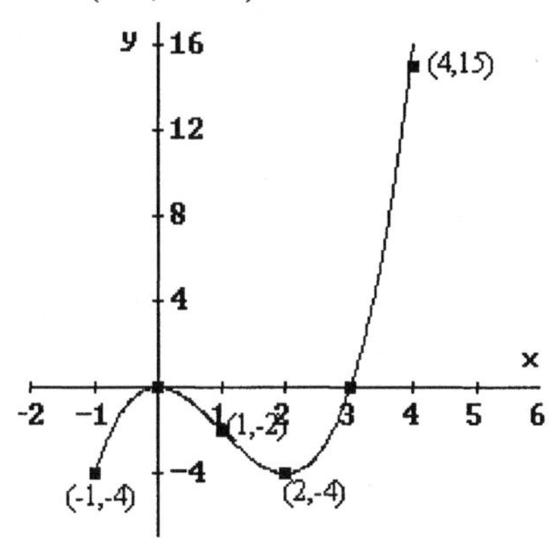

Figure B

15. The graph of $y=2x-4$ is a straight line.

x	y
−1	−6
0	−4
1	−2
2	0

17. The following graph corresponds to the equation $y=-x^2+9$, in the viewing window Xmin$=-10$, Xmax=10, Ymin$=-10$, Ymax=10.

x	y
−4	−7
−3	0
−2	5
−1	8
0	9
1	8
2	5
3	0
4	−7

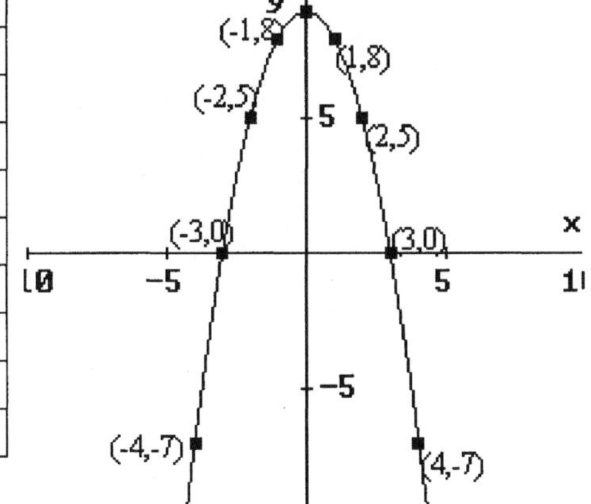

19. The graph of $y = \sqrt{x-2}$, in the viewing window Xmin=−1, Xmax=7, Ymin=−1, Ymax=7:

x	y
2	0
3	1
6	2

21. The circle of equation $x^2 + y^2 = 25$ has its center at (0,0) and its radius is 5.

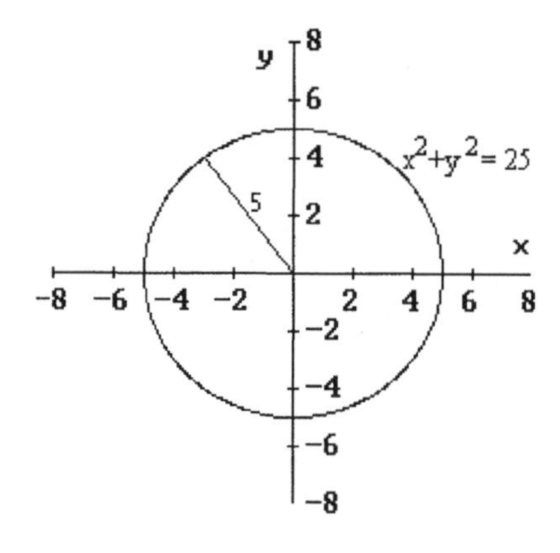

23. The circle of equation $2x^2 + 2y^2 = 16$ has its center at (0, 0) and its radius is $2\sqrt{2}$.

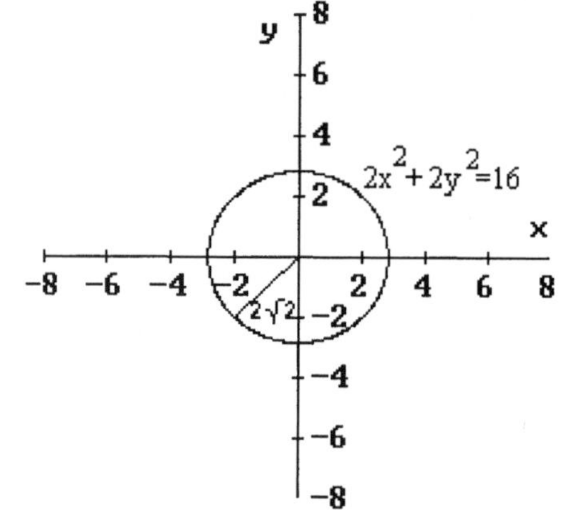

25. The equation $(x+4)^2 + (y-1)^2 = 4$ corresponds to a circle of radius 2 and center at $(-4,1)$.

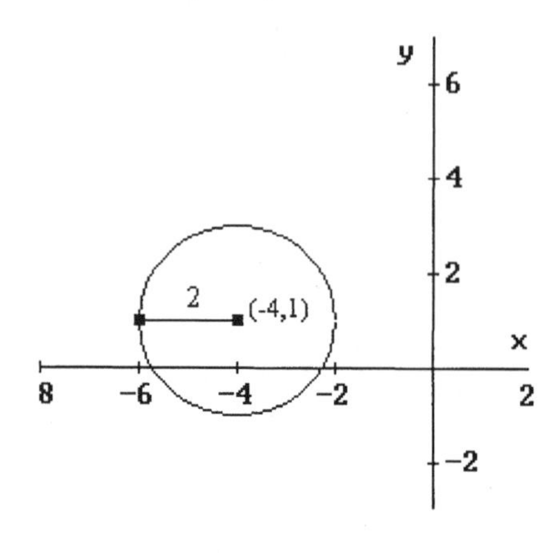

27. The equation $x^2 - 4x + y^2 + 6y + 13 = 4$ corresponds to a circle of radius 2 and center at $(2,-3)$.

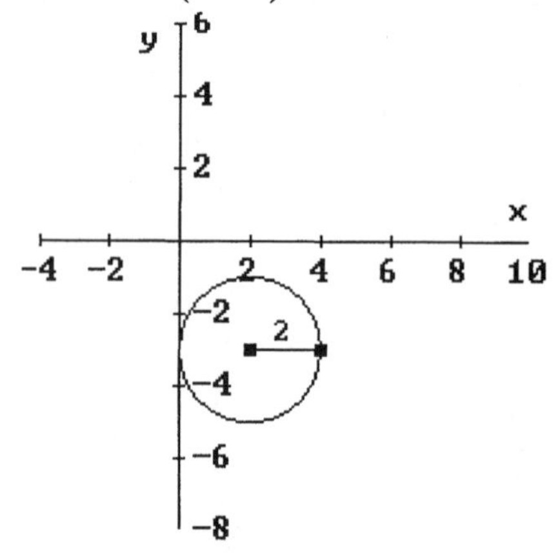

29. The equation $4x^2+12x+4y^2+8y=3$ is equivalent to $\left(x+\frac{3}{2}\right)^2+(y+1)^2=4$, so its graph is a circle of radius 2, and center at $\left(-\frac{3}{2},-1\right)$.

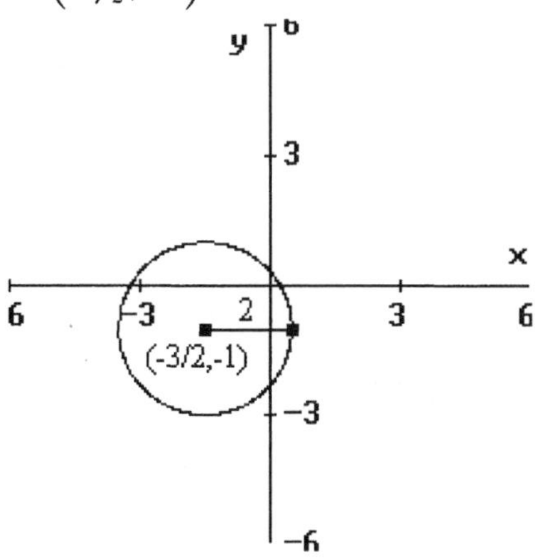

31. The equation of the circle with center $(-1,3)$ and radius 4, in standard form, is: $(x+1)^2+(y-3)^2=16$. In general form, the equation is: $x^2+y^2+2x-6y-6=0$.

33. The circle with center $(4,-3)$ that passes through the origin has radius 5 (the distance from $(4,-3)$ to the origin.) The equation of this circle is: $(x-4)^2+(y+3)^2=25$.

35. The circle that has as one of its diameters the segment with endpoints $(-4,2)$ and (2, 8) has center $(-1,5)$, the midpoint of this segment. The radius is $3\sqrt{2}$, half the length of the diameter. The equation of this circle is: $(x+1)^2+(y-5)^2=18$.

Exercises 37-42 match equations with their graphs.

37. $y=\dfrac{x^4}{108000}-\dfrac{x^3}{1800}+x$

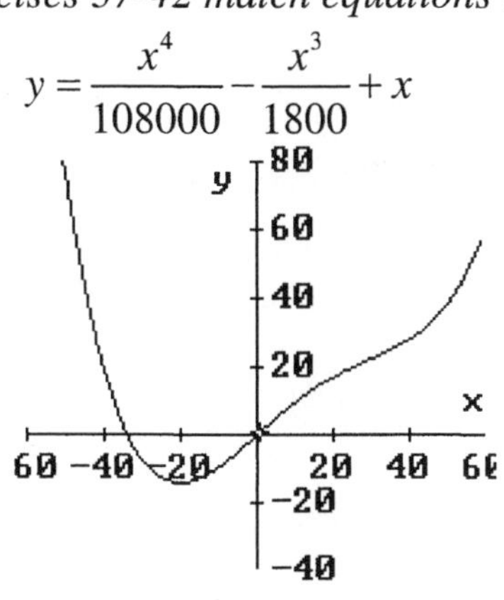

39. $y=\dfrac{x^2}{3}$

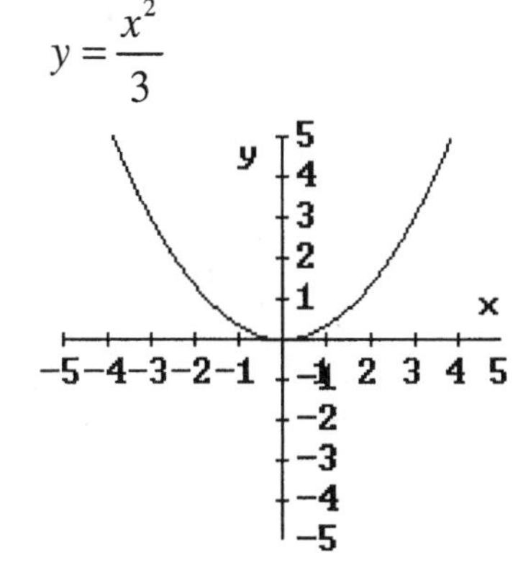

41. $y = \frac{9x^2}{32} - \frac{x^4}{1600}$

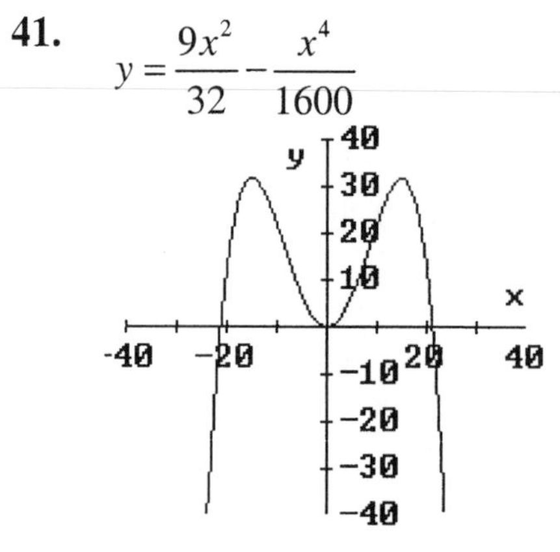

43. **(a)**●▸ **(ii)**, **(b)**●▸ **(i)**, **(c)**●▸ **(iv)**, **(d)**●▸ **(iii)**

45. **(a)**●▸ **(i)**, **(b)**●▸ **(iii)**, **(c)**●▸ **(iv)**, **(d)**●▸ **(ii)**

47. For the equation $y = -\frac{1}{2}x^3 - 4$

(a) there is only one x-intercept, $x = -2$

(b) $y > 0$ for $x < -2$

(c) $y < 0$ for $x > -2$

49. For the equation $y = -x^2 - x + 6$

(a) the x-intercepts are $x = -3$, and $x = 2$

(b) $y > 0$ if $-3 < x < 2$

(c) $y < 0$ if $x < -3$, or $x > 2$

51. The x-intercepts of $y = \frac{x^3}{24} - \frac{x^2}{10} - \frac{22x}{5} + 12$ are $x \approx -10.41$, $x \approx 2.75$, and $x \approx 10.05$. The y-intercept is 12.

53. The lowest point of the graph of $y = \frac{1}{2}x^2 + 4x - 4$ is $(-4, -12)$.

55. The x-coordinates of the points where $y = \frac{10}{x^2 - 16x + 64}$ equals 2 are $x \approx 5.76$, and $x \approx 10.24$.

Applications

57. The predicted minimum hourly wage for the year 2000 by the line of best fit is \$4.70. This is \$0.45 less than the actual value of \$5.15.

59. The shortest route from $A(-6, -2)$ to $B(5, 4)$ is to go from A to (3,4), a distance of 10.82 units, plus the two blocks from (3, 4) to (5, 4), for a total distance of 12.82. If Lloyd is closed, the shortest distance using the north-south and east-west streets is 17.

61. The approximate year when the U.S. population reached 260 million is 1994.

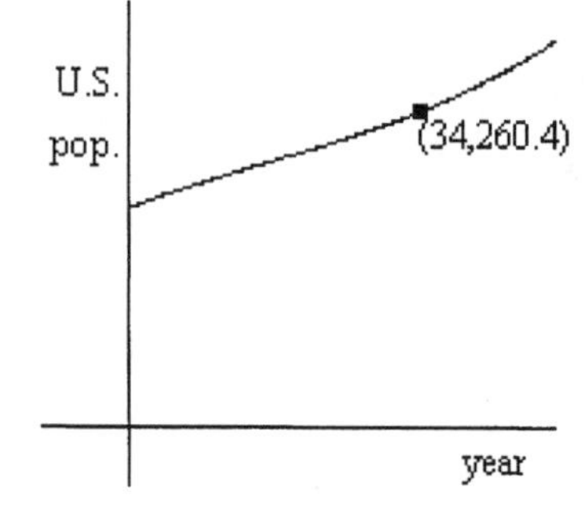

63. The model establishes 1993 as the year with highest number of new AIDS cases.

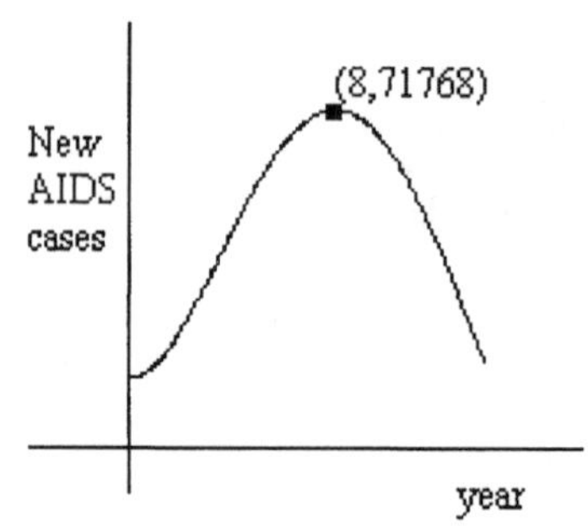

65. The shortest ladder that would reach the stranded passenger if $x=10$ should be 63.28 ft long. The equation of the circular edge is $x^2+(y-35)^2=900$.

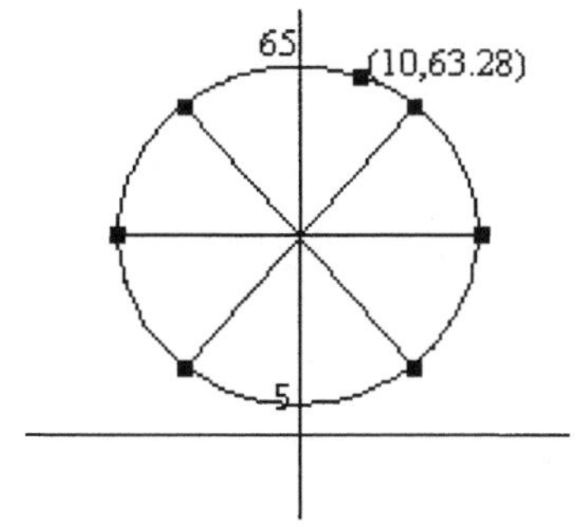

Concepts and Critical Thinking.

67. The statement: "the distance between two points is the sum of the differences of the *x*- and *y*-coordinates" is false.

69. The statement: "the *x*-intercepts of the graph of an equation can be found by setting $x=0$" is false. In order to find the *x*-intercepts, we set $y=0$.

71. Answers may vary. An example of a point in the third quadrant is $(-3,-4)$.

73. Answers may vary. Examples of points lying on the graph of $x^2+y^2=16$ are (4, 0), (0, 4), $(-4,0)$, and $(0,-4)$.

75. The equation $(x-2)^2+(y+3)^2=16$ expresses "the distance from (*x*, *y*) to $(2,-3)$ is 4."

77. The distance d_1 between $P(x_1,y_1)$ and $(\bar{x},\bar{y})$ is $\sqrt{\left(x_1-\frac{x_1+x_2}{2}\right)^2+\left(y_1-\frac{y_1+y_2}{2}\right)^2}$, or

$d_1=\sqrt{\left(\frac{x_1-x_2}{2}\right)^2+\left(\frac{y_1-y_2}{2}\right)^2}$. Similarly, the distance d_2 between $Q(x_2,y_2)$ and $(\bar{x},\bar{y})$ is

$\sqrt{\left(x_2-\frac{x_1+x_2}{2}\right)^2+\left(y_2-\frac{y_1+y_2}{2}\right)^2}$, or $d_2=\sqrt{\left(\frac{x_2-x_1}{2}\right)^2+\left(\frac{y_2-y_1}{2}\right)^2}$. Hence, $d_1=d_2$.

Chapter 1. Section 1.5 Techniques for Solving Equations

Exercise	Solution(s)
1. $3x-4=2$	$x=2$
3. $-2(m+3)=9$	$m=-\frac{15}{2}$
5. $5(-2w+9)=3(w-18)+5$	$w=\frac{94}{13}$
7. $(a+3)(a-4)(a+6)=0$	$a=-6,\ a=-3,\ a=4$
9. $(x-2)(x^2+5x+6)=0$	$x=-3,\ x=-2,\ x=2$
11. $s^2=3s$	$s=0,\ s=3$
13. $(y+1)^2=3$	$y=-1-\sqrt{3},\ y=-1+\sqrt{3}$
15. $u^4-16=0$	$u=-2,\ u=2$
17. $x^2+6x=-9$	$x=-3$
19. $2x^2+4x+1=0$	$x=\frac{-2-\sqrt{2}}{2},\ x=\frac{-2+\sqrt{2}}{2}$
21. $w^2-4w+5=0$	$w=2-i,\ w=2+i$
23. $\frac{7}{y+4}=\frac{3}{y-4}$	$y=10$
25. $\frac{s}{s^2-1}+\frac{2}{s+1}=\frac{4}{s-1}$	$s=-6$
27. $\frac{1}{x+2}+\frac{3}{x}=\frac{2x+2}{x(x+2)}$	*No solution* (Note: $x=-2$ is not a solution.)
29. $\frac{7}{x+5}-\frac{3}{x+1}=-1$	$x=-5+2\sqrt{7},\ x=-5-2\sqrt{7}$
31. $\sqrt{x}=3$	$x=9$
33. $\sqrt{8x-7}=0$	$x=\frac{7}{8}$
35. $\sqrt{4t+1}=-\sqrt{4t+1}$	$t=-\frac{1}{4}$
37. $\sqrt{x+5}-\sqrt{x}=1$	$x=4$

39. If $E=mc^2$, then $m=\frac{E}{c^2}$.

41. If $A=P(1+r)$, then $r=\frac{A-P}{P}$.

43. If $d_1(t+1)=d_2(t-1)$, then $t=\frac{d_1+d_2}{d_2-d_1}$.

45. If $w^2 + pw + q = 0$, then $w = \dfrac{-p - \sqrt{p^2 - 4q}}{2}$, $w = \dfrac{-p + \sqrt{p^2 - 4q}}{2}$.

47. If $\dfrac{l}{h} = \dfrac{s+d}{s}$, then $s = \dfrac{hd}{l-h}$.

51. The solutions of the equation $\sqrt{3x+7} = \sqrt{x+2} + 1$ are $x = -2$ and $x = -1$.

53. To the nearest hundredth, the approximate solution of $x^5 + x = 1$ is $x \approx 0.75$.

55. The approximate solutions (to the nearest hundredth) of the equation $x^4 - 143x^2 = 100$ are $x \approx -11.99$ and $x \approx 11.99$.

57. Solving for *y* in the equation $y^2 + 3x - 4 = 0$ gives two solutions, $y = \sqrt{4 - 3x}$ and $y = -\sqrt{4 - 3x}$. The *x*-intercept of the graph of $y^2 = 4 - 3x$ is $x = \dfrac{4}{3}$.

59. Solving for *y* in the equation $4x^2 - 15 = 8x + 4y$ gives the equivalent equation $y = x^2 - 2x - \dfrac{15}{4}$. The approximate *x*-intercepts, to the nearest hundredth, of the graph of the equation $4x^2 - 15 = 8x + 4y$, are $x \approx -1.18$ and $x \approx 3.18$.

61. Solving for *y* in the equation $x = 2y^2 + 4y + 5$ gives two possibilities for *y*: $y = -1 + \dfrac{1}{2}\sqrt{2(x-3)}$ and $y = -1 - \dfrac{1}{2}\sqrt{2(x-3)}$. The second equation doesn't have any *x*-intercepts, but the first has an *x*-intercept, $x = 5$.

63. In order to get an *A*, the average of 85, 88 and *x* needs to be greater than or equal to 90, that is $\dfrac{85 + 88 + x}{3} \geq 90$, or $x \geq 97$.

65. For the first car, at an average speed of 55 miles per hour, after *t* hours the distance traveled is $d = 55t$. For the second car, at the average speed of 65 miles per hour, after $t - \dfrac{1}{2}$ hours, the distance traveled is $65\left(t - \dfrac{1}{2}\right)$ hours. The second car catches up with the first when the traveled distances are equal, or after $t = 3\frac{1}{4}$ hours, or 3 hours and 15 minutes.

67. The solutions of $-16t^2 + 1454 = 0$ are $t = \pm\dfrac{\sqrt{1454}}{4}$. In this situation, only the positive solution is relevant, approximately $t \approx 9.53$ seconds.

69. The solution of the equation $\frac{1}{6}\left(\frac{4}{3}\pi R^3\right)=5440$ is, approximately, $R=19.83$ miles.

71. The Intel stock is worth \$100 when $21.7x^4-111x^3+162x^2-29.2x+28.6=100$, according to this model. That is, when $x\approx 2.71$, which corresponds to August, 1998.

73. There are two values of t for which $-192t^2+16\sqrt{72}t=12$, $t\approx 0.1$ sec, and $t\approx 0.6$ sec. (One of the solutions on the way up, the other on the return.)

75. **(a)** The equation $\frac{l}{h}=\frac{s+d}{s}$, when solved for s, gives: $s=\frac{hd}{l-h}$.

(b) If $h=5.5ft$, $d=30ft$, and $h=40ft$, we get the shadow length $s\approx 4.78ft$.

77. **(a)** If we assume that each roofer works at the rate of 12 hours per roof, it will take them 6 hours to complete one roof working together.

(b) The solution of $\frac{1}{x}=\frac{1}{8}+\frac{1}{16}$ is $x=\frac{16}{3}$ hours, or $x=5$ hours $20\,\text{min}$.

(c) The contractor is overestimating the number of hours required to complete 20 roofs by $20\left(\frac{2}{3}\right)$, or 13 hours 20 min.

79. The average speed when driving 40 miles per hour for 50 miles and cycling 20 miles per hour for another 50 miles is given by *distance traveled/total time* $=\frac{100}{3.75}\approx 26.67$ mph.

We can expect the average speed to be lower than the simple average, since we spend more time going 20 miles per hour than going 40 miles per hour.

81. The statement: "The x-intercepts of an equation of the form $y=\square$ correspond to solutions of $\square=0$" is true.

83. The statement: "If $x(x+2)=0$, then we can conclude that $x=0$ or $x+2=0$" is true.

85. "$d=\text{dimes}$" is not an appropriate variables definition because "*dimes*" doesn't correspond to a real number, but to a category. Appropriate definitions could be "$d=\text{number of dimes}$", or "$d=\text{value of dimes}$", depending on the situation.

87. Answers may vary. An example of a quadratic equation that has 3 as its only solution is: $2(x-3)^2=0$.

89. We can't divide by 0, so the sequence of steps: $x=0,\ x+x=0+x,\ 2x=x$ is correct, but at this point we may not divide by x, since $x=0$.

91. If $x=-1$, $x=1$, and $x=3$ are solutions of $x^3-4x^2+x+6=0$, then $x^3-4x^2+x+6=a(x+1)(x-1)(x-3)$, where a is a constant. Hence $a=1$, and $x^3-4x^2+x+6=(x+1)(x-1)(x-3)$. The polynomial $2x^2-5x+2$ factors as $2\left(x-\frac{1}{2}\right)(x-2)$, or as $(2x-1)(x-2)$.

Chapter 1. Section 1.6 Inequalities

Exercises 1-17 correspond to inequalities, with their solution as explicit inequalities, in interval notation, and including the graph of the solution set.

	Inequality	Solution	Interval Notation	Graph of the Solution Set
1.	$x-8<5$	$x<13$	$(-\infty,13)$	 0, 13
3.	$23-4v<27$	$v>-1$	$(-1,\infty)$	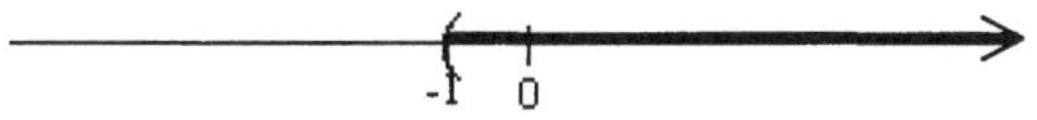 -1, 0
5.	$2z+7\geq 5-6z$	$z\geq -\frac{1}{4}$	$\left[-\frac{1}{4},\infty\right)$	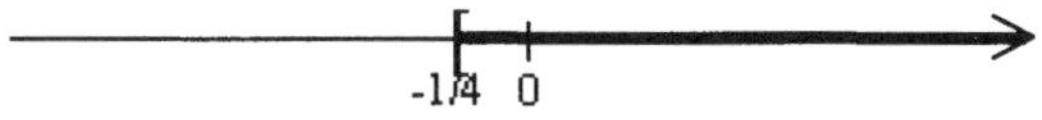 -1/4, 0
7.	$3(x-1)\leq 17-(8-3x)$	All real values of x	$(-\infty,\infty)$	 0
9.	$\frac{1}{4}w+1<w$	$w>\frac{4}{3}$	$\left(\frac{4}{3},\infty\right)$	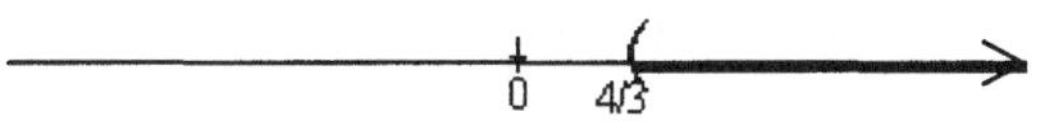 0, 4/3
11.	$x+3>1$ and $x-2<4$	$-2<x<6$	$(-2,6)$	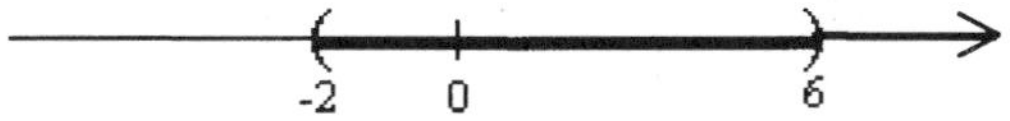 -2, 0, 6
13.	$5\leq 8t-3\leq 9$	$1\leq t\leq \frac{3}{2}$	$\left[1,\frac{3}{2}\right]$	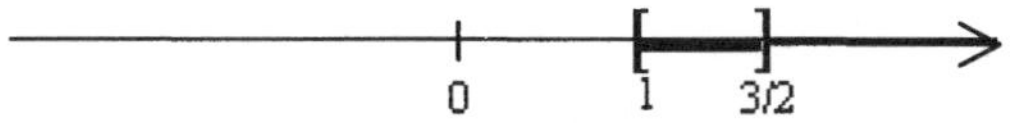 0, 1, 3/2
15.	$-2x+7\leq 1$ and $\frac{1}{2}x\leq \frac{9}{4}$	$3\leq x\leq \frac{9}{2}$	$\left[3,\frac{9}{2}\right]$	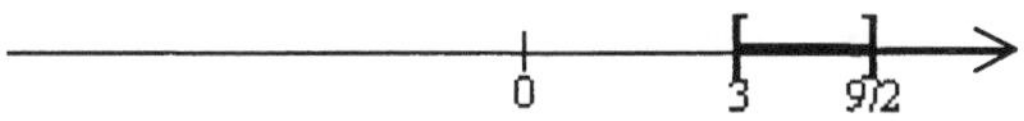 0, 3, 9/2
17.	$\begin{cases} 2z\geq 5 \text{ or} \\ \frac{1}{4}(z-1)\geq 1 \end{cases}$	$z\geq 5$	$[5,\infty)$	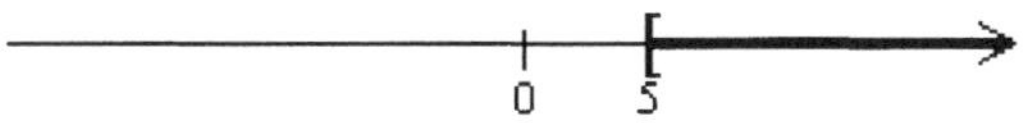 0, 5

Exercises 19-23 correspond to Absolute Value Equations, and their solutions.

	Absolute Value Equation	Solution(s)		Absolute Value Equation	Solution(s)
19.	$\lvert x\rvert=6$	$x=-6,\ x=6$	**21.**	$\lvert 2u-5\rvert=1$	$u=2,\ u=3$
23.	$\left\lvert 4-\frac{1}{2}z\right\rvert=-3$	No solutions			

Exercises 25-35 are Absolute Value Inequalities, and their solutions in interval notation.

Absolute Value Inequality	Solution in Interval Notation	Absolute Value Inequality	Solution in Interval Notation
25. $\lvert 2w\rvert < 6$	$(-3,3)$	**27.** $\frac{1}{3}\lvert z\rvert > 4$	$(-\infty,-12)\cup(12,\infty)$
29. $\lvert s-5\rvert \le 2$	$[3,7]$	**31.** $\lvert -2r+1\rvert < 8$	$\left(-\frac{7}{2},\frac{9}{2}\right)$
33. $\left\lvert \frac{3z-2}{5}\right\rvert \le 1$	$\left[-1,\frac{7}{3}\right]$	**35.** $\lvert 6t-4\rvert + 3 < 8$	$\left(-\frac{1}{6},\frac{3}{2}\right)$

Exercises 37-39 translate a verbal statement to an absolute value inequality, and its solution.

Verbal Statement	Absolute Value Inequality	Solution
37. The distance between x and 1 is less than 5.	$\lvert x-1\rvert < 5$	$-4 < x < 6$
39. The distance between twice a number and -3 is greater than 6.	$\lvert 2x+3\rvert > 6$	$x < -\frac{9}{2}$, or $x > \frac{3}{2}$

41. **(a)** $\lvert 3x-2\rvert - 5 = 0$

The solutions are $x=-1$, and $x = 7/3$.

(b) $\lvert 3x-2\rvert - 5 > 0$

The solution set is: $(-\infty,-1)\cup(7/3,\infty)$.

(c) $\lvert 3x-2\rvert - 5 < 0$

The solution set is: $(-1, 7/3)$.

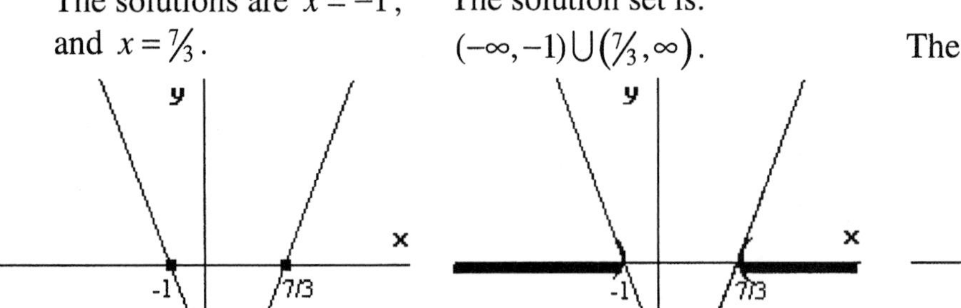

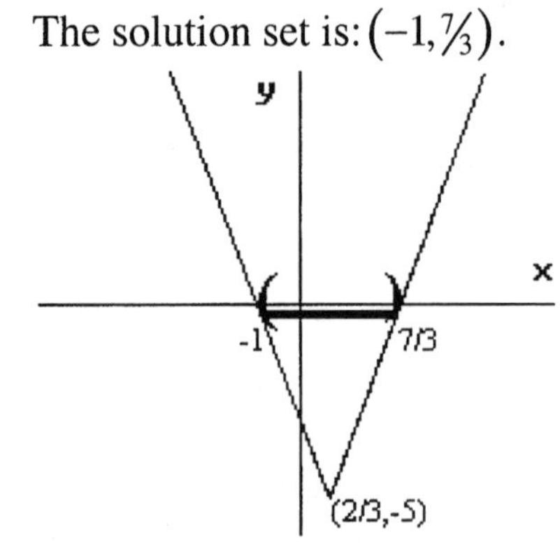

43. The solutions of $\lvert x-1\rvert + \lvert x-2\rvert = 3\lvert x-3\rvert$ are $x = \frac{12}{5}$, and $x = 6$.

Exercises 45-57 correspond to polynomial inequalities and their solutions.

Polynomial Inequality	Solution	Polynomial Inequality	Solution
45. $(t+3)(t-1) \le 0$	$-3 \le t \le 1$	**47.** $x^2+4x \le 5$	$-5 \le x \le 1$
47. $x^2+4x \le 5$	$-5 \le x \le 1$	**49.** $u^2 \le 8$	$-2\sqrt{2} \le u \le 2\sqrt{2}$
51. $x^4+x>1$	$x<-1.22$, or $x>0.72$	**53.** $(x-5)^2(x+1)^3<0$	$x<-1$
55. $(7x-2)^4(2x+3)^6 \le 0$	$x=\frac{2}{7}$, $x=-\frac{3}{2}$	**57.** $2v^3-8v \ge 0$	$-2 \le v \le 0$, or $v>2$

Applications

59. For the cost for one day at \$0.39 per mile plus a flat fee of \$19.95 per day, the number x of miles traveled should be in the range: $141.15 \text{ miles} < x < 205.26 \text{ miles}$.

61. If the total distance covered by the car with an average of 750 revolutions per minute is between 59.8 and 60.69 miles, then the radius r of the tires is between 13.4 and 13.6 inches.

63. If the Celsius temperature C satisfies $37^\circ < C < 39^\circ$, then the Fahrenheit temperature F satisfies $100.4^\circ < F < 102.2^\circ$.

65. The height $h=-16t^2+80t$ will be above 96 feet, when $2 \text{ sec} < t < 3 \text{ sec}$.

67. The height $h=-16t^2+4000t$ will be above 158,400 feet, when $49.34 \text{ sec} < t < 200.66 \text{ sec}$.

Concepts and Critical Thinking

69. The statement: "If $-2x<4$ then $x<-2$" is false.

71. The statement: "The solution set to an absolute value inequality may be the union of two disjoint intervals" is true.

73. Answers may vary. An absolute value equation with empty solution set is: $|x|=-3$.

75. Answers may vary. $|x|<-5$ is an absolute value inequality having the empty set as its solution.

Chapter 1. Section 1.7 Lines

Exercises 1-11 consist in finding the slope of the line that passes through the given pair of points.

	Pair of Points	Slope
1.	$(1,3)$ and $(2,7)$	$m=4$
5.	$(0,32)$ and $(100,212)$	$m=\frac{9}{5}$
9.	$(2,3.5)$ and $(4.6,7.9)$	$m=\frac{22}{13}$

	Pair of Points	Slope
3.	$(-2,6)$ and $(3,-5)$	$m=-\frac{11}{5}$
7.	$\left(\frac{1}{2},2\right)$ and $(-1,3\frac{1}{2})$	$m=-1$
11.	$(35,1.5\times10^7)$ and $(40,2.0\times10^7)$	$m=1.0\times10^6$

13. The line contains (0,0) and (1,3); the slope is 3.

15. The line contains $(0,4)$ and $(1,0)$; the slope is -4.

17. If the line l is parallel to a line with slope -3, the slope of l is -3 also.

19. If the line l is perpendicular to a line with slope $3/5$, the slope of l is $-5/3$.

21. l_1 has slope -2, l_2 has slope $2/3$, l_3 has slope 3, and l_4 has slope 0.

23. The line l_1 passes through $(-5,13/7)$ and $(-2,22/7)$, so the slope of l_1 is $3/7$. l_2 is perpendicular to l_1, so l_2 has slope $-7/3$. l_3 is parallel to l_1, so the slope of l_3 is $3/7$.

Exercises 25-33 consist in finding the equation of the line in slope-intercept form when possible, knowing the slope and one point P on the line, or knowing two points that belong to the line.

	Slope *m* and point *P* on the line	Slope-intercept form of equation
25.	$m=1,\ P=(2,4)$	$y=x+2$
29.	m is undefined, $P=(3,7)$	Impossible; the equation is $x=3$
33.	$m=0,\ P=(2,1)$	$y=1$

	Slope *m* and point *P* on the line	Slope-intercept form of equation
27.	$m=-1,\ P=(2,8)$	$y=-x+10$
31.	$m=5,\ P=\left(\frac{1}{2},\frac{2}{3}\right)$	$y=5x-\frac{11}{6}$

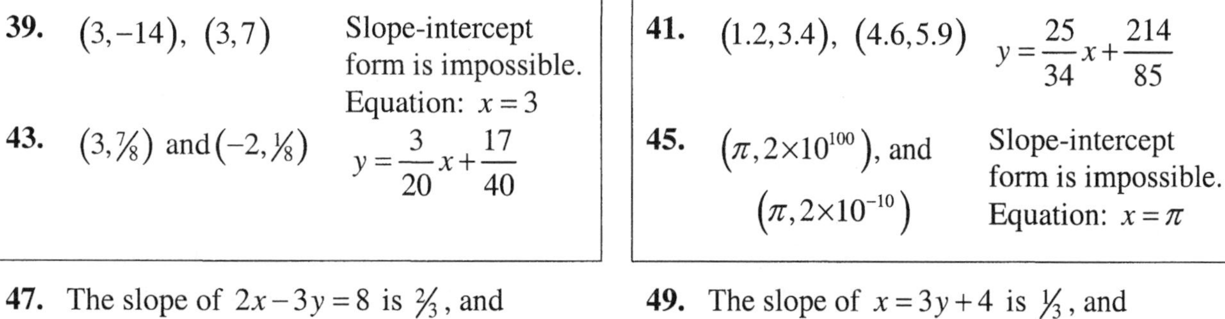

	Pair of points	Slope-intercept form of equation
35.	$(1,4),\ (-2,3)$	$y=\frac{1}{3}x+\frac{11}{3}$
37.	$(2,0),\ (0,4)$	$y=-2x+4$
39.	$(3,-14),\ (3,7)$	Slope-intercept form is impossible. Equation: $x=3$
41.	$(1.2,3.4),\ (4.6,5.9)$	$y=\frac{25}{34}x+\frac{214}{85}$
43.	$(3,7/8)$ and $(-2,1/8)$	$y=\frac{3}{20}x+\frac{17}{40}$
45.	$(\pi,2\times10^{100})$, and $(\pi,2\times10^{-10})$	Slope-intercept form is impossible. Equation: $x=\pi$

47. The slope of $2x-3y=8$ is $2/3$, and its y-intercept is $-8/3$.

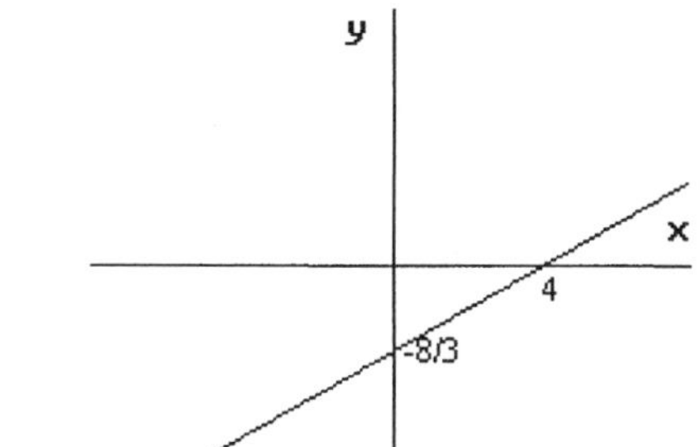

49. The slope of $x=3y+4$ is $1/3$, and its y-intercept is $-4/3$.

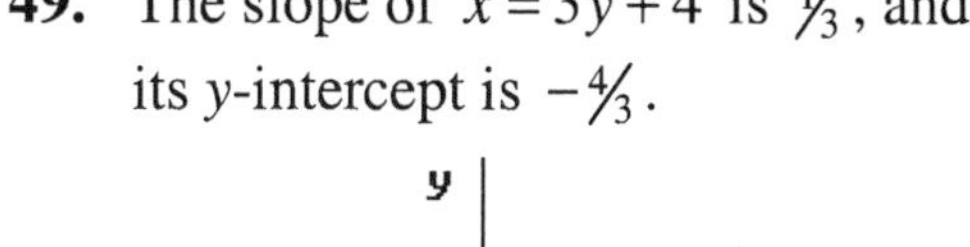

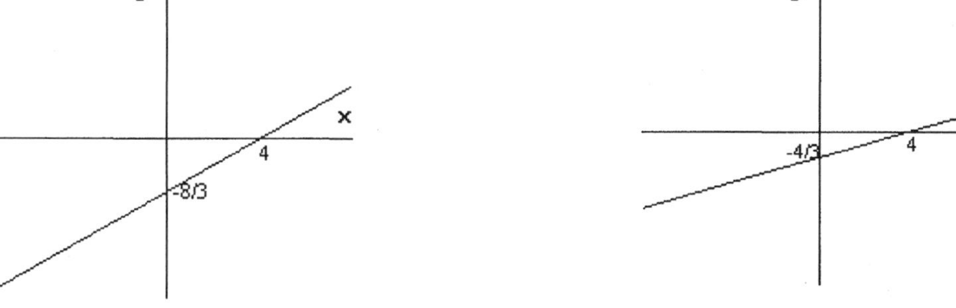

51. The slope of $y-3=-2(x+3)$ is -2, and its y-intercept is -3.

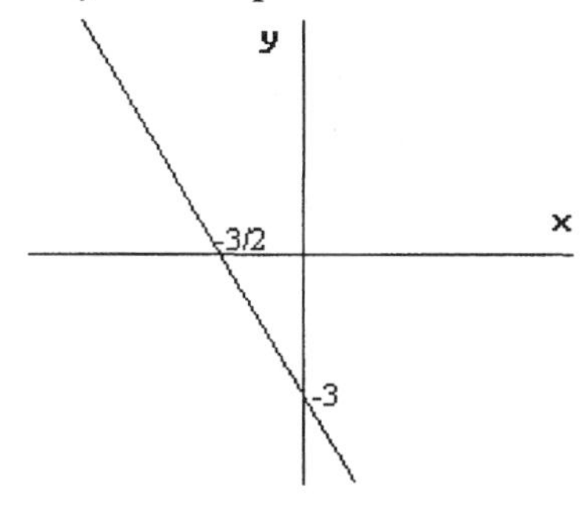

53. If the slope is $2/3$ and the y-intercept is 2, the equation of the line is: $y=\frac{2}{3}x+2$.

55. The equation of the line parallel to $y=4x+2$, with y-intercept 5, is: $y=4x+5$.

57. The equation of the line that contains the point $(2,-4)$ that is perpendicular to $x-3y=8$ is: $y=-3x+2$.

59. The equation of the line parallel to $x=6$ that contains the point (3,8) is: $x=3$.

61. The standard form of $y=3x+4$ is: $3x-y=-4$.

63. The standard form of $y-\frac{2}{3}=2\left(x-\frac{1}{2}\right)$ is: $6x-3y=1$.

65. The point of intersection of $y=3x+5$ and $y=-2x-8$ is $\left(-\frac{13}{5},-\frac{14}{5}\right)$

67. The point of intersection of $x=2$ and $y=5$ is $(2,5)$.

69. The slope of the line connecting $(1,0)$ and $(7,4)$ is $2/3$; the slope of the line connecting (7,4) and $(15,9)$ is $5/8$. Then, the points are not collinear.

71. The line joining (2,3) and (3,5) is parallel to the line joining (6,3) and (7,5), since the slope of each of these is 2. The line joining (3,5) and (7,5) is horizontal, as is the line joining (2,3) and (6,3). So, the quadrilateral with vertices (2,3), (3,5), (6,3) and (7,5) is a parallelogram.

73. If N is the number of orders received, at \$5 per order, plus \$1000 overload, the monthly cost is: $C=5N+1000$.

75. **(a)** If h is the height in inches, $h=0$ at conception and $h=76$ inches at 16 years of age, and if we assume constant rate of growth, the equation relating h and t is $h=4.75t$.
(b) If the birth of the baby happens at 9 months after conception, then $t=0.75$ and $h=4.75(0.75)=3.5\ in$.
(c) The height of the child at 12 years of age is obtained by substituting $t=12.75$ in the equation found in (a), or $h=60.56in$ (5.05 *ft*). This model is, of course, very inadequate.

77. The value of the machinery is given by $V=-18{,}000t+200{,}000$, where t is the time in years, and $t=0$ corresponds to 2003. The value in 2006 will be \$146,000.

79. **(a)** Assuming a constant rate of growth in the U.S. population between 1990 and 2000, the population is given by $P=249{,}000{,}000+3{,}200{,}000t$ where t is the time in years since 1990. In 1997, the population predicted by this equation would have been 271,400,000.
(b) Extrapolation to the year 2010 gives the predicted population of 313,000,000.

Concepts and Critical Thinking.

81. The statement: "Slope isn't defined for horizontal lines" is false. For horizontal lines, the slope is 0.

83. The statement: "If the y-coordinates of points on a line increases as the x-coordinates decrease, then the line has negative slope" is true.

85. The statement: "A line with slope 3 is steeper than a line with slope -4" is false. The line with slope -4 is steeper, only decreasing instead of increasing, as the other line.

87. Answers may vary. Any vertical line has undefined slope, for example the line $x=3$.

89. Answers may vary. An example of a line that does not intersect $2x+7y=11$ is $2x+7y=5$.

91. The quotient $\frac{y_2-y_1}{x_2-x_1}$ is equivalent to $\frac{y_1-y_2}{x_1-x_2}$: $\frac{y_2-y_1}{x_2-x_1}=\frac{(-1)(y_1-y_2)}{(-1)(x_1-x_2)}=\frac{y_1-y_2}{x_1-x_2}$

Chapter 1. Review

1. -2.33 is rational, real and complex.

3. A number that when squared gives -11 is complex.

5. $\{x \mid -5 \le x < -1\}$ is the interval $[-5,-1)$

-5 0 1

7. $\left\{x \mid x \le \frac{1}{2}\right\}$ is the interval $\left(-\infty, \frac{1}{2}\right]$

0 1/2

9. $(-3)^4 = 81$

11. $\dfrac{(a^3 b)^2}{a^4 b^6} = \dfrac{a^2}{b^4}$

13. $31,400,000 = 3.14 \times 10^7$

15. $\sqrt[3]{a^6 b^4} = a^2 b \sqrt[3]{b}$

17. $\dfrac{9}{\sqrt{3}} = 3\sqrt{3}$

19. $\dfrac{1}{x-\sqrt{2}} = \dfrac{x+\sqrt{2}}{x^2-2}$

21. $\sqrt{6s^5 t} \cdot \sqrt{2st} = 2\sqrt{3}s^3 t$

23. $\sqrt{28} + 9\sqrt{7} = 11\sqrt{7}$

25. $\sqrt[4]{t^2} = t^{1/2}$

27. $x^{2/3} = \sqrt[3]{x^2}$

29. $\left(x^{1/3}\right)^{1/4} = x^{1/12}$

31. $\dfrac{a^{2/3} b^{4/3}}{a^{5/3} b^{-2/3}} = \dfrac{b^2}{a}$

33. $(2t^3 + 4t) - (3t^3 + t^2 - t) = -t^3 - t^2 + 5t$

35. $(2-w)(4-3w) = 3w^2 - 10w + 8$

37. $9u^2 - 4 = (3u-2)(3u+2)$

39. $t^2 + 4t - 32 = (t+8)(t-4)$

41. $9x^2 + 6x + 1 = (3x+1)^2$

The simplified answers in Exercises 43-51 are valid where the original expression is defined.

43. $\dfrac{x+3}{x^2-9} = \dfrac{1}{x-3}$

45. $\dfrac{t^3-1}{t^3-t} = \dfrac{t^2+t+1}{t^2+t}$

47. $\dfrac{3}{y+2} - \dfrac{1}{y} = \dfrac{2y-2}{y^2+2y}$

49. $\dfrac{x^2}{x^2-9} \cdot \dfrac{3-x}{x^2+x} = -\dfrac{x}{x^2+4x+3}$

51. $\dfrac{\frac{1}{t}+\frac{1}{2}}{t+2} = \dfrac{1}{2t}$

53. The distance between the points (4,5) and $(-2,-3)$ is 10. The midpoint of the segment connecting (4,5) and $(-2,-3)$ is (1,1).

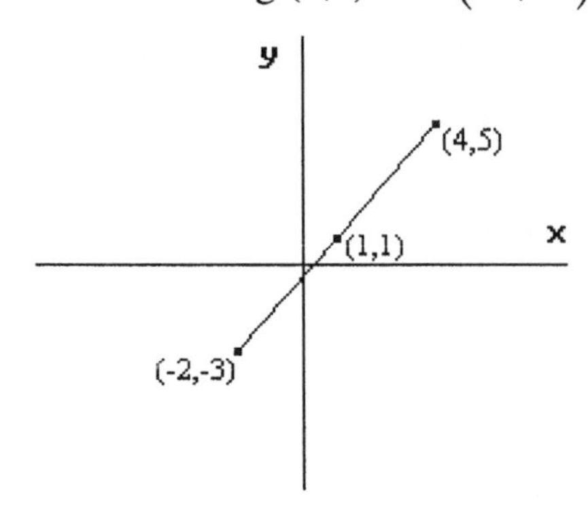

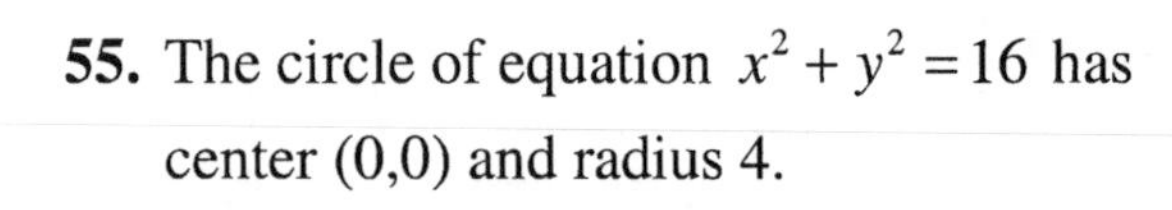

55. The circle of equation $x^2+y^2=16$ has center (0,0) and radius 4.

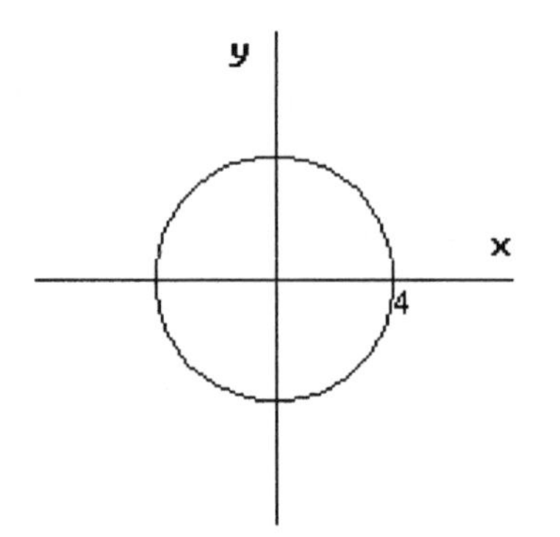

57. The circle of equation $x^2+y^2+10x-8y+32=0$ has center $(-5,4)$ and radius 3.

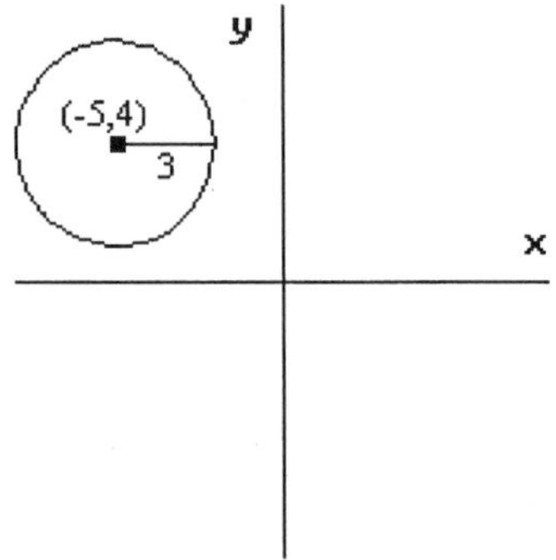

59. The equation of the circle with center (2,5) and radius 3 is $(x-2)^2+(y-5)^2=9$.

61. The equation $y=\dfrac{x^3}{48}-\dfrac{x^2}{2}+4$ corresponds to graph (iv).

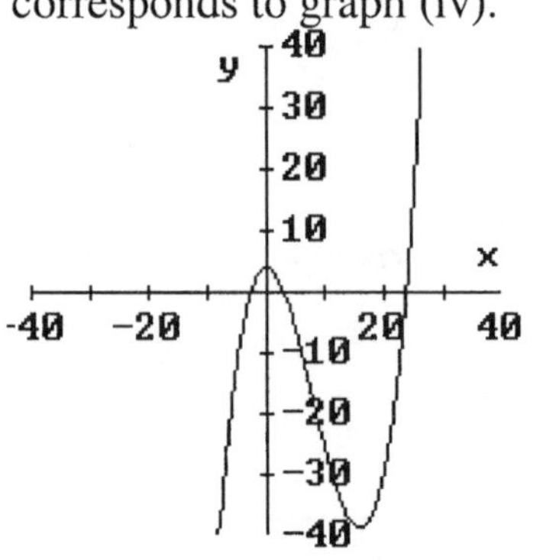

63. The equation $y=\dfrac{x^3}{4}-\dfrac{33x^2}{32}$ corresponds to graph (ii).

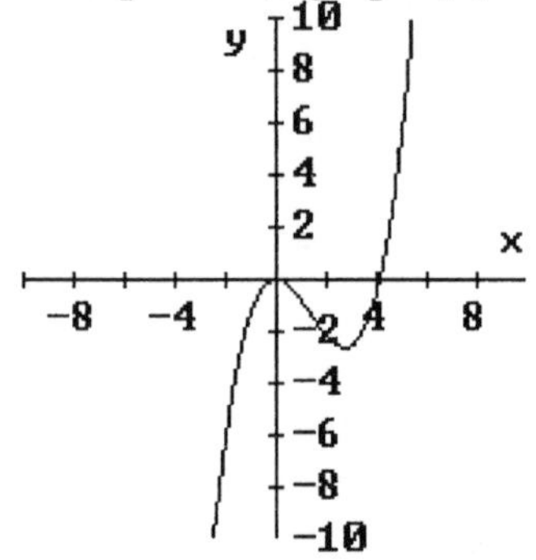

65. The x-intercepts of $y=\dfrac{x^3}{8}-\dfrac{15x^2}{16}-\dfrac{129x}{32}+5$ are, to the nearest hundredth, -3.79, 1.03, and 10.26; the y-intercept is 5.

67. (2,3) is the highest point of the graph of $y=\sqrt{-x^2+4x+5}$.

	Equation	Solution(s)		Equation	Solution(s)
69.	$3x+4=9$	$x=5/3$	**71.**	$2(x+1)-3(2x-2)=1$	$x=7/4$
73.	$x(2x-1)(x+3)=0$	$x=-3,\ 0,\ 1/2$	**75.**	$x^2+x-12=0$	$x=-4,\ 3$
77.	$w^2+4w+2=0$	$w\approx-3.41,\ -0.59$	**79.**	$6y^2-y-2=0$	$y=-1/2,\ 2/3$
81.	$\frac{2}{x}=\frac{3}{x+1}$	$x=2$	**83.**	$\frac{x}{2x+1}=\frac{x+2}{5x}$	$x=-1/3,\ 2$
85.	$\sqrt{3x-5}=1$	$x=2$	**86.**	$\sqrt{2x-3}+7=5$	No solution
87.	$\sqrt{3a+10}-a=2$	$a=2$; $a=-3$ is not a solution.			

89. The solution of $x^7-x^3+1=0$ is $x\approx-1.14$.

91. The solutions of $x^4-3x=1$ are $x\approx-0.33,\ 1.54$.

93. If $2x+3y=12$, then $y=4-\frac{2}{3}x$.

The x-intercept is 6.

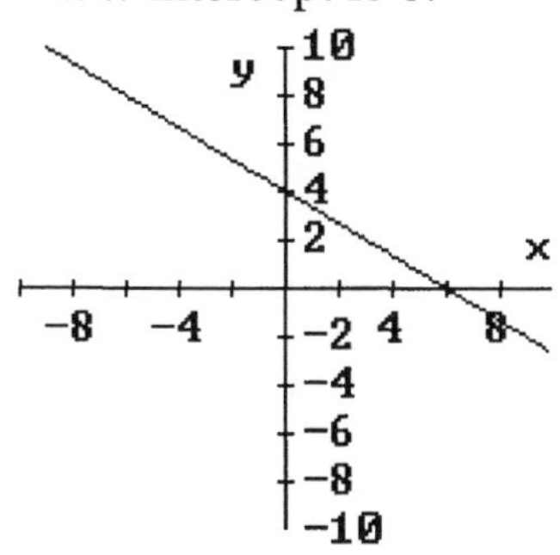

95. If $y^4-x=5$, then $y=-\sqrt[4]{x+5}$,

or $y=\sqrt[4]{x+5}$. The x-intercept is -5.

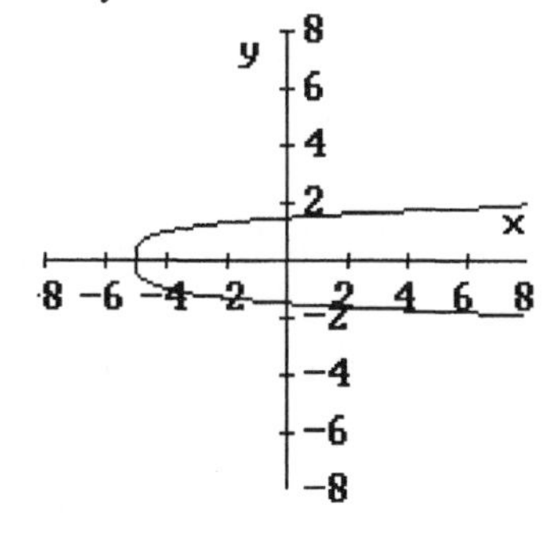

97. If $x+4y=y^2+5$, then $y=2-\sqrt{x-1}$, or $y=2+\sqrt{x-1}$. The x-intercept is 5.

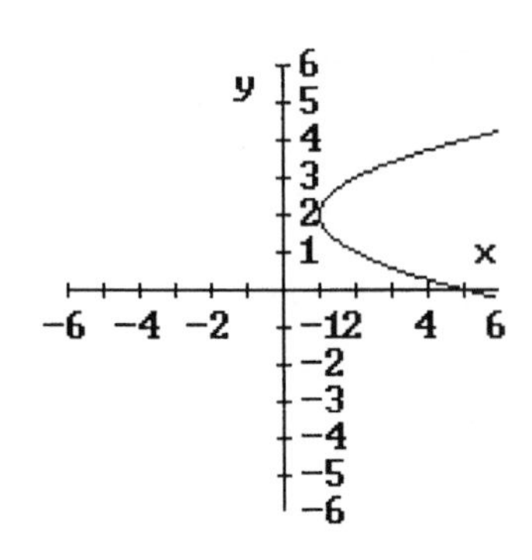

Inequality	Solution and Graph	Inequality	Solution and Graph
99. $-3x-4<8$	$x>-4$	**101.** $\frac{1}{3}(x+2)-\frac{1}{2}(x-2)>0$	$x<10$

-4 0

0 10

103.

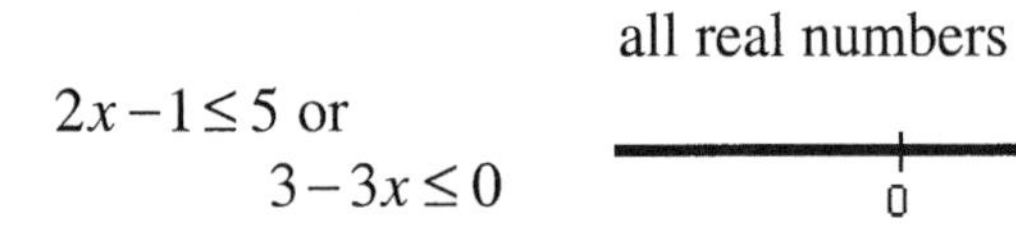

$2x-1\le 5$ or $3-3x\le 0$ — all real numbers

105. $|-3x+4|<2$ — $\frac{2}{3}<x<2$

0 2/3 4

107. The solution of $(s+1)(s+5)>0$ is: $s<-5$, or $s>-1$.

109. The solution of $x^4-2x<1$ is, approximately: $-0.47<x<1.4$.

111. The slope of the line that passes through (2,5) and (−1,4) is $\frac{1}{3}$

113. The slope of the line with equation $y=4x-7$ is 4.

115. The line of slope 3 and passing through (1,5) has equation $y=3x+2$.

117. The line that passes through $(-1,6)$ and $(3,-2)$ has equation $y=-2x+4$.

119. The line with slope -2 and y-intercept 2 has equation $y=-2x+2$.

121. The line parallel to $y=-3x+1$ that passes through (7,1) has equation $y=-3x+22$.

123. The line perpendicular to $2x-3y=1$ that passes through $(-4,2)$ has equation $y=-\frac{3}{2}x-4$.

125. (a) $(2x-3y)^2=4x^2-12xy+9y^2$; **(b)** if $2x\geq 3y$, then $\sqrt{4x^2-12xy+9y^2}=2x-3y$.

127. (a) The midpoint of the line segment connecting $(-2,3)$ and $(2,11)$ is $(0,7)$;
(b) The point P such that the line segment connecting $(-2,3)$ and $(0,7)$ is half of the line segment connecting $(-2,3)$ and P is $(2,11)$.

129. (a) $x^2-x-6=(x-3)(x+2)$;
(b) The solutions of $\dfrac{x^2-x-6}{x^{14}+7x^{12}+6x^3+192.35x+8}=0$ are $x=3$, and $x=-2$.

131. (a) $(x-3)(x+3)(x+9)=x^3+9x^2-9x-81$;
(b) The solutions of $Z^3+9Z^2-9Z=81$ are $Z=-9,\ -3$, and 3.

133. The volume of the Great Pyramid of Giza is 84,375,000 ft^3, or 1055,35 times the volume of an Olympic-size swimming pool.

135. If the cost of a *CD* including 6% sales tax is \$12.19, the price of the *CD* before taxes is \$11.50.

137. The revenue is at least 625 million dollars if the number of units x sold is between 25 million and 75 million units.

139. Daniel takes approximately (to the nearest minute) 2 h 18 min , and Roberto takes 1 h 18 min to wax the car.

Chapter 1. Test

1. The statement: "All rational numbers are real" is true.

3. The statement: "The principal root of any positive number is positive" is true.

5. The statement: "If $(x+1)(x-4)=0$, then either $x=-1$ or $x=4$" is true.

7. The statement: "If $|x|>1$, then $-1>x>1$" is false. No real number satisfies the compound inequality $-1>x>1$, which is read x less than -1 and greater than 1. A correct statement is: "If $|x|>1$, then $x<-1$ or $x>1$."

9. Answers may vary. An example of an irrational number is $\pi/2$.

11. Answers may vary. An example of a quadratic equation with 3 as its only solution is: $2x^2-12x+18=0$.

13. $\left(\dfrac{a^{-3}b^4}{a^2b^{-1}}\right)^2=\dfrac{b^{10}}{a^{10}}$

15. $\dfrac{2}{\sqrt[3]{4}}=\sqrt[3]{2}$

17. $\dfrac{1}{x^2+2x}+\dfrac{x}{x^2-4}=\dfrac{x-1}{x^2-2x}$

19. $u^3+8v^3=(u+2v)(u^2-2uv+4v^2)$

21. The solutions of the equation $x^4-2x^2+x-1=0$ are: $x\approx-1.71,\ 1.35$.

23. The solution of the inequality $x^2-5\leq 4x$ is: $-1<x<5$.

25. The equation of the line that passes through $(-2,3)$ and $(-2,7)$ is $x=-2$.

27. The following graphs correspond to the equation $y=x(x-1.5)(x+1.5)$:

Viewing rectangle:
Xmin = −10, Xmax = 10
Ymin = −10, Ymax = 10

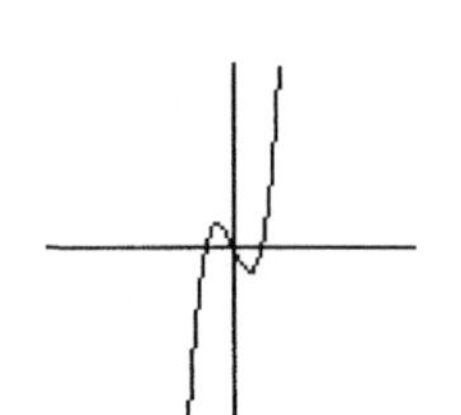

(a) Viewing rectangle:
Xmin = −2, Xmax = 2
Ymin = −2, Ymax = 2
(ii)

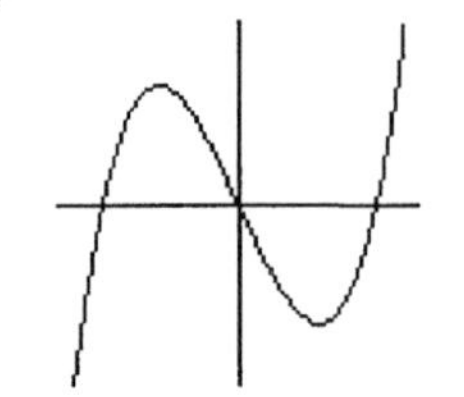

(b) Viewing rectangle:
Xmin = −2, Xmax = 2
Ymin = −10, Ymax = 10
(i)

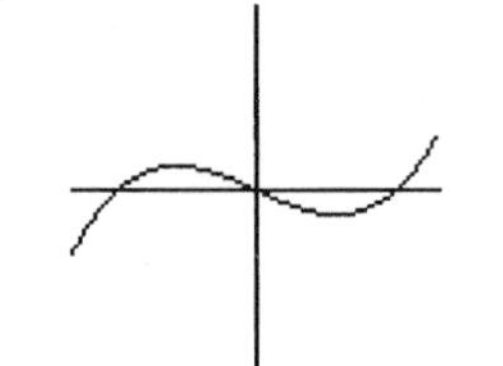

(c) Viewing rectangle:
Xmin = −10, Xmax = 10
Ymin = −2, Ymax = 2
(iii)

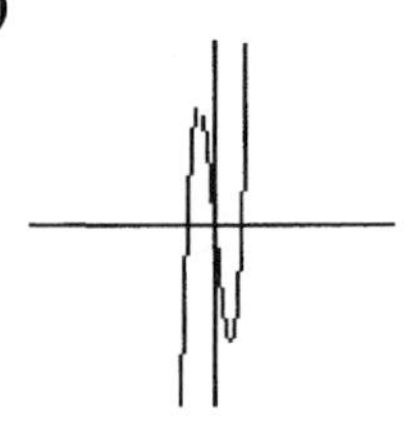

Chapter 2. FUNCTIONS AND THEIR GRAPHS

Section 2.1 Functions

1.

D	-2	-1	0	1	2
R	4	1	0	1	4

The correspondence from D to R is a function.

3. The correspondence from the indicated group of Presidents D to the group of their Vice presidents R is not a function. James Madison in D has two Vice Presidents associated, George Clinton and Elbridge Gerry.

5. The correspondence from D (Year 1991:2002) to R (Internet Hosts) is a function.

7. The equation $3x+4y=12$ defines y as a function of x, $y=\dfrac{12-3x}{4}$ (for x any real number.)

9. The equation $x^2+y^2=1$ doesn't define y as a function of x. For example, $x=\dfrac{\sqrt{3}}{2}$ has two y-values associated, $y=-\dfrac{1}{2}$ and $y=\dfrac{1}{2}$.

11. The equation $0=x-y^3$ defines y as a function of x, $y=x^{1/3}$ (x any real number.)

13. The equation $y=-x^2+1$ defines y as a function of x (any real number.)

15. The equation $x=y^3-4y$ doesn't define y as a function of x (for example, the value $x=0$ has three y-values associated, $y=-2$, $y=0$, and $y=2$.)

17. The equation $xy=4$ defines y as a function of x (for x any real number except 0.)

19. If $f(x)=x^2+3x+2$, we have: **(a)** $f(2)=12$ **(b)** $f(-3)=2$ **(c)** $f(0)=2$.

21. If $f(x)=\dfrac{3x+2}{4x-5}$: **(a)** $f(4)=\dfrac{14}{11}$ **(b)** $f(5)=\dfrac{17}{15}$ **(c)** $f(-t)=\dfrac{-3t+2}{-4t-5}$.

23. If $x(t)=3t^2$: **(a)** $x(t^2)=3t^4$ **(b)** $x(t-1)=3t^2-6t+3$ **(c)** $x\left(\dfrac{5}{t}\right)=\dfrac{75}{t^2}$.

25. If $g(x)=(x+1)^2$: **(a)** $g(x-1)=x^2$ **(b)** $g(x^2)=x^4+2x^2+1$ **(c)** $g\left(\dfrac{1-a}{a}\right)=\dfrac{1}{a^2}$.

27. If $f(x)=\dfrac{1}{x-1}$, $g(x)=x^2$: **(a)** $f(g(3))=\dfrac{1}{8}$ **(b)** $f(g(t))=\dfrac{1}{t^2-1}$ **(c)** $g(f(x+1))=\dfrac{1}{x^2}$.

29. The function $h(x)=|x|$ assigns to every real number the absolute value of the number.

31. The function $f(x)=\frac{1}{x+3}$ assigns to every real number $x \neq 3$ the reciprocal of $x+3$.

Function	Domain	Function	Domain
33. $f(x)=x^2$	All real numbers	**35.** $h(x)=\dfrac{3}{x-4}$	All real numbers except 4
37. $r(t)=\dfrac{5t^2+3t+1}{t^2-9}$	All real numbers except $t=-3,\ 3$	**39.** $f(x)=\sqrt{x+1}$	$x \geq -1$
41. $f(x)=\dfrac{\sqrt{x+2}}{x^2-4x+3}$	All real numbers except $x=1,\ 3$	**43.** $f(x)=\dfrac{2}{\lvert x\rvert-3}$	All real numbers except $x=-3,\ 3$

45. The function $f(x)=2x+1$ satisfies: $f(0)=1, f(1)=3, f(2)=5,$ and $f(3)=7$.

47. The function $f(x)=\frac{1}{x+1}$ satisfies: $f(-1)$ is undefined, $f(0)=1$, $f(1)=\frac{1}{2}$, and $f(2)=\frac{1}{3}$.

49. Answers may vary. The function $f(x)=\sqrt{7-x}$ has domain $(-\infty,7]$.

51. Answers may vary. The function $p(x)=\frac{1}{x(x-3)}$ has domain $(-\infty,0)\cup(0,3)\cup(3,\infty)$.

53. $f(x)=-2x+5$

x	$f(x)$
0	5
2	1
4	−3

55. $f(x)=(x+1)^2-3$

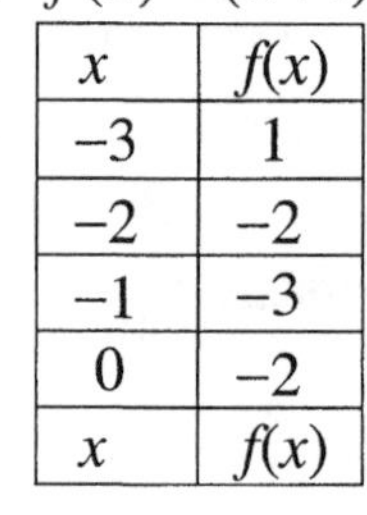

x	$f(x)$
−3	1
−2	−2
−1	−3
0	−2
x	$f(x)$

57. **(a)** $f(x)=x^2$

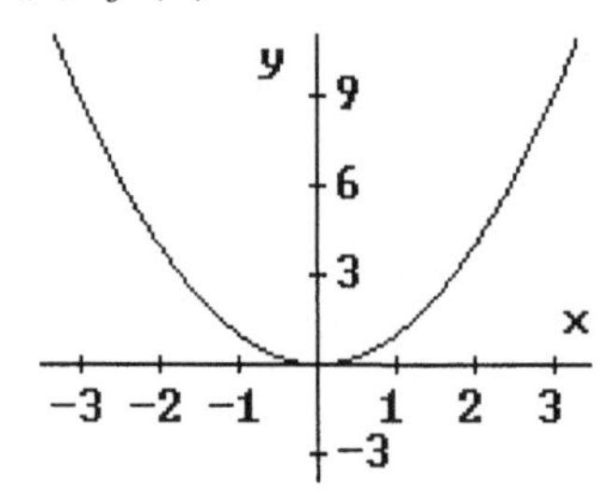

Domain: $(-\infty,\infty)$; range: $[0,\infty)$

(b) $f(x)=(x+1)^2$

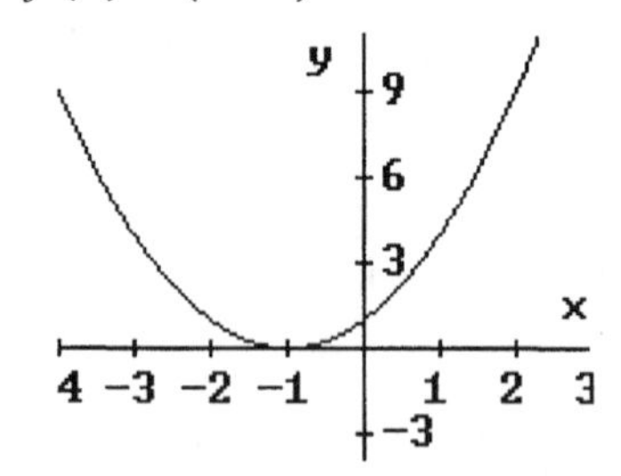

Domain: $(-\infty,\infty)$; range $[0,\infty)$

(c) $f(x) = x^2 + 1$

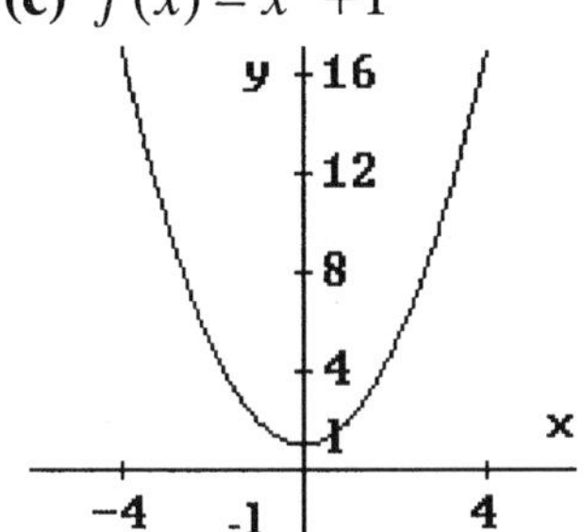

Domain: $(-\infty, \infty)$; range: $[1, \infty)$

(d) $f(x) = 1/(x^2 + 1)$

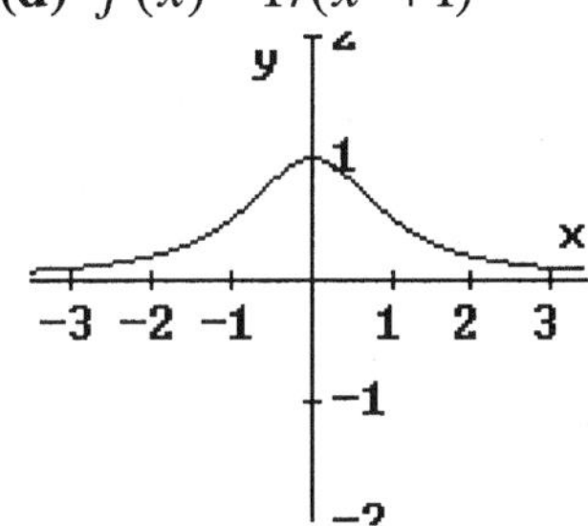

Domain: $(-\infty, \infty)$; range: $(0, 1]$

59. (a) $f(x) = \sqrt{x^2 + 1}$

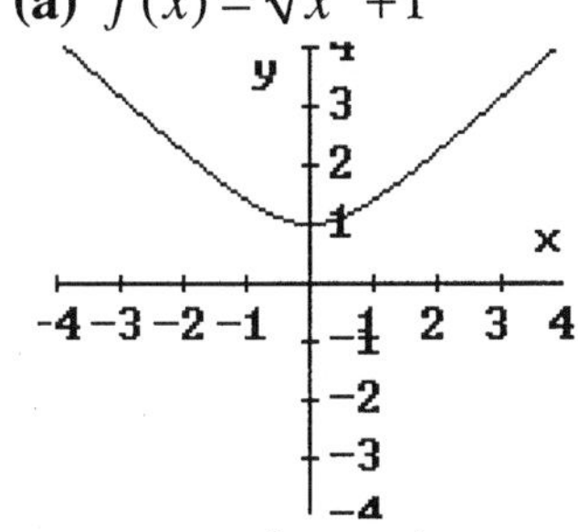

Domain: $(-\infty, \infty)$; range: $[1, \infty)$

(b) $f(x) = x + \sqrt{x^2 + 1}$

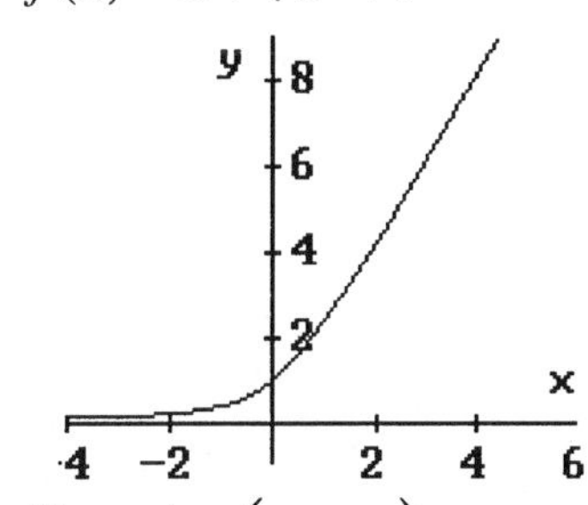

Domain: $(-\infty, \infty)$; range: $(0, \infty)$

(c) $f(x) = x\sqrt{x^2 + 1}$

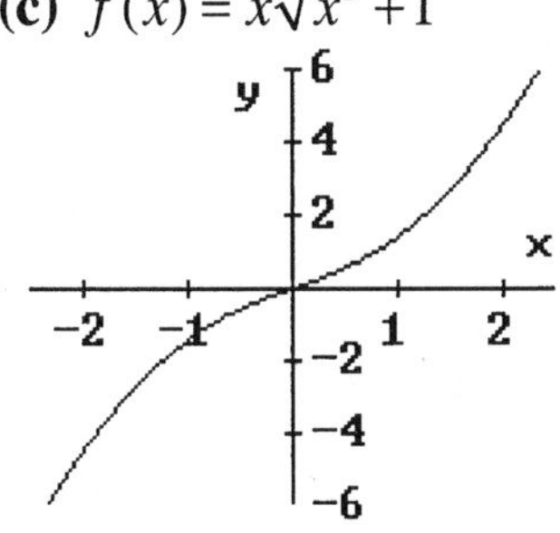

Domain: $(-\infty, \infty)$; range: $(-\infty, \infty)$

(d) $f(x) = x / \sqrt{x^2 + 1}$

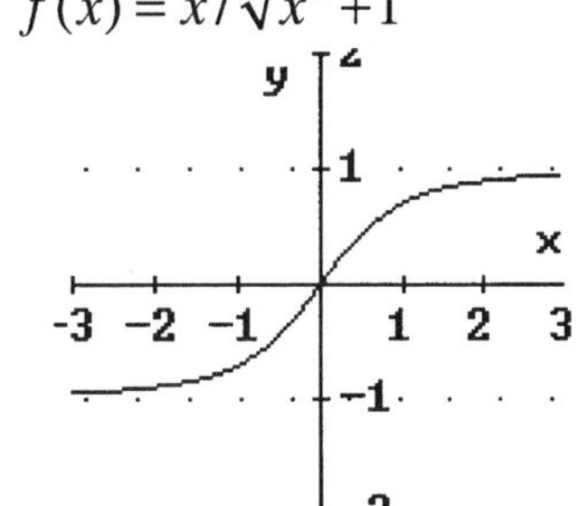

Domain: $(-\infty, \infty)$; range: $(-1, 1)$

61. The average rate of change of $f(x) = 2x^2$ over [0,2] is 4.

63. The average rate of change of $f(x) = 7x + 5$ over [2.3,6.8] is 7

65. (a) Solving for *x* in the equation $y = x^2 + 4x + 6$, we obtain two possibilities: $x = -2 - \sqrt{y - 2}$, and $x = -2 + \sqrt{y - 2}$.

(b) If $y \geq 2$, the solutions obtained in (a) are real.

(c) The range of $g(x) = x^2 + 4x + 6$ is the set of *y* such that $y \geq 2$, or $[2, \infty]$.

Applications

67. The volume of a sphere with diameter d is given by $V = \dfrac{\pi d^3}{6}$.

69. The area of a square inscribed in a circle with radius r is $2r^2$. The diagonal of the square is equal to $2r$, so if s is the side of the square, we have the equation $s^2 + s^2 = 4r^2$, or $s^2 = 2r^2$.

71. The area of the triangle is $\dfrac{ab}{2} = \dfrac{a^2}{2a-4}$. The value of b in terms of a is obtained by computing the slope of the line in two different ways, using the pair of points $(0,b)$ and $(2,1)$, and the pair of points $(2,1)$ and $(a,0)$. From the equation $\dfrac{1-b}{2} = \dfrac{1}{2-a}$, we get $b = \dfrac{a}{a-2}$.

73. $V = x(16-2x)^2$, where $0 < x < 8$, since we need x and $16-2x$ to represent lengths.

75. $V \approx 4.36 \times 10^{40}\ mi^3$

77.

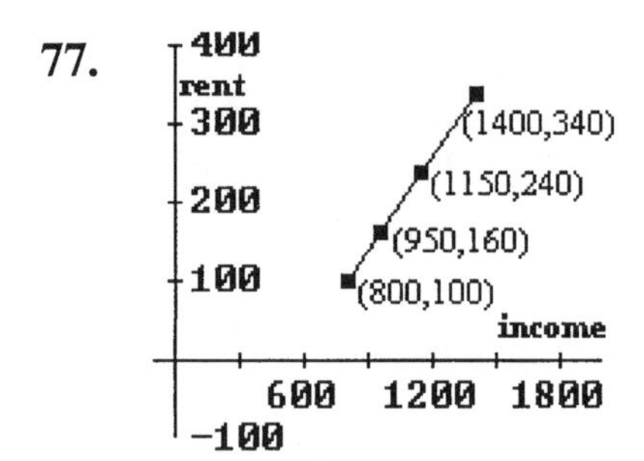

If x represents income, and $R(x)$ the rent, then $R(x) = 100 + 0.4(x - 800)$, $800 \le x \le 1400$.
For incomes between \$800 and \$1400, the rent will be \$100 plus 40% of the income in excess of \$800.

Concepts and Critical Thinking

79. The statement: "The domain of a function is the set of all output values" is false.

81. The statement: "If there is a vertical line that crosses the graph of an equation once, then the graph is the graph of a function" is false.

83. Answers may vary. An example of a function f with range $[5,\infty)$ is $f(x) = \sqrt{x-5}$.

85. Answers may vary. The function $f(x) = \sqrt{x}$ has domain $[0,\infty)$ and range $[0,\infty)$.

87. If the range of $g(x)$ is $(-\infty,4]$ and its domain is $(-\infty,\infty)$, the domain of $\dfrac{1}{g(x)-5}$ is $(-\infty,\infty)$, since $g(x) \ne 5$ for all x in $(-\infty,\infty)$.

89. The functions $f(x) = \dfrac{x^2-1}{x+1}$ and $g(x) = x-1$ are equal for every $x \ne -1$. $f(-1)$ is not defined, and $g(-1) = -2$. Note that $\dfrac{x}{x} = 1$ for all $x \ne 0$. If $x = 0$, $\dfrac{x}{x} \ne 1$ ($\dfrac{0}{0}$ is not defined.)

Chapter 2. Section 2.2 Graphs of Functions

1. $f(x)=3x+1$ has only one zero: $x=-\frac{1}{3}$.

3. $x=-2,\ 4$ are zeros of $g(x)=x^2-2x-8$.

5. $h(x)=|2x+5|$ has one zero: $x=-\frac{5}{2}$.

7. $f(x)=\dfrac{x^2-4}{x^{345}+345x^{23}-172{,}896}$ has two zeros: $x=-2$, and $x=2$.

9. $f(a)=2-\frac{6}{a}$ has one zero: $a=3$.

11. $f(x)=2x^3-3x^2-20x+2$ has three zeros: $x\approx-2.56,\ x\approx0.10$, and $x\approx3.96$.

13. $f(x)=\dfrac{15}{x^2+1}-5x-14$ has three zeros: $x\approx-2.34,\ x\approx-0.61$, and $x\approx0.14$.

15. $f(x)=\sqrt{2x+5}-\sqrt{x^2+1}$ has two zeros: $x\approx-1.24$, and $x\approx3.24$.

17. **(a)** $g(x)=|x-2|$

(b) $g(x)=|x|+1$

(c) $g(x)=|x+2|-2$

19. **(a)** $g(x)=-\frac{1}{4}x^4+\frac{1}{3}x^3+x^2$

(b) $g(x)=\frac{1}{4}x^4+\frac{1}{3}x^3-x^2$

21. Given the graph of the function f:

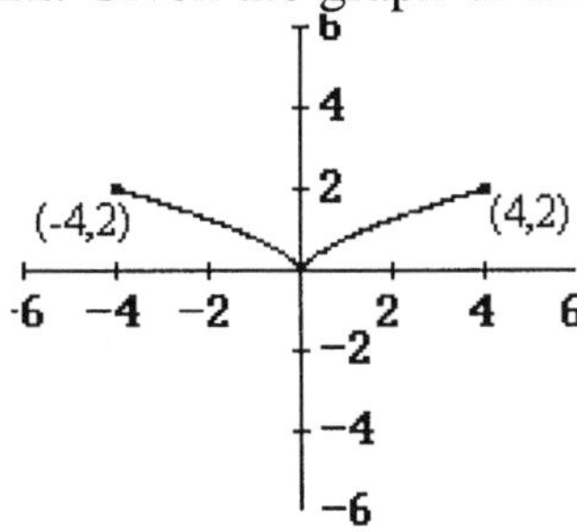

We have the following graphs of transformations of f:

(a) $h(x)=f(x-2)$

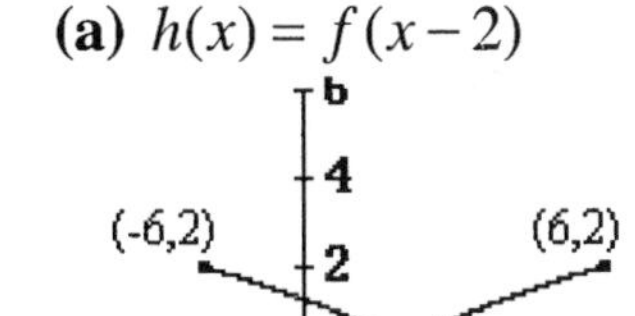

(b) $h(x)=f(x)+3$

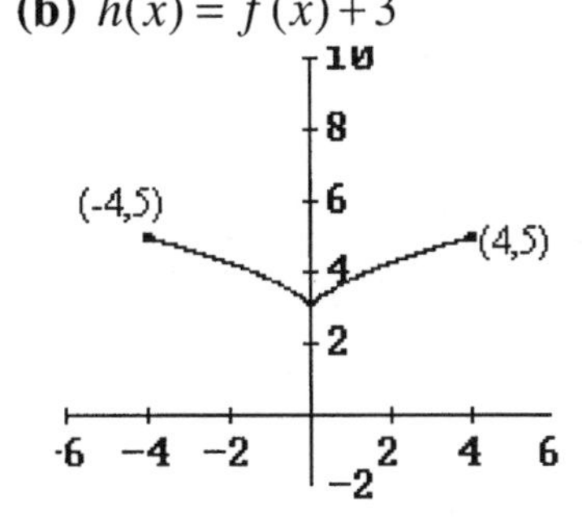

(c) $h(x)=f(x+1)-4$

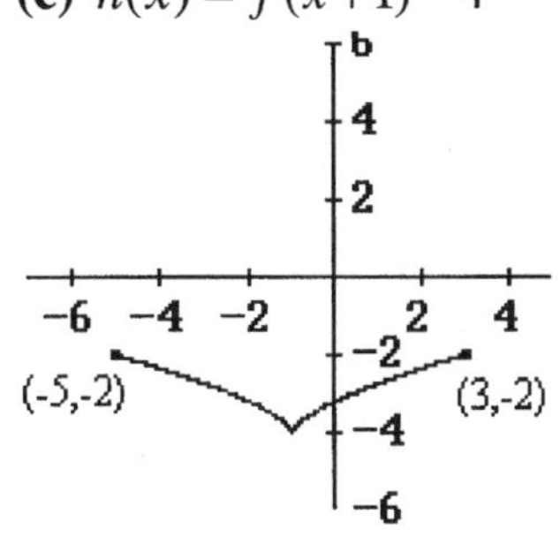

(d) $h(x) = f(-x)$

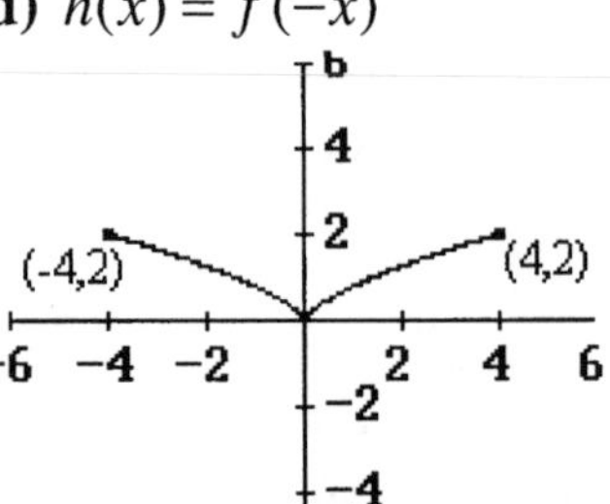

(e) $h(x) = -f(x) + 2$

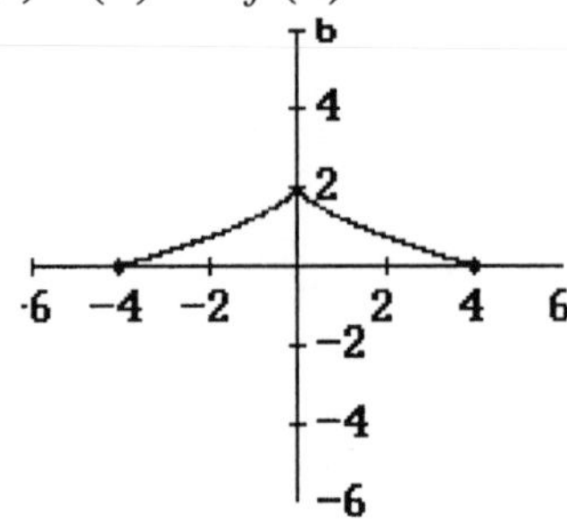

(f) $h(x) = -f(x-1)$

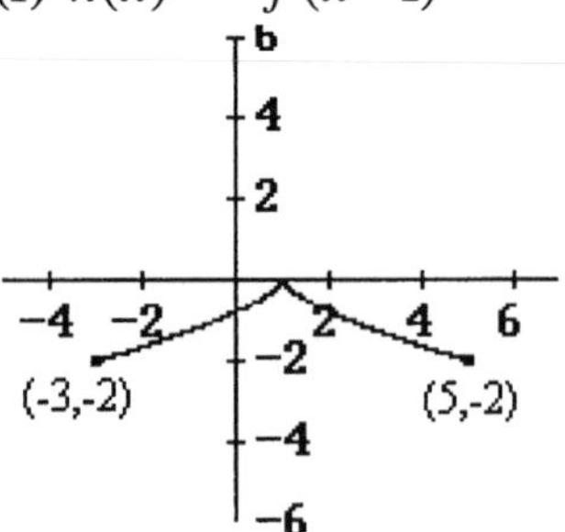

23. Given the graph of the function f:

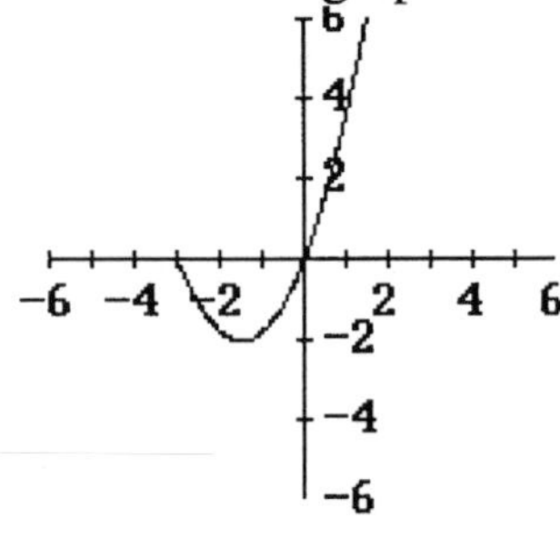

We have the following graphs of transformations of f:

(a) $h(x) = f(x-3) + 2$

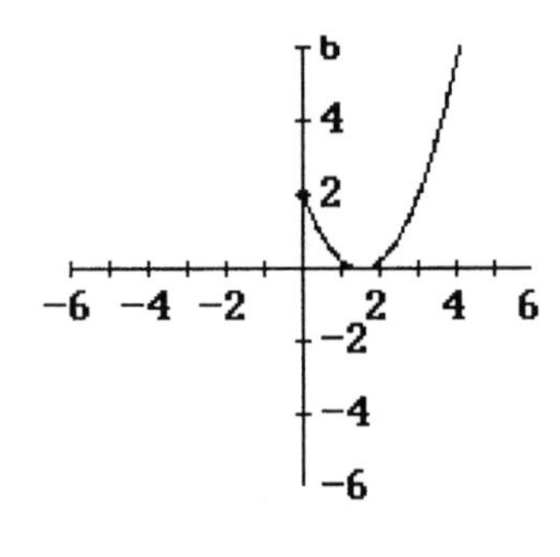

(b) $h(x) = f(-x) - 1$

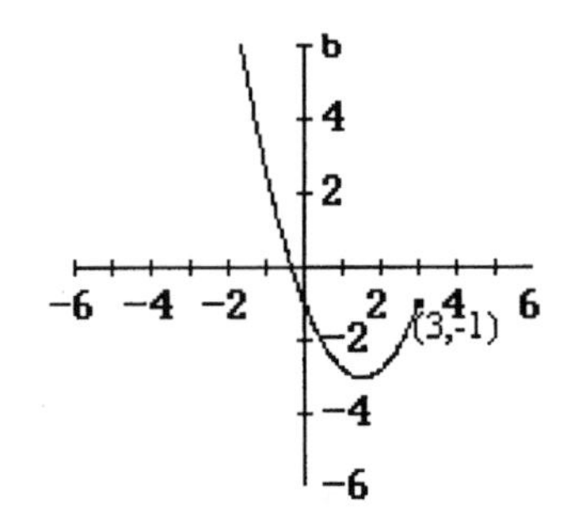

(c) $h(x) = -f(x+1)$

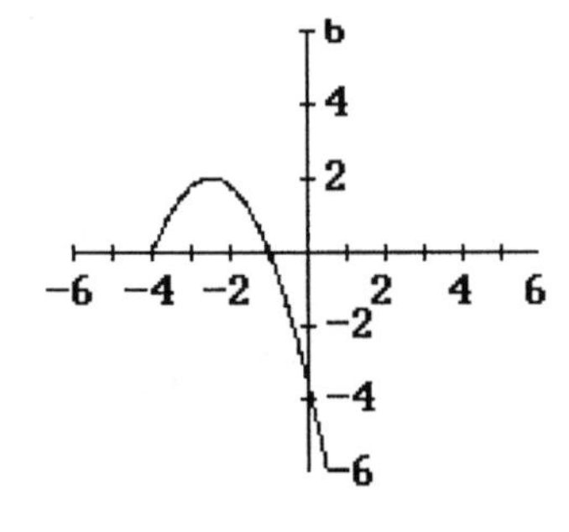

(d) $h(x) = f(-x+2) - 3$

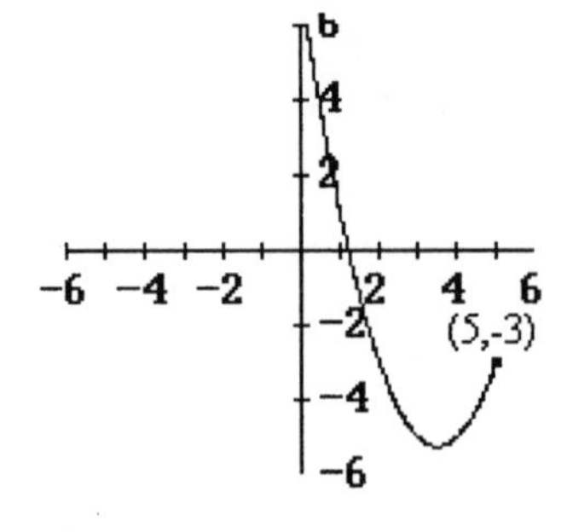

25. The function $f(x) = x^4 - 4x^2$ is even.

27. The function $f(x) = x\sqrt{x^2 + 1}$ is odd.

29. $f(x) = x^4 - \dfrac{x}{12}$ is neither even nor odd.

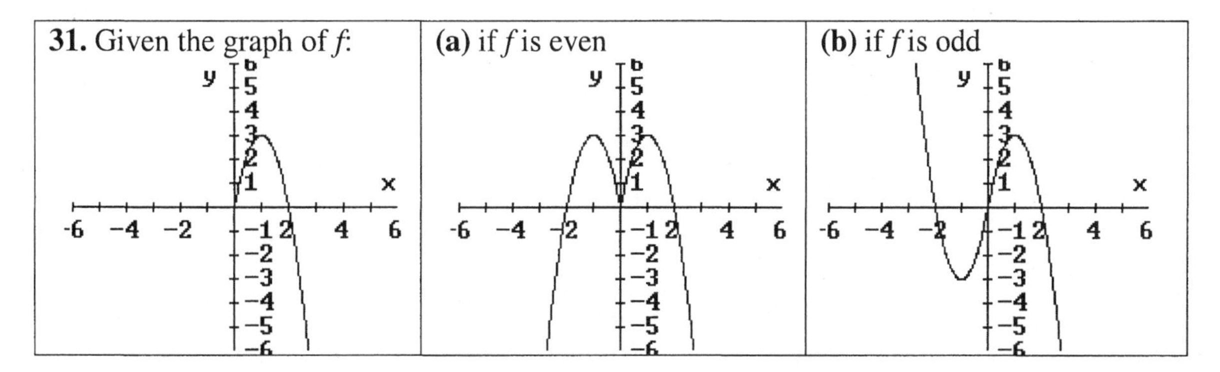

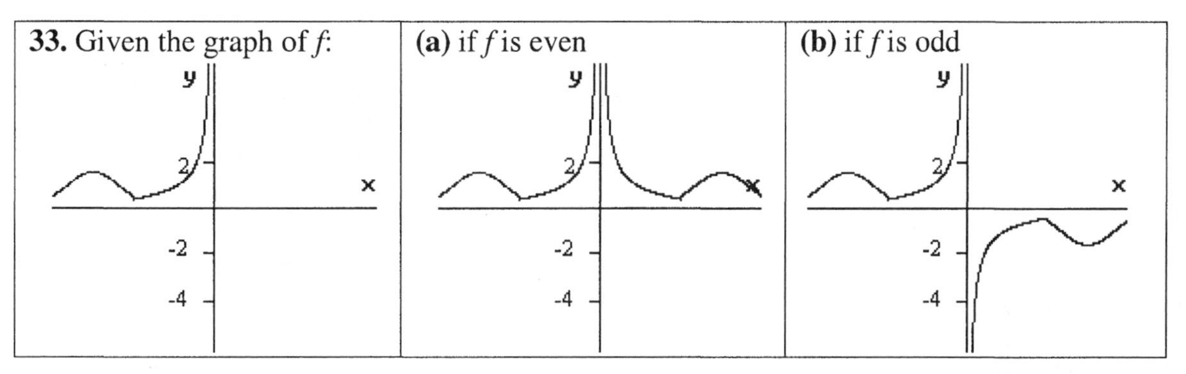

35. The function $f(x) = x^2 - 4x + 5$ is decreasing on $(-\infty, 2)$, increasing on $(2, \infty)$, and has a local minimum at (2, 1).

37. The function $h(x) = 2x^3 - 3x^2 - 6x + 5$ is increasing on the intervals $(-\infty, -0.62)$ and $(1.62, \infty)$. It is decreasing on the interval $(-0.62, 1.62)$. There is a local maximum at the point $(-0.62, 7.09)$ and a local minimum at the point $(1.62, -4.09)$

39. The function $q(x) = \dfrac{5}{x^2 - 4x + 5}$ is increasing on the interval $(-\infty, 2)$ and decreasing on the interval $(2, \infty)$. There is a local maximum at the point (2,5).

41. The function $f(x) = x^4 - 3x^2 + x$ is decreasing on the intervals $(-\infty, -1.3)$ and $(0.17, 1.13)$. It is increasing on the intervals $(-1.3, 0.17)$ and $(1.13, \infty)$. There is a local minimum at the point $(-1.3, -3.51)$, a local maximum at $(0.17, 0.08)$, and a local minimum at $(1.13, -1.07)$.

Applications

43. If $P(x)=x^2+14x+100$ represents the number of deer in hundreds, where x is the number of years since 1995 ($x=0$ represents 1995), then the function $Q(x)=P(x-5)=x^2+4x+55$ represents the same population approximations, with $x=0$ corresponding to the year 2000.

45. If $f(t)=-16t^2+80t$ represents the height in feet of a ball thrown upwards from ground level with an initial velocity of 80 ft/sec, then $g(t)=-16t^2+80t+20$ represents the height in feet of a ball thrown upwards from the top of a 20-foot building, with an initial velocity of 80 ft/sec. The graph of g is a vertical shift of the graph of f, 20 units up.

47. The break-even point occurs when $\text{Revenue}=\text{Cost}$, that is $200x-0.5x^2=50x+4050$. There are two values of x for which this happens: $x=30$, and $x=270$. The maximum profit occurs when $P(x)=R(x)-C(x)=-0.5x^2+150x-4050$ has a maximum, which happens when $x=150$ items sold each month.

49. The rectangle formed by the lines parallel to the axes from the point $P\left(2,\frac{3}{2}\right)$ on the line $y=-\frac{3}{4}x+3$ and the coordinate axes has the maximum area. This maximum area is 3.

51. The area of the enclosed pasture is given by: $A(x)=x(200-2x)=200x-2x^2,\ 0<x<100$. This area is maximum when $x=50$. So, the dimensions of the pasture are 100×50, with an area of 5,000 ft^2.

53. **(a)** The car's maximum distance from the checkpoint is the maximum of the function $f(t)=\dfrac{4}{33}(33t^2-t^3)$, a distance of 645.33 *ft*, 22 seconds after going past the checkpoint.

(b) The graph of $g(t)=33t$ is a ray starting at (0, 0).

(c) The racer and the car meet 16.5 seconds after the racer goes past the checkpoint.

(d) The velocity $v(t)=\dfrac{4}{11}\left(22t-t^2\right)$ when $t=16.5$ sec is 33 *ft/sec,* which is the velocity of the racer. This is what the bottle handler should aim for, to insure a smooth delivery.

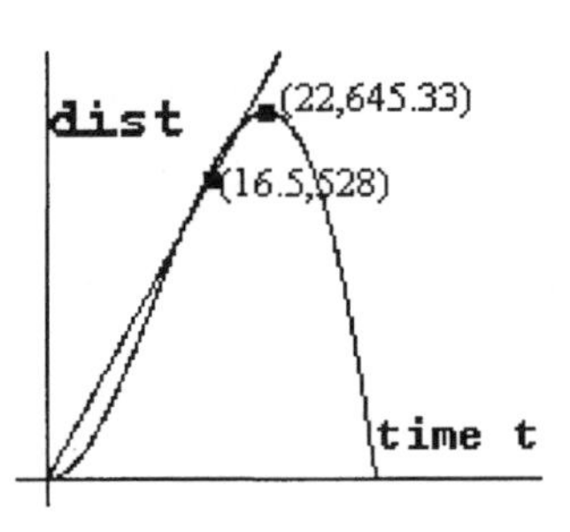

55. The turning points of $f(x)=-14.3x^5+84x^4-175x^3+153.4x^2-49.4x+371$ are: $x\approx 0.25,\ 0.85,\ 1.42$, which correspond to August 1999 (local minimum), April 2000 (local maximum, and October 2000 (minimum), using the fact that $x=0$ corresponds to May 1999. The CO_2 concentration seems to be lowest at the end of the summer, beginning of the fall, and highest in April (midspring.)

57. The model $f(x)=0.0005x^4-0.026x^3+0.48x^2-3.9x+275$ for the U.S. gold reserves in millions of fine troy ounces between 1975 and 1995 says that the reserves have been decreasing between 1975 and 1994, with a small increase in 1995. The lowest gold reserve according to the model was 261 million fine troy ounces in 1994.

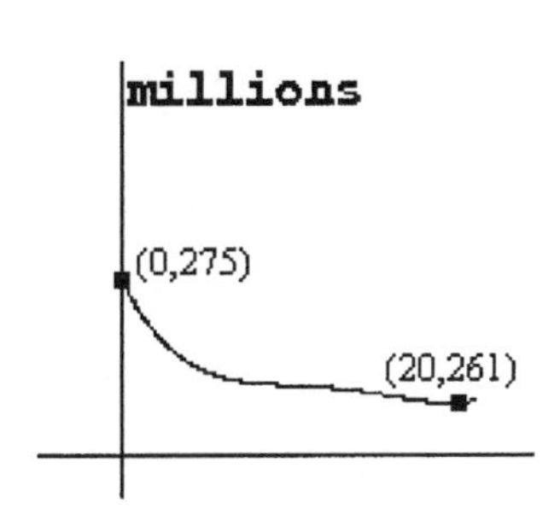

Concepts and Critical Thinking

59. The statement: "If the graph of a function *f* lies entirely in the third quadrant, then 2 is in neither the domain nor the range of *f*" is true.

61. The statement: "If a function *f* satisfies $f(-x)=f(x)$ for all *x*, then its graph will be symmetric with respect to the *y*-axis" is true.

63. Answers may vary. An example of a function for which 2 and -2 are zeros: $f(x)=x^2-4$.

65. Answers may vary. An example of an even function whose graph is a line is $f(x)=5$.

67. If *f* is an even function, and $g(x)=f(x-3)$, then the graph of *g* is symmetric with respect to the line $x=3$. If *f* is odd, then the graph of *g* is symmetric with respect to the point (3, 0).

69. If *n* is even, say $n=2k$, then: $(-x)^n=(-1)^{2k}x^n=\left((-1)^2\right)^k x^n=x^n$, for all *x*, so the function $f(x)=x^n$ is even. If *n* is odd, say $n=2k+1$ then:
$(-x)^n=(-1)^{2k+1}x^n=(-1)^{2k}(-1)x^n=(-1)x^n=-x^n$, for all *x*, so the function $f(x)=x^n$ is odd.

71. It is impossible to have a function symmetric with respect to the *y* axis (other than $f(x)=0$) because there may be only one output value for any *x*, so if $(a,f(a))$ is on the graph of *f*, with $f(a)\neq 0$, then $(a,-f(a))$ may not be on the graph of *f* (otherwise there would be two different output values for the input value *a*.

73. Answers may vary. The graphical concepts of translation, reflection, and symmetry are very valuable to determine properties of functions based on their "parent" function. A graphing calculator is a great tool to help us confirm our hunches, guesses, or inferences, as well as to explore the behavior of computationally involved functions.

Chapter 2. Section 2.3 Combinations of Functions

1. If $f(x)=x$, $g(x)=5$:

(a) $(f+g)(x)=x+5$

(b) $(f-g)(x)=x-5$

(c) $(fg)(x)=5x$

(d) $\left(\frac{f}{g}\right)(x)=\frac{x}{5}$

3. Given $f(x)=x^2$, $g(x)=3x+1$:

(a) $(f+g)(x)=x^2+3x+1$

(b) $(f-g)(x)=x^2-3x-1$,

(c) $(fg)(x)=3x^2+x$

(d) $\left(\frac{f}{g}\right)(x)=\frac{x^2}{3x+1}$, for $x\neq-\frac{1}{3}$

5. Given $f(x)=x$, $g(x)=x$, we have:

(a) $(f+g)(x)=2x$

(b) $(f-g)(x)=0$

(c) $(fg)(x)=x^2$

(d) $\left(\frac{f}{g}\right)(x)=1$, for $x\neq 0$

7. If $f(x)=2x-2$, $g(x)=\frac{2}{x+5}$, we have:

(a) $(f+g)(x)=2x-2+\frac{2}{x+5}$, for $x\neq-5$

(b) $(f-g)(x)=2x-2-\frac{2}{x+5}$, for $x\neq-5$

(c) $(fg)(x)=\frac{4x-4}{x+5}$, if $x\neq-5$

(d) $\left(\frac{f}{g}\right)(x)=x^2+4x-5$, for $x\neq-5$

9. If $f(x)=\sqrt{x-2}$, $g(x)=x-4$, we have:

(a) $(f+g)(x)=\sqrt{x-2}+x-4$, for $x\geq 2$

(b) $(f-g)(x)=\sqrt{x-2}-x+4$, for $x\geq 2$

(c) $(fg)(x)=(x-4)\sqrt{x-2}$, if $x\geq 2$

(d) $\left(\frac{f}{g}\right)(x)=\frac{\sqrt{x-2}}{x-4}$, for $x\geq 2$, $x\neq 4$

11. f and g are given by the first three columns of the table; the rest follow from them:

x	$f(x)$	$g(x)$	**(a)** $(f+g)(x)$	**(b)** $(f-g)(x)$	**(c)** $(fg)(x)$	**(d)** $(f/g)(x)$
0	7	0	7	7	0	undefined
1	0	4	4	−4	0	0
2	−2	8	6	−10	−16	$-\frac{1}{4}$

13. Given $f(x)=2x+3$, and $g(x)=4x-5$, we have: $(fog)(x)=2(4x-5)+3$, or $(fog)(x)=8x-7$, and $(gof)(x)=4(2x+3)-5$, or $(gof)(x)=8x+7$.

15. If $f(x)=x^2$, and $g(x)=2x+7$, then: $(fog)(x)=(2x+7)^2$, or $(fog)(x)=4x^2+28x+49$; on the other hand, $(gof)(x)=2x^2+7$.

17. If $f(x)=\frac{3}{x}$, and $g(x)=\frac{1}{3x}$, then $(fog)(x)=\frac{3}{\left(\frac{1}{3x}\right)}$, or

$(fog)(x)=9x,\ x\neq 0$; on the other hand, $(gof)(x)=\frac{1}{3\left(\frac{3}{x}\right)}$, or $(gof)(x)=\frac{x}{9},\ x\neq 0$.

19. Given $f(x)=x+1$, and $g(x)=x^3-3x$, we obtain $(fog)(x)=x^3-3x+1$. The other order for combining f and g by composition gives $(gof)(x)=(x+1)^3-3(x+1)$, or $(gof)(x)=x^3+3x^2-2$.

21. For $f(x)=7,\ g(x)=x^2+3x+1$: $(fog)(x)=7$, and $(gof)(x)=71$.

23. For $f(x)=\sqrt{x}$, $g(x)=x^4$, we have $(fog)(x)=x^2$. On the other hand, $(gof)(x)=x^2,\ x\geq 0$. (*Note*: the algebraic expression for $(fog)(x)$ and $(gof)(x)$ is the same, but the domain of fog is the set of all real numbers, while the domain of gof is the set of x such that $x\geq 0$.

25. Given $f(x)=3x+5$, and $g(x)=\sqrt{x-2}$, it follows that $(fog)(x)=3\sqrt{x-2}+5$, for $x\geq 2$, and $(gof)(x)=\sqrt{3x+3},\ x\geq -1$.

27. For $f(x)=\frac{2}{x^3+8}$, and $g(x)=\sqrt[3]{\frac{2}{x}-8}$, we obtain $(fog)(x)=\frac{2}{\left(\sqrt[3]{\frac{2}{x}-8}\right)^3+8}$, or

$(fog)(x)=x$, for $x\neq 0$, and $(gof)(x)=\sqrt[3]{\frac{2}{\left(\frac{2}{x^3+8}\right)}-8}$, or $(gof)(x)=x$, for $x\neq -2$.

29. The functions f and g are given by the first three columns on the left of the table, the last two columns correspond to fog, and gof:

f(g(1))

x	$f(x)$	$g(x)$	$(fog)(x)$	$(gof)(x)$
1	2	1	2	2
2	4	2	4	8
3	6	4	8	undefined
4	8	8	undefined	undefined

31. We use the estimates from the graph given in the second and third columns to complete the table:

x	$f(x)$	$g(x)$	$(fog)(x)$	$(gof)(x)$
0	2	1	0	0
1	0	2	1	1
2	1	0	2	2

33. If $f(x)=x+5$, the third iterate of f is:
$$\begin{aligned} f^3(x) &= f(f(f(x))) \\ &= f(f(x+5)) \\ &= f(x+10) \\ &= x+15 \end{aligned}$$

35. If $f(x)=2x$, the fourth iterate of f is:
$$\begin{aligned} f^4(x) &= f\left(f\left(f\left(f(x)\right)\right)\right) \\ &= f\left(f\left(f(2x)\right)\right) \\ &= f\left(f(4x)\right) \\ &= f(8x) \\ &= 16x \end{aligned}$$

37. If $f(x)=x^2$, then $f^{10}(0.98)\approx 1.04\times 10^{-9}$, and $f^{10}(1.02)\approx 640{,}583{,}700.88$.
The iterates $f^n(0.98)$ approach 0 as n gets larger, while the iterates $f^n(1.02)$ grow out of bounds as n gets larger.

39. Given $f(x)=3x+4$,

(a)
$$\begin{aligned} \frac{f(x)-f(2)}{x-2} &= \frac{(3x+4)-10}{x-2} \\ &= \frac{3x-6}{x-2} \\ &= \frac{3(x-2)}{x-2} \\ &= 3, \qquad x\neq 2 \end{aligned}$$

(b)
$$\begin{aligned} \frac{f(x+h)-f(x)}{h} &= \frac{(3(x+h)+4)-(3x+4)}{h} \\ &= \frac{3h}{h} \\ &= 3, \qquad h\neq 0 \end{aligned}$$

41. If $f(x)=x^3$,

(a)
$$\begin{aligned} \frac{f(x)-f(2)}{x-2} &= \frac{x^3-8}{x-2} \\ &= \frac{(x-2)(x^2+2x+4)}{x-2} \\ &= x^2+2x+4, \qquad x\neq 2 \end{aligned}$$

(b)
$$\begin{aligned} \frac{f(x+h)-f(x)}{h} &= \frac{(x+h)^3-x^3}{h} \\ &= \frac{3x^2h+3xh^2+h^3}{h} \\ &= 3x^2+3xh+h^2, \qquad h\neq 0 \end{aligned}$$

Applications

43. The production in thousands of barrels of oil by each of two refineries can be modeled by the functions $B_1(t) = 75t^2 - 450t + 1800$, and $B_2(t) = -85t^2 + 510t + 900$, where t is measured in months.

(a) Sketch the graphs of both functions. For the First refinery: the production slows down from maximum of 1,800,000 barrels the first month, to minimum of 1,125,000 the third month; then production increases to go back to maximum of 1,800,000 barrels the sixth month. Second refinery: output increases from minimum of 900,000 the first month, to maximum of 1,665,000 the third month, then slowing down to minimum production of 900,000 the sixth month.

(b) The function that gives the total production is $B(t) = B_1(t) + B_2(t)$

$$= -10t^2 + 60t + 2700$$

The total production increases from a minimum of 2,700,000 barrels on the first month, to a maximum of 2,900,000 on the third month, and then slows down to reach its minimum again on the sixth month.

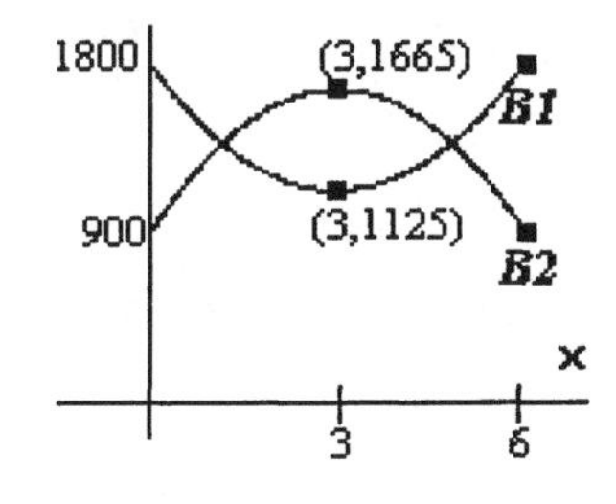

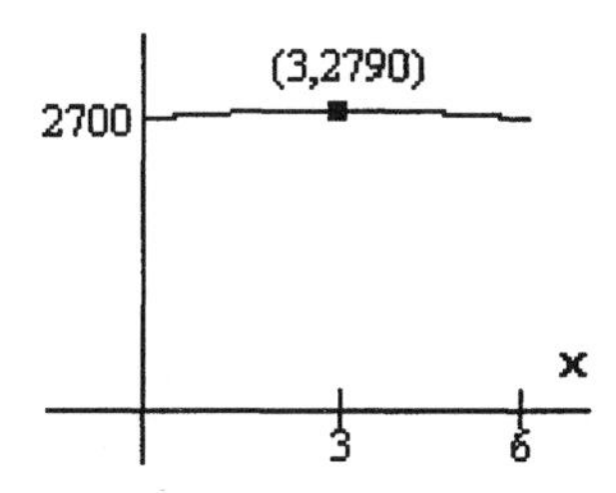

45. **(a)** Kobayashi eats at a constant rate 50 hot dogs in 12 minutes, the function that gives the number of hot dogs eaten by him after t minutes, is $f(t) = \frac{25}{6}t$ (the points (0,0) and (12,50) belong to this linear function, so the slope is $\frac{50-0}{12-0} = \frac{25}{6}$; moreover, the y-intercept is 0, hence $y = \frac{25}{6}t$.)

(b) Kobayashi's weight before eating is 113 lbs, and if we assume that each hot dog and bun weighs 1/3 lb, then the function that models his weight after he eats n hot dogs is $g(n) = 113 + \frac{1}{3}n$.

(c) $(f o g)(t) = \frac{25}{6}\left(113 + \frac{1}{3}t\right)$; $(g o f)(t) = 113 + \frac{1}{3}\left(\frac{25}{6}t\right)$, or $(g o f)(t) \approx 113 + 1.39t$. The combination fog doesn't have any real-world interpretation. The combination gof models Kobayashi's weight as a function of the time t in minutes while he eats the 50 hot dogs. This expression indicates that Kobayashi ingests 1.39 lbs per minute.

Concepts and Critical Thinking

47. The statement: "$(fog)(x) = f(x)g(x)$ for all real numbers *x*" is false. Take, for example, $f(x) = 3x+1$, $g(x) = 2x$. We have $(fog)(x) = 6x+1$, and $f(x)g(x) = 6x^2 + 2x$.

49. The statement: "If the domain of each of *f* and *g* is the set of all real numbers, then the domain of f/g is the set of all real numbers" is false. Consider, for instance, $f(x) = x^2$, $g(x) = x^2$. We have $(f/g)(x) = 1$ for $x \neq 0$, since x^2/x^2 is undefined for $x = 0$.

51. Answers may vary. An example of a function f such that $f(x+c) = f(x)$, for all real numbers *c*, is given by $f(x) = 5$. Any constant function will be such an example.

53. Answers may vary. An example of a pair of functions f and g such that $fog = gof$ is: $f(x) = 2x$, $g(x) = 3x$. We have $(fog)(x) = 2(3x) = 6x$, and $(gof)(x) = 3(2x) = 6x$.

55. $(fo(goh))(x) = f((goh)(x)) = f(g(h(x)))$, and $((fog)oh)(x) = (fog)(h(x)) = f(g(h(x)))$, therefore $fo(goh) = (fog)oh$, so the composition of functions is associative.

57. Answers may vary. A reflection about the *x*-axis followed by a counterclockwise rotation by 90° yields a reflection about the line $y = x$.

Chapter 2. Section 2.4 Inverses of Functions

1. If $f(x)=x+2$, $g(x)=x-2$: $(f o g)(x)=f(x-2)=(x-2)+2=x$, and $(g o f)(x)=g(x+2)=(x+2)-2=x$; then f and g are inverses of one another.

3. For $f(x)=2x+1$, and $g(x)=\frac{1}{2}x-1$: $(f o g)(x)=2\left(\frac{1}{2}x-1\right)+1=x-1$. Then, f and g are not inverse functions.

5. If $f(x)=(x+1)^3$, and $g(x)=\sqrt[3]{x}-1$, then $(f o g)(x)=\left(\left(\sqrt[3]{x}-1\right)+1\right)^3=\left(\sqrt[3]{x}\right)^3=x$, and $(g o f)(x)=\sqrt[3]{(x+1)^3}-1=(x+1)-1=x$; then f and g are inverses of one another.

7. Given $f(x)=\frac{x+3}{2x-1}$, and $g(x)=\frac{x+3}{2x-1}$, then:

$$(f o g)(x)=\frac{\left(\frac{x+3}{2x-1}\right)+3}{2\left(\frac{x+3}{2x-1}\right)-1}=\frac{\frac{(x+3)+3(2x-1)}{2x-1}}{\frac{2x+6-(2x-1)}{2x-1}}=\frac{7x}{7}=x$$; since $f=g$, we also have $(g o f)(x)=x$; then f and g are inverses of one another.

9. $f(x)=\sqrt{x-6}$; $g(x)=x^2+6$, $x\geq 0$. Then $(f o g)(x)=\sqrt{\left(x^2+6\right)-6}=\sqrt{x^2}=|x|=x$, and $(g o f)(x)=\left(\sqrt{x-6}\right)^2+6=(x-6)+6=x$. So, f and g are inverses of one another.

In Exercises 11-19, the Horizontal Line Test may be used to determine if a function is 1-1. An algebraic argument is provided here in each case.

11. $f(x)=-2x+5$ is 1-1, since $-2a+5=-2b+5$ implies $-2a=-2b$, and $a=b$.

13. $f(x)=x^2$ is not 1-1. For example, $f(-2)=4$, and $f(2)=4$.

15. $h(x)=x^3+2x+1$ is 1-1.

To see this, we first note that if $a<b$, then $a^3<b^3$:

(i) if $a<0<b$, then $a^3<0$, and $b^3>0$, so $a^3<b^3$;

(ii) if $0<a<b$, or $a<b<0$, then $ab>0$, and $b^3-a^3=(b-a)\left(b^2+ab+b^2\right)>0$, since the two factors in the last expression are positive, then in both cases $a^3<b^3$.

We apply this fact to show that $h(x)=x^3+2x+1$ is 1-1: if $a\neq b$, we can assume $a<b$, then $a^3+2a+1<b^3+2b+1$, so $h(a)\neq h(b)$.

17. $g(x) = \frac{x^4}{12} - x^3$ is not 1-1. For example, $g(0) = 0$, and $g(12) = 0$.

19. $h(x) = |x-3|$ is not 1-1. For example, $|-4+3| = 1$, and $|-2+3| = 1$.

21. $f(x) = \frac{1}{3}x - 1$ is 1-1. To find the inverse of *f*: start with $y = \frac{1}{3}x - 1$, interchange *x* and *y*, and solve for *y*: $x = \frac{1}{3}y - 1$, so $y = 3x + 3$, and $f^{-1}(x) = 3x + 3$.

23. $f(x) = x^4$ is not 1-1. For example, $f(-1) = 1$, and $f(1) = 1$.

25. $f(x) = x^2 + 1$, $x \geq 0$, is 1-1. To find the inverse, start with $y = x^2 + 1$. Interchange *x* and *y*, $x = y^2 + 1$, and solve for *y*: $y^2 = x - 1$, so $y = \sqrt{x-1}$, or $y = -\sqrt{x-1}$, but this last option has been eliminated since the domain of *f* was stipulated to be the set of real numbers *x*, $x \geq 0$. So, $f^{-1}(x) = \sqrt{x-1}$.

27. $f(x) = \sqrt{2x+5}$ is 1-1. To find the inverse: start with $y = \sqrt{2x+5}$, interchange x and y, $x = \sqrt{2y+5}$, and solve for *y*, $x^2 = 2y + 5$, so $y = \frac{x^2-5}{2}$. Then, $f^{-1}(x) = \frac{x^2-5}{2}$.

29. $f(x) = \frac{x^2}{x^2+1}$ is not 1-1. For example: $f(-1) = \frac{1}{2}$, and $f(1) = \frac{1}{2}$.

31. The function *f* given by the table on the left is 1-1. Its inverse is given by the table next to it.

x	$f(x)$
0	4
1	7
2	10
3	13
4	16

x	$f^{-1}(x)$
4	0
7	1
10	2
13	3
16	4

33. The function *f* given by the table is not 1-1: $f(-1) = 1$, and $f(1) = 1$.

x	$f(x)$
−2	5
−1	1
0	0
1	1
3	10

35. The graph of f passes the horizontal line test, so it is one to one and it has an inverse. The graph of the inverse function is the reflection of the graph of f about the line $y = x$.

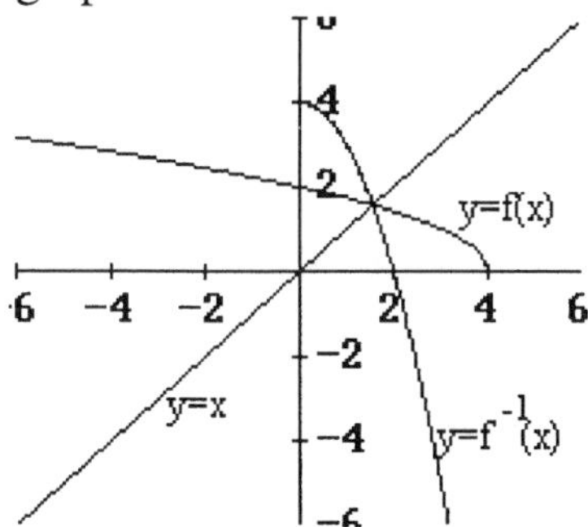

37. The graph of f doesn't pass the horizontal line test. For example, $f(-1) = f(1)$. f is not 1-1.

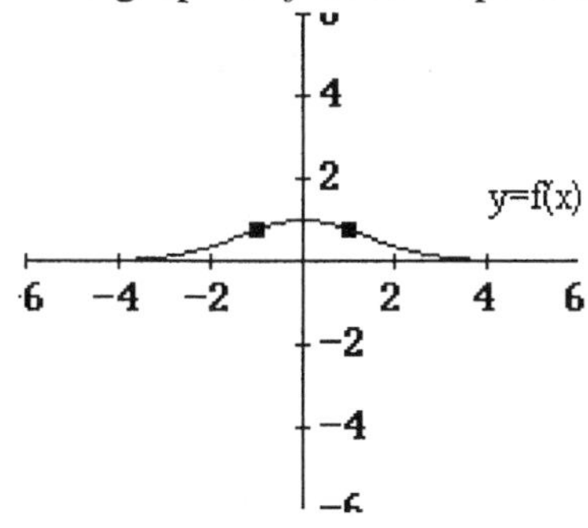

39. The graph of f passes the horizontal line test, so it is one to one and it has an inverse. The graph of the inverse function is the reflection of the graph of f about the line $y = x$,

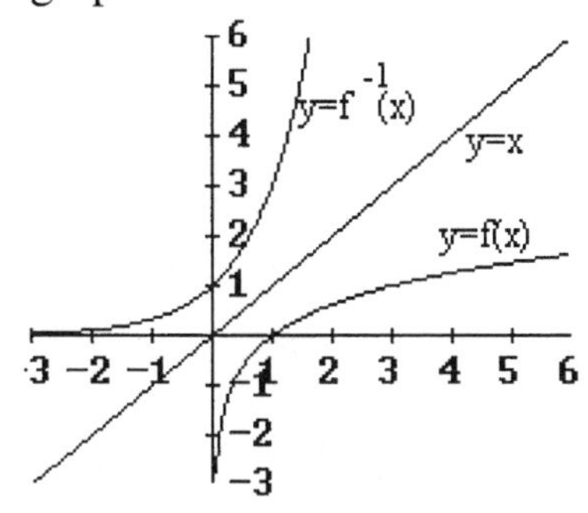

Applications

41. M is the function that assigns to every year since 1935 the maximum temperature in Los Angeles, assuming a constant rate of increase of $0.12°\text{F}$ per year, starting at $98°\text{F}$ for $t = 0$ (where $t = 0$ corresponds to 1935.)

(a) The algebraic expression for M is $M(t) = 98 + 0.12t$.

(b) The year 2035 corresponds to $t = 100$; the model in (a) predicts a maximum temperature in Los Angeles of $110°\text{F}$.

(c) Solving for t in the equation $M = 98 + 0.12t$, we have $t = \dfrac{M - 98}{0.12}$. So, $M^{-1}(t) = \dfrac{t - 98}{0.12}$.

(d) M^{-1} takes the expected maximum temperature in Los Angeles assuming an increase of $0.12°\text{F}$ per year, and giving as output the year when that temperature will be reached, in years since 1935.

43. The *Rule of* 72 gives the time t in years that it would take an investment to double if it is earning interest at an annual rate of $i\%$: $t = f(i) = \dfrac{72}{i}$.

(a) For a debt of \$400 to grow to \$800 if the annual interest rate is 18% and no payments are made, it will take $t = f(18) = \frac{72}{18} = 4$ years.

(b) The inverse of f takes as input the doubling time, and gives as output the annual interest rate necessary to achieve the given doubling time: $i = g(t) = \frac{72}{t}$ (g is f^{-1}.)

(c) The annual interest rate needed for an investment to double in 8 years is $i = \frac{72}{8}$, or 9%.

45. The described maximal healthy weight in pounds as a function of the height in inches is: $w = f(h) = 0.0355264h^2$.

(a) At a height of 6' 2", or 74 inches, Arnold Scharzenegger should have had a maximal healthy weight of $f(74) = 194$ lbs. So, by this standard, Arnold would have been considered obese.

(b) For f to have an inverse, we need to restrict its domain: $h \geq 0$, which is after all the set of values of the height that make sense (height may not be negative.) The inverse of $w = f(h) = 0.0355264h^2$, $h \geq 0$, is $h = 5.30547655\sqrt{w}$.

(c) For a 1400 lbs individual to be healthy under this model, he should have been 198 inches tall (obtained using the result in (c), $h = 5.30547655\sqrt{1400}$), that is 16' 5".

Concepts and Critical Thinking

47. The statement: "If $g(x) = \frac{1}{f(x)}$, then $g = f^{-1}$" is false.

49. "If f is 1-1 function, then so is $g(x) = f(x-1)$" is a true statement. The graphs of f and g are horizontal translations of one another, so if f passes the horizontal line test, g does also.

51. Answers may vary. An example of the graph of a function that is not a 1-1 function corresponds to $f(x) = \sqrt{9 - x^2}$, or any other even function with more than one point.

53. Answers may vary. For $f(x) = -x^2 + 2x$, and $g(x) = x^2$, we have $(f + g)(x) = 2x$. $f + g$ is a 1-1 function, and neither f nor g are 1-1.

55. We agree on the definition: $(fogoh)(x) = f\left(g\left(h(x)\right)\right)$.

(a) $\left(f^{-1}ofop\right)(x) = f^{-1}\left(f(p(x))\right) = p(x)$, so $f^{-1}ofop = p$.

(b) $\left(pofof^{-1}\right)(x) = p\left(f(f^{-1}(x))\right) = p(x)$, so $pofof^{-1} = p$.

(c) Assume that g and h are inverses of f. Because of (a), $gofoh = h$ (g in the place of f^{-1}.)

(d) Assume that g and h are inverses of f. Because of (b), $gofoh = h$ (h in the place of f^{-1}.)

(e) Parts (c) and (d) imply that there is at most one inverse function for a function f.

57. Answers may vary. An example of a function that is its own inverse is $f(x) = \frac{1}{x}$.

Chapter 2. Section 2.5 Selected Functions

1. $f(x) = 3x^2 + 4$ defines a quadratic function.

3. $h(x) = \dfrac{x^2 - 5x + 3}{x+1}$ is neither linear, nor quadratic, nor a piecewise-defined function.

5. $q(x) = 2^x$ is neither linear, nor quadratic, nor a piecewise-defined function.

7. $f(x) = 3x + 9$ is a linear function, with slope 3 and y-intercept 9.

9. $f(t) = 4^{13}t + 7$ is a linear function, with slope 4^{13} and y-intercept 7.

11. $f(x) = \begin{cases} x & x < 0 \\ x^2 & x > 0 \end{cases}$ is a piecewise-defined function.

13. $f(x) = 2x - 1$ is a linear function, with slope 2 and y-intercept -1.

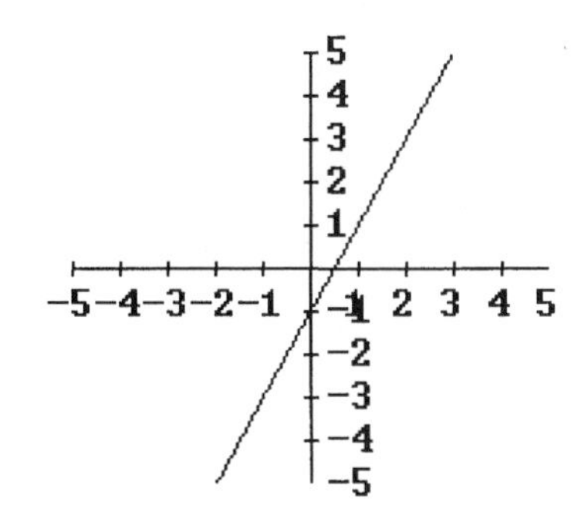

15. $h(x) = \dfrac{-3x+4}{4}$ is a linear function, with slope $-\frac{3}{4}$ and y-intercept 1.

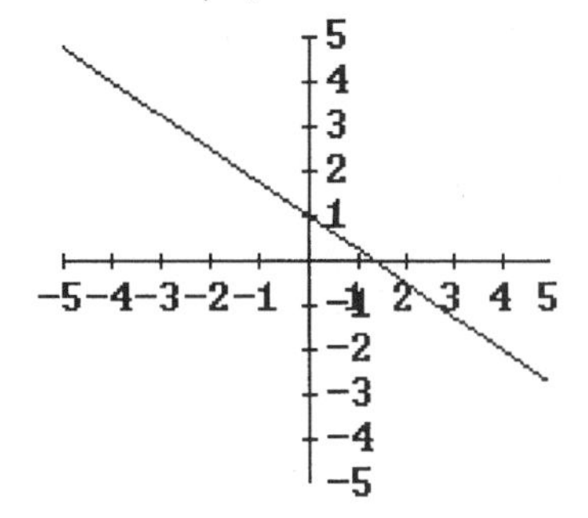

17. If the graph of the linear function h has slope $-\frac{1}{2}$, and $h(2) = 4$, then $y = -\dfrac{1}{2}x + b$, and $4 = -\dfrac{1}{2}(2) + b$, so $b = 5$ and $h(x) = -\dfrac{1}{2}x + 5$.

19. If g is a linear function, the graph of g has y-intercept -4 and $g(3) = 0$, then the slope of g is $m = \dfrac{0-(-4)}{3-0}$, or $m = \dfrac{4}{3}$. Then $g(x) = \dfrac{4}{3}x - 4$.

21. If $f(x) = 3x^2 - 4$, the graph of f has its vertex at $(0, -4)$.

23. Given $f(x) = 2x^2 - 6x + 10$, we have $f(x) = 2\left(x^2 - 3x + \frac{9}{4}\right) + 10 - \frac{9}{2} = 2\left(x - \frac{3}{2}\right)^2 + \frac{11}{2}$.

The vertex of the graph of f is $\left(\frac{3}{2}, \frac{11}{2}\right)$.

25. For $m(x) = x^2 + 16x + 10$, we can rewrite m as: $m(x) = \left(x^2 + 16x + 64\right) + 10 - 64$, or $m(x) = \left(x + 8\right)^2 - 54$. The vertex of the graph of m is the point $\left(-8, -54\right)$.

27. If $P(x) = 1 - 2x + x^2$, then $P(x) = \left(x - 1\right)^2$. The vertex of the graph of P is the point (1,0).

29. If f is a quadratic function, with vertex (1,3), and y-intercept 5, then an equation for f is $f(x) = a(x-1)^2 + 3$, and $5 = a(0-1)^2 + 3$, from where it follows that $a = 2$, so $f(x) = 2(x-1)^2 + 3 = 2x^2 - 4x + 5$.

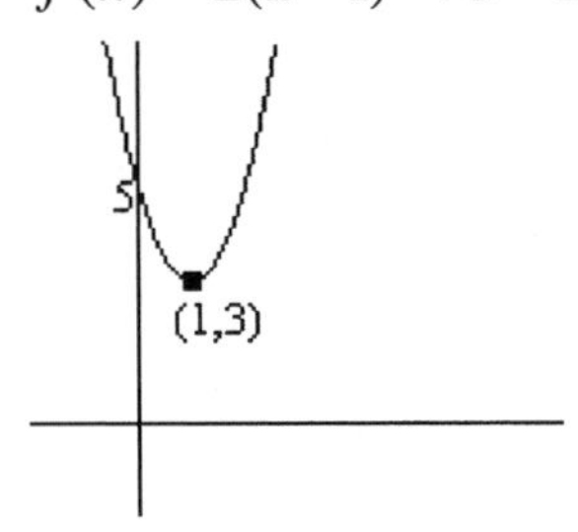

31. If g is a quadratic function, with axis of symmetry $x = 3$, $g(3) = -1$, and $g(2) = -2$, then the vertex of the graph of g is $\left(3, -1\right)$, so $g(x) = a(x-3)^2 - 1$, $-2 = a(2-3)^2 - 1$, so $a = -1$, and $g(x) = -(x-3)^2 - 1$.

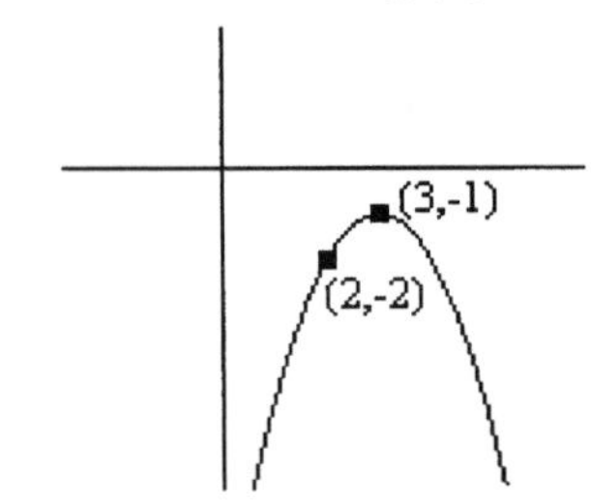

33. $f(x) = \begin{cases} -x - 4, & x \le 0 \\ 2x - 4, & x > 0 \end{cases}$

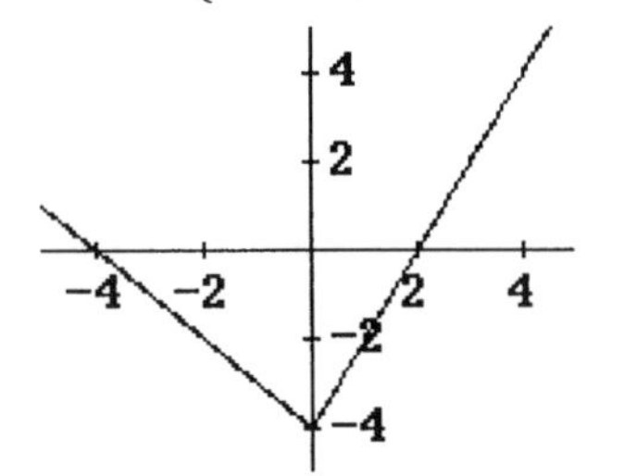

$f(-2) = -2$,
$f(0) = -4$,
$f(3) = 2$

35. $h(x) = \begin{cases} x^2 - 1, & x < 2 \\ \frac{1}{2}x + 1, & x \ge 2 \end{cases}$

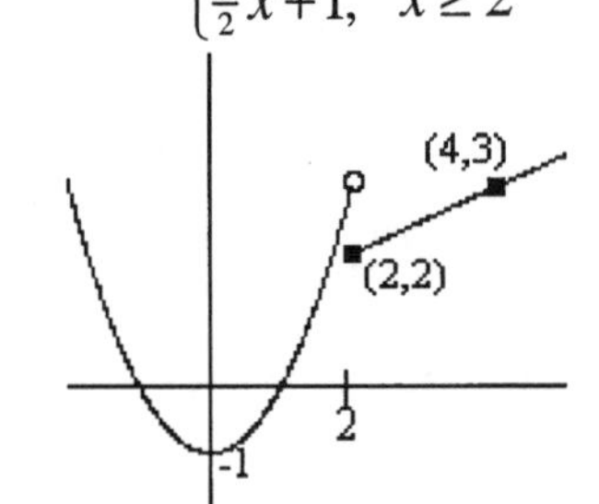

$h(0) = -1$,
$h(2) = 2$,
$h(4) = 3$

37. $f(x)=\begin{cases} x, & x<-1 \\ 1, & -1\le x\le 1 \\ x+2, & x>1 \end{cases}$

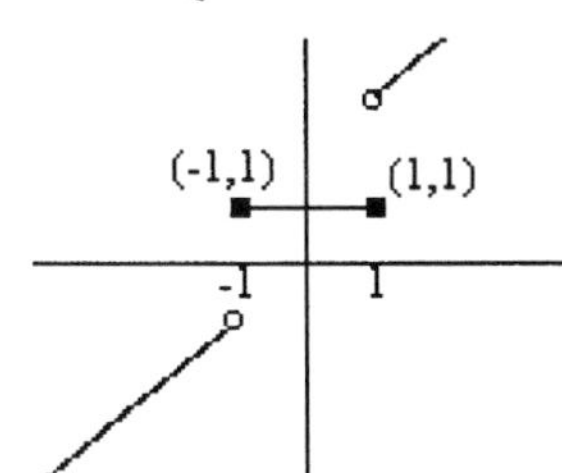

$f(-3)=-3,$
$f(-1)=1,$
$f(4)=6$

In Exercises 39-42, we use the graphs of f and g given below:

$y=f(x)$

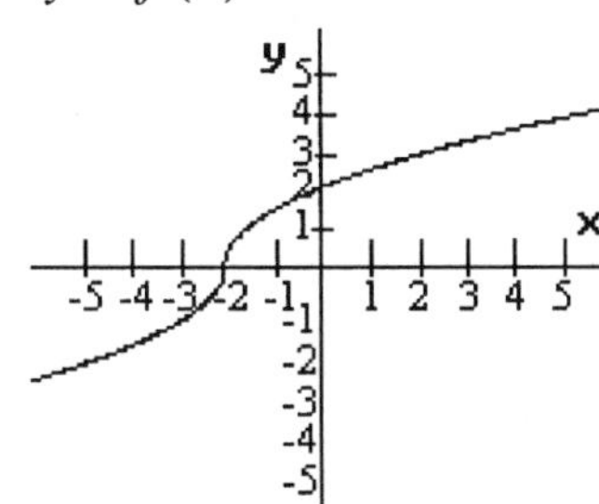

$y=g(x)$

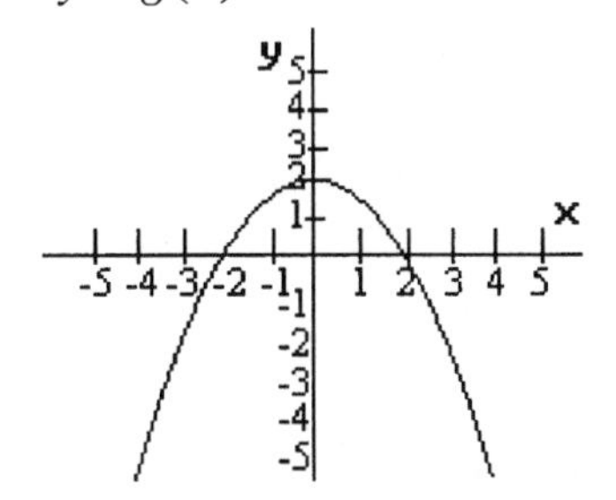

39. $h(x)=\begin{cases} f(x), & x\le -2 \\ g(x), & x>-2 \end{cases}$

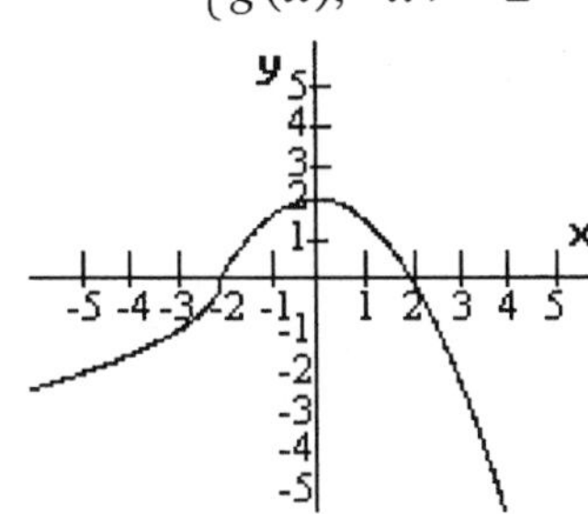

41. $h(x)=\begin{cases} g(x), & x\le -2 \\ x, & -2<x<0 \\ f(x), & x\ge 0 \end{cases}$

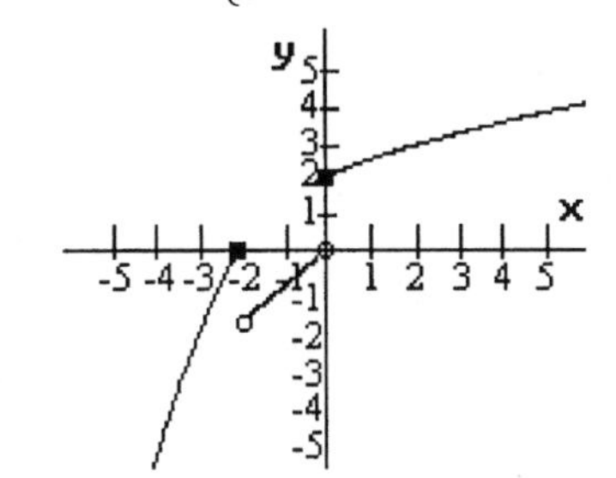

43. $f(x)=\begin{cases} -x^2+2x, & x\le 1 \\ (x-2)^2, & x>1 \end{cases}$

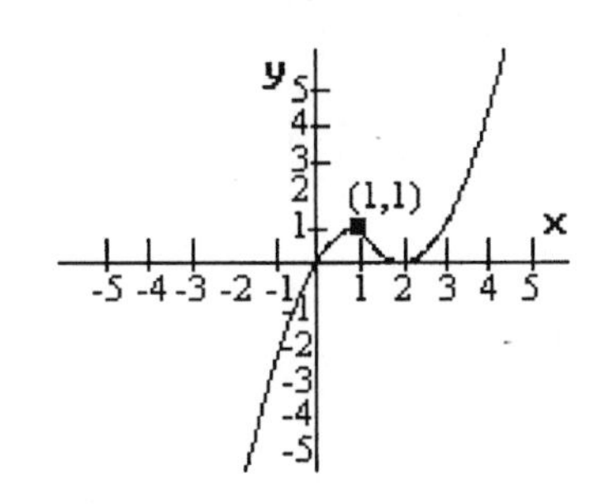

45. $h(x)=\begin{cases} x^3+2x^2, & x<-1 \\ 1, & -1\le x\le 1 \\ x^3-2x^2+2, & x>1 \end{cases}$

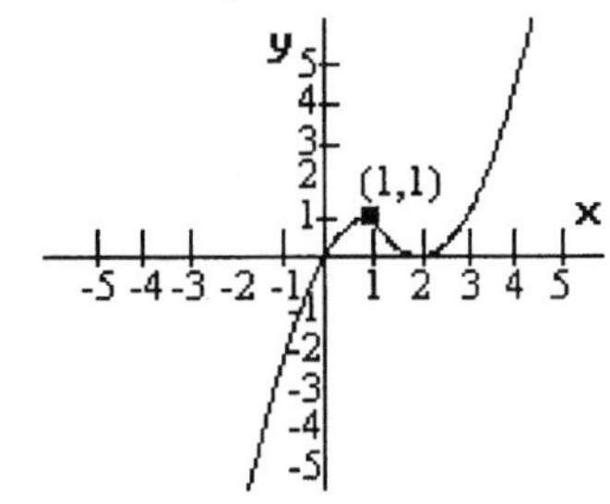

47. If $f(x)=\begin{cases} x+1, & x\le 1 \\ ax^2, & x>1 \end{cases}$, and the graph of f is to be unbroken at $x=1$, we need $a(1^2)=2$, so $a=2$, and

$$f(x)=\begin{cases} x+1, & x\le 1 \\ 2x^2, & x>1 \end{cases}.$$

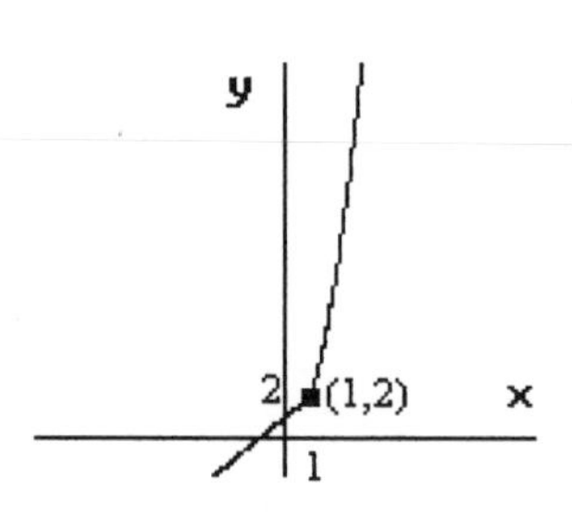

49. If $f(x)=\begin{cases} \dfrac{x^2-1}{x-1}, & x\ne 1 \\ a, & x=1 \end{cases}$, and the graph of f is to be unbroken at $x=1$, we need $a=2$, since $\dfrac{x^2-1}{x-1}=x+1$ for $x\ne 1$, and a has to be the value that this expression approaches as x approaches 1.

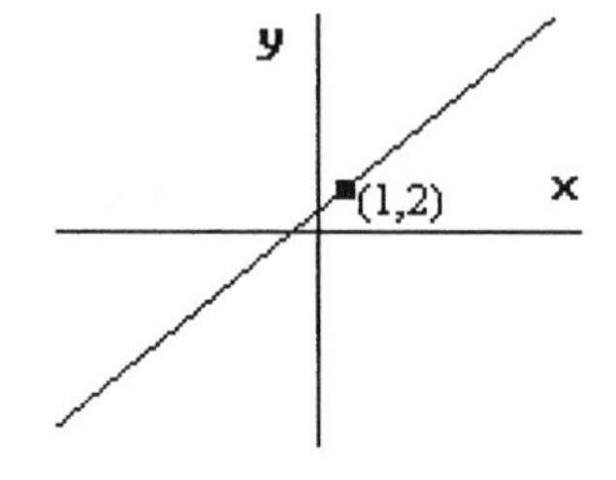

51. Sketch of the graph of $f(x)=\lfloor x+2 \rfloor$: the graph "jumps" at every integer value of x by one unit.

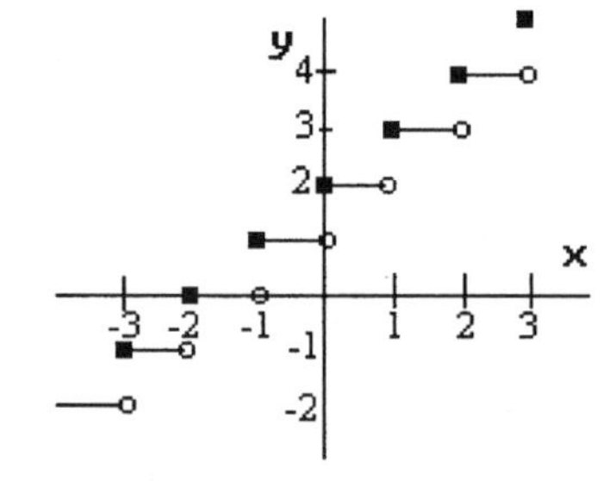

53. For an interval $[a,b]$, define $f_{[a,b]}(x)=\begin{cases} 1, & a\le x\le b \\ 0, & \text{otherwise} \end{cases}$.

(a) The graph of $f_{[0,\,1]}$

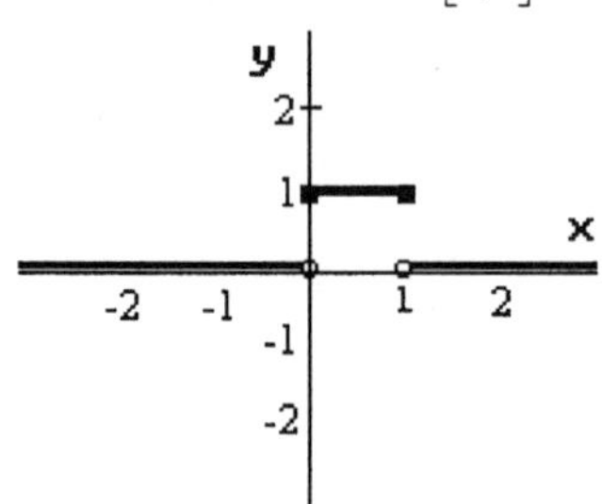

(b) The graph of $f_{[0,\,2]}+f_{[1,\,3]}$

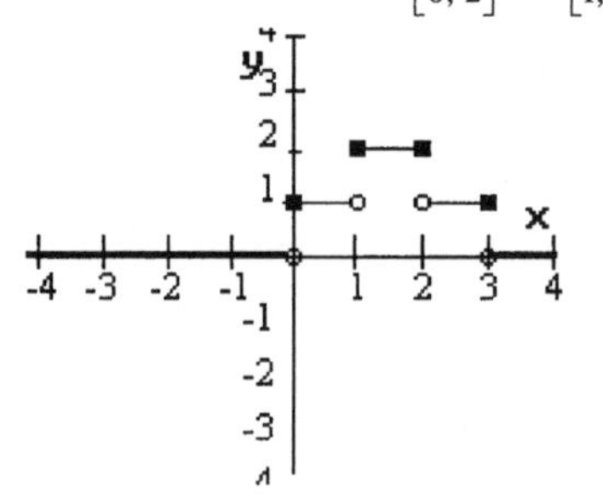

(c) The graph of $f_{[0,\,1]}+f_{[1,\,2]}+f_{[2,\,3]}+\ldots$

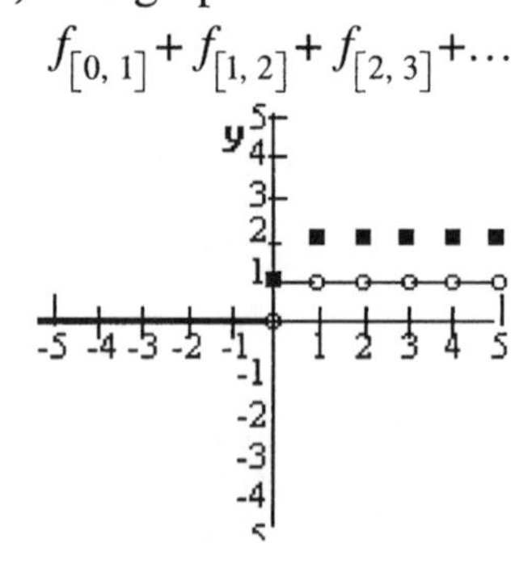

55. If $g(x) = ((x))$ is the function that assigns to each real number x its fractional part, then:

(a) (i) $g(2.8) = 0.8$ (ii) $g(4.7) = 0.7$

(iii) $g\left(\frac{23}{5}\right) = \frac{3}{5}$

(b) Sketch of the graph of g, for $x \geq 0$

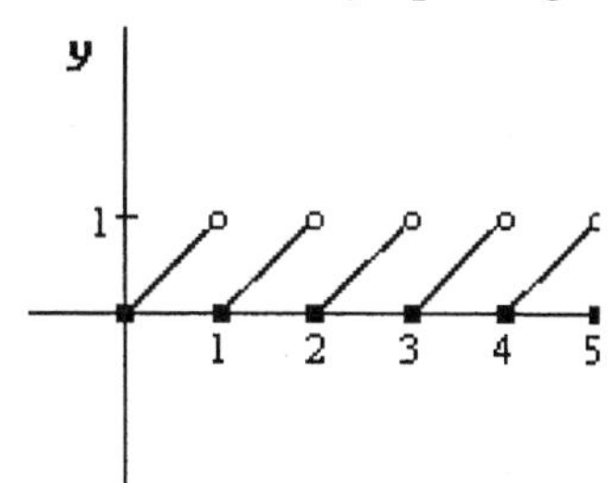

(c) If $x \geq 0$, then $\lfloor x \rfloor + ((x)) = x$.

Applications

57. Given that the fixed overhead costs are \$5000 per month, and there is an added cost of \$80 per month for every apartment that is rented, plus the information that if the rent is \$300 per month then all 50 apartments owned by this landlord will be rented, and that every \$10 increase in the monthly rent will result in two fewer apartments being rented, we have:

(a) $C(x) = 5000 + x$, if x is the number of apartments rented.

(b) To find the equation that gives the rental price p as a function of the number x of apartments rented, we note that $p = 300$ when $x = 50$, and $p = 310$ when $x = 48$. The line passing through the points $(48, 310)$ and $(50,300)$ has slope $m = \frac{310-300}{48-50}$, that $m = -5$. Then $p(x) = -5x + b$. To find b, we use $300 = -5(50) + b$, so $b = 550$ and $p(x) = -5x + 550$.

(c) The revenue function R in terms of the number of apartments rented x is the product of the monthly rent times x, that is: $R(x) = p(x) \cdot x$, so $R(x) = (-5x + 550) \cdot x$. That is, $R(x) = -5x^2 + 550x$.

(d) The profit function P is the difference between the revenue and the total cost, $P(x) = R(x) - C(x)$, or

$P(x) = -5x^2 + 470x - 5000$

(e) The profit P is maximum when $x = 47$, at a monthly rent of \$315 per apartment.

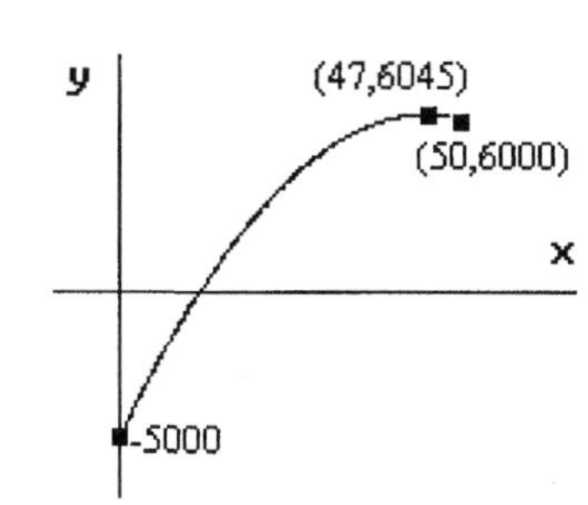

59. The height of a baseball hit from a height of 3 feet with an initial velocity of 100 feet per second, at an angle of 45°, if we ignore air resistance, is given by $f(x) = -\frac{2}{625}x^2 + x + 3$, where x is the distance in feet from home plate.

(a) The maximum height of the baseball is 81.125 *ft*, attained when $x = 156.25\ ft$.

(b) When $x = 300\ ft$ from home plate, $f(300) = 15\ ft$, so the ball clears a 12-ft high fence.

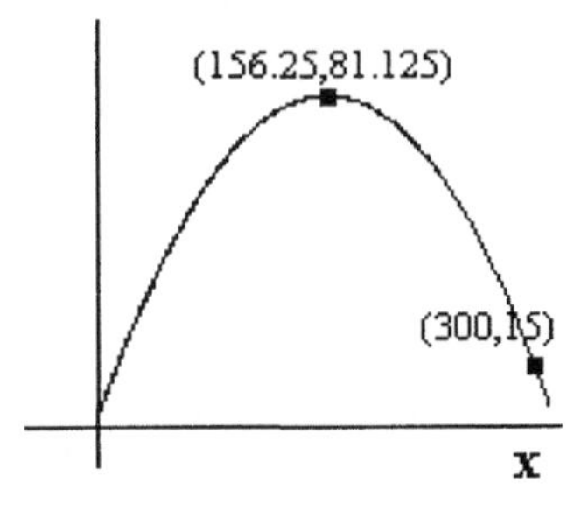

61. If $f(x) = -0.148x^2 + 8.885x + 1004.5$ models the acreage in millions of U.S. farm land as a function of the year x since 1930 (up to 1997), the highest is 1137.85 million acres, when $x = 30$, corresponding to 1960.

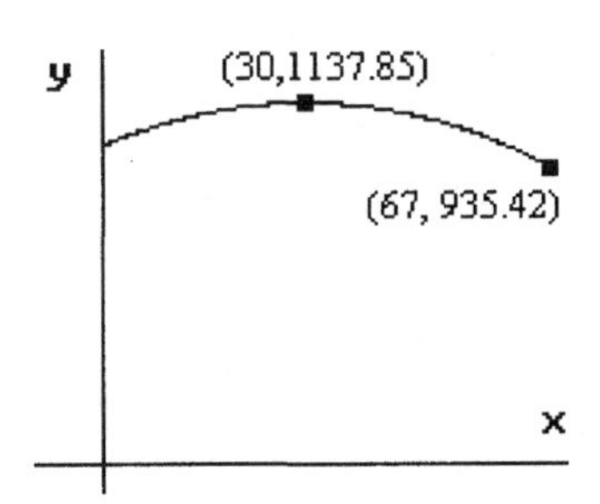

63. For the cost of a cell phone call from New York to Los Angeles, the charges are \$0.25 for the first minute, and \$0.15 for each additional minute. The cost function, in terms of the duration x in minutes of the call is $C(x) = 0.25 - 0.15\lfloor 1 - x \rfloor,\ x \geq 0$.
The longest call that can be made for no more than \$3.00 is 19 minutes long, with a cost of $C(19) = 0.25 - 0.15\lfloor 1 - 19 \rfloor = \$2.95$.

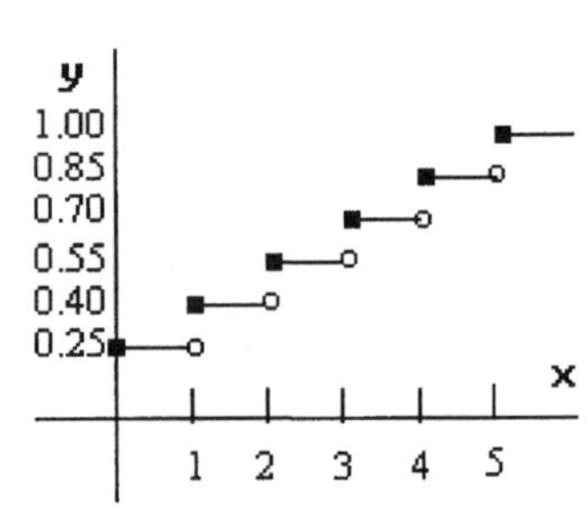

65. If the the tax functions for filing singly and jointly are *S* (singly) and *J* (jointly) are given by:

$$S(x)=\begin{cases} 0.1x, & 0\le x\le 6000 \\ 600+0.15(x-6000), & 6,000<x\le 23,350 \\ 3,202.50+0.27(x-23,350), & 23,350<x\le 56,425 \\ 12,132.75+0.30(x-56,425), & 56,425<x\le 85,975 \\ 20,997.75+0.35(x-85,975), & 85,975<x\le 153,325 \\ 44,640.25+0.386(x-153,325), & x\ge 153,325 \end{cases}$$

$$J(x)=\begin{cases} 0.1x, & 0\le x\le 12,000 \\ 1,200+0.15(x-12,000), & 12,000<x\le 46,700 \\ 6,405+0.27(x-46,700), & 46,700<x\le 112,850 \\ 24,265.50+0.30(x-112,850), & 112,850<x\le 171,950 \\ 41,995.50+0.35(x-171,950), & 171,950<x\le 307,050 \\ 89,280.50+0.386(x-307,050), & x\ge 307,050 \end{cases}$$

If Fred files singly, his taxes are $S(60,000)=13,205.25$. If Irma files singly, her taxes are $S(86,000)=21,006.50$. So, the total of their taxes filing singly is \$34,211.75. Filing singly, their taxes would be $J(146,000)=\$34,210.50$. Filing singly is slightly more expensive than filing jointly for Fred and Irma.

67. The message EFPSNKXEFL would get translated to TWA JULY TWO if the relative position of each letter (from 1 to 26) is considered input of the function $f(x)=3x+5-26\left\lfloor \dfrac{3x+5}{26} \right\rfloor$, and output corresponds to a letter in order in the alphabet.

Concepts and Critical Thinking

69. The statement: "If x^2 appears in the formula defining a function, then the function is said to be quadratic" is false. For example, $f(x)=3x^4+x^2-5$ is not a quadratic function.

71. The statement: "The floor function is an example of a step function" is true.

73. Answers may vary. An example of a piecewise-defined function that is linear on one piece and quadratic on another is $f(x)=\begin{cases} -3x-2, & x\le 0 \\ x^2-2, & x>0 \end{cases}$.

75. Answers may vary. An example of two linear functions whose graphs do not intersect is the pair $f(x)=2x-3$, and $g(x)=2x+7$. Any pair of parallel lines will be an example of this.

77. Answers may vary. An example of a quadratic function that has one zero is $f(x)=x^2-2x+1$.

1. If *y* varies directly as *x*, and $y = 8$ when $x = 2$, then $k = 4$ is the constant of proportionality, and $y = 4x$.

3. If *z* varies inversely as *x*, and $z = 12$ when $x = 4$, then the constant of proportionality is $k = 48$, and $z = \frac{48}{x}$.

5. If *w* varies jointly as *x* and *y*, and $w = 15$ when $x = \frac{1}{2}$ and $y = 3$, then $k = 10$ is the constant of proportionality, and $w = 10xy$.

7. If *w* varies directly as *x*, and $w = 9$ when $x = 3$, then $w = 12$ when $x = 4$.
($9 = 3k$, therefore $k = 3$, and $w = 3x$).

9. If *q* varies jointly as *x* and *y*, and $q = 90$ when $x = 3$ and $y = 5$, then $k = 6$ is the constant of proportionality, and $q = 6xy$. Therefore, if $x = 2$ and $y = 5$, it follows that $q = 60$.

11. If *V* varies directly as the square of *x*, inversely as the cube of *y*, then $V = \frac{kx^2}{y^3}$. When $x = 4$ and $y = 3$, $V = 2$: $2 = \frac{16k}{27}$, or $k = \frac{27}{8}$, and $V = \frac{27x^2}{8y^3}$. If $x = 3$, and $y = 2$, then $V = \frac{243}{64}$.

13. $f(x) = \frac{k}{x^2}$, and $(-3, 2)$ belongs to the graph of *f*. Then $2 = \frac{k}{9}$, so $k = 18$ and $f(x) = \frac{18}{x^2}$.

15. $f(x) = ax^2 + bx$; the graph of *f* passes through the points $(2, -4)$ and $(-1, 5)$. Then:

$\begin{cases} -4 = 4a + 2b \\ 5 = a - b \end{cases}$; from the second equation: $a = b + 5$, substitution in the first equation then gives $-4 = 4(b+5) + 2b$, $6b = -24$, so $b = -4$, and hence $a = 1$. That is: $f(x) = x^2 - 4x$.

17. Since the points $(0,5)$, $(1,3)$, and $(2,3)$ belong to the graph of $h(x) = ax^2 + bx + c$, we obtain three equations involving *a*, *b*, and *c*:

$\begin{cases} 5 = c \\ 3 = a + b + c \\ 3 = 4a + 2b + c \end{cases}$; substituting $c = 5$ in the last two equations gives: $\begin{cases} 3 = a + b + 5 \\ 3 = 4a + 2b + 5 \end{cases}$, so

$\begin{cases} -2 = a + b \\ -2 = 4a + 2b \end{cases}$ $\begin{cases} 8 = -4a - 4b \\ -2 = 4a + 2b \end{cases}$. Add these last two equations, to obtain $6 = -2b$, so $b = -3$; and since $-2 = a + b$, we obtain $a = 1$. So, $h(x) = x^2 - 3x + 5$.

19. The line with the least squares best fit for the points $(1,3)$, $(4,4)$, and $(7,6)$ is $y \approx 0.5x + 2.33$

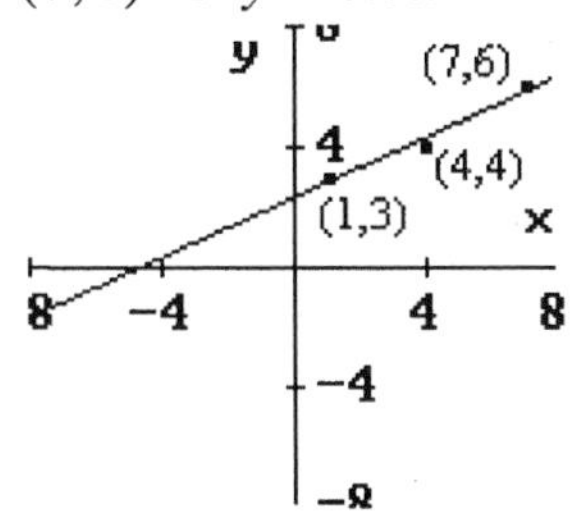

21. The line with the least squares best fit for the points $(-1,-6)$, $(0,-5)$, $(1,-4)$, and $(2,-1)$ is $y \approx 1.6x - 4.8$

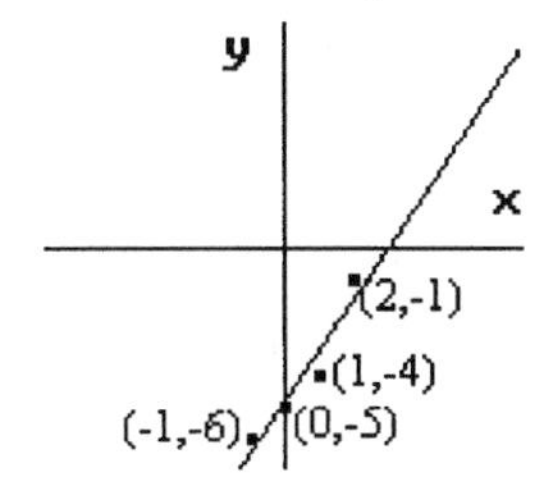

23. For the points $(0,90),(5,101),(10,116),(15,130)$, and $(20,138)$, the line with the least squares best fit is $y = 2.5x + 90$. The predicted value of y when $x = 25$ is 152.5

25. The line with the least squares best fit for the points $(0,80),(1,45),(2,12),(4,-50)$, and $(-5,-78)$ is $y = -31.55x + 77.51$. The predicted value of y when $x = 3$ is -17.14.

Applications

27. The number B of board feet of lumber in a ponderosa pine varies directly as the cube of the circumference c of the pine at waist height, that is: $B = kc^3$. When $c = 100$ inches, $B = 1500$ board feet, *i.e.* $1500 = k(100^3)$. Therefore, $k = 0.0015$, and $B = 0.0015c^3$. Then, if the circumference of a ponderosa pine at waist height is 120 inches, $B = 2592$ board feet.

29. The kinetic energy E of a body in motion varies jointly as the mass m of the body and the square of its velocity v. So, $E = kmv^2$. The kinetic energy E of a 4000-pound vehicle traveling at a velocity of 30 mph is 200,000 joules. That is, $200{,}000 = k(4000)(30^2)$, or $200{,}000 = 3{,}600{,}000k$. Therefore, $k = \frac{1}{18}$, and $E = \frac{mv^2}{18}$. The kinetic energy of the same vehicle traveling at a velocity of 60 mph is 800,000 joules. If the velocity is doubled, the kinetic energy is quadrupled.

31. According to Coulomb's Law, the electric force between two charged particles at rest varies jointly as the charges q_1 and q_2, of the particles and inversely as the square of the distance r between the particles, that is $F = k\frac{q_1 q_2}{r^2}$. Two electrons, each with a charge of 1.6×10^{-19} coulomb, are 1 centimeter apart and exert a force of 2.3×10^{-24} newtons. Therefore, $2.3\times10^{-24} = k\frac{(1.6\times10^{-19})(1.6\times10^{-19})}{(1)^2}$, that is $k = \frac{2.3\times10^{-24}}{2.56\times10^{-38}} = 8.98\times10^{13}$. If the electrons are 2 centimeters apart, the force that they exert is $F = (8.98\times10^{13})\frac{(1.6\times10^{-19})(1.6\times10^{-19})}{(2)^2}$ $= 5.75\times10^{-25}$. (We could have done this directly by dividing 2.3×10^{-24} by 4.)

33. If an ant weighing 0.01 gram can lift an object weighing 0.2 gram, and if strength S varies directly as weight w, then $S = kw$, and in our known example $0.2 = k(0.01)$, so $k = 20$, and $S = 20w$. Under this model, a 150 pound man ought to be able to lift 3000 pounds!

35. Based on the data for weight classes and bench press records:

Weight class (kg)	52	60	75	90
Bench press record (kg)	173	205.5	236	255

the line with the least-squares best fit is $y = 2.08x + 73.18$. The predicted record for the 100-kg class is 281.18 kg, much more than the actual record of 265 kilograms.

37. The line of least-squares best fit to the data on concentration of CFC-11, is $y = 9.21x + 178.1$ ($x = 0$ corresponds to 1980.)

Year	1980	1982	1984	1986	1988	1990
Concentration of CFC-11 (parts per trillion)	180	195	215	230	255	270

The linear model predicts a concentration of $y = 9.21(22) + 178.1 \approx 381$ parts per trillion of CFC-11. The actual concentration was 262 parts per trillion. This indicates that the efforts to reduce the use of CFCs have been successful in curbing the use of CFCs.

39. The number of inmates in state and federal prisons in 1980, 1990, and 2000:

Year	1980	1990	2000
Number of inmates (thousands)	320	743	1,316

The quadratic regression model for these data is $y = 0.75x^2 + 34.8x + 320$ ($x = 0$ corresponds to 1980.) According to this model, the number of inmates in 2005 will be 1,658,750 people.

Concepts and Critical Thinking

41. The statement: "If y varies directly as x, then doubling x will cause y to double as well" is true. (If $y = kx$, then $k(2x) = 2(kx) = 2y$.)

43. "If y varies inversely to x, then doubling x will cause y to double as well" is a false statement.

45. Answers may vary. For example, the three points (1, 2), (3, 4), and (7, 8) produce the regression line $y = x + 1$, and this line passes through all three points.

47. Answers may vary. An example of a collection of three points and a quadratic model that fits the data points with no error may be given with three points that satisfy a quadratic equation. For example, $g(x) = x^2 + 1$ fits the points $(1,2)$, $(2,5)$, and $(3,10)$ with no error.

Chapter 2. Review

Exercises 1-3 refer to the Los Angeles Lakers 2002-2003 Roster displayed below on the right.

Player	Height	Weight
Kobe Bryant	6-7	215
Derek Fisher	6-1	205
Rick Fox	6-7	235
Devean George	6-8	225
A.J. Guyton	6-2	185
Robert Horry	6-10	238
Mark Madsen	6-9	245
Stanislav Medvedenko	6-10	255
Tracy Murray	6-7	228
Shaquille O'Neal	7-1	335
Jannero Pargo	6-2	170
Guy Rucker	6-9	270
Kareem Rush	6-6	215
Soumalia Samake	6-10	240
Brian Shaw	6-6	200
Nick Sheppard	6-11	200
Jefferson Sobral	6-8	210
Samaki Walker	6-9	255

1. The correspondence from D = the set of players to R = the set of heights is a function.

3. The correspondence from D=the set of weights to R=the set of heights is not a function. For example, 215 is associated to the height values 6-6 and 6-7.

5. $y - x^2 + 1 = 0$ defines y as a function of x.

7. $x^2 + y^2 = 5$ does not define y as a function of x. For example, $x = 0$ is associated to two values of y, $-\sqrt{5}$ and $\sqrt{5}$.

9. If $g(x) = 2x^2 - 3x + 1$, then:

(a) $g(4) = 21$ **(b)** $g(-3) = 28$

11. If $f(x) = \frac{x^2}{x^2+1}$, then:

(a) $f(-x) = \frac{x^2}{x^2+1}$

(b) $f(x+1) = \frac{x^2+2x+1}{x^2+2x+2}$

13. If $h(x) = \begin{cases} 4 - x^2, & x \le 2 \\ -x+4, & x > 2 \end{cases}$, then:

(a) $h(0) = 4$

(b) $h(5) = -1$

(c) $h(a) = -a + 4$ if $a > 2$

15. The domain of $f(x) = x^3 + 1$ is the set of all real numbers.

17. The domain of $h(t) = \frac{t}{\sqrt{2t+1}}$ is $\left\{t \mid t > -\frac{1}{2}\right\}$.

19. The domain of $h(x) = \sqrt{x^2 - 4}$ is $(-\infty, -2] \cup [2, \infty)$. The range of this function is $[0, \infty)$.

21. The domain of $f(x) = \frac{|1-x|}{1-x}$ is the set of all real numbers except 1. The range of *f* is $\{-1, 1\}$.

23. Given: $f(x) = 2x^3 - 6x + 1$ **(a)** 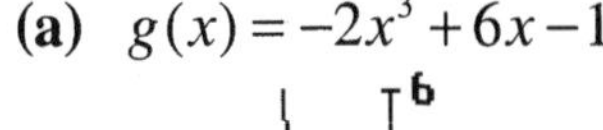 $g(x) = -2x^3 + 6x - 1$ **(b)** 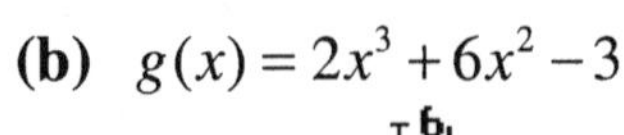 $g(x) = 2x^3 + 6x^2 - 3$

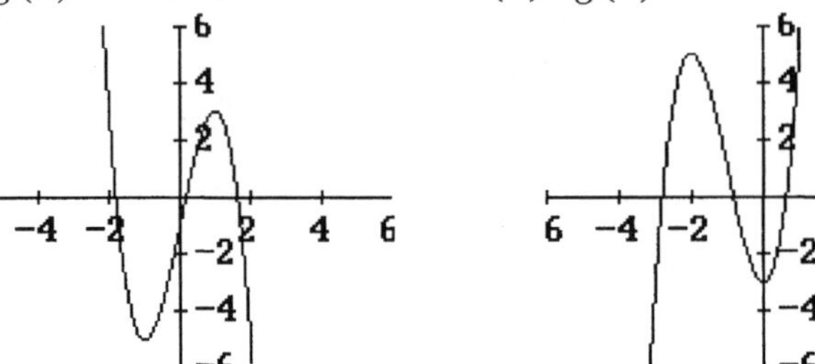

25. Given the graph of $y = f(x)$:

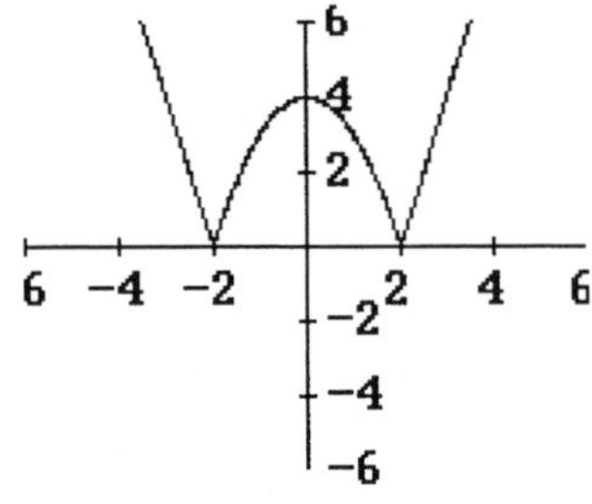

(a) $h(x) = f(x+3)$ **(b)** $h(x) = f(x-2)+1$ **(c)** $h(x) = -f(x) - 2$

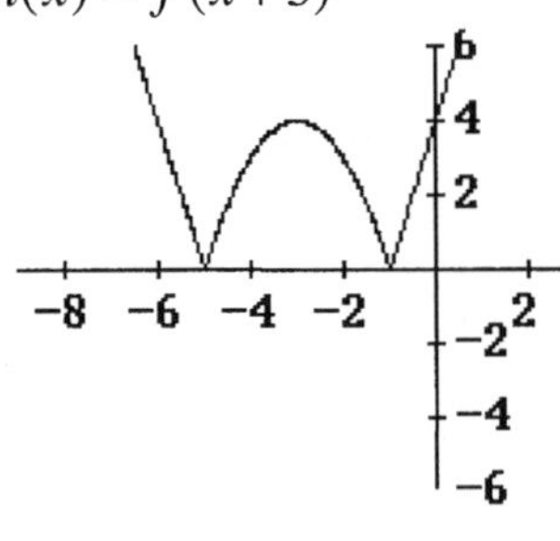

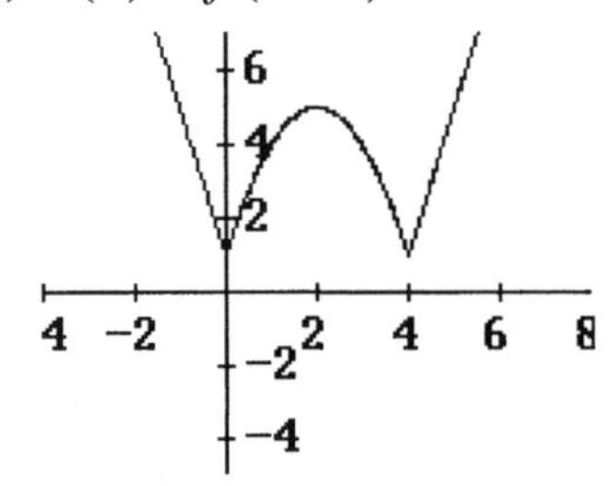

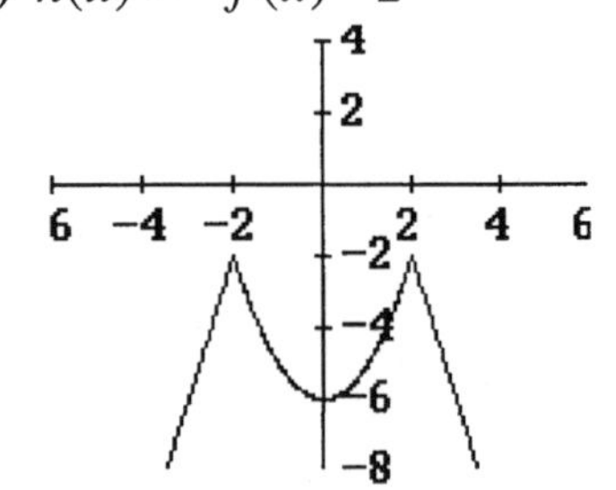

27. The function $f(x) = x(4x^2 - 1)$ is odd.

29. The function $f(x) = |x+1|$ is neither odd nor even.

31. The function $h(x) = 4x - 9$ has a zero at $x = \frac{9}{4}$. (If $4x - 9 = 0$, then $4x = 9$, so $x = \frac{9}{4}$.)

33. The function $g(x) = 3 - \frac{8}{x}$ has a zero at $x = \frac{8}{3}$. (If $3 - \frac{8}{x} = 0$, then $3 = \frac{8}{x}$, so $3x = 8$.)

35. $f(x) = x^3 - 3x + 1$

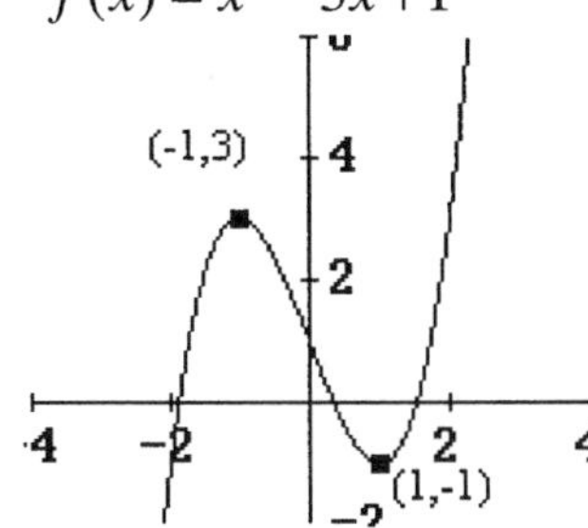

(a) The zeros of *f*, approximate to the nearest hundredth, are -1.88, 0.35, and 1.53.

(b) The turning points of the graph of *f* are: $(-1,3)$ (local maximum), and $(1,-1)$ (local maximum.)

(c) The function f is increasing on $(-\infty,-1)$, and $(1,\infty)$, and decreasing on $(-1,1)$.

37. $h(x) = \frac{x^2-3}{2x-4}$

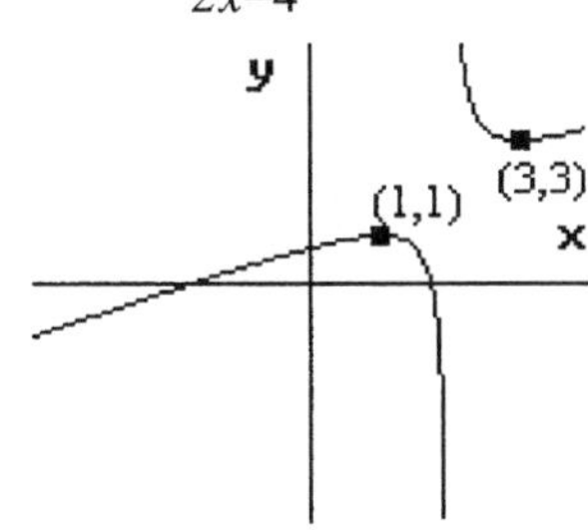

(a) The zeros of *f* are $x = -\sqrt{3}$, and $x = \sqrt{3}$, or -1.73, and 1.73 approximate to the nearest hundredth.

(b) The turning points of the graph of *f* are: $(1,1)$ (local maximum), and $(3,3)$ (local minimum.)

(c) The function f is increasing on $(-\infty,1)$ and $(3,\infty)$, and decreasing on the intervals $(1,2)$ and $(2,3)$.

39. If $f(x) = x^2$, and $g(x) = 2x - 1$:

(a) $(f+g)(x) = x^2 + 2x - 1$

(b) $(fg)(x) = 2x^3 - x^2$

(c) $\left(\frac{f}{g}\right)(x) = \frac{x^2}{2x-1}$, if $x \neq \frac{1}{2}$

(d) $(fog)(x) = 4x^2 - 4x + 1$

(e) $(gof)(x) = 2x^2 - 1$

41. If $f(x) = \frac{1}{x-4}$, and $g(x) = x^3$:

(a) $(f+g)(x) = \frac{x^4-4x^3+1}{x-4}$, if $x \neq 4$

(b) $(fg)(x) = \frac{x^3}{x-4}$, if $x \neq 4$

(c) $\left(\frac{f}{g}\right)(x) = \frac{1}{x^4-4x^3}$, if $x \neq 0, 4$

(d) $(fog)(x) = \frac{1}{x^3-4}$, if $x \neq \sqrt[3]{4}$

(e) $(gof)(x) = \frac{1}{(x-4)^3}$, if $x \neq 4$

43. If $f(x) = 2x^2$, and $g(x) = \sqrt{2x+4}$:

(a) $(f+g)(x) = 2x^2 + \sqrt{2x+4}$, if $x \geq -2$

(b) $(fg)(x) = 2x^2\sqrt{2x+4}$, if $x \geq -2$

(c) $\left(\frac{f}{g}\right)(x) = \frac{2x^2}{\sqrt{2x+4}}$, if $x > -2$

(d) $(fog)(x) = 4x + 8$, if $x \geq -2$

(e) $(gof)(x) = 2\sqrt{x^2+1}$

45. If f and g are given by the first three columns in the following table:

x	$f(x)$	$g(x)$	**(a)** $(f+g)(x)$	**(b)** $(fg)(x)$	**(c)** $\left(\frac{f}{g}\right)(x)$	**(d)** $(f \circ g)(x)$	**(d)** $(g \circ f)(x)$
0	1	3	4	3	$\frac{1}{3}$	0	2
1	2	2	4	4	1	3	0
2	3	0	3	0	undefined	1	1
3	0	1	1	0	0	2	3

47. If $f(x)=x-5$, and $g(x)=x+5$, then $(f \circ g)(x)=(x+5)-5=x$, and $(g \circ f)(x)=(x-5)+5=x$, for all real values of x. Then f and g are inverses of each other.

49. If $f(x)=x^3-4$, and $g(x)=\sqrt[3]{x}+4$, then $(f \circ g)(x)=\left(\sqrt[3]{x}+4\right)^3-4$. So, for example, $(f \circ g)(1)=\left(\sqrt[3]{1}+4\right)^3-4=121$. f and g are not inverses of one another.

51. The function $f(x)=3x-2$ is 1-1. The inverse of f is $f^{-1}(x)=\frac{x+2}{3}$.
(Start with $y=3x-2$; interchange x and y: $x=3y-2$, and solve for y: $y=\frac{x+2}{3}$.)

53. The function $h(x)=x^3-x+1$ is not 1-1. For example: $h(-1)=1$, $h(0)=1$, and $h(1)=1$.

55. The function $f(x)=\sqrt{x-4}$, $x \geq 4$, is 1-1. The inverse of f is $f^{-1}(x)=x^2+4$, $x \geq 0$.
(Start with $y=\sqrt{x-4}$; interchange x and y, $x=\sqrt{y-4}$ – note how $x \geq 0$ – and solve for y: $y-4=x^2$, so $y=x^2+4$, if $x \geq 0$.)

57. The function f (table on the left) is 1-1.

x	$f(x)$
1	2
2	5
3	10
4	17
5	26

x	$f^{-1}(x)$
2	1
5	2
10	3
17	4
26	5

59. The graphs of $y=f(x)$, $y=f^{-1}(x)$, $y=x$, are sketched below.

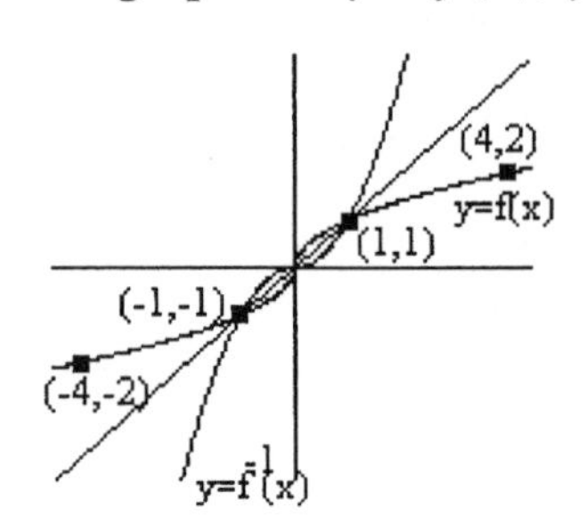

61. The function $h(x) = -(x-3)^2 + 1$ is a quadratic function, with vertex at (3, 1). The x-intercepts of the graph of h are $x = 2$, and $x = 4$.

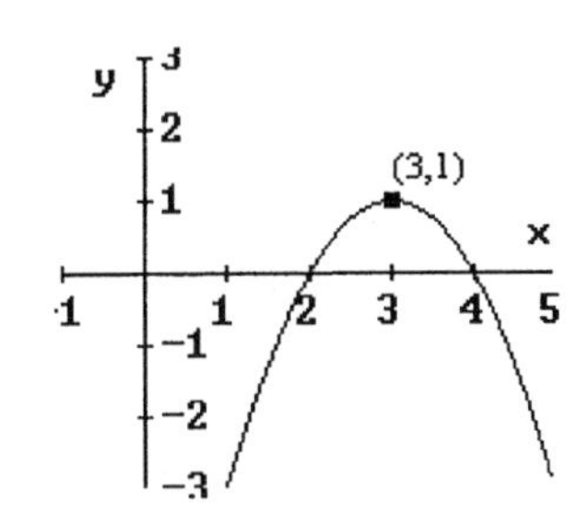

63. The function $f(x) = \dfrac{x+2}{x-2}$ is not linear, quadratic, or piecewise-defined. There is one x-intercept, $x = -2$.

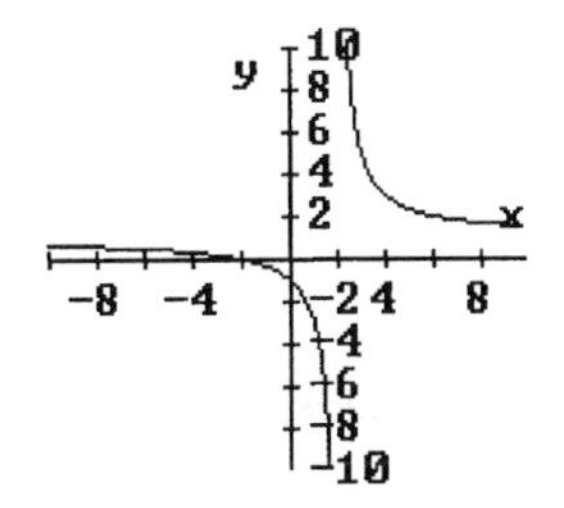

65. The function $g(x) = \begin{cases} 2x+3, x \le 1 \\ 5-x^2, x > 1 \end{cases}$ is a piecewise-defined function. There are two x-intercepts, $x = -\frac{3}{2}$, and $x = \sqrt{5}$.

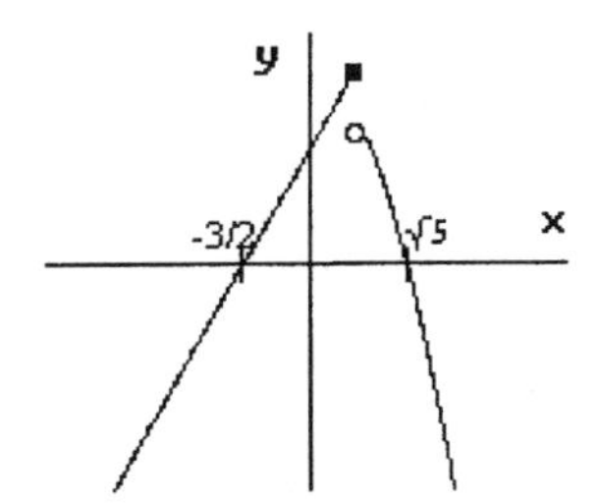

67. The point $(-1, -4)$ is the vertex of the graph of $f(x) = (x+1)^2 - 4$ The x-intercepts are -3 and $x = 1$.

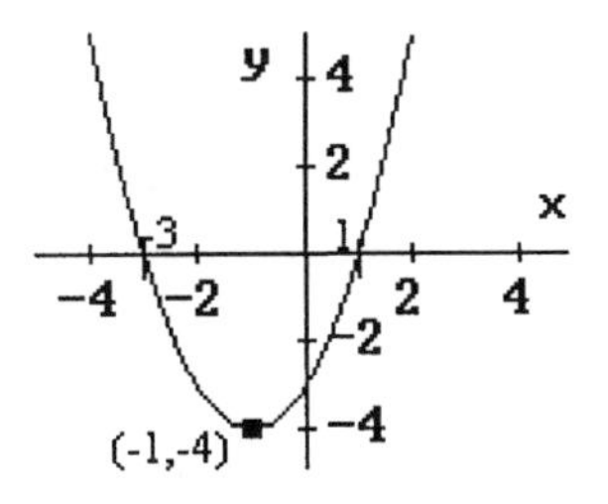

69. The graph of $g(x) = -x^2 + 8x - 17$ has its vertex at $(4, -1)$.
(Complete squares: $-x^2 + 8x - 17$
$= -(x^2 - 8x) - 17$
$= -(x^2 - 8x + 16) - 1 = -(x-4)^2 - 1$.)
There are no x-intercepts.

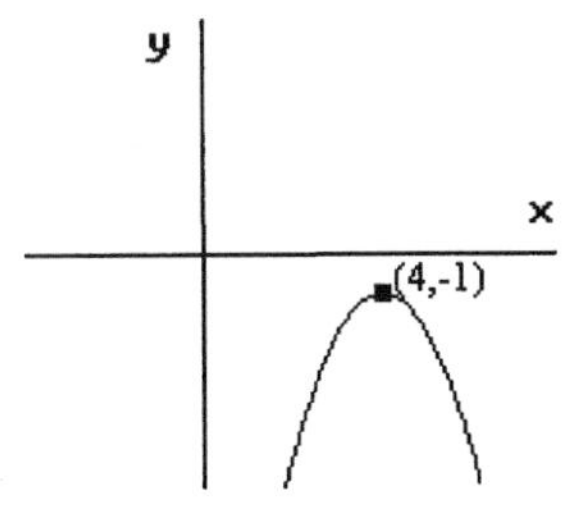

71. If f is linear, the graph of f has slope $-\frac{1}{2}$, and $f(0) = 3$, then $f(x) = -\frac{1}{2}x + 3$.

73. If h is quadratic, the graph of h has vertex $(2, -1)$, and $h(4) = 3$, then $h(x) = x^2 - 4x + 3$. ($h(x) = a(x-2)^2 - 1$ (the vertex-form of h), and $3 = a(4-2)^2 - 1$. So, $3 = 4a - 1$, or $a = 1$. Then $h(x) = (x-2)^2 - 1$, or $h(x) = x^2 - 4x + 3$.)

75. If z varies directly as a, and $z = 20$ when $a = 4$, then $z = ka$, $20 = k(4)$, so $k = 5$ and $z = 5a$. Then, if $a = 3$, $z = 15$.

77. T varies jointly as x and y, that is $T = kxy$. When $x = \frac{1}{2}$, and $y = \frac{1}{3}$, we know that $T = 1$. So, $1 = k\left(\frac{1}{2}\right)\left(\frac{1}{3}\right)$, or $k = 6$, and $T = 6xy$. If $x = 2$, and $y = 3$, then $T = 6(2)(3)$, or $T = 36$. (*Note*: increasing x by a factor of 4, and y by a factor of 9, makes T increase by 36.)

79. $f(x) = ax^3 + b$, and the graph of f passes through the points $(1,2)$, and $(-1,4)$. Then, $\begin{cases} a(1^3)+b=2 \\ a(-1)^3+b=4 \end{cases}$, so $\begin{cases} a+b=2 \\ -a+b=4 \end{cases}$. Add these last two equations, to obtain $2b = 6$, of $b = 3$. Then, $a = -1$, and $f(x) = -x^3 + 3$.

81. The equation of the line with the least squares fit to the points (1, 2), (3, 3), and (6, 5) is $y = 0.61x + 1.32$.

83. The equation of the line with the least squares fit to the five known data points:

x	0	10	20	30	40	50
y	42	67	88	118	135	?

is $y = 2.37x + 42.6$. The predicted value of y when $x = 50$ is 161.1.

85. **(a)** To solve the quadratic equation $x^2 + 6 = -5x$, we first add $5x$ to both sides: $x^2 + 5x + 6 = 0$, and factor to obtain the equivalent equation $(x+3)(x+2) = 0$; we conclude that $x + 3 = 0$, or $x + 2 = 0$, so $x = -3$, or $x = -2$.

(b) Finding the zeros of $f(x) = x^2 + 5x + 6$ is equivalent to solving $x^2 + 5x + 6 = 0$, which was done in part (a): $x = -3$, or $x = -2$.

87. **(a)** To solve $y = \frac{1}{x+2}$ for x, we multiply both sides of the equation by $x + 2$, to obtain $xy + 2y = 1$; then, $xy = 1 - 2y$, and $x = \frac{1-2y}{y}$.

(b) If $f(x) = \frac{1}{x+2}$, using (a) we conclude that $f^{-1}(x) = \frac{1-2x}{x}$.

89. **(a)** Expanding $(2x-1)^3$: $(2x-1)^3 = 8x^3 + 3(2x)^2(-1) + 3(2x)(-1)^2 + (-1)^3$
$= 8x^3 - 12x^2 + 6x - 1.$

(b) If $f(x) = x^3$, and $g(x) = 2x - 1$, then $(fog)(x) = (2x-1)^3$
$= 8x^3 - 12x^2 + 6x - 1.$

91. **(a)** Simplifying, $\frac{3}{x+5} - \frac{4}{x-2} = \frac{3(x-2)-4(x+5)}{(x+5)(x-2)} = -\frac{x+26}{x^2+3x-10}$

(b) If $f(x) = \frac{3}{x+5}$, and $g(x) = \frac{4}{x-2}$, the domain of $f - g$ is the set of all real numbers x, $x \neq -5$, $x \neq 2$. Since $\frac{3}{x+5} - \frac{4}{x-2} = -\frac{x+26}{x^2+3x-10}$, we can easily see that $x = -26$ is the only zero of $f - g$.

93. **(a)** $\left(k\sqrt{x}\right)^4 = k^4x^2$.

(b) If y is proportional to the square root of x, and z is proportional to the fourth power of y, then: $y = k\sqrt{x}$, $z = ay^4$, where both k and a are real constants. Then, $z = a(k\sqrt{x})^4$, so $z = a(k^4x^2)$ (as seen in (a)). Then $z = (ak^4)x^2$. So, z varies directly as the square of x.

95. The circumference C of a circle of radius r is given by $C = 2\pi r$. So, $r = \frac{C}{2\pi}$. Then, the area of such a circle is $A = \pi r^2$, or in terms of C: $A = \frac{C^2}{4\pi}$.

97. The segment connecting the observer to the balloon is the hypotenuse of the right triangle formed by the balloon B, the point P directly below the balloon on the ground, and the point O on the ground where the observer is. One leg of the triangle is the segment from O to P, which measures 200 *ft*. By the Pythagorean Theorem for right triangles: $d^2 = 200^2 + h^2$. Solving for h: $h = \sqrt{d^2 - 40{,}000}$, $d \geq 200$.

99. **(a)** The maximum height of the ball is attained when $f(x) = -\frac{1}{128}x^2 + x$ is a maximum, that is, a height of $32\ ft$ when $x = 64\ ft$ from where the ball is kicked.

(b) If $x = 30$ yds, or $90\ ft$, then $f(90) = 26.72\ ft$, well over the 10-foot goal post. Solving the inequality $-\frac{1}{128}x^2 + x > 10$, gives $10.93 < x < 117.07$. So, the furthest distance from which the ball could be kicked to clear the 10-*ft* goal post is 117.07 *ft*, or 39.02 yds.

101. The line with the least-squares best fit to (10, 8.3), (12, 18.8), (14, 28.1), and (16, 38.4), is $y = 4.98x - 41.34$ (Here, x represents the number of years of education, and y the percent of the population with x years of education that do volunteer work.) This predicts that the percent of the population with 18 years of education that do volunteer work is 48.34%.

103. The quadratic equation with the least-squares best fit to $(1985, 23)$, $(1991, 339)$, and (1999,995) is $y = 2.095238095x^2 - 8278x + 8{,}176{,}143.476$ (Here, x represents the year and y the number of CDs sold, in millions.) For the year 2010, this quadratic model predicts 2,334.9 million CDs to be sold.

Chapter 2. Test

1. The statement: "If a horizontal line intersects the graph of an equation in more than one point, the equation does not define y as a function of x" is false. A true statement would be: "If a vertical line intersects the graph of an equation in more than one point, the equation does not define y as a function of x."

3. The statement: "If $h(x)$ is defined by $h(x) = f(x-2)+3$, then the graph of h can be obtained by translating the graph of f 2 units to the right and 3 units upward" is true.

5. The statement: "Only functions satisfying the horizontal line test are 1-1, and only 1-1 functions have inverse functions" is true.

7. The statement: "If y varies directly as x, then y is also a linear function of x" is true. The converse statement: "If y is a linear function of x, then y varies directly as x" is false. Take, for example, $y = 2x+1$ is a linear function of x but y does not vary directly as x.

9. Answers may vary. An example of a linear function f and its inverse: $f(x) = 2x-1,\ f^{-1}(x) = \frac{1}{2}x + \frac{1}{2}$.

11. Answers may vary. An example of a piecewise-defined function that is linear over each of its three pieces is: $p(x) = \begin{cases} x+2 & x \le -2 \\ 2x+4 & -2 < x \le 0 \\ x+4 & x > 0 \end{cases}$

13. Answers may vary. An example of a function whose graph is symmetric with respect to the origin is the piecewise-defined function $f(x) = \begin{cases} -\sqrt{-x} & x < 0 \\ \sqrt{x} & x \ge 0 \end{cases}$.

15. The equation $\frac{1}{y} + x = 2$ defines y as a function of x: $y = \frac{1}{2-x}$, with domain $\{x \mid x \ne 2\}$.

17. If $f(x) = \begin{cases} 3x+1 & x \le 2 \\ 4x-5 & x > 2 \end{cases}$, and $g(x) = x^2$, then:

(a) $f(3) = 7$

(b) $(f \circ g)(\sqrt{2}) = f(g(\sqrt{2})) = f(2) = 7$

(c) $(g \circ f)(3) = g(f(3)) = g(7) = 49$

19. Starting with the graph of *f*:

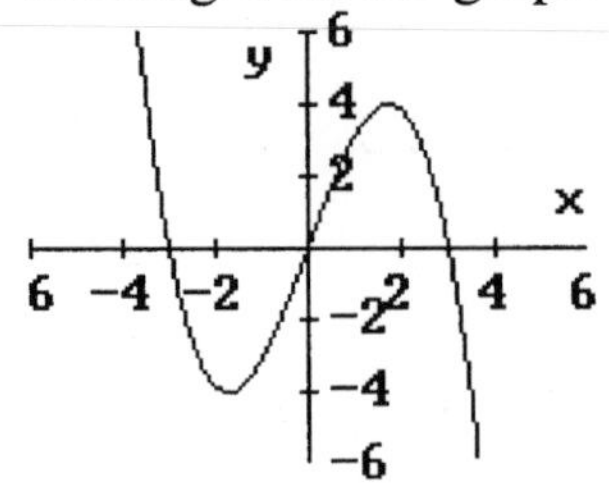

the graph of $h(x) = f(x-1) - 3$ is obtained by translating the graph of *f* 1 unit to the right and 3 units downward:

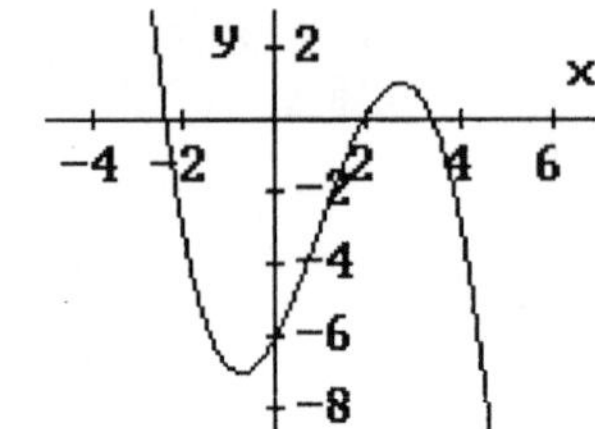

21. The vertex of the graph of $f(x) = 3x^2 - 6x + 10$ is (1, 7).

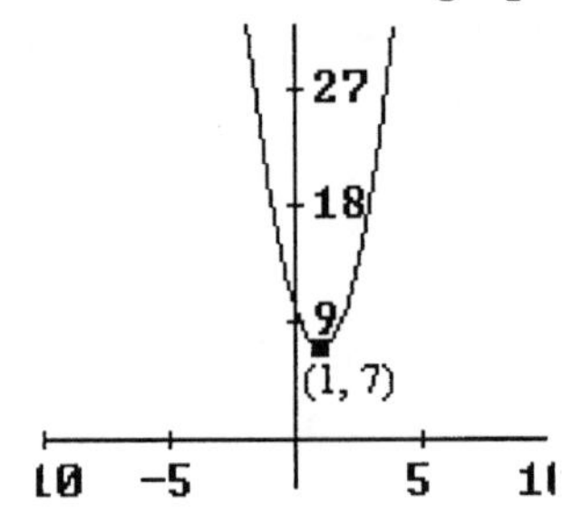

23. If $f(x) = ax^2 + bx$, and $(2, -4)$ and $(-1, 5)$ are points on the graph of *f*, then: $\begin{cases} 4a + 2b = -4 \\ a - b = 5 \end{cases}$;

the solution to this system is $a = 1,\ b = -4$. So, $f(x) = x^2 - 4x$, and $f(3) = -3$.

Chapter 3. POLYNOMIAL AND RATIONAL FUNCTIONS

Section 3.1 Polynomial Functions

1. For the function $f(x) = x^6$, we have: as $x \to -\infty$, the values $f(x) \to \infty$, and as $x \to \infty$, the values $f(x) \to \infty$ also. The graph of *f* matches (ii).

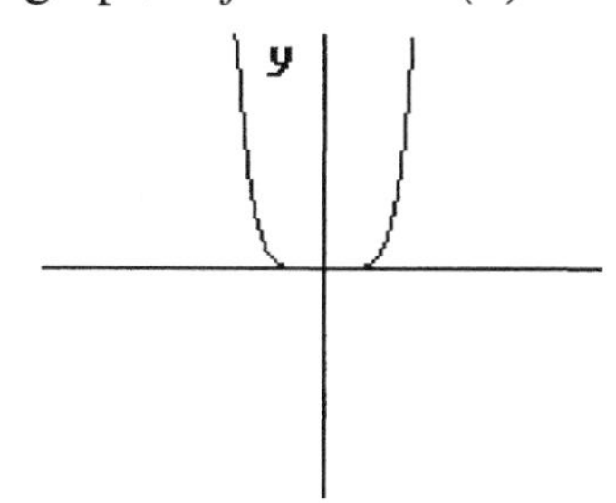

3. For the function $p(x) = -2x^4$, we have: as $x \to -\infty$, the values $p(x) \to -\infty$, and as $x \to \infty$, the values $p(x) \to -\infty$ also. The graph of *p* matches (i).

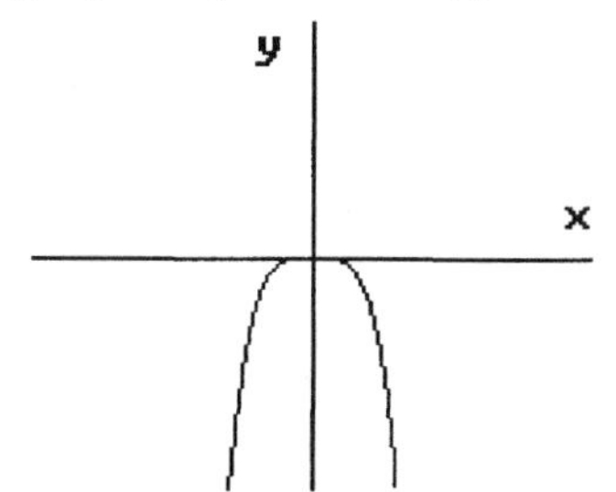

5. The function $f(x) = x^2 - 3x + 7$ has end behavior like that of $g(x) = x^2$. As $x \to \infty$ or $x \to -\infty$, the values $f(x) \to \infty$.

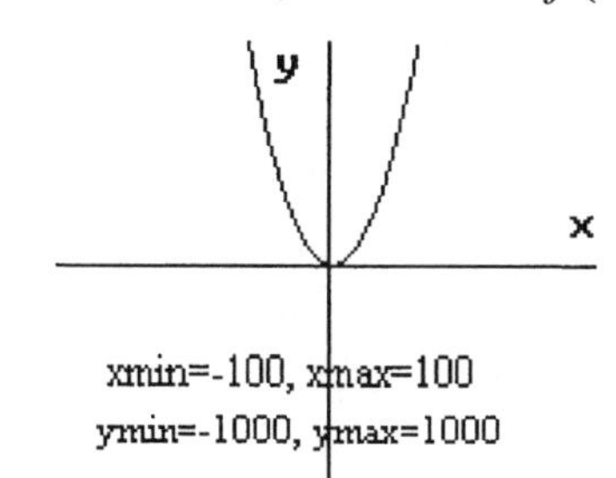

7. The function $p(x) = -4x^4 + 7x - 1$ has end behavior like that of $-4x^4$. As $x \to \infty$ or $x \to -\infty$, the values $f(x) \to -\infty$.

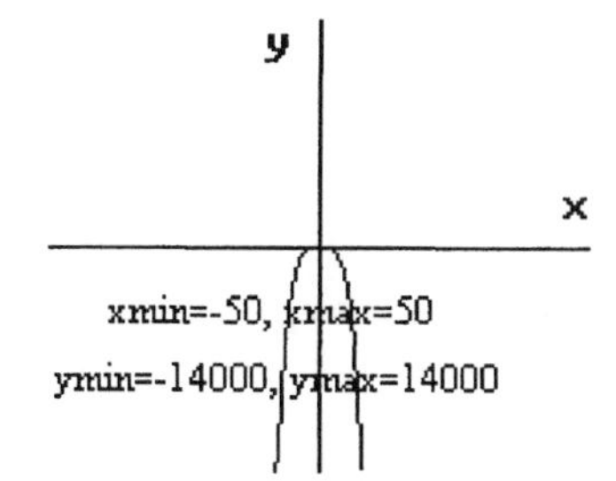

9. The function $h(x) = 10{,}000x^3 - 0.001x^5$ has end behavior like that of $-0.001x^5$. As $x \to \infty$, $h(x) \to -\infty$, while $h(x) \to \infty$ as $x \to -\infty$.

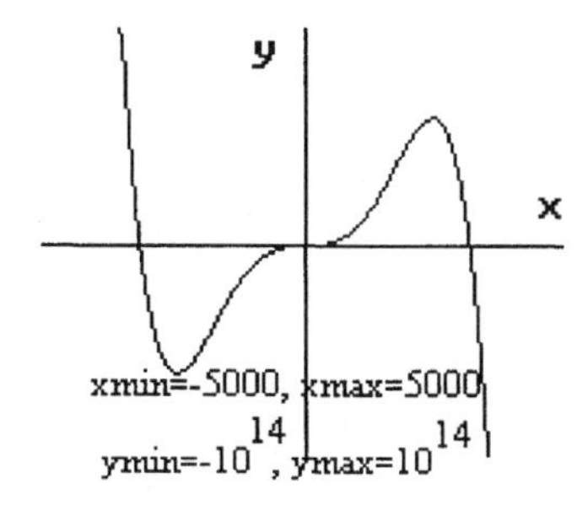

11. The function $g(x) = 10^7 + x^3$ has end behavior like that of x^3. As $x \to \infty$, the values $f(x) \to \infty$, and as $x \to -\infty$ the values

$f(x) \to -\infty$.

13. A polynomial function of the form $P(x) = mx + b$ (that is, a linear function) has at most one zero, and it has no turning points.

15. If $P(x) = ax^5 + bx^4 + cx^3 + dx^2 + ex + f$, a polynomial function of degree 5, then P has at most 5 zeros, and at most 4 turning points.

17. If $f(x) = x^2 - 2x - 8$, then $f(x) = (x-4)(x+2)$. The zeros of *f* are $x = -2$ and $x = 4$.

19. $f(x) = 3x^3 - 3x$. Since $3x^3 - 3x = 3x(x^2 - 1) = 3x(x+1)(x-1)$. The zeros of *f* are $-1, 0,$ and 1.

21. $f(x) = -x^4 - 12x^3 - 36x^2 = -x^2(x^2 + 12x + 36) = -x^2(x+6)^2$. The zeros of *f* are -6 and 0.

23. The graph of $f(x) = x^3 - 6x^2 + 5x + 13$ has two turning points, approximately a local maximum at $(0.47, 14.13)$ and a local minimum at $(3.53, -0.13)$.

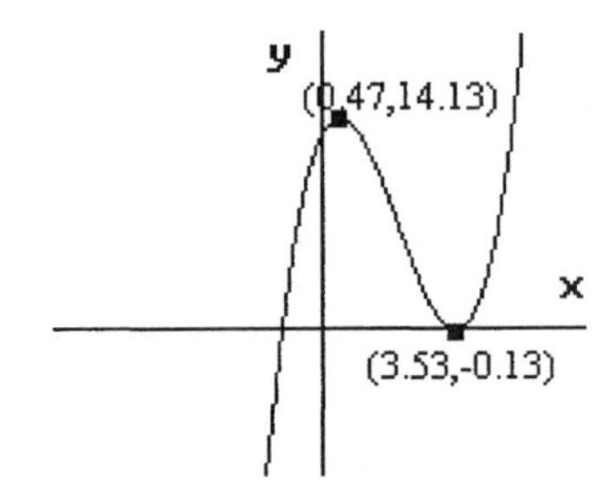

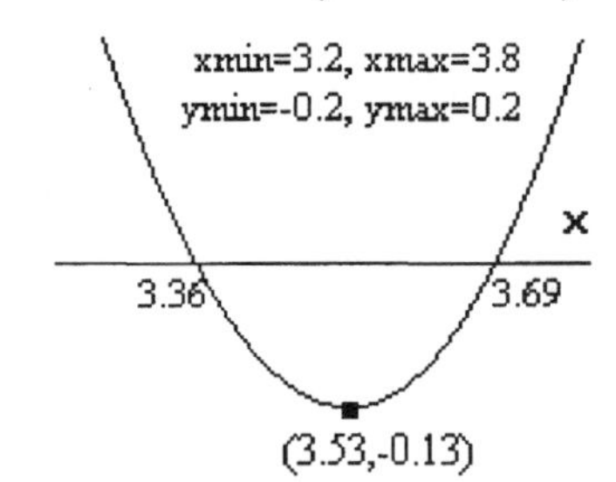

There are three zeros, approximately at $x \approx -1.05$, 3.36, and 3.69. (*Note*: the positive zeros are "hidden;" the graph on the right shows a window where the zeros can be actually seen.) There are three zeros, $x \approx -4.69$, -4.36, and 0.05 (approximate to the nearest hundredth.)

25. The graph of $h(x) = x^3/8 + 2x^2 + 2x + 3$ has two turning points: a local maximum at $(-10.14, 58.04)$ and a local minimum at $(-0.53, 2.48)$. There is only one zero, $x \approx -15.04$.

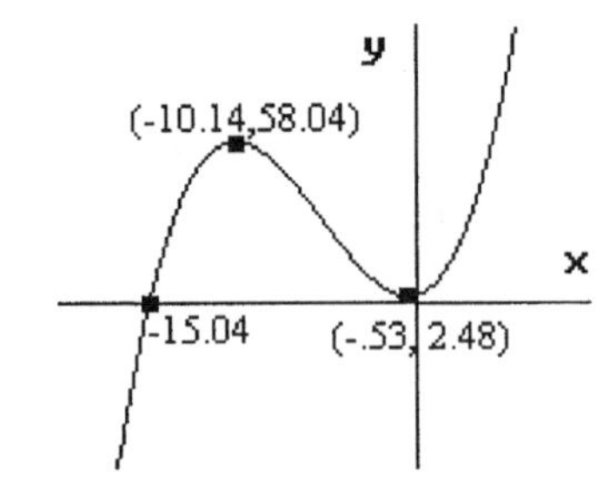

27. The graph of $p(x) = -x^4/4 + 3x^3 + x^2/2 - 9x$ has three turning points: a local maximum at $(-1, 6.25)$, a local minimum at $(1, -5.75)$, and a local maximum at $(9, 506.25)$. There are four zeros: $x \approx -1.7,\ 0,\ 1.78$ and 11.91.

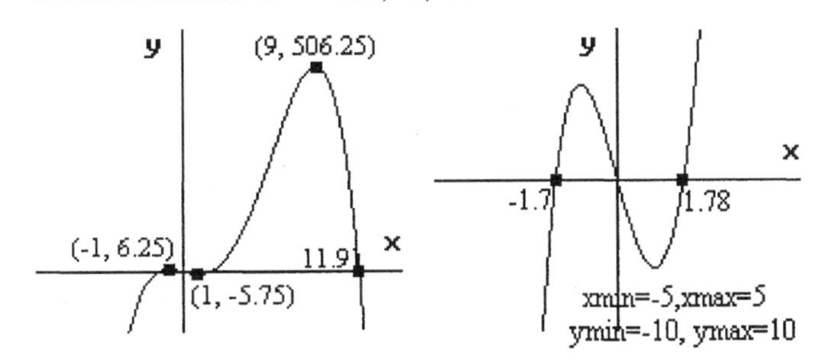

29. The graph of $h(x) = 0.1x^3 - 0.7x^2 - 5.6x + 4$ has two turning points: a local maximum at $(-2.58, 12.07)$, and a local minimum at $(7.24, -35.29)$.

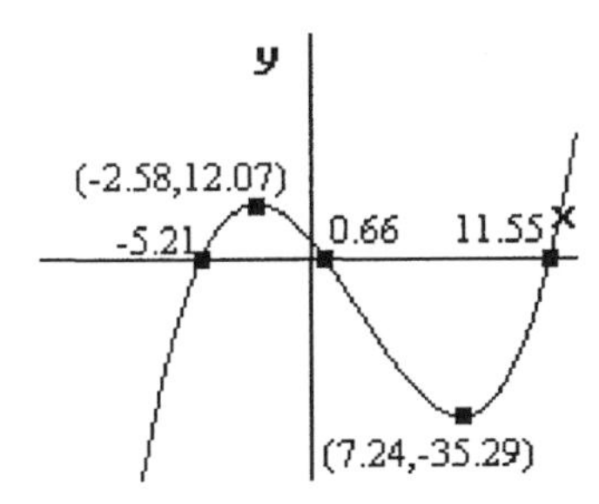

h has three zeros, $x \approx -5.21, 0.66,$ and 11.55 (approximate to the nearest hundredth.)

31. The graph of $g(x) = 0.1x^5 - 0.01x^6 - 1$ has one turning point, a local maximum, at (8.33, 668.8). *f* has two zeros, 1.64 and 10 (approximate to the nearest hundredth.)

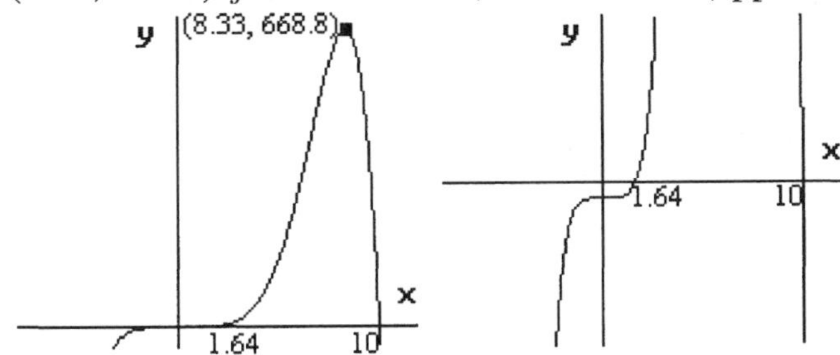

33. The polynomial function $f(x) = 1.25x^4 - 73x^3 + 980x^2 + 130$ models the number of AIDS deaths reported in the year *x*, for the years 1981-2001, where $x = 0$ corresponds to 1981. This function has a maximum at (12, 41195), which corresponds to 1993, with 41195 AIDS deaths reported. The first year after 1990 when the number of deaths is less than 13,000 is 2001, or $x = 20$ with a number of reported deaths of 8130 according to the model. If $f(1)$ represents the number of AIDS deaths in 1982 (from January 1 through December 31), then $f(1.5)$ can be interpreted as the number of deaths reported during the period Jan 1-June 30 of 1983.

35. For the period 1960-2000, the percentage of paper waste remaining after recycling is modeled by $f(x) = 0.0000002x^4 - 0.0003x^3 + 0.012x^2 - 0.23x + 18.1$, where $x = 0$ corresponds to 1960. Points that belong to the graph of *f*: (0, 18.1), (10, 16.7), (20, 15.9) (30, 14.1), (40, 9.4). Recycling seems to be increasing during 1960 to 2000, since the percentage of paper waste remaining after recycling has been steadily decreasing.

37. **(a)** The end behavior of $f(t) = -1.2t^3 + 16.1t^2 + 90.9t + 2349.3$ is like that of $y = -t^3$, so it eventually takes only negative values with increasingly large absolute values, as $t \to \infty$. The end behavior of $g(t) = 7.2t^2 + 107.2t + 2345.8$ is like that of $y = t^2$, so it eventually takes only positive values, and as $t \to \infty$, $g(t) \to \infty$ also.

(b) The quadratic model is more likely to provide the best fit for the next century, since the model $f(t) = -1.2t^3 + 16.1t^2 + 90.9t + 2349.3$ produces negative values in the long run.

(c) The two models are very close in the domain [0, 5], with a difference in their values < 6.

(d) The maximum amount by which the models differ for the years 1990-1993 is 5.32.
(The graph on the right corresponds to $y = |f(x) - g(x)|$, on the interval [0, 5].)

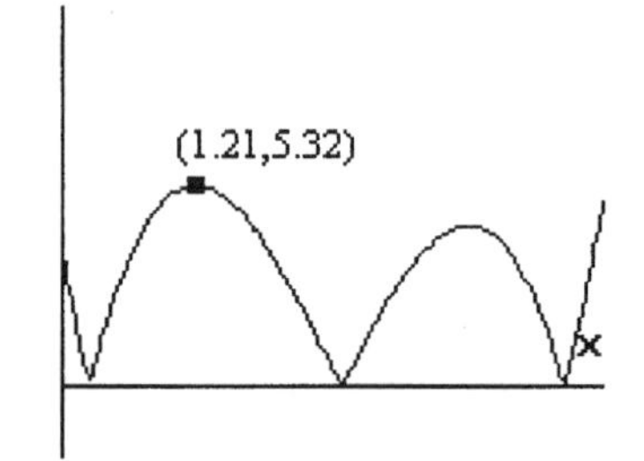

Concepts and Critical Thinking

39. The statement: "If *f* is a fifth-degree polynomial and $f(x) \to \infty$ as $x \to \infty$, then $f(x) \to -\infty$ as $x \to -\infty$" is true.

41. The statement: "Between any two turning points of a polynomial there must be at least one zero" is false.

43. The statement: "Only the term of highest degree has any effect on the end behavior of the graph of a polynomial" is true.

45. Answers may vary. A cubic polynomial with no turning points is $f(x) = x^3 + 1$.

47. Answers may vary. One *f*(*x*) such that $f(x) \to \infty$ as $x \to -\infty$ is $f(x) = -x^3$.

49. Polynomials of odd degree have at least one zero, since their end behavior as $x \to \infty$ is opposite to the one as $x \to -\infty$. The graph has to cross the *x* axis.

Chapter 3. Section 3.2 Division of Polynomials

1. If we divide $f(x)=2x^2-5x-12$ by $d(x)=x-4$, the quotient is $q(x)=2x+3$, and the remainder is $r(x)=0$.

$$
\begin{array}{r}
2x+3 \\
x-4\overline{)2x^2-5x-12} \\
\underline{2x^2-8x} \\
3x-12 \\
\underline{3x-12} \\
0
\end{array}
$$

3. When we divide $f(x)=4x^3-3x^2+20x-15$ by $d(x)=x^2+5$, the quotient is $q(x)=4x-3$ and the remainder is $r(x)=0$.

$$
\begin{array}{r}
4x-3 \\
x^2+5\overline{)4x^3-3x^2+20x-15} \\
\underline{4x^3+20x} \\
-3x^2-15 \\
\underline{-3x^2-15} \\
0
\end{array}
$$

5. Dividing $f(x)=2x^4-7x^3-6x+9$ by $d(x)=2x^2-x+3$, we obtain the quotient $q(x)=x^2-3x-3$ and the remainder $r(x)=18$.

$$
\begin{array}{r}
x^2-3x-3 \\
2x^2-x+3\overline{)2x^4-7x^3+0x^2-6x+9} \\
\underline{2x^4-x^3+3x^2} \\
-6x^3-3x^2-6x+9 \\
\underline{-6x^3+3x^2-9x} \\
-6x^2+3x+9 \\
\underline{-6x^2+3x-9} \\
18
\end{array}
$$

7. The division of $f(x)=x^2+1$ by $d(x)=x+1$ produces the quotient $q(x)=x-1$ and the remainder $r(x)=2$.

$$\begin{array}{r|l} & x-1 \\ x+1 & x^2+0x+1 \\ & \underline{x^2+x} \\ & -x+1 \\ & \underline{-x-1} \\ & 2 \end{array}$$

9. When dividing $f(x)=3x^3-4x+2$ by $d(x)=x^2-3$, the quotient is $q(x)=3x$ and the remainder is $r(x)=5x+2$.

$$\begin{array}{r|l} & 3x \\ x^2-3 & 3x^3-4x+2 \\ & \underline{3x^3-9x} \\ & 5x+2 \end{array}$$

11. The quotient of dividing $f(x)=x^5$ by $d(x)=x^3+x+1$ is $q(x)=x^2-1$, and the remainder is $r(x)=-x^2+x+1$.

$$\begin{array}{r|l} & x^2-1 \\ x^3+x+1 & x^5 \\ & \underline{x^5+x^3+x^2} \\ & -x^3-x^2 \\ & \underline{-x^3 \qquad -x-1} \\ & -x^2+x+1 \end{array}$$

13. Dividing $f(x)=2x^2+9x-18$ by $d(x)=x+6$ gives the quotient $q(x)=2x-3$ and the remainder $r(x)=0$.

$$\begin{array}{r|rrr} -6 & 2 & 9 & -18 \\ & & -12 & 18 \\ \hline & 2 & -3 & 0 \end{array}$$

15. Dividing $f(x)=x^3-2x^2-2x+4$ by $d(x)=x-2$ gives the quotient $q(x)=x^2-2$ and the remainder $r(x)=0$.

$$\begin{array}{r|rrrr} 2 & 1 & -2 & -2 & 4 \\ & & 2 & 0 & -4 \\ \hline & 1 & 0 & -2 & 0 \end{array}$$

17. Dividing $f(x)=3x^4-8x^3-10x+3$ by $d(x)=x-3$ gives the quotient $q(x)=3x^3+x^2+3x-1$ and the remainder $r(x)=0$.

$$\begin{array}{r|rrrrr} 3 & 3 & -8 & 0 & -10 & 3 \\ & & 9 & 3 & 9 & -3 \\ \hline & 3 & 1 & 3 & -1 & 0 \end{array}$$

19. Dividing $f(x)=x^2-2$ by $d(x)=x+1$ gives the quotient $q(x)=x-1$ and the remainder $r(x)=-1$.

$$\begin{array}{r|rrr} -1 & 1 & 0 & -2 \\ & & -1 & 1 \\ \hline & 1 & -1 & -1 \end{array}$$

21. Dividing $f(x)=x^3+6x^2-16x+2$ by $d(x)=x-2$ gives the quotient $q(x)=x^2-2x$ and the remainder $r(x)=2$.

$$\begin{array}{r|rrrr} -8 & 1 & 6 & -16 & 2 \\ & & -8 & 16 & 0 \\ \hline & 1 & -2 & 0 & 2 \end{array}$$

23. Dividing $f(x)=1+x^3-4x^4$ by $d(x)=x-⅓$ gives the quotient $q(x)=-4x^3-\frac{1}{3}x^2-\frac{1}{9}x-\frac{1}{27}$ and the remainder $r(x)=\frac{80}{81}$.

$$\begin{array}{r|rrrrr} \frac{1}{3} & -4 & 1 & 0 & 0 & 1 \\ & & -\frac{4}{3} & -\frac{1}{9} & -\frac{1}{27} & -\frac{1}{81} \\ \hline & -4 & -\frac{1}{3} & -\frac{1}{9} & -\frac{1}{27} & \frac{80}{81} \end{array}$$

25. We divide $f(x)=x^3+6x^2+3x-10$ by the given factor $x+2$, to obtain $f(x)=(x+2)(x^2+4x-5)$. Therefore, $f(x)=(x+2)(x+5)(x-1)$. The zeros of f are $x=-5,\ -2,$ and 1.

$$\begin{array}{r|rrrr} -2 & 1 & 6 & 3 & -10 \\ & & -2 & -8 & 10 \\ \hline & 1 & 4 & -5 & 0 \end{array}$$

27. The division of $p(x)=2x^3-9x^2+7x+6$ by the given factor $x-3$, gives the quotient $2x^2-3x-2$, which in turn factors as $(2x+1)(x-2)$. So, $p(x)=(2x+1)(x-2)(x-3)$ and the zeros of p are $x=-½,\ 2,$ and 3.

$$\begin{array}{r|rrrr} 3 & 2 & -9 & 7 & 6 \\ & & 6 & -9 & -6 \\ \hline & 2 & -3 & -2 & 0 \end{array}$$

29. The division of $g(x)=x^4+9x^3+22x^2-32$ by the given factors $x+4$ and $x-1$ gives the quotient x^2+6x+8, which in turn factors as $(x+4)(x+2)$. Therefore, $g(x)=(x+4)^2(x+2)(x-1)$ and the zeros of g are $x=-4,\ -2,$ and 1.

$$\begin{array}{r|rrrrr} -4 & 1 & 9 & 22 & 0 & -32 \\ & & -4 & -20 & -8 & 32 \\ \hline 1 & 1 & 5 & 2 & -8 & 0 \\ & & 1 & 6 & 8 & \\ \hline & 1 & 6 & 8 & 0 & \end{array}$$

31. The division of $h(x) = 6x^4 - 47x^3 - 13x^2 + 38x + 16$ by $2x^2 - 15x - 8$ (which is the product of the given factors $2x+1$ and $x-8$) gives the quotient $3x^2 - x - 2$, which in turn factors as $(3x+2)(x-1)$. Therefore, $h(x) = (3x+2)(x-1)(2x+1)(x-8)$ and the zeros of h are $x = -\frac{2}{3},\ -\frac{1}{2},\ 1$, and 8.

$$\begin{array}{r} 3x^2 - x - 2 \\ 2x^2 - 15x - 8 \overline{\big)\, 6x^4 - 47x^3 - 13x^2 + 38x + 16} \\ \underline{6x^4 - 45x^3 - 24x^2 } \\ -2x^3 + 11x^2 + 38x + 16 \\ \underline{-2x^3 + 15x^2 + \ \ 8x } \\ -4x^2 + 30x + 16 \\ \underline{-4x^2 + 30x + 16} \\ 0 \end{array}$$

33. **(a)** The division of $f(x) = 2x^3 - x^2 - 4x + 6$ by $x-3$ leaves a remainder of 39. Then, $f(3) = 39$.

3	2	−1	−4	6
		6	15	33
	2	5	11	39

(b) The division of $f(x) = 2x^3 - x^2 - 4x + 6$ by $x+2$ leaves a remainder of -6. Then, $f(-2) = -6$.

−2	2	−1	−4	6
		−4	10	−12
	2	−5	6	−6

(b) The division of $f(x) = 2x^3 - x^2 - 4x + 6$ by $x - \frac{1}{2}$ leaves a remainder of 4. Then, $f\left(\frac{1}{2}\right) = 4$.

½	2	−1	−4	6
		1	0	−2
	2	0	−4	4

35. **(a)** The division of $h(x) = 3x^4 - 2x^3 - 4x + 3$ by $x+1$ leaves a remainder of 12. Then, $h(-1) = 12$.

−1	3	−2	0	−4	3
		−3	5	−5	9
	3	−5	5	−9	12

(b) The division of $h(x) = 3x^4 - 2x^3 - 4x + 3$ by $x-5$ leaves a remainder of 1608. Then, $h(5) = 1608$.

5	3	−2	0	−4	3
		15	65	325	1605
	3	13	65	321	1608

(c) The division of $h(x) = 3x^4 - 2x^3 - 4x + 3$ by $x+2.1$ leaves a remainder of 88.2663. Then, $h(-2.1) = 88.2663$.

−2.1	3	−2	0	−4	3
		−6.3	17.43	−36.603	85.2663
	3	−8.3	17.43	−40.603	88.2663

37. **(a)** The division of $g(x)=\frac{1}{2}x^4-\frac{5}{3}x^3+\frac{2}{3}x^2-\frac{9}{2}x$ by $x-4$ leaves a remainder of 14. Then, $g(4)=14$.

4	$\frac{1}{2}$	$-\frac{5}{3}$	$\frac{2}{3}$	$-\frac{9}{2}$	0
		2	$\frac{4}{3}$	8	14
	$\frac{1}{2}$	$\frac{1}{3}$	2	$\frac{7}{2}$	14

(b) The division of $g(x)=\frac{1}{2}x^4-\frac{5}{3}x^3+\frac{2}{3}x^2-\frac{9}{2}x$ by $x+2$ leaves a remainder of 33. Then, $g(-2)=33$.

−2	$\frac{1}{2}$	$-\frac{5}{3}$	$\frac{2}{3}$	$-\frac{9}{2}$	0
		−1	$\frac{16}{3}$	−12	33
	$\frac{1}{2}$	$-\frac{8}{3}$	6	$-\frac{33}{2}$	33

Concepts and Critical Thinking

39. The statement: "When one polynomial is divided by another, the degree of the remainder is always greater than that of the quotient" is false.

41. The statement: "If the remainder obtained when dividing the polynomial $f(x)$ by $x-3$ is 8, then $f(3)=8$" is true.

43. Answers may vary. A third-degree polynomial having $x+2$ as a factor is $f(x)=x^3+2x^2$.

45. Answers may vary. An example of a fourth-degree polynomial having $x+1$ and $x-2$ as factors is $f(x)=x^2(x+1)(x+2)$, that is $f(x)=x^4+3x^3+2x^2$.

Chapter 3. Section 3.3 Zeros and Factors of Polynomials

1. -3 is a zero of $f(x)=x^3+3x^2-4x-12$. In fact, we can factor $f(x)$ by grouping:

$$\begin{aligned} f(x) &= x^2(x+3)-4(x+3) \\ &= \left(x^2-4\right)\left(x+3\right) \\ &= \left(x-2\right)\left(x+2\right)\left(x+3\right) \end{aligned}$$

Then, the zeros of *f* are $x=-3,\ -2$, and 2.

3. 5 is a zero of $f(x)=2x^3-7x^2-14x-5$. We divide $f(x)$ by $x-5$:

$$\begin{array}{r|rrrr} 5 & 2 & -7 & -14 & -5 \\ & & 10 & 15 & 5 \\ \hline & 2 & 3 & 1 & 0 \end{array}$$

Therefore, $f(x)=(x-5)(2x^2+3x+1)$
$=(x-5)(2x+1)(x+1)$

The zeros of *f* are $x=-1,\ -\frac{1}{2}$, and 5.

5. $-\frac{1}{3}$ and -4 are zeros of $f(x)=3x^4+7x^3-19x^2+5x+4$. We divide by $x+\frac{1}{3}$ and $x+4$:

$$\begin{array}{r|rrrrr} -\frac{1}{3} & 3 & 7 & -19 & 5 & 4 \\ & & -1 & -2 & 7 & -4 \\ \hline -4 & 3 & 6 & -21 & 12 & 0 \\ & & -12 & 24 & -12 & \\ \hline & 3 & -6 & 3 & 0 & \end{array}$$

$$\begin{aligned} \text{That is, } f(x) &= \left(x+\frac{1}{3}\right)(x+4)\left(3x^2-6x+3\right) \\ &= \left(x+\frac{1}{3}\right)(x+4)\left(3(x^2-2x+1)\right) \\ &= \left(3x+1\right)(x+4)\left(x-1\right)^2 \end{aligned}$$

The zeros of f are $x=-4,\ -\frac{1}{3}$, and 1.

7. If $f(x)=5x^3+3x^2-12x+4$, then $f(1)=0$. Dividing $f(x)$ by $x-1$, we obtain the quotient $5x^2+8x-4$. So, $f(x)=(x-1)(5x^2+8x-4)$, and factoring this quadratic factor we finally get $f(x)=(x-1)(5x-2)(x+2)$. The zeros of *f* are $x=-2,\ \frac{2}{5}$, and 1.

9. Given $f(x)=6x^3-13x^2+x+2$, we have $f(2)=0$. When we divide $f(x)$ by $x-2$:

$$\begin{array}{r|rrrr} 2 & 6 & -13 & 1 & 2 \\ & & 12 & -2 & -2 \\ \hline & 6 & -1 & -1 & 0 \end{array}$$

so, $f(x)=(x-2)(6x^2-x-1)$

$=(x-2)(3x+1)(2x-1)$

and the zeros of f are $x=-\frac{1}{3}, \frac{1}{2}$, and 2.

11. If $f(x)=x^4-x^3-5x^2+3x+6$, we observe that $f(-1)=(-1)^4-(-1)^3-5(-1)^2+3(-1)+6=0$ and $f(2)=(2)^4-(2)^3-5(2)^2+3(2)+6=0$. After dividing by $x+1$ and $x-2$, we obtain $f(x)=(x+1)(x-2)(x^2-3)$. The zeros of f are $x=-\sqrt{3}, \sqrt{3}, -1$, and 2.

13. Answers may vary. An example of a polynomial of degree 2, with zeros -3 and 4, is $f(x)=a(x+3)(x-4)$, where a is any real number. If $a=2$, $f(x)=2x^2-2x-24$.

15. A polynomial of degree 3, with zeros -2, 1, and 5, is of the form $f(x)=a(x+2)(x-1)(x-5)$. That is, $f(x)=a(x^3-4x^2-7x+10)$. The extra information $f(0)=5$ allows us to determine the value of a, since $f(0)=10a$, therefore $10a=5$, $a=\frac{1}{2}$ and $f(x)=\frac{1}{2}x^3-2x^2-\frac{7}{2}x+5$.

17. A polynomial of degree 4 with zeros 0 and 2, both of multiplicity 2, is of the form $f(x)=ax^2(x-2)^2$. Since $f(-1)=3$, we have $a(-1)^2(-1-2)^2=3$, then $a=\frac{1}{3}$, and

$$f(x)=\frac{1}{3}x^2(x-2)^2=\frac{1}{3}x^2(x^2-4x+4)=\frac{1}{3}x^4-\frac{4}{3}x^3+\frac{4}{3}x^2$$

19. Answers may vary. A polynomial of degree 5 with zeros -1 (of multiplicity 3) and $\frac{1}{2}$ (of multiplicity 2), is of the form $f(x)=a(x+1)^3(x-\frac{1}{2})^2$, or equivalently $f(x)=b(x+1)^3(2x-1)^2$, where b is any real number. That is, $f(x)=b(4x^5+8x^4+x^3-5x^2-x+1)$, with b any real number.

21. Given the graph of $y=f(x)$: f is of degree 2 and has zeros -1 and 3, so it has the form $f(x)=a(x+1)(x-3)$. The value of f when $x=1$ is 4, so $a(1+1)(1-3)=4$, and $a=-1$.

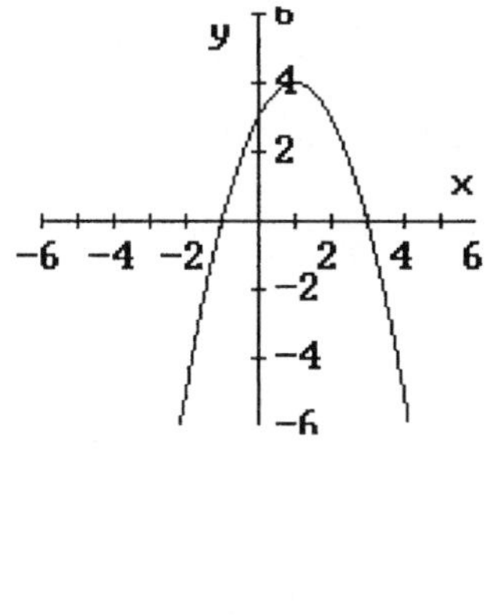

Then, $f(x)=-(x+1)(x-3)$

$=-x^2+2x+3.$

y
6
4
2
x
−6 −4 −2 2 4 6
−2
−4
−6

23. Given the graph of $y = f(x)$: f is of degree 3 and has zeros -4 (of multiplicity 1), and 2 (of multiplicity 2), so it has the form $f(x) = a(x+4)(x-2)^2$.

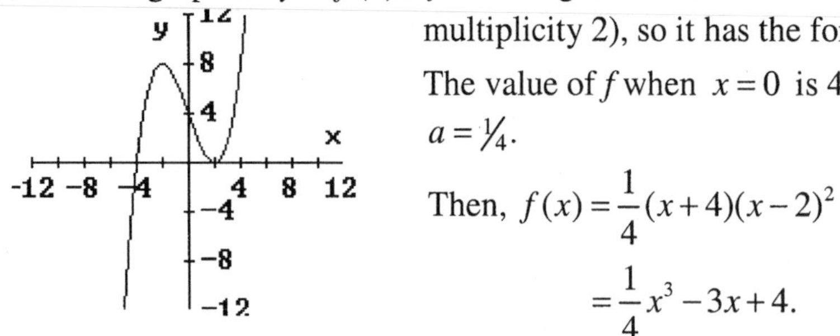

The value of f when $x = 0$ is 4, so $a(0+4)(0-2)^2 = 4$, and $a = \frac{1}{4}$.

$$\text{Then, } f(x) = \frac{1}{4}(x+4)(x-2)^2 = \frac{1}{4}x^3 - 3x + 4.$$

25. Given the graph of $y = f(x)$: f is of degree 4 and has zeros -2, -1, 1, and 2, so f is of the form $f(x) = a(x+2)(x+1)(x-1)(x-2)$.

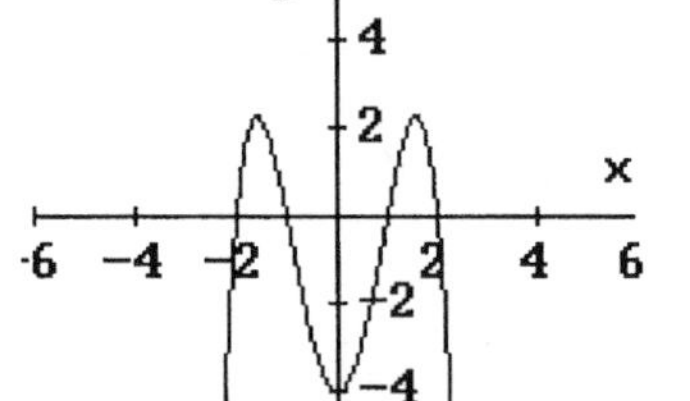

The value of f when $x = 0$ is -4, so $4a = -4$, and $a = -1$.

$$\text{Then, } f(x) = -1(x+2)(x+1)(x-1)(x-2) = -x^4 + 5x^2 - 4.$$

27. Since $-2i$ is a zero of the polynomial with real coefficients $f(x) = x^3 - 3x^2 + 4x - 12$, the conjugate of $-2i$ is also a zero. Hence, $(x+2i)(x-2i)$ is a factor of $f(x)$. In fact

$$\begin{array}{r|rrrr} -2i & 1 & -3 & 4 & -12 \\ & & -2i & -4+6i & 12 \\ \hline & 1 & -3-2i & 6i & 0 \end{array} \qquad \begin{array}{r|rrr} 2i & 1 & -3-2i & 6i \\ & & 2i & -6i \\ \hline & 1 & -3 & 0 \end{array}$$

That is, $f(x) = (x+2i)(x-2i)(x-3)$, and the zeros of f are $-2i$, $2i$, and 3.

29. Since $1+i$ is a zero of the polynomial with real coefficients $f(x) = x^4 - 5x^3 + 4x^2 + 2x - 8$, the conjugate of $1+i$ is also a zero. Hence, $(x-(1+i))(x-(1-i))$ is a factor of $f(x)$. In fact

$$\begin{array}{r|rrrrr} 1+i & 1 & -5 & 4 & 2 & -8 \\ & & 1+i & -5-3i & 2-4i & 8 \\ \hline & 1 & -4+i & -1-3i & 4-4i & 0 \end{array} \qquad \begin{array}{r|rrrr} 1-i & 1 & -4+i & -1-3i & 4-4i \\ & & 1-i & -3+3i & -4+4i \\ \hline & 1 & -3 & -4 & 0 \end{array}$$

That is, $f(x) = (x-(1+i))(x-(1-i))\left(x^2 - 3x - 4\right) = (x-(1+i))(x-(1-i))(x-4)(x+1)$

The zeros of f are $1+i$, $1-i$, -1, and 4.

31. If $f(x) = x^2 + 1$, the zeros of f are $-i$ and i, and $f(x) = (x+i)(x-i)$.

33. We can factor $f(x) = x^3 - 2x^2 + 5x$ as $f(x) = x\left(x^2 - 2x + 5\right)$. So $x = 0$ is a zero of f. To find the other zeros of f, we observe that they satisfy $x^2 - 2x + 5 = 0$. Using the quadratic formula: $x = \dfrac{-2 \pm \sqrt{-16}}{2}$, that is $x = \dfrac{2 \pm 4i}{2}$, or $x = 1+2i$, $x = 1-2i$. The factored expression for $f(x)$ is $f(x) = x(x-(1-2i))(x-(1+2i))$.

35. We can factor $f(x) = x^4 + 3x^2 - 4$ as $f(x) = (x^2+4)\left(x^2-1\right)$; therefore, $f(x) = (x-2i)(x+2i)(x-1)(x+1)$. The zeros of *f* are $x = -2i,\ 2i,\ -1,$ and 1.

37. The polynomial $f(x) = x^5 - 5x^4 + 9x^3 - 5x^2$ factors as $f(x) = x^2(x^3 - 5x^2 + 9x - 5)$. One of the zeros of $x^3 - 5x^2 + 9x - 5 = 0$ is $x = 1$. Dividing $x^3 - 5x^2 + 9x - 5$ by $x - 1$, we obtain

$$\begin{array}{r|rrrr} 1 & 1 & -5 & 9 & -5 \\ & & 1 & -4 & 5 \\ \hline & 1 & -4 & 5 & 0 \end{array}$$

Then, $f(x) = x^2(x-1)\left(x^2 - 4x + 5\right)$, so $f(x) = x^2(x-1)\left(x-(2+i)\right)\left(x-(2-i)\right)$, and the full list of zeros of *f* is $x = 0,\ 1,\ 2-i,$ and $2+i$.

39. A polynomial with real coefficients with zeros $x = -2$ and $2i$ also has the zero $x = -2i$. The least degree that such a polynomial has is 3, and it is of the form:

$$\begin{aligned} f(x) &= a(x+2)(x-2i)(x+2i) \\ &= a(x+2)(x^2+4) \\ &= a(x^3 + 2x^2 + 4x + 8) \end{aligned}$$

(*a* is any real number, $a \neq 0$.)

41. A polynomial with real coefficients with zeros $x = i$ and $1-i$ also has the zeros $x = -i$ and $1+i$. The least degree that such a polynomial has is 4, and it is of the form:

$$\begin{aligned} f(x) &= a(x+i)(x-i)(x-(1+i))(x-(1-i)) \\ &= a(x^2+1)(x^2-2x+2) \\ &= a(x^4 - 2x^3 + 3x^2 - 2x + 2) \end{aligned}$$

(*a* is any real number, $a \neq 0$.)

43. A polynomial with real coefficients with zeros $x = 0,\ \pm\sqrt{2}$ and $-2+2i$ also has the zero $x = -2-2i$. The least degree that such a polynomial has is 5, and it is of the form:

$$\begin{aligned} f(x) &= ax(x-\sqrt{2})(x+\sqrt{2})(x-(-2+2i))(x-(-2-2i)) \\ &= ax(x^2-2)(x^2+4x+8) \\ &= ax(x^4 + 4x^3 + 6x^2 - 8x - 16) \\ &= a(x^5 + 4x^4 + 6x^3 - 8x^2 - 16x) \end{aligned}$$

(*a* is any real number, $a \neq 0$.)

Applications

45. A box is constructed by cutting squares out square corners of a rectangular piece of cardboard, and then folding up the sides, and its volume is $V(x) = 4x^3 - 72x^2 + 320x$, then

$$V(x) = 4x(x^2 - 18x + 320)$$
$$= 4x(10 - x)(8 - x)$$
$$= x(20 - 2x)(16 - 2x)$$

The zeros of V are $x = 0$, 8, and 10. $V(x)$ is nonnegative for $0 \le x \le 8$, and $x > 10$. All three dimensions x, $20 - 2x$, and $16 - 2x$ must be nonnegative, so the domain that is relevant in this situation is $0 \le x \le 8$. The original piece of cardboard was $16'' \times 20''$.

47. If the monthly profit of selling x bicycle frames is $P(x) = -0.005x^3 + 1.5x^2 + 50x - 15000$, we have

$$P(x) = -0.005(x^3 - 300x^2 - 10000x + 3000000)$$
$$= -0.005(x^2(x - 300) - 10000(x - 300))$$
$$= -0.005(x - 300)(x^2 - 10000)$$
$$= -0.005(x - 300)(x - 100)(x + 100)$$

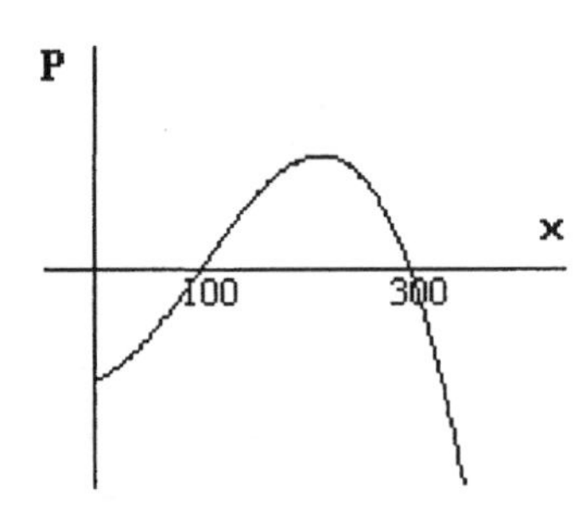

The zeros of P are $x = -100$, 100, and 300. Negative values of x are irrelevant; the values that make sense for the profit to be zero are 100 and 300. The profit is positive for $100 < x < 300$.

49. Knowing that the height $s(t)$ is a quadratic function, we must have $s(t) = at^2 + bt + c$. If a ball is projected upward from 64 feet below ground level, the value of the height at the time $t = 0$ is -64. Then, $s(t) = at^2 + bt - 64$. Since we know that the ball passes ground level when $t = 1$ and $t = 4$, we must have $s(t) = a(t - 1)(t - 4)$, so:

$$at^2 + bt - 64 = a(t - 1)(t - 4)$$
$$= a(t^2 - 5t + 4)$$
$$= at^2 - 5at + 4a$$

that is, $4a = -64$, and $b = -5a$. Therefore, $a = -16$, $b = 80$, and $s(t) = -16t^2 + 80t - 64$. The maximum height is obtained after 2.5 seconds, $s(2.5) = -16(2.5^2) + 80(2.5) - 64$, or 36 feet.

Concepts and Critical Thinking

51. The statement: "If -4 is a zero of a polynomial $f(x)$, then $x - 4$ is a factor if $f(x)$" is false.

53. The statement: "All nth-degree polynomials have exactly n distinct complex zeros" is false". For example, the polynomial $f(x) = x^2 - 2x + 1$ has only one zero.

55. Answers may vary. An example of a second-degree polynomial with no real zeros is $f(x) = x^2 + 3$.

57. Answers may vary. An example of a fourth-degree polynomial with exactly two x-intercepts is $f(x) = x^2(x-2)^2$, that is $f(x) = x^4 - 4x^2 + 4x^2$.

59. According to the Linear Factorization Theorem, a polynomial of degree $n \geq 1$, say $f(x) = a_n x^n + a_{n-1}x^{n-1} + ... + a_1 x + a_0$ can be expressed as $f(x) = a_n(x-c_1)(x-c_2)...(x-c_n)$. If the leading coefficient is 1, we have $f(x) = x^n + a_{n-1}x^{n-1} + ... + a_1 x + a_0$, and $f(x) = (x-c_1)(x-c_2)...(x-c_n)$. If $x = 0$, we get $f(0) = a_0$, and also $f(0) = (-1)^n c_1 c_2 ... c_n$; that is, $(-1)^n c_1 c_2 ... c_n = a_0$ and $c_1 c_2 ... c_n$ has to be a_0 or its opposite.

Chapter 3. Section 3.4 Real Zeros of Polynomials

1. The possible rational zeros of the function $f(x) = x^4 - 12x^2 + 27$ are of the form $\frac{a}{b}$, with a a factor of 27 and b a factor of 1, namely $\pm 1, \ \pm 3, \ \pm 9, \ \pm 27$.

The rational zeros of f are -3 and 3, as we can see using synthetic division.

$$\begin{array}{r|rrrrr}
3 & 1 & 0 & -12 & 0 & 27 \\
 & & 3 & 9 & -9 & -27 \\
\hline
-3 & 1 & 3 & -3 & -9 & 0 \\
 & & -3 & 0 & 9 & \\
\hline
 & 1 & 0 & -3 & &
\end{array}$$

We have $f(x) = (x-3)(x+3)(x^2-3)$, so the real zeros of f are $-3, \ 3, \ -\sqrt{3}$, and $\sqrt{3}$.

3. The polynomial function $g(x) = x^3 - 3x^2 - 4x + 12$ has as possible zeros those of the form $\frac{a}{b}$, where a is a factor of 12 and b is a factor of 1, namely $\pm 1, \ \pm 2, \ \pm 3, \ \pm 4, \ \pm 6, \ \pm 12$.

$$\begin{array}{r|rrrr}
3 & 1 & -3 & -4 & 12 \\
 & & 3 & 0 & -12 \\
\hline
 & 1 & 0 & -4 & 0
\end{array}$$

We obtain $g(x) = (x-3)(x^2-4)$, or $g(x) = (x-3)(x-2)(x+2)$. The real zeros of f are $-2, \ 2$, and 3.

5. If $\frac{a}{b}$ is a rational zero of $f(x) = 4x^4 - 4x^3 - 9x^2 + x + 2$, then a is a factor of 2 and b a factor of 4. So, the possible rational zeros of f are $\pm\frac{1}{4}, \ \pm\frac{1}{2}, \ \pm 1, \ \pm 2$.

$$\begin{array}{r|rrrrr}
2 & 4 & -4 & -9 & 1 & 2 \\
 & & 8 & 8 & -2 & -2 \\
\hline
\frac{1}{2} & 4 & 4 & -1 & -1 & 0 \\
 & & 2 & 3 & 1 & \\
\hline
 & 4 & 6 & 2 & 0 &
\end{array}$$

and, lastly:

$$\begin{array}{r|rrr}
-\frac{1}{2} & 4 & 6 & 2 \\
 & & -2 & -2 \\
\hline
 & 4 & 4 & 0
\end{array}$$

Then, $f(x) = (4x+4)(x-2)\left(x-\frac{1}{2}\right)\left(x+\frac{1}{2}\right)$, or $f(x) = (x+1)(x-2)\left(x-\frac{1}{2}\right)\left(x+\frac{1}{2}\right)$, and the real zeros of f are $-1, \ -\frac{1}{2}, \ \frac{1}{2}$, and 2.

7. For $g(x)=2x^5+3x^4-2x-3$, the possible rational zeros are of the form $\frac{a}{b}$, where a is a factor of -3, and b a factor of 2, that is $\pm 3,\ \pm\frac{3}{2},\ \pm 1,\ \pm\frac{1}{2}$.

We have
$$\begin{aligned} g(x) &= 2x^5+3x^4-2x-3 \\ &= x^4(2x+3)-(2x+3) \\ &= (x^4-1)(2x+3) \\ &= (x^2+1)(x-1)(x+1)(2x+3) \end{aligned}$$
The real zeros of g are $-3/2,\ -1$, and 1.

9. The zeros of the polynomial function $f(x)=x^3-\frac{9}{2}x^2+\frac{1}{2}x+6$ are also zeros of the polynomial with integer coefficients obtained multiplying $f(x)$ by 2, $g(x)=2x^3-9x^2+x+12$. The possible rational zeros of these two polynomials, then, are of the form a/b, with a a factor of 12 and b a factor of 2. That is, the possible rational zeros are $\pm 12,\ \pm 6,\ \pm 4,\ \pm 3,\ \pm 2,\ \pm 3/2,\ \pm 1,\ \pm 1/2$. Using synthetic division for $g(x)$,

$$\begin{array}{r|rrrr} -1 & 2 & -9 & 1 & 12 \\ & & -2 & 11 & -12 \\ \hline 4 & 2 & -11 & 12 & 0 \\ & & 8 & -12 & \\ \hline & 2 & -3 & 0 & \end{array}$$

So, we obtain $g(x)=(2x-3)(x+1)(x-4)$, and the zeros of g and f are $x=-1,\ 3/2$, and 4.

11. $f(x)=x^3+4$ has no variations in sign, so there are no positive real zeros, and $f(-x)=(-x)^3+4$, that is $f(-x)=-x^3+4$, has one variation in sign, so there is only one negative real zero, $x=-\sqrt[3]{4}$, approximately -1.59.

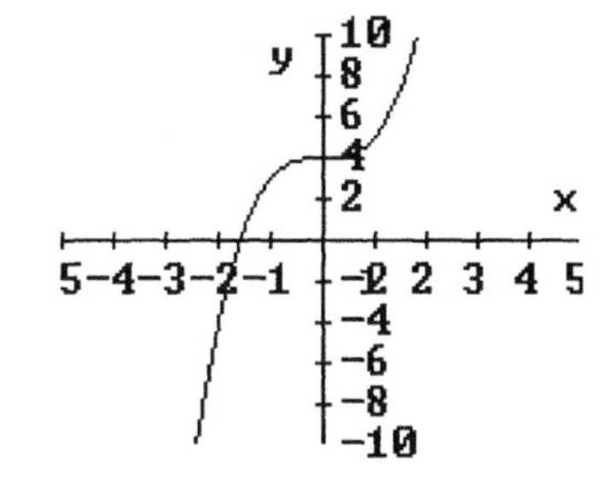

13. $g(x)=2x^4+x^2+3$ has no variations in sign, so there are no positive real zeros, and $g(-x)=2(-x)^4+(-x)^2+3$, that is $g(-x)=2x^4+x^2+3$, has no variations in sign either, so there are no negative real zeros.

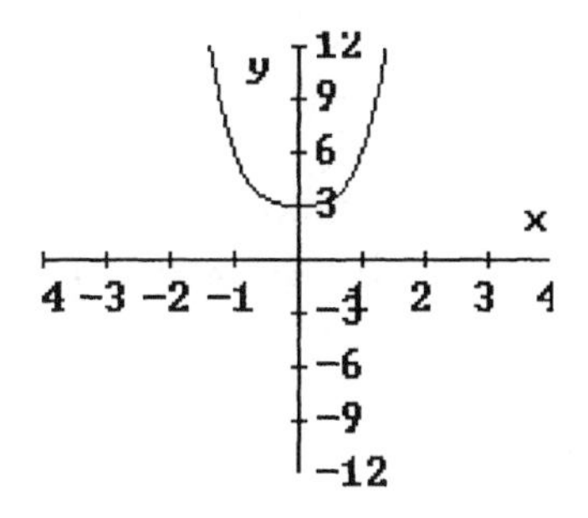

15. $f(x)=x^3-4x^2+7x-2$ has three variations in sign, so there is one or three positive real zeros; $f(-x)=-x^3-4x^2-7x-2$ has no variations in sign, so there are no negative real zeros. There is, in fact, only one real zero, $x \approx 0.35$

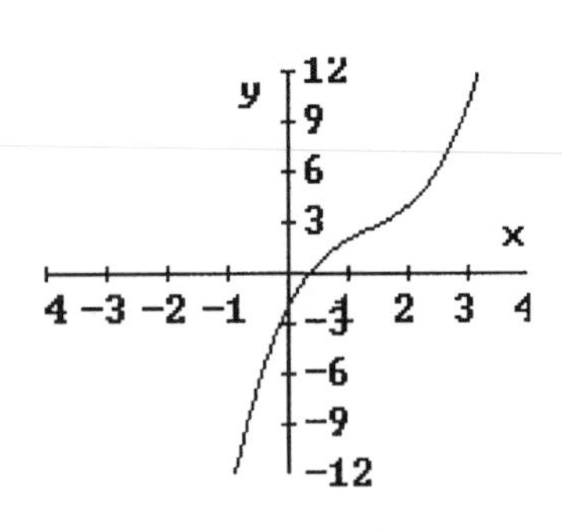

17. $f(x)=x^5-10x^4-11x^3-5$ has one variation in sign, so there is exactly one positive real zero.
$f(-x)=(-x)^5-10(-x)^4-11(-x)^3-5$, that is $f(-x)=-x^5-10x^4+11x^3-5$, has two variations in sign, so f has either two negative real zeros or none. There are, in fact, no negative real zeros. The positive real zero is $x \approx 11$.

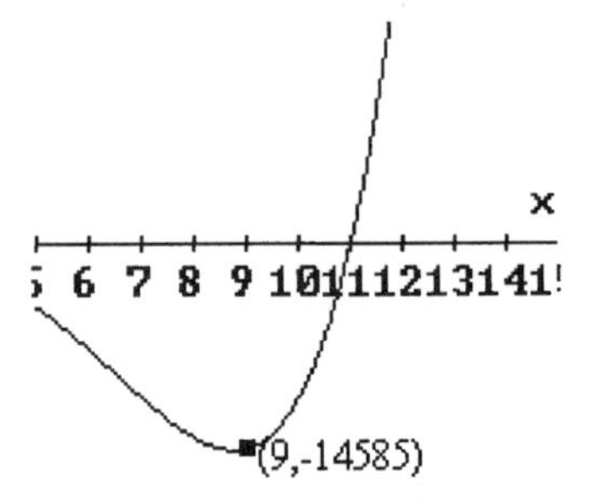

19. $h(x)=x^3-x^2-\dfrac{101}{100}x+\dfrac{99}{100}$ has two variations in sign, so there are two positive real zeros, or none. Now,
$h(-x)=-x^3-x^2+\dfrac{101}{100}x+\dfrac{99}{100}$, has one variation in sign, so h has one negative real zero. The zeros of h are $x=-1,\ 0.9,\ 1.01$.

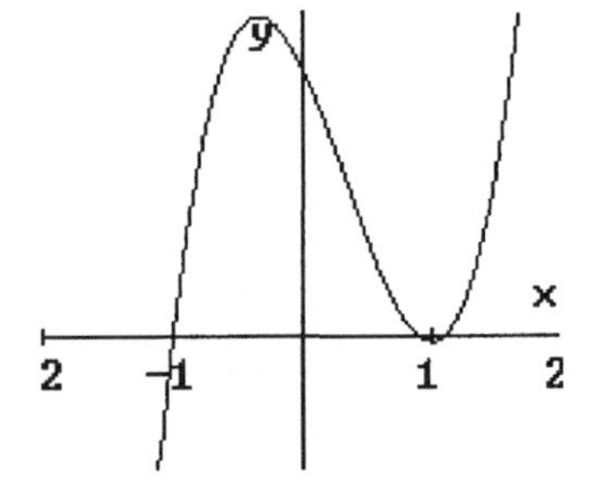

21. If we divide $f(x)=x^4-5x^3-11x^2+33x-18$ by $x-7$ and by $x+4$ using synthetic division:

7	1	−5	−11	33	−18
		7	14	21	378
	1	2	3	54	360

−4	1	−5	−11	33	−18
		−4	36	−100	268
	1	−9	25	−67	250

all numbers in the last row of the first division are positive, therefore 7 is an upper bound for the real zeros of *f*, and the signs of the numbers in the last row of the second division alternate, so -4 is a lower bound for the real zeros of *f*. So, all real zeros of *f* are in $[-4,\ 7]$.

23. We divide $f(x)=x^4-10x^3-5$ by $x-11$ and by $x+1$:

11	1	−10	0	0	−5
		11	11	121	1331
	1	1	11	121	1326

−1	1	−10	0	0	−5
		−1	11	−11	11
	1	−11	11	−11	6

all numbers in the last row of the first division are positive, therefore 11 is an upper bound for the real zeros of *f*, and the signs of the numbers in the last row of the second division alternate, so -1 is a lower bound for the real zeros of *f*. So, all real zeros of *f* are in $[-1,\ 11]$.

25. We divide $f(x) = x^4 - 62x^3 + 962x^2 - 62x + 960$ by $x - 62$ and by $x + 1$:

$$\begin{array}{r|rrrrr} 62 & 1 & -62 & 962 & -62 & 960 \\ & & 62 & 0 & 59644 & 3694084 \\ \hline & 1 & 0 & 962 & 59582 & 3695044 \end{array} \qquad \begin{array}{r|rrrrr} -1 & 1 & -62 & 962 & -62 & 960 \\ & & -1 & 63 & -1025 & 1087 \\ \hline & 1 & -63 & 1025 & -1087 & 2047 \end{array}$$

62 is an upper bound and -1 a lower bound for the real zeros of *f*, since in the first case all numbers in the last row are positive, and in the second case, the signs alternate.

27. If $h(x) = x^3 + 6x^2 - x - 6$, we have $h(x) = x^2(x+6) - (x+6)$, or $h(x) = (x^2 - 1)(x+6)$, so $h(x) = (x-1)(x+1)(x+6)$. The zeros of *h* are $x = -6,\ -1,$ and 1.

29. For $f(x) = x^3 - 14x^2 + 25x - 12$, we apply synthetic division by $x - 1$ twice:

$$\begin{array}{r|rrrr} 1 & 1 & -14 & 25 & -12 \\ & & 1 & -13 & 12 \\ \hline & 1 & -13 & 12 & 0 \end{array} \qquad \begin{array}{r|rrr} 1 & 1 & -13 & 12 \\ & & 1 & -12 \\ \hline & 1 & -12 & \end{array}$$

That is, $f(x) = (x-1)^2(x-12)$, and the zeros of *f* are $x = 12$ and a double zero $x = 1$.

31. The polynomial function $h(x) = x^3 + \frac{7}{3}x^2 - \frac{23}{12}x + \frac{1}{4}$ has the same zeros as the multiple of *h* obtained multiplying by 12, $f(x) = 12x^3 + 28x^2 - 23x + 3$. The possible rational zeros of *f* are of the form a/b, where *a* is a factor of 3, and *b* a factor of 12, namely $\pm 12,\ \pm 6,\ \pm 4,\ \pm 3,\ \pm 2,\ \pm 1$. Applying synthetic division,

$$\begin{array}{r|rrrr} -3 & 12 & 28 & -23 & 3 \\ & & -36 & 24 & -3 \\ \hline & 12 & -8 & 1 & 0 \end{array}$$

$f(x) = (x+3)(12x^2 - 8x + 1)$, so $f(x) = (x+3)(6x-1)(2x-1)$. The zeros are $-3,\ 1/6,$ and $1/2$.

33. If $h(x) = 8x^3 + x^2 - 16x - 2$, we have $h(x) = x^2(8x+1) - 2(8x+1)$, or $h(x) = (x^2 - 2)(8x+1)$, and the real zeros of *h* are $x = -\sqrt{2},\ -1/8,$ and $\sqrt{2}$.

35. The possible rational zeros of $g(x) = x^4 + x^3 - 120x^2 - 121x - 121$ are $\pm 1,\ \pm 11,\ \pm 121$. In fact, we have $g(x) = (x+11)(x-11)(x^2 + x + 1)$. The rational zeros are $x = -11$ and $x = 11$.

37. To solve $x^3 + x^2 - 4x - 4 = 0$, we factor the left hand side: $x^2(x+1) - 4(x+1) = 0$, so $(x^2 - 4)(x+1) = 0$, and finally $(x+2)(x-2)(x+1) = 0$. The solutions are $-2,\ -1,$ and 2.

39. To solve $x^4 - 27 = 6x^2$, we first subtract $6x^2$ from both sides so that one of the sides is 0: $x^4 - 6x^2 - 27 = 0$. This equation can be factored as $(x^2 - 9)(x^2 + 3) = 0$, or $(x-3)(x+3)(x^2 + 3) = 0$. The equation has only two real solutions, $x = -3,\ 3$.

41. To solve $8x^5 - 8x^3 = 1 - x^2$, we add $x^2 - 1$ to both sides, to obtain the equivalent equation

$$8x^5 - 8x^3 + x^2 - 1 = 0$$
$$8x^3(x^2 - 1) + (x^2 - 1) = 0$$
$$\left(8x^3 + 1\right)\left(x^2 - 1\right) = 0$$
$$(2x+1)\left(4x^2 - 2x + 1\right)(x-1)(x+1) = 0$$

The real solutions of $8x^5 - 8x^3 = 1 - x^2$ are $x = -1,\ -\frac{1}{2}$, and 1.

43. The possible rational zeros of $f(x) = x^3 + 3x + 1$ are of the form a/b, where a is a factor of 1, and so is b. Then, the only candidates to be rational zeros of f are -1 and 1. We have $f(-1) = -3$, and $f(1) = 5$. f has one real zero, an irrational one, $x \approx -0.32$.

45. The possible rational zeros of $h(x) = x^4 + 2x - 2$ are of the form a/b, where a is a factor of -2 and b a factor of 1. So, the only possibilities are $\pm 1,\ \pm 2$. By inspection of the graph of $y = h(x)$, we can see that neither one of these is a zero. h has two irrational zeros, $x \approx -1.49,\ 0.8$

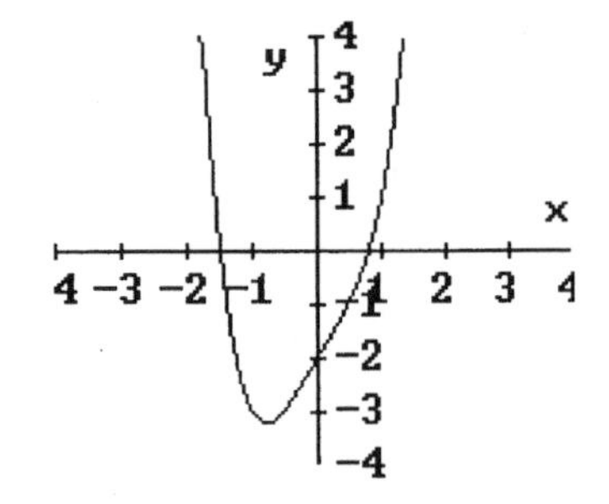

47. The possible rational zeros of $h(x) = x^5 - 10x^4 - 11x^3 - 5$ are of the form $\dfrac{a}{b}$, where a is a factor of -5 and b a factor of 1. The only possible rational zeros are $\pm 1,\ \pm 5$. Neither one of these is a zero. h has one irrational zero, $x \approx 11$.

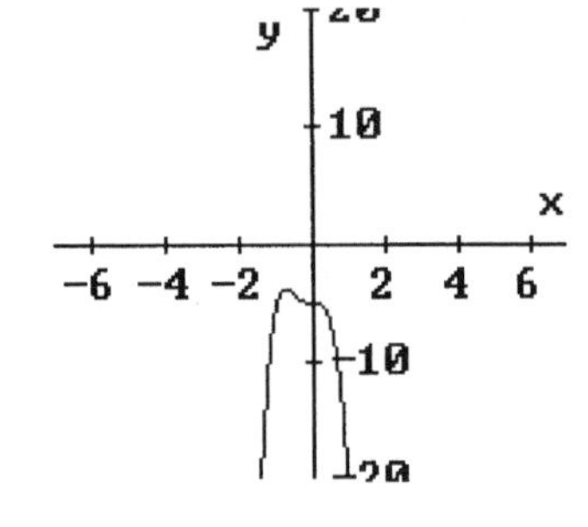

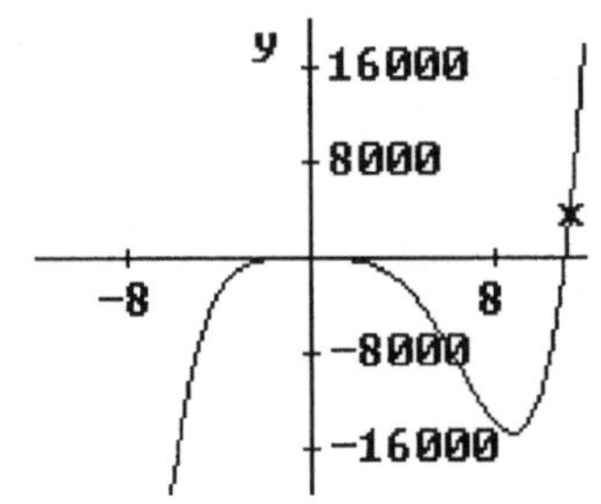

Applications

49. A box with square base of side length x and height h has volume $V = x^2 h$. Given that the volume is 9 cubic inches, and that the height is one inch less than twice the side length of the base, we have $x^2(2x - 1) = 9$, that is $2x^3 - x^2 - 9 = 0$, for which the solution is $x \approx 1.84$. So the side length of the base is $x \approx 1.84$ inches and the height is $h \approx 2.68$ inches.

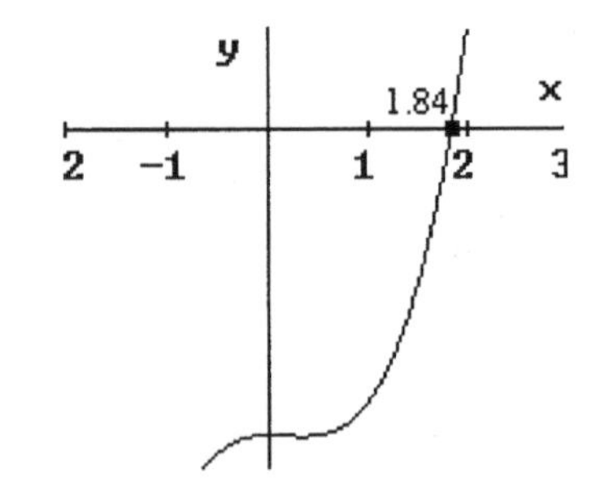

51. If the side length of the squares that are cut out of the board is x, the box obtained by folding up the sides will have height x, and a rectangular base of width $5-2x$ and length $10-2x$. The volume of the box is to be 18 cubic inches, so $4x^3-30x^2+50x=18$, and there are two relevant solutions ($0<x<2.5$), $x=0.5$, $x\approx 1.7$.

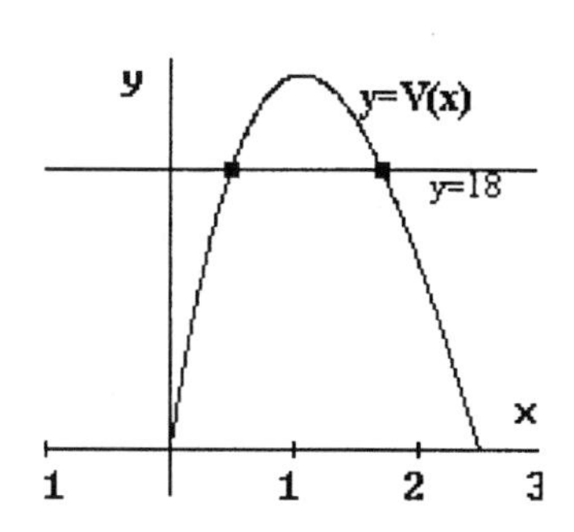

53. For $p(x)=-0.001x^3+0.07x^2-1.4x+17.5$ to be 10.3, there are three solutions $x\approx 7.93$, $x\approx 23.59$, and $x\approx 38.48$. Since x is the year since the year 1960, this model says that the poverty level was 10.3 in 1968, 1984 and 1998.

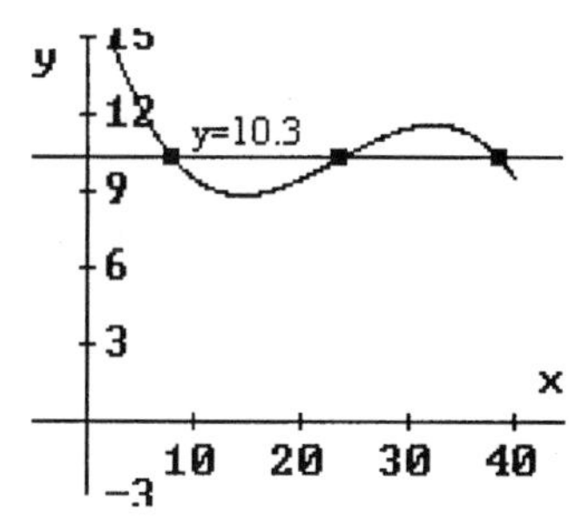

55. The interest rate on a loan of \$3000 is r, and $v=\dfrac{1}{1+r}$.

(a) Present value of payments of \$1100 each (at times 1, 2, 3)$=1100v+1100v^2+1100v^3$.

(b) $1100v+1100v^2+1100v^3=3000$, so $1100v+1100v^2+1100v^3-3000=0$. There is only one variation in sign in this equation, therefore only one positive solution.

(c) The solution for $1100v+1100v^2+1100v^3-3000=0$ is, approximately, $v\approx 0.9531$, that is, $\dfrac{1}{1+r}\approx 0.9531$, so $r\approx 0.0492$, or a rate of 4.92% per period.

57. **(a)-(c)** If a line-of-credit loan is initiated with a distribution of \$1000 at time 0, payments at times 1 and 2, a second distribution d at time 3, and a final payment at time 4, we must have the present value of the payments be equal to the present value of the distributions: $p_1v+p_2v^2+p_3v^4=1000+dv^3$, so $p_3v^4-dv^3+p_2v^2+p_1v-1000=0$. The number of variations in sign is 3, so there are one or three positive solutions for v.

(d) If the second distribution is $d=500$, and the payments are \$400 each, we obtain $400v^4-500v^3+400v^2+400v-1000=0$. The solution of this equation is, approximately, $v\approx 1.19$, so $\dfrac{1}{1+r}\approx 1.19$. From here, $r\approx -0.16$. The lender would be losing money under this scheme, since the distributions total \$1500, and the payments total \$1200. For a lender to make money on a loan, it is necessary that the payments total more than the distributions.

Concepts and Critical Thinking

59. The statement: "A polynomial of the form $f(x) = x^3 + ax^2 + bx + 1$ (where a and b are integers) can have at most 2 rational zeros" is true. A rational zero must be of the form a/b, where a is a factor of 1 and b a factor of 1 as well. The only possibilities are -1 and 1.

61. The statement: "It is possible that $1/3$ is a zero of a polynomial of the form $f(x) = x^4 + ax^2 + x + 3$, where a is an integer" is false.

63. Answers may vary. An example of a polynomial that, according to Descartes' Rule of Signs, has either 1, 3, or 5 positive zeros is $f(x) = x^5 - x^4 + x^2 - 4x^3 + 10x^2 - x - 1$.

65. Answers may vary. An example of a fifth-degree polynomial that, according to the Rational Zero Theorem, has ± 1 as its only possible rational zeros is $f(x) = x^5 - 3x - 1$.

67. Answers will not vary, even though the polynomials will be different for different people. In any case, the polynomial will be of the form $P(x) = x^{20} - ax^{19} + bx^5 + 2x + 27$, where $a > 0$, and $b > 0$. The number of variations in sign is 2, so the number of positive real zeros is 2 or 0. We have $P(-x) = x^{20} + ax^{19} - bx^5 - 2x + 27$. The number of variations in sign in $P(-x)$ is 2, so the number of negative real zeros is 2 or 0.

Chapter 3. Section 3.5 Rational Functions

1. The domain of the function $f(x) = \dfrac{1}{x-2}$ is the set of all real numbers $x \neq 2$. f doesn't have any zeros. Its graph matches (vii).

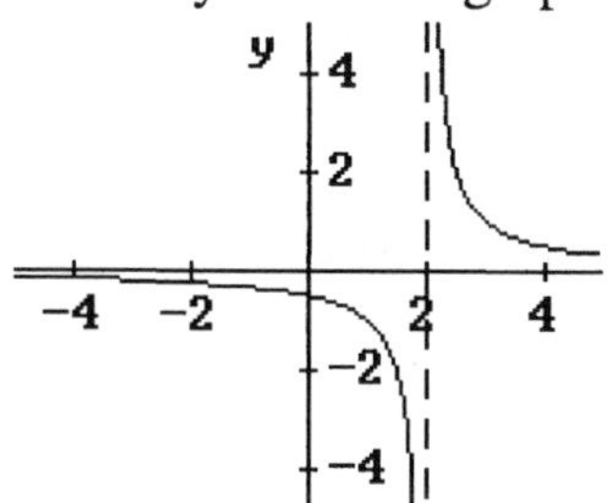

3. The domain of the function $f(x) = \dfrac{x+1}{x}$ is the set of all real numbers $x \neq 0$. f has one zero, $x = -1$. Its graph matches (iii).

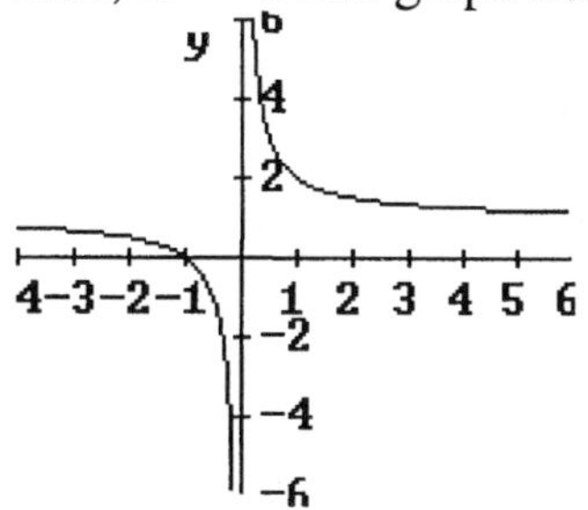

5. The domain of the function $f(x) = \dfrac{1}{x+2}$ is the set of all real numbers $x \neq -2$. f doesn't have any zeros. Its graph matches (iv).

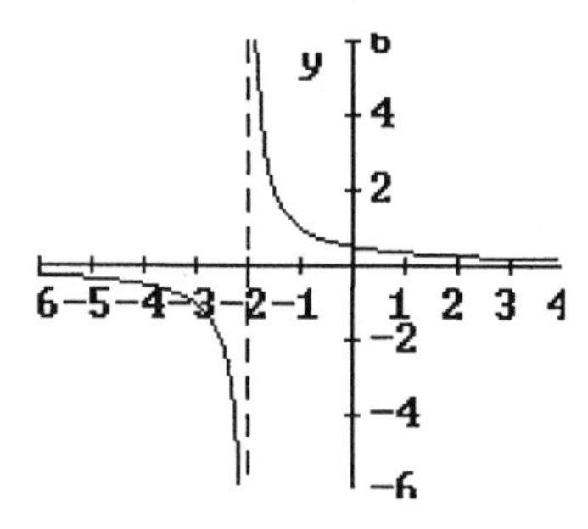

7. The domain of the function $f(x) = \dfrac{x^2-4}{x^2-1}$ is the set of all real numbers $x \neq -1,\ 1$. f has two zeros, $x = -2$ and $x = 2$. Its graph matches (ii).

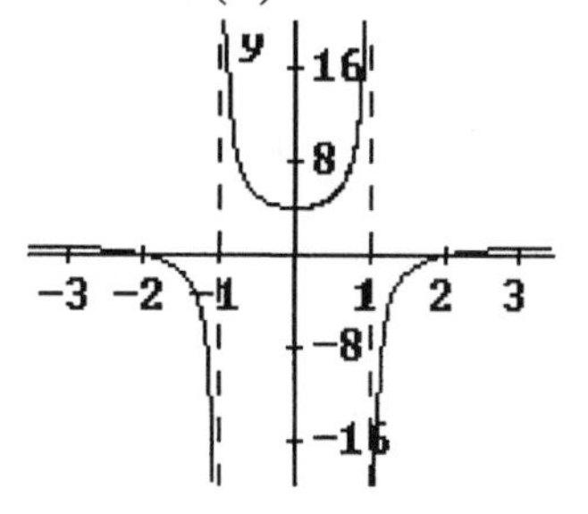

9. Based on the graph of $f(x) = \dfrac{1}{x}$:

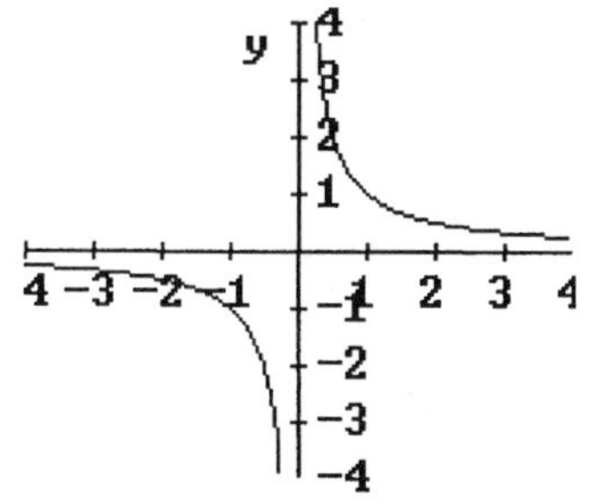

(a) Sketch of the graph of $g(x) = \dfrac{1}{x-3}$

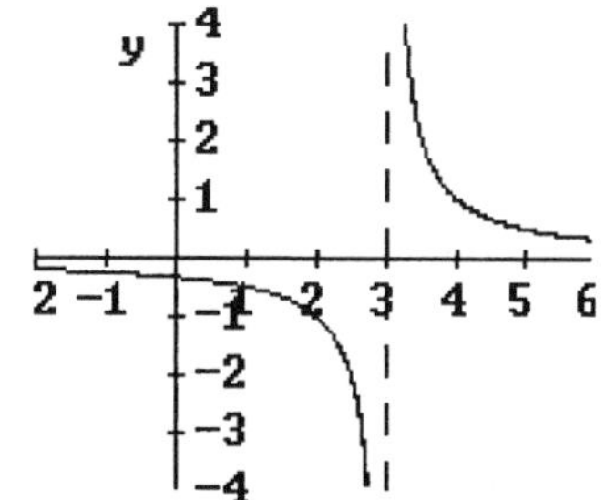

(b) Sketch of the graph of $g(x) = \dfrac{1}{x} + 2$

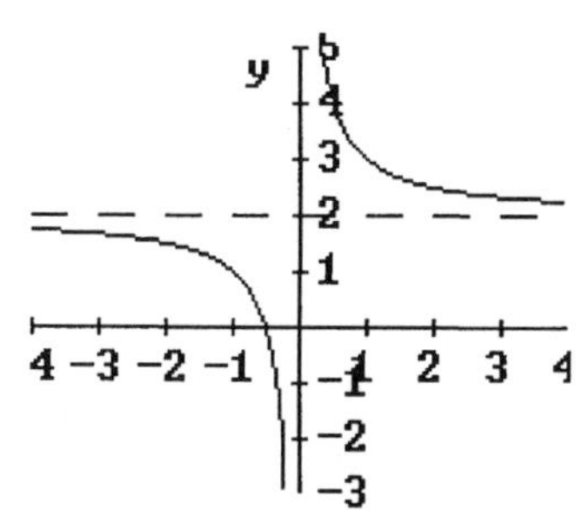

(c) Sketch of the graph of $g(x) = -\dfrac{1}{x}$

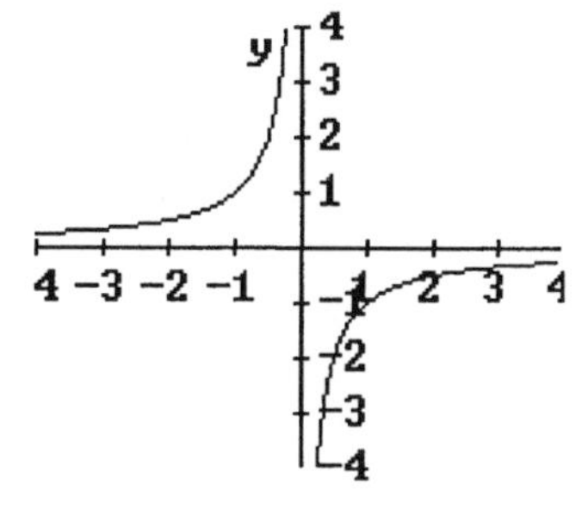

11. Based on the graph of $y = f(x)$:

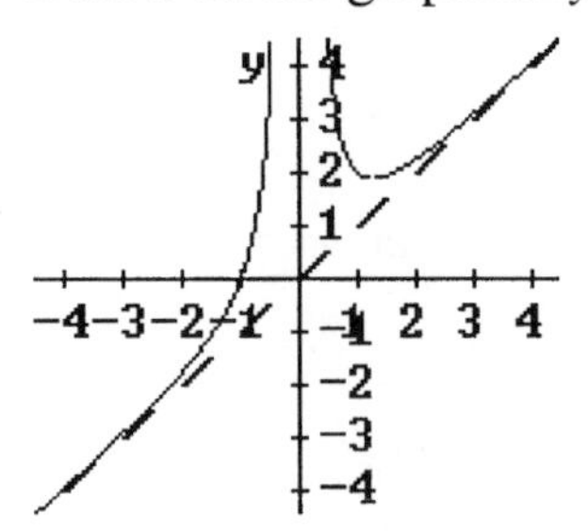

(a) Sketch of the graph of $g(x) = f(x-1)$

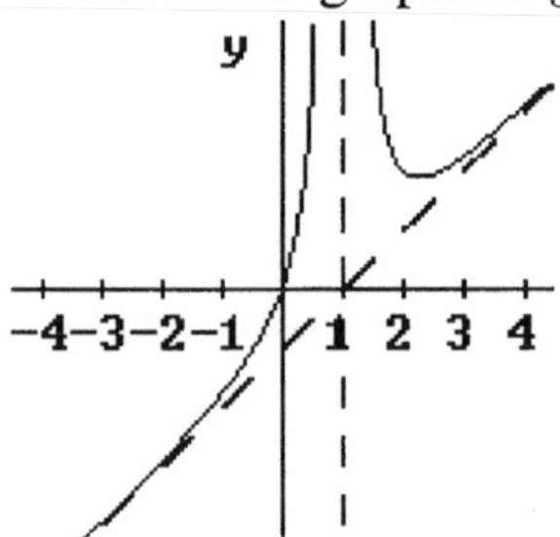

(b) Sketch of the graph of $g(x) = f(x) - 2$

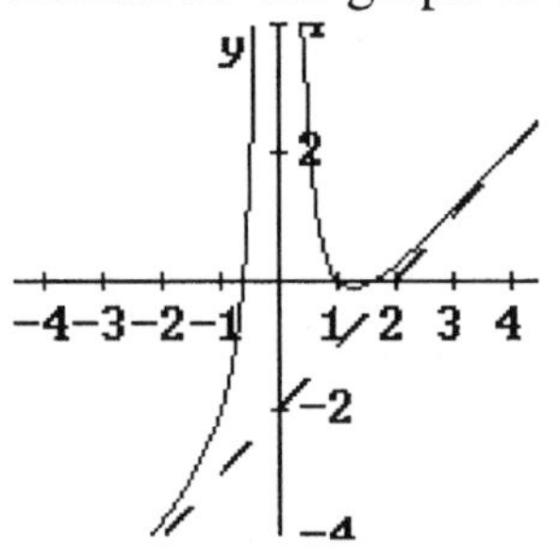

(c) Sketch of the graph of $g(x) = -f(x)$

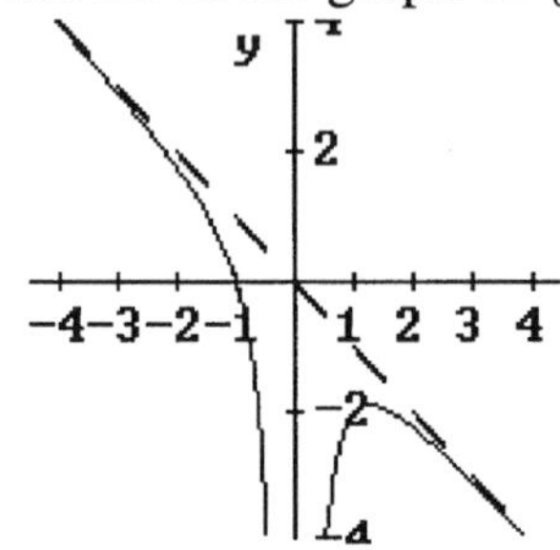

13. Answers may vary. The function $f(x) = -\dfrac{1}{(x-4)^2}$ has a vertical asymptote at $x = 4$, horizontal asymptote $y = 0$, and $f(x) < 0$ for all $x \neq 4$.

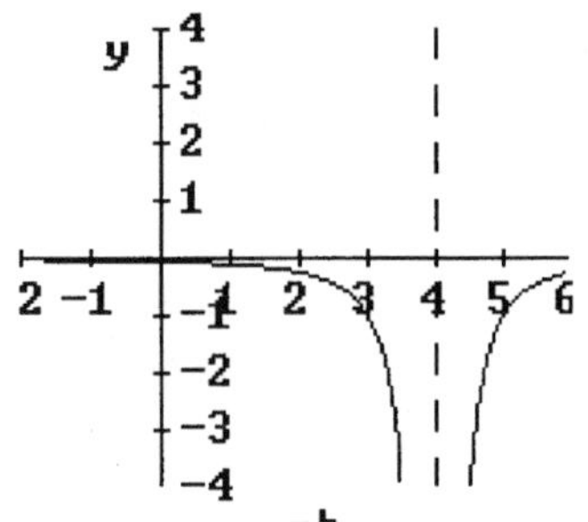

15. Answers may vary. The function $g(x) = \dfrac{2x+3}{x+1}$ has a vertical asymptote at $x = -1$, horizontal asymptote $y = 2$, an x-intercept at $\left(-\tfrac{3}{2}, 0\right)$, and a y-intercept at (0, 3).

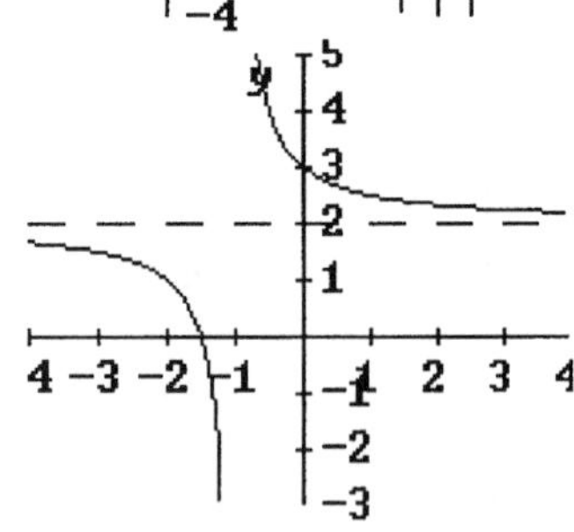

17. The function $f(x) = \dfrac{1}{x+3}$ has a vertical asymptote at $x = -3$, and horizontal asymptote $y = 0$.

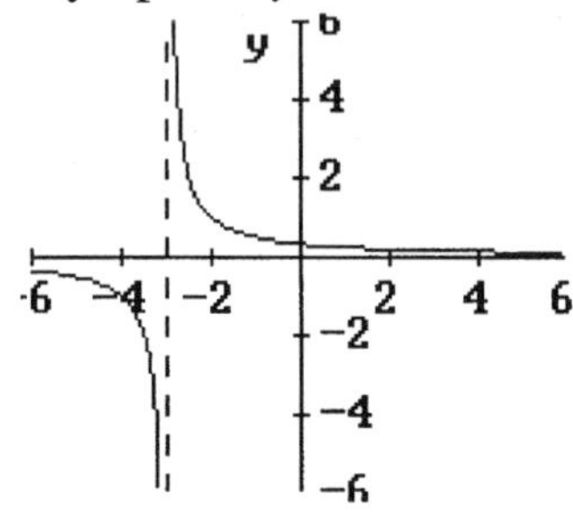

19. The function $h(x) = \dfrac{x}{x+2}$ has a vertical asymptote at $x = -2$, and horizontal asymptote $y = 1$.

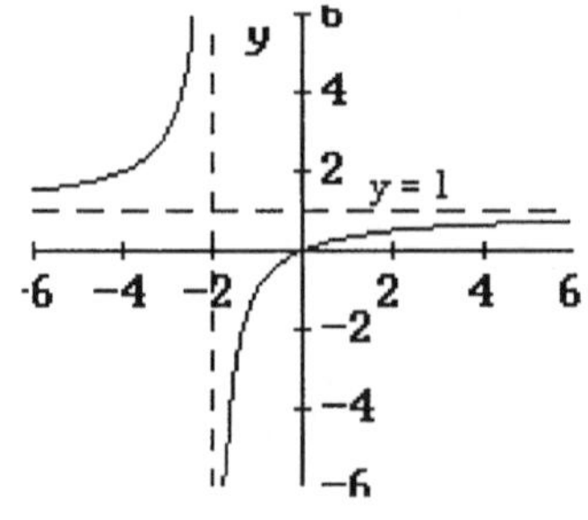

21. The function $f(x)=\dfrac{3x}{x-2}$ has a vertical asymptote at $x=2$, and horizontal asymptote $y=3$.

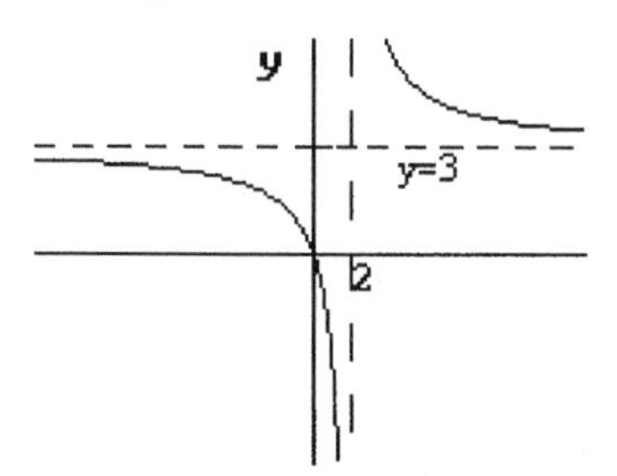

23. The function $g(x)=\dfrac{x}{x^2-9}$ has vertical asymptotes at $x=-3,\ x=3$, and horizontal asymptote $y=0$.

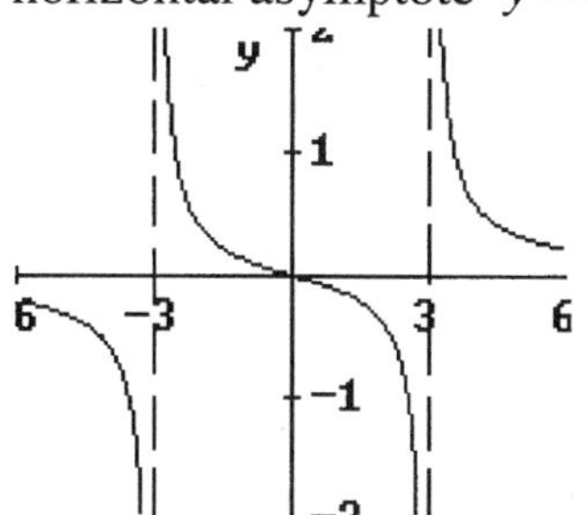

25. The function $g(x)=\dfrac{45x^2}{9x^2+1}$ has no vertical asymptotes. It has a horizontal asymptote $y=5$.

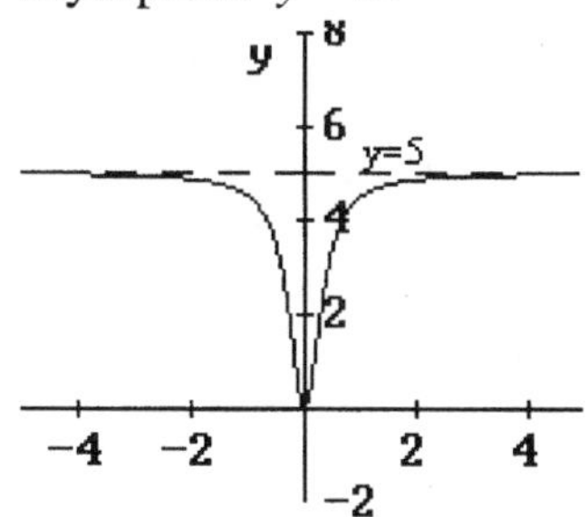

27. The function $h(x)=\dfrac{1}{(x+1)^2}$ has a vertical asymptote at $x=-1$, and horizontal asymptote $y=0$.

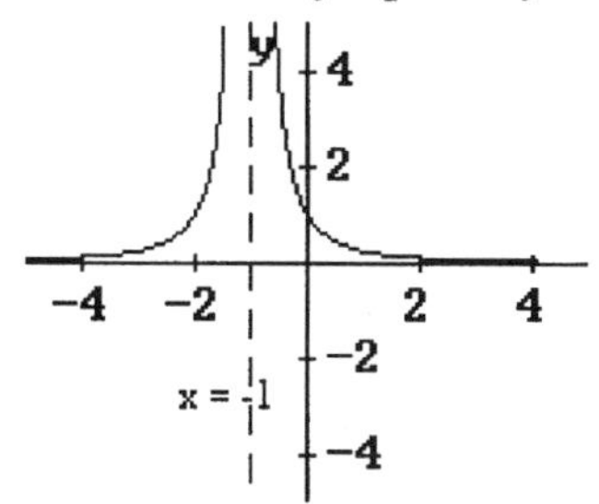

29. The function $f(x)=\dfrac{x^2}{(x-2)^2}$ has a vertical asymptote at $x=2$, and horizontal asymptote $y=1$.

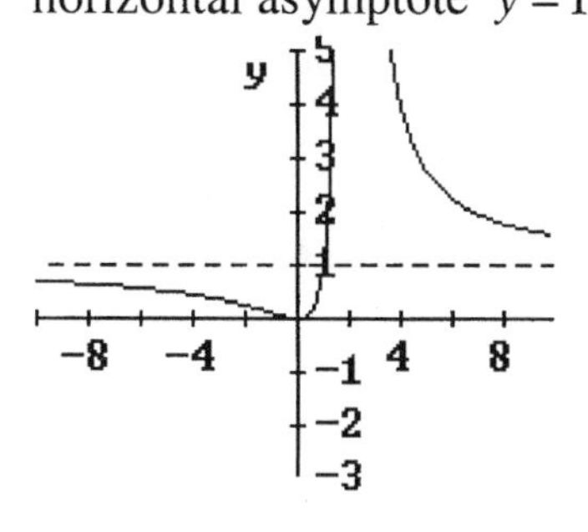

31. The function $g(x)=2x+\dfrac{1}{x}$ has a vertical asymptote at $x=0$, and an inclined asymptote $y=2x$.

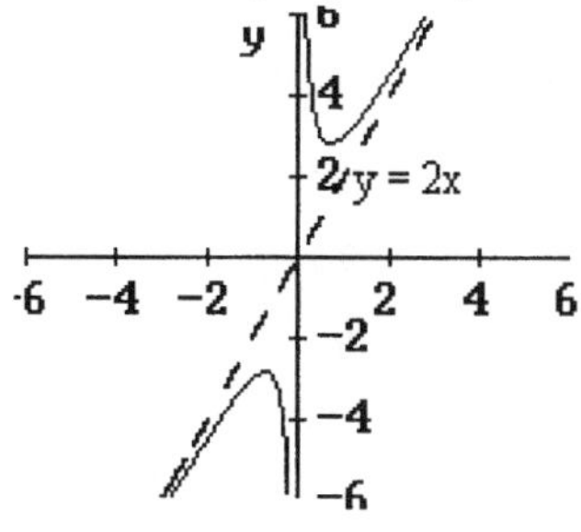

33. The function $f(x)=\dfrac{x^2}{x-1}$ has a vertical asymptote at $x=1$, and an inclined asymptote $y=x+1$.

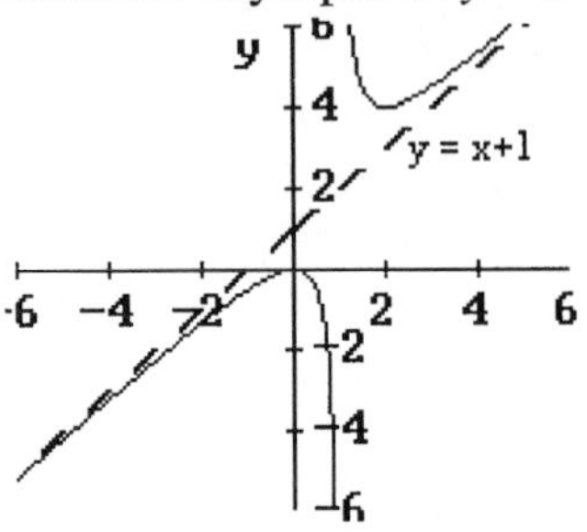

35. The function $g(x)=\dfrac{-3x^2-6x+1}{x+2}$ has a vertical asymptote at $x=-2$, and an inclined asymptote $y=-3x$.

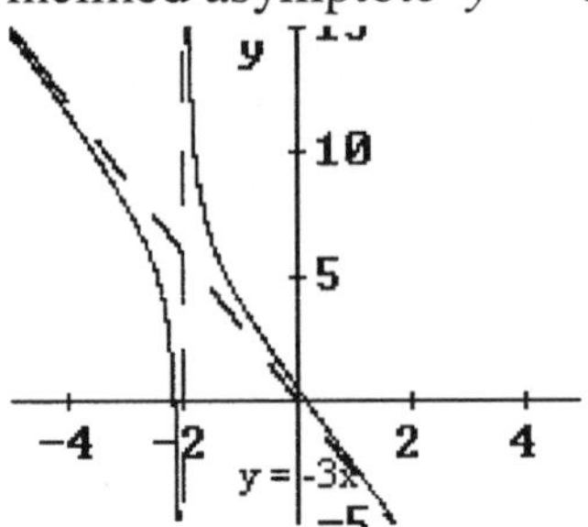

37. The function $f(x)=\dfrac{2}{x^2+1}$ doesn't have vertical asymptotes. It has a horizontal asymptote $y=0$.

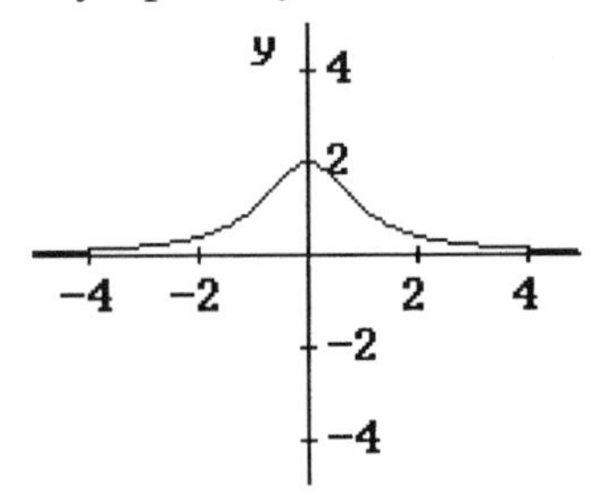

39. The function $h(x)=\dfrac{12x+24}{x-15}$ has a vertical asymptote at $x=15$. It has a horizontal asymptote $y=12$.

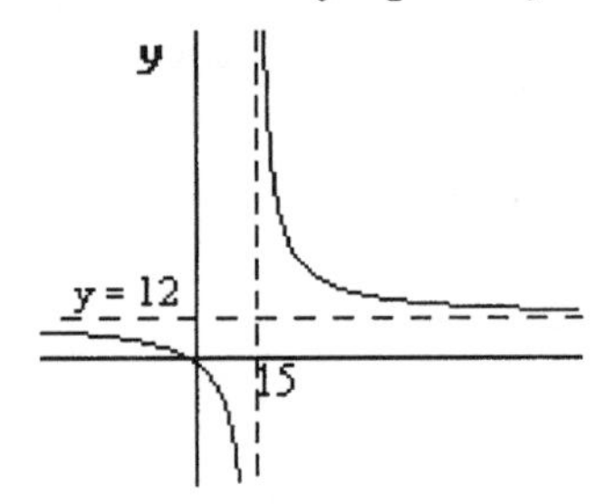

41. The function $g(x)=\dfrac{x^2-4x+7}{x-3}$ has a vertical asymptote at $x=3$. It has an inclined asymptote $y=x-1$.

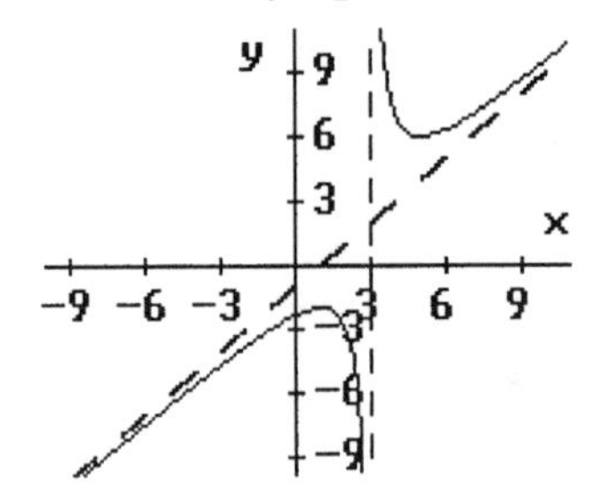

43. The function $h(x)=-\dfrac{1}{x^2-9}$ has vertical asymptotes at $x=-3$ and $x=3$. It has a horizontal asymptote $y=0$.

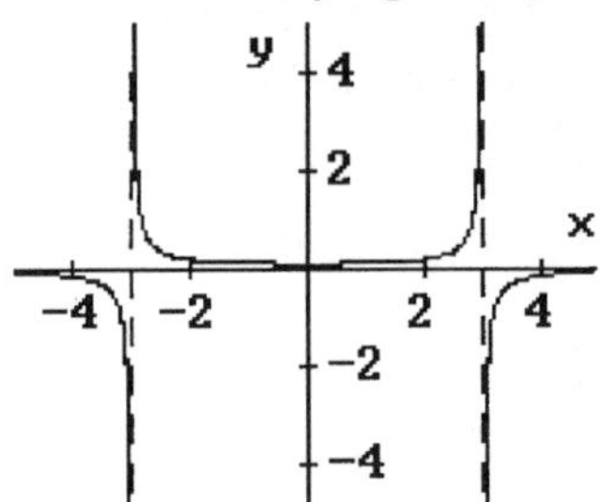

43. The function $h(x) = -\dfrac{1}{x^2-9}$ has vertical asymptotes at $x = -3$ and $x = 3$. It has a horizontal asymptote $y = 0$.

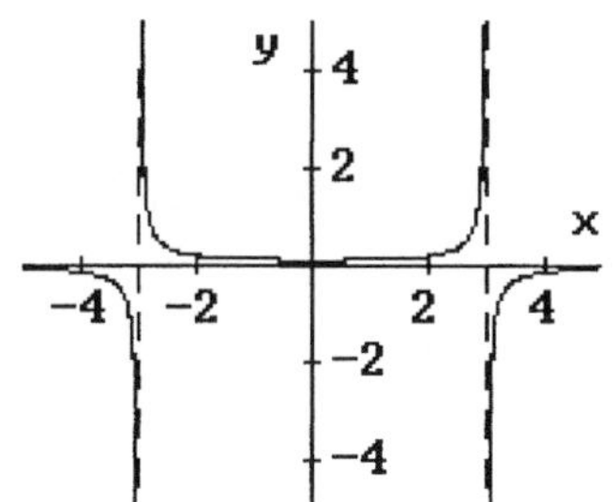

45. The function $f(x) = \dfrac{x^3}{x-3}$ has a vertical asymptote at $x = 3$. It has no horizontal or inclined straight-line asymptotes.

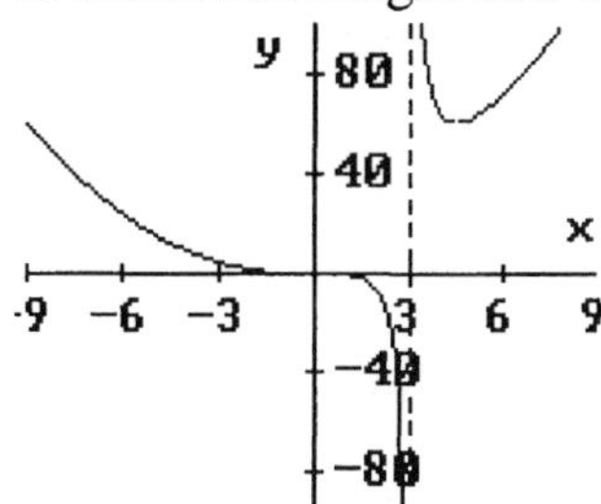

47. The function $f(x) = \dfrac{\sqrt{1+4x^2}}{4+x}$ has a vertical asymptote at $x = -4$. It has two horizontal asymptotes, $y = -2$, $y = 2$.

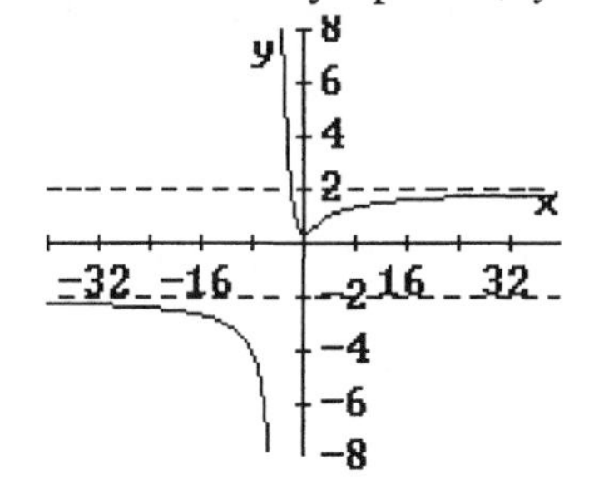

49. The function $f(x) = \dfrac{3x}{\sqrt{x^2+1}}$ has two horizontal asymptotes $y = 3$ and $y = -3$.

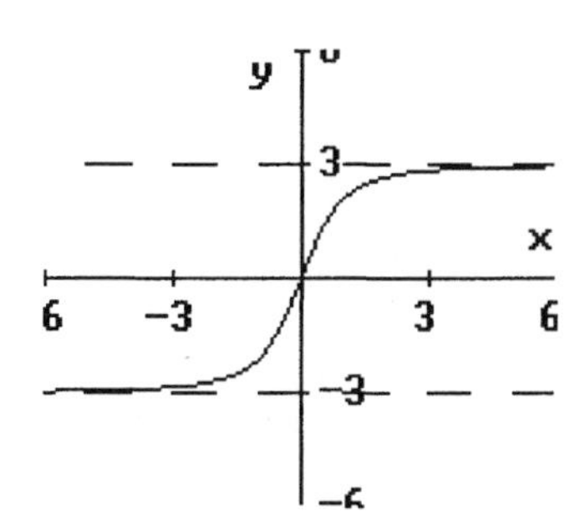

Applications

51. $f(x) = \dfrac{x}{1000(100-x)}$ gives the increase in refinery costs (per barrel) as a function of the percent reduction x in the refinery discharge. The function has a vertical asymptote at $x = 100$; in fact, the relevant domain of this function is $[0,\ 100]$. With an increase of \$0.003 per barrel, a 75% reduction in the refinery discharge is possible. A 100% reduction is not attainable ($x = 100$ is not in the domain of f.)

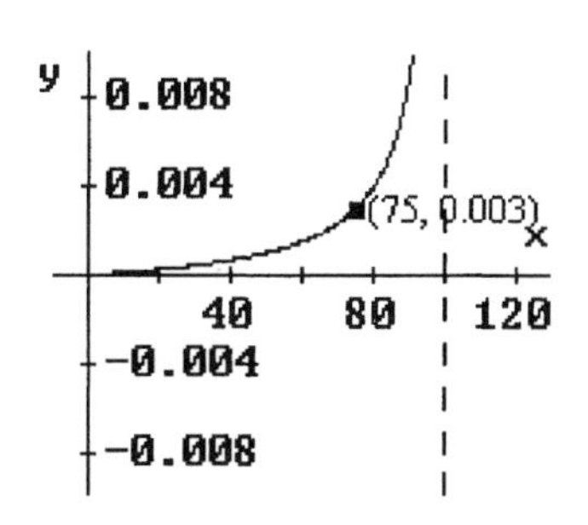

53. $f(x) = \dfrac{77}{(x-10)^2+7}$ gives the percentage rate of unemployment in the year 2000 as a function of the years of education x. The function has a horizontal asymptote $y = 0$. According to this model, the group of people who had 4.9% rate of unemployment that year was those with $x \approx 12.95$ years of education (that is, one incomplete year of college, if we count years of education starting at elementary school.)

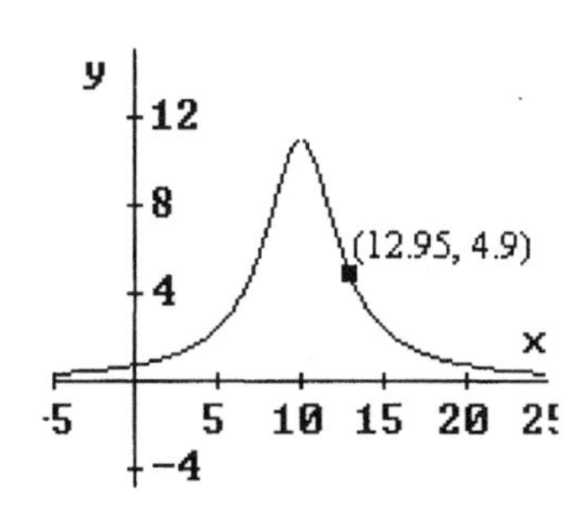

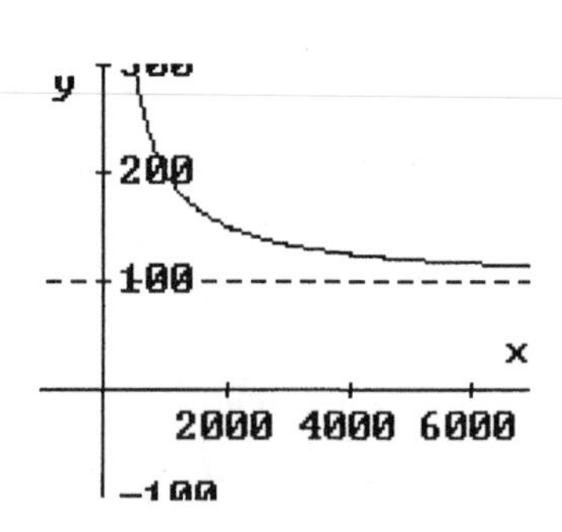

55. $\overline{C}(x) = \dfrac{100{,}000 + 100x}{x}$ is the average cost if x bicycles are manufactured. The horizontal asymptote is $y = 100$. As more bicycles are produced, the average cost gets closer to \$100 per bicycle.

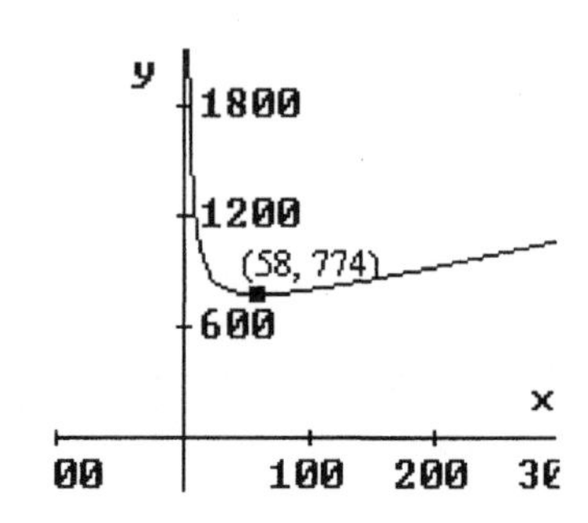

57. $C(x) = \dfrac{3}{2}x + 600 + \dfrac{5046}{x}$ is the yearly cost for inventory if x balls are ordered at a time. The minimum cost of \$774 is achieved when $x = 58$ basketballs per order. There should be 10 orders during the year, in order to sell a total of 580 basketballs. So, orders should be made about every 5 weeks. The inclined asymptote is $y = \dfrac{3}{2}x + 600$, with a slope of $3/2$.

Concepts and Critical Thinking

59. The statement: "The graphs of all rational functions have vertical asymptotes" is false.

61. The statement: "The zeros of a rational function in lowest terms are the zeros of its numerator" is true.

63. Answers may vary. An example of a rational function with horizontal asymptote $y = 3$ and vertical asymptotes $x = 1$ and $x = -1$ is $f(x) = \dfrac{3x^2}{x^2 - 1}$.

Chapter 3. Review

1. The end behavior of $g(x) = -3x^6$ is: $g(x) \to -\infty$ as $x \to \infty$, and $g(x) \to -\infty$ as $x \to -\infty$.

3. The end behavior of $f(x) = 4x^3 - x^2 + 9x + 3$ is like that of $4x^3$: $f(x) \to \infty$ as $x \to \infty$, and $f(x) \to -\infty$ as $x \to -\infty$.

5. The end behavior of $h(x) = x^3 + 100 - 0.00005x^4$ is like that of $-0.00005x^4$: $h(x) \to -\infty$ as $x \to -\infty$, and $h(x) \to \infty$ as $x \to \infty$.

7. The polynomial $f(x) = \dfrac{x^3}{3} - 3x^2 - 3$ has one zero, $x \approx 9.11$, and two turning points, a maximum at $(0, -3)$ and a minimum at $(6, -39)$.

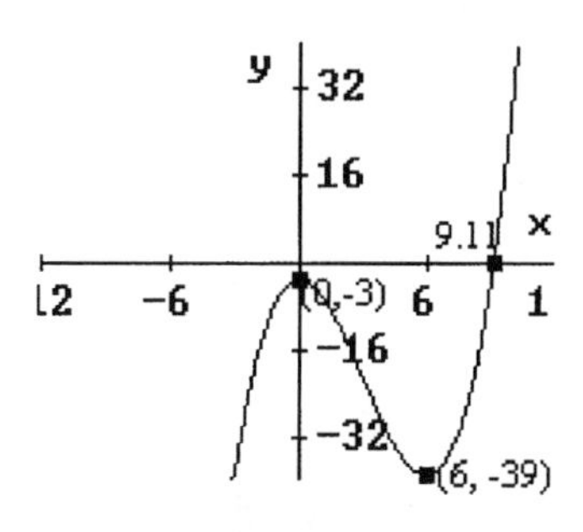

9. The polynomial $g(x) = 0.1x^4 + x^3 - 4x^2 + x - 4$ has two zeros, $x \approx -13.12$ and $x \approx 3.12$. There is a maximum at $(0.13, -3.93)$, minima at $(-9.61, -417.63)$ and $(1.98, -8.4)$.

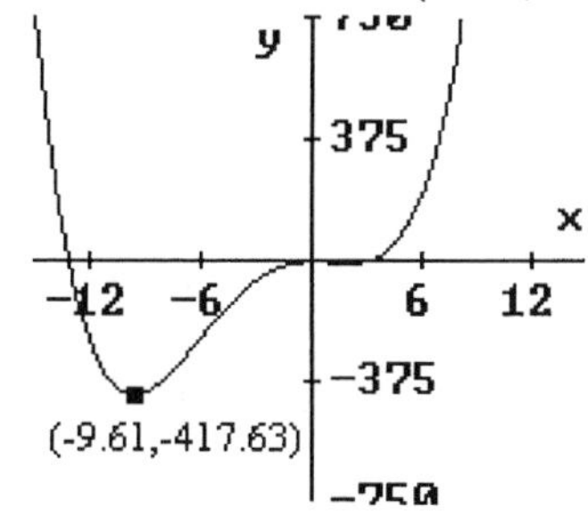

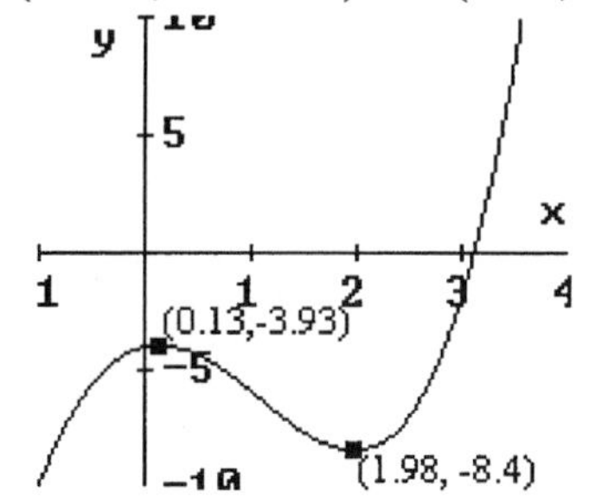

11. When we divide $f(x) = 2x^2 + 3x - 2$ by $d(x) = x + 2$, the quotient is $q(x) = 2x - 1$ and the remainder is $r(x) = 0$.

$$\begin{array}{r} 2x - 1 \\ x + 2 \overline{\smash{\big)} 2x^2 + 3x - 2} \\ \underline{2x^2 + 4x} \\ -x - 2 \\ \underline{-x - 2} \\ 0 \end{array}$$

13. When we divide $f(x) = x^4 - 5x^2 + 2$ by $d(x) = x^2 + 4x$, the quotient is $q(x) = x^2 - 4x + 11$ and the remainder is $r(x) = -44x + 2$.

$$
\begin{array}{r}
x^2 - 4x + 11 \\
x^2 + 4x \overline{\big) x^4 \qquad -5x^2 \qquad +2} \\
\underline{x^4 + 4x^3 \qquad\qquad\qquad} \\
-4x^3 - 5x^2 \qquad +2 \\
\underline{-4x^3 - 16x^2 \qquad\quad} \\
11x^2 \qquad +2 \\
\underline{11x^2 + 44x \quad} \\
-44x + 2
\end{array}
$$

15. Dividing $f(x) = 3x^3 - 17x^2 + 22x - 8$ by $d(x) = x - 4$ using synthetic division we obtain:

$$
\begin{array}{r|rrrr}
4 & 3 & -17 & 22 & -8 \\
 & & 12 & -20 & 8 \\
\hline
 & 3 & -5 & 2 & 0
\end{array}
$$

That is, the quotient is $q(x) = 3x^2 - 5x + 2$, and the remainder is $r(x) = 0$.

17. Dividing $f(x) = -2x^4 + 4x^2 + x - 3$ by $d(x) = x + 3$ using synthetic division we obtain:

$$
\begin{array}{r|rrrrr}
-3 & -2 & 0 & 4 & 1 & -3 \\
 & & 6 & -18 & 42 & -129 \\
\hline
 & -2 & 6 & -14 & 43 & -132
\end{array}
$$

That is, the quotient is $q(x) = -2x^3 + 6x^2 - 14x - 43$, and the remainder is $r(x) = -132$.

19. Dividing $f(x) = 3x^3 + 11x^2 + 2x - 6$ by $d(x) = x - 3/2$, $d(x) = x + 2$ using synthetic division:

$$
\begin{array}{r|rrrr}
3/2 & 3 & 11 & 2 & -6 \\
 & & 9/2 & 93/4 & 303/8 \\
\hline
 & 3 & 31/2 & 101/4 & 255/8
\end{array}
\qquad
\begin{array}{r|rrrr}
-2 & 3 & 11 & 2 & -6 \\
 & & -6 & -10 & 16 \\
\hline
 & 3 & 5 & -8 & 10
\end{array}
$$

That is, $f(3/2) = 255/8$ and $f(-2) = 10$.

21. Dividing $h(x) = -2x^4 + 5x^3 + 3x$ by $d(x) = x - 5$, $d(x) = x - 1/2$ using synthetic division:

$$
\begin{array}{r|rrrrr}
5 & -2 & 5 & 0 & 3 & 0 \\
 & & -10 & -25 & -125 & -610 \\
\hline
 & -2 & -5 & -25 & -122 & -610
\end{array}
\qquad
\begin{array}{r|rrrrr}
1/2 & -2 & 5 & 0 & 3 & 0 \\
 & & -1 & 2 & 1 & 2 \\
\hline
 & -2 & 4 & 2 & 4 & 2
\end{array}
$$

That is, $h(5) = -610$ and $h(1/2) = 2$.

23. The real zeros of $f(x) = 2x^2 + 5x - 3$ are $x = 1/2$ and $x = -3$, since $f(x) = (2x-1)(x+3)$.

25. The polynomial $h(x)=x^4-11x^2+18$ can be expressed in factored form:
$h(x)=\left(x^2-9\right)\left(x^2-2\right)$, or $h(x)=\left(x-\sqrt{3}\right)\left(x+\sqrt{3}\right)\left(x-\sqrt{2}\right)\left(x+\sqrt{2}\right)$. The real zeros of h are $x=-\sqrt{3},\ -\sqrt{2},\ \sqrt{2},$ and $\sqrt{3}$.

27. The polynomial function $h(x)=6x^3+7x^2-1$ has a zero at $x=-1$, in fact:

$$\begin{array}{r|rrrr} -1 & 6 & 7 & 0 & -1 \\ & & -6 & -1 & 1 \\ \hline & 6 & 1 & -1 & 0 \end{array}$$

That is, $h(x)=(x+1)(6x^2+x-1)$. Now, $6x^2+x-1=(3x-1)(2x+1)$. So, h has zeros $x=-1,\ -\frac{1}{2},$ and $\frac{1}{3}$.

29. Answers may vary. Any real multiple of the polynomial $f(x)=(x+2)(x-3)(x-5)$ is a polynomial of least degree with real coefficients that has zeros -2, 3, and 5.

31. The least degree for a polynomial f with real coefficients to have zeros 1 and -1 (both with multiplicity 2) is 4. That is $f(x)=a(x-1)^2(x+1)^2$. We also need $f(0)=4$, therefore , $a(0-1)^2(0+1)^2=4$, so $a=4$ and $f(x)=4(x-1)^2(x+1)^2$.

33. If the graph of $y=f(x)$ appears as:

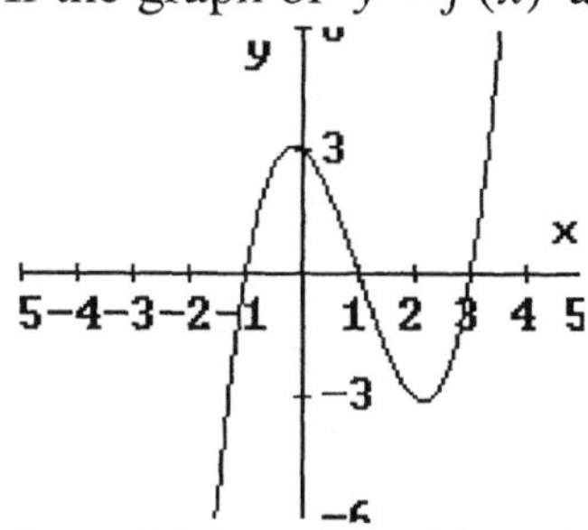

then $f(x)=a(x+1)(x-1)(x-3)$. Since $f(0)=3$, we get $a(0+1)(0-1)(0-3)=3$, or $3a=3$ and $a=1$. So, $f(x)=(x+1)(x-1)(x-3)$.

35. If f is a polynomial function with real coefficients, and -3 and $4i$ are zeros of f, then $-4i$ is also a zero of f. Then the least degree of a polynomial function f with real coefficients is 3. Any real multiple of $f(x)=(x+3)(x^2+16)$ is an example of such a function.

37. $f(x)=x^2-4x+5$ can be factored as $f(x)=\left(x-(2+i)\right)\left(x-(2-i)\right)$. Its zeros are $x=2\pm i$.

39. If $h(x)=x^4+7x^2-144$, then $h(x)=\left(x^2+16\right)\left(x^2-9\right)$, which in turns may be factored: $h(x)=\left(x-4i\right)\left(x+4i\right)\left(x-3\right)\left(x+3\right)$. The complex zeros of h are $x=\pm 3,\ \pm 4i$.

41. A possible rational zero of $f(x) = x^3 - 2x^2 - 5x + 6$ is of the form a/b, where a is a factor of 6 and b a factor of 1, namely $\pm 1, \ \pm 2, \ \pm 3, \ \pm 6$.

$$\begin{array}{r|rrrr}
-2 & 1 & -2 & -5 & 6 \\
 & & -2 & 8 & -6 \\ \hline
1 & 1 & -4 & 3 & 0 \\
 & & 1 & -3 & \\ \hline
 & 1 & -3 & 0 &
\end{array}$$

That is, $f(x) = (x+2)(x-1)(x-3)$; f has the zeros $x = -2$, 1, and 3.

43. The possible rational zeros of $g(x) = 4x^4 - 8x^3 - x^2 + 8x - 3$ are of the form a/b, where a is a factor of 3, and b a factor of 4, namely $\pm 3, \ \pm 3/2, \ \pm 1, \ \pm 3/4, \ \pm 1/2, \ \pm 1/4$.

$$\begin{array}{r|rrrrr}
-1 & 4 & -8 & -1 & 8 & -3 \\
 & & -4 & 12 & -11 & 3 \\ \hline
1 & 4 & -12 & 11 & -3 & 0 \\
 & & 4 & -8 & 3 & \\ \hline
 & 4 & -8 & 3 & 0 &
\end{array}
\qquad
\begin{array}{r|rrrr}
1/2 & 4 & -8 & 3 & 0 \\
 & & 2 & -3 & \\ \hline
3/2 & 4 & -6 & 0 & \\
 & & 6 & & \\ \hline
 & 4 & 0 & &
\end{array}$$

That is, $g(x) = 4\left(x - 3/2\right)\left(x - 1/2\right)\left(x-1\right)(x+1)$, and g has zeros at $x = -1, \ 1/2$, 1, and $3/2$.

45. There are no variations in sign in $h(x) = x^3 + 2x + 1$, so there are no positive real zeros. $h(-x) = -x^3 - 2x + 1$, so h has exactly one negative real zero, approximately $x \approx -0.45$.

47. We can factor x in $g(x) = -x^4 - 5x^3 + 2x^2 + 8x$: $g(x) = x\left(-x^3 - 5x^2 + 2x + 8\right)$. There is one variation in sign in $h(x) = -x^3 - 5x^2 + 2x + 8$, so there is one positive real zero of $-x^3 - 5x^2 + 2x + 8 = 0$ (therefore one positive real zero of g.) $h(-x) = x^3 - 5x^2 - 2x + 8$ has two variations in sign, so there are two negative real zeros or none. In fact, g has four real zeros, $x = 0$, two negative zeros, $x \approx -5.08$, $x \approx -1.21$, and one positive zero, $x \approx 1.3$.

49. If $f(x) = x^4 - 2x^3 - 49x^2 - 2x + 20$, we divide $f(x)$ by $x - 9$ and by $x + 7$, to obtain:

$$\begin{array}{r|rrrrr}
9 & 1 & -2 & -49 & -2 & 20 \\
 & & 9 & 63 & 126 & 1116 \\ \hline
 & 1 & 7 & 14 & 124 & 1136
\end{array}
\qquad
\begin{array}{r|rrrrr}
-7 & 1 & -2 & -49 & -2 & 20 \\
 & & -7 & 63 & -91 & 651 \\ \hline
 & 1 & -9 & 14 & -93 & 671
\end{array}$$

All numbers in the last row of the first division are positive, so 9 is an upper bound for the real zeros of *f*. The numbers in the last row of the second division have alternating signs, then -7 is a lower bound for the real zeros of *f*. So, all the real zeros of *f* are in $(-7, \ 9)$.

51. The equation $x^3+4x+12=7x^2$ is equivalent to $x^3-7x^2+4x+12=0$. We divide $x^3-7x^2+4x+12$ by $x+1$:

$$\begin{array}{r|rrrr} -1 & 1 & -7 & 4 & 12 \\ & & -1 & 8 & -12 \\ \hline & 1 & -8 & 12 & 0 \end{array}$$

Then, $x^3-7x^2+4x+12=0$ is equivalent to $(x+1)(x^2-8x+12)=0$, or $(x+1)(x-6)(x-2)=0$. The solutions to $x^3+4x+12=7x^2$ are, then $x=-1$, 2, and 6.

53. The equation $x^3-\frac{37}{12}x^2-\frac{7}{2}x-\frac{2}{3}=0$ is equivalent to $12x^3-37x^2-42x-8=0$. We divide $12x^3-37x^2-42x-8$ by $x-4$, $x+\frac{2}{3}$, and $x+\frac{1}{4}$:

$$\begin{array}{r|rrrr} 4 & 12 & -37 & -42 & -8 \\ & & 48 & 44 & 8 \\ \hline -\frac{2}{3} & 12 & 11 & 2 & 0 \\ & & -8 & -2 & \\ \hline -\frac{1}{4} & 12 & 3 & 0 & \\ & & -3 & & \\ \hline & 12 & 0 & & \end{array}$$

Then, $12x^3-37x^2-42x-8=0$ is equivalent to $12(x-4)(x+\frac{2}{3})(x+\frac{1}{4})=0$. The solutions of this equation, and of our original equation $x^3-\frac{37}{12}x^2-\frac{7}{2}x-\frac{2}{3}=0$, are $x=-\frac{2}{3}$, $-\frac{1}{4}$, and 4.

55. The possible rational zeros of $f(x)=x^3-5x-1$ are ± 1. Neither one of these is a zero. There are three irrational zeros, $x\approx -2.13$, -0.2, and 2.33.

57. The possible rational zeros of $g(x)=0.05x^4+x^3-x^2+x-3$ are rational zeros of $h(x)=20g(x)$, that is $h(x)=x^4+20x^3-20x^2+20x-60$. Any possible rational zero of this polynomial must be an integer, a factor of 60. Inspecting the graph, we conclude that none of the factors of 60 is a zero of h, so its two zeros are irrational, approximately $x\approx -21$ and $x\approx 1.52$.

59. The domain of the rational function $f(x)=\frac{1}{x-3}$ is the set of all real numbers $x\neq 3$. f has a vertical asymptote at $x=3$ and horizontal asymptote $y=0$. There are no zeros of f.

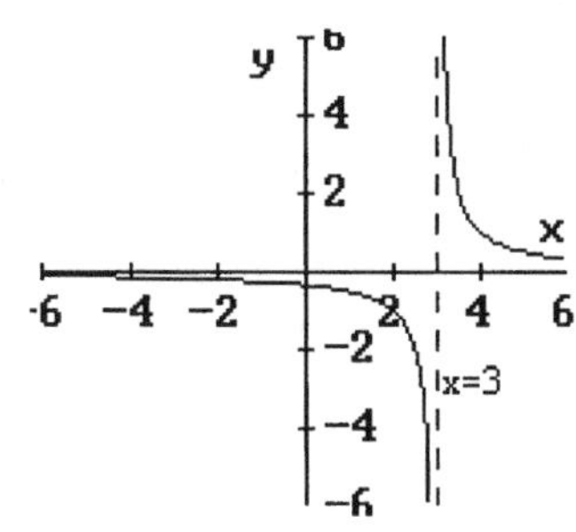

61. The domain of the rational function $f(x)=\dfrac{x+2}{x^2-9}$ is the set of all real numbers $x\neq -3,\ 3$. f has vertical asymptotes at $x=-3$ and $x=3$, and horizontal asymptote $y=0$. f has a zero at $x=-2$.

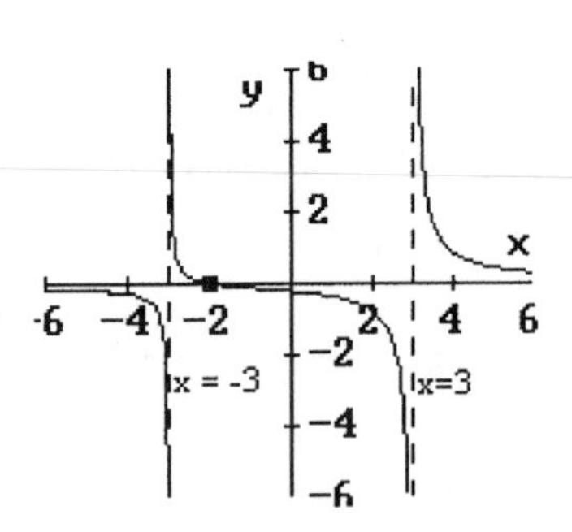

63. The rational function $f(x)=\dfrac{1}{x+2}$ has a vertical asymptote at $x=-2$ and horizontal asymptote $y=0$.

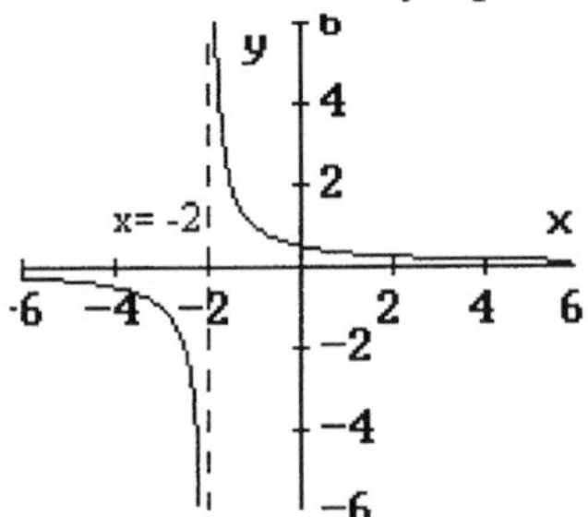

65. The rational function $g(x)=\dfrac{2x+5}{3x-1}$ has a vertical asymptote at $x=\frac{1}{3}$ and horizontal asymptote $y=\frac{2}{3}$.

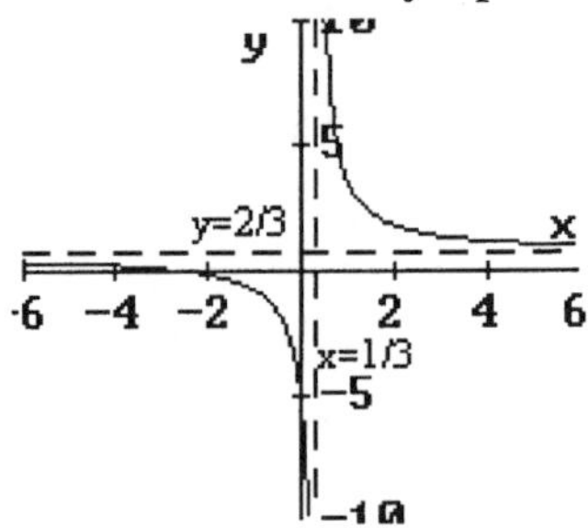

67. The rational function $f(x)=\dfrac{3-x}{x^2}$ has a vertical asymptote at $x=0$ and horizontal asymptote $y=0$.

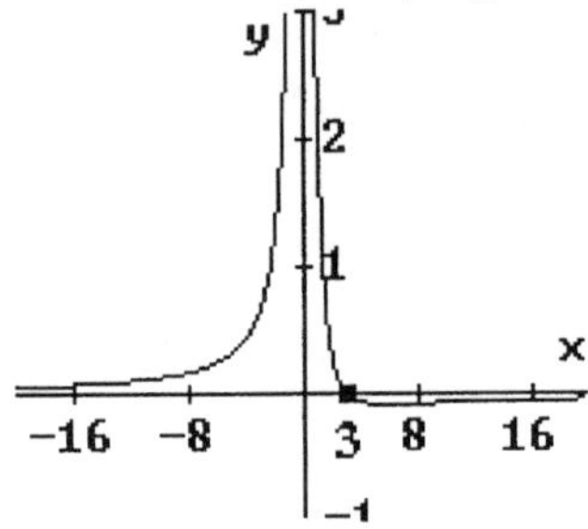

69. The rational function $g(x)=\dfrac{2x^2}{x^2+x+2}$ has no vertical asymptotes. g has horizontal asymptote $y=2$.

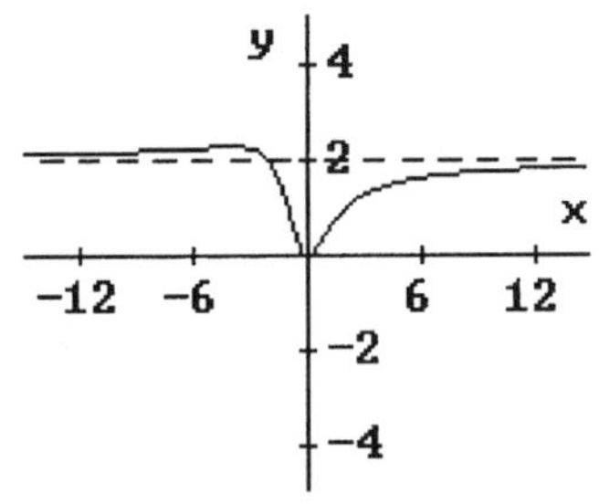

71. The rational function $f(x)=\dfrac{2}{(1-x)^2}$ has a vertical asymptote at $x=1$. f has horizontal asymptote $y=0$.

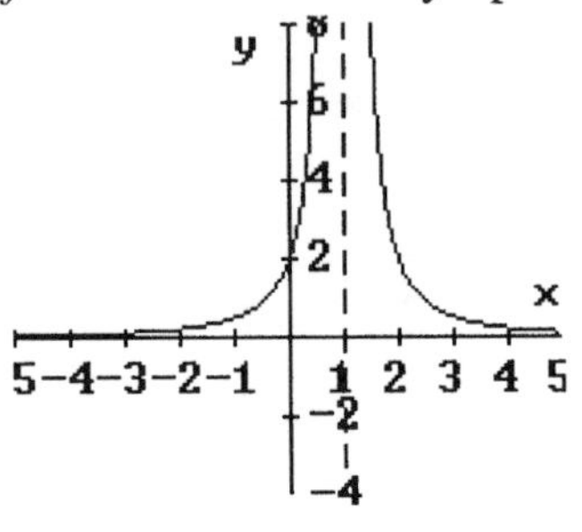

73. The rational function $h(x)=-2x-\dfrac{1}{x}$ has a vertical asymptote at $x=0$. h has inclined asymptote $y=-2x$.

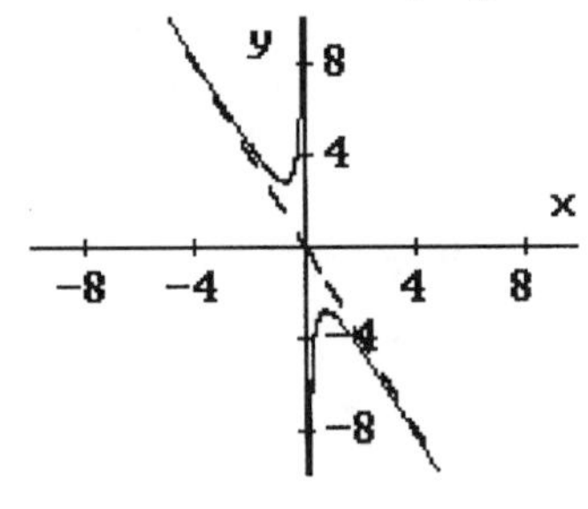

75. The rational function $h(x) = \dfrac{3x^2 + 8x - 9}{x+3}$
has a vertical asymptote at $x = -3$.
h has inclined asymptote $y = 3x - 1$,
which can be seen using the fact that
$h(x) = 3x - 1 - \dfrac{6}{x+3}$ (by long division.)

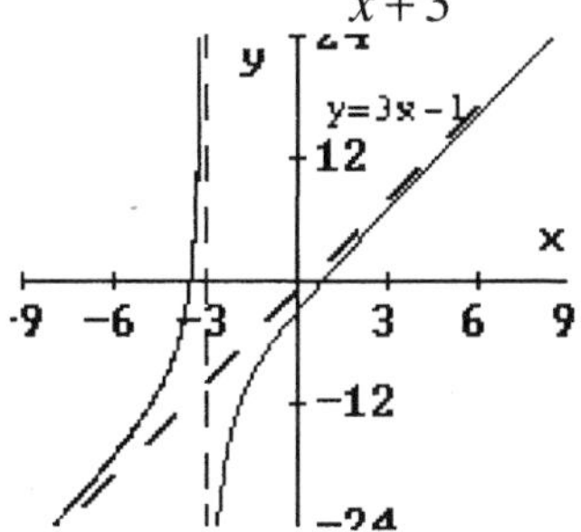

77. **(a)** $x - 3 + \dfrac{2}{x-1} = \dfrac{x^2 - 4x + 5}{x-1}$

(b) The inclined asymptote of $f(x) = \dfrac{x^2 - 4x + 5}{x-1}$ is $y = x - 3$, according to part (a).

79. **(a)** The solutions of $x^2 + 3x - 1 = 0$ are $x = -\dfrac{3}{2} \pm \dfrac{\sqrt{13}}{2}$.

(b) If $f(x) = x^3 + 3x^2 - x$, we have $f(x) = x\left(x^2 + 3x - 1\right)$ and $f(x) = 0$ for $x = 0,\ -\dfrac{3}{2} \pm \dfrac{\sqrt{13}}{2}$.

81. **(a)** $(x-1)(x+5) + 9 = x^2 + 4x + 4$

(b) $\dfrac{x^2 + 4x + 4}{x-1} = \dfrac{(x-1)(x+5) + 9}{x-1}$, therefore $\dfrac{x^2 + 4x + 4}{x-1} = x + 5 + \dfrac{9}{x-1}$.

83. **(a)** If f has horizontal asymptote $y = 0$, and a vertical asymptote at $x = 0$ (the y-axis), then $h(x) = f(x-3) - 2$ has horizontal asymptote $y = -2$ and a vertical asymptote at $x = 3$.

(b) Answers may vary. An example of a rational function with vertical asymptote $x = 3$ and horizontal asymptote $y = -2$ is $h(x) = \dfrac{1}{x-3} - 2$, or $h(x) = \dfrac{5-2x}{x-3}$.

85. **(a)** $(2 - 3i) + (2 + 3i) = 4$, and $(2 - 3i)(2 + 3i) = 13$.

(b) Answers may vary. A polynomial with zeros $2 \pm 3i$ is $f(x) = (x - (2+3i))(x - (2-3i))$, that is $f(x) = x^2 - 4x + 13$.

87. If $f(x)=ax^2+bx+c$ is the profit when x units are sold, and it is known that if $x=0$ there is a loss of \$15,000, and that the profit is zero when $x=1000$ and $x=5000$, we have $f(0)=-15{,}000$, $f(1000)=0$, and $f(15000)=0$. That is, $a(0)^2+b(0)+c=-15{,}000$, so $c=-15{,}000$, and $1{,}000{,}000a+1000b-15000=0$, $25{,}000{,}000a+5000b-15000=0$. Multiplying the first of these last two equations by -5, and combining it with the second, we obtain $20{,}000{,}000a+6000=0$, so $a=-0.003$. Finally, we get b, $1{,}000{,}000(-0.003)+1000b-15000=0$, or $1000b-18000=0$, and $b=18$. Then $f(x)=-0.003x^2+18x-15{,}000$, which has a maximum of \$12,000, when $x=3000$ units.

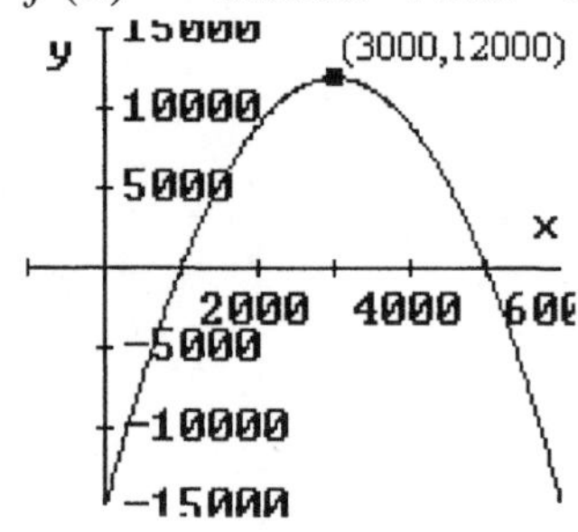

89. If $f(x)=0.0134x^4-0.33x^3+2.07x^2-0.456x-5.04$ models the average monthly Celsius temperature in Juneau, Alaska, where x is the month, $0\le x\le 12$, the zeros of this function are the months when the average Celsius temperature for the month is 0^0 Celsius. The zeros are, approximately, $x\approx 2.02$ and $x\approx 10.32$. This model says that February and October are the months with 0^0 Celsius average temperature for the month in Juneau.

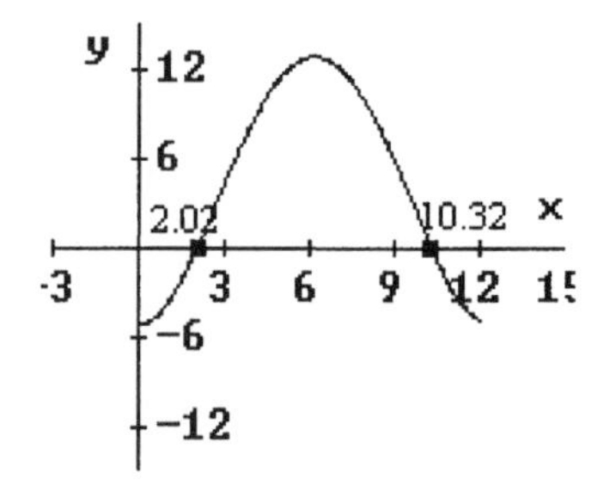

91. The function $\overline{C}(x)=\dfrac{15{,}000+20x}{x}$, which gives the average cost per calculator produced at a company, approaches 20 when x increases. As the number of calculators produced increases, the average cost per calculator approaches 20.

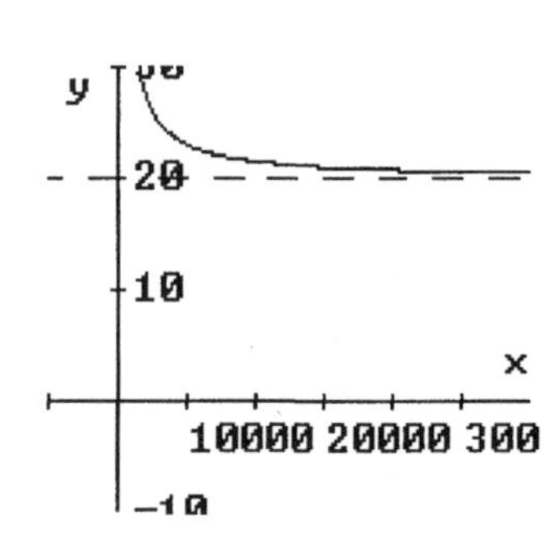

Chapter 3. Test

1. The statement: "A polynomial of degree 3 with real coefficients must have at least one real zero" is true.

3. The statement: "According to Descartes' Rule of Signs, if a polynomial function has three variations in sign, then it must have three positive zeros" is false. (There has to be at least one but there may be only one.)

5. The statement: "If a polynomial of degree 2 divides evenly into one of degree 5, the quotient will have degree 3" is true.

7. The statement: "A rational function must have a vertical asymptote" is false. For example, the function $f(x)=\dfrac{x}{x^2+4}$ doesn't have a vertical asymptote. In fact, any polynomial function is a rational function (with denominator 1), and polynomial functions don't have vertical asymptotes.

9. Answers may vary. An example of a polynomial function *f* for which $f(x)\to-\infty$ as $x\to\infty$ is $f(x)=-x^2+3$.

11. Answers may vary. An example of a polynomial function of least degree with real coefficients and with zeros 2 and $-3i$ is $f(x)=(x-2)(x^2+9)$ (the other possible answers are real multiples of this polynomial function.)

13. Answers may vary. An example of a rational function with vertical asymptote $x=-1$ and horizontal asymptote $y=1$ is $f(x)=\dfrac{x}{x+1}$.

15. The polynomial $f(x)=2x^4-7x^3-4x^2$ factors as $f(x)=x^2\left(2x^2-7x-4\right)$, or further, $f(x)=x^2\left(2x+1\right)\left(x-4\right)$. The zeros of *f* are $x=-\dfrac{1}{2}$, 0, and 4.

17. Dividing $f(x)=3x^4-12x^2+5x+14$ by $d(x)=x+2$, we obtain

$$\begin{array}{r|rrrrr} -2 & 3 & 0 & -12 & 5 & 14 \\ & & -6 & 12 & 0 & -10 \\ \hline & 3 & -6 & 0 & 5 & 4 \end{array}$$

that is, the quotient is $q(x)=3x^3-6x^2+5$ and the remainder is $r(x)=4$.

19. A possible rational zero of $f(x)=2x^3+x^2-12x+9$ are of the form $\frac{a}{b}$, where a is a factor of 9, and b a factor of 2, namely $\pm 9,\ \pm 9/2,\ \pm 3,\ \pm 3/2,\ \pm 1,\ \pm 1/2$. -3 seems to be a zero of f:

$$\begin{array}{r|rrrr} -3 & 2 & 1 & -12 & 9 \\ & & -6 & 15 & -9 \\ \hline & 2 & -5 & 3 & 0 \end{array}$$

So, $f(x)=(x+3)\left(2x^2-5x+3\right)$, and further, $f(x)=(x+3)(2x-3)(x-1)$. The zeros of f are $x=-3$, 1, and $3/2$.

21. $f(x)=x^3-x-2$ has possible rational zeros ± 1, and ± 2. There is only one real zero, an irrational one, $x \approx 1.52$.

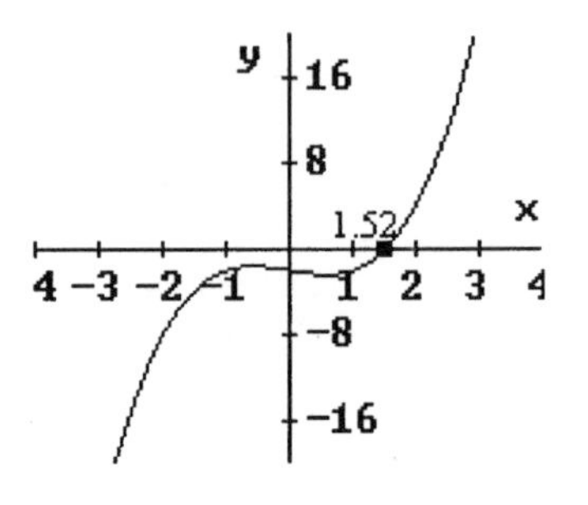

23. If the number of homicides per 100,000 people in the U.S., from 1985 to 2000 is modeled by
$f(x)=0.001x^4-0.03x^3+0.23x^2-0.29x+8.2$.

(a) The years between 1985 and 2000 when the homicide rate was 8.7 in 1988 and 1994.

(b) The homicide rate peaked during 1991, at about 9.57 homicides per 100,000 people.

(c) The period when the homicide rate is declining is 1992-2000, according to this model.

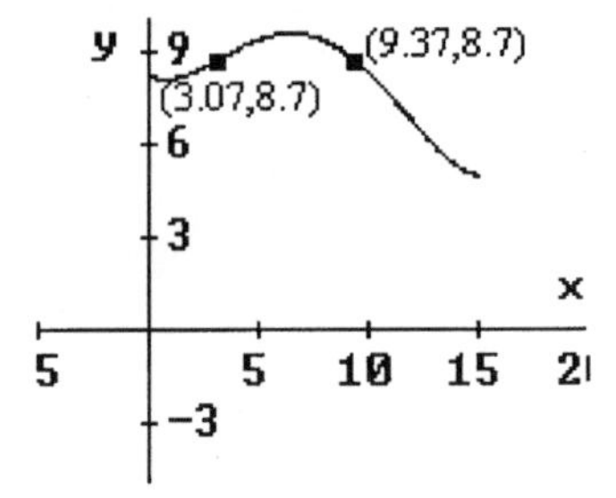

Chapter 4. EXPONENTIAL AND LOGARITHMIC FUNCTIONS

Section 4.1 Exponential Functions

1. Starting with the graph of $f(x) = 3^x$:

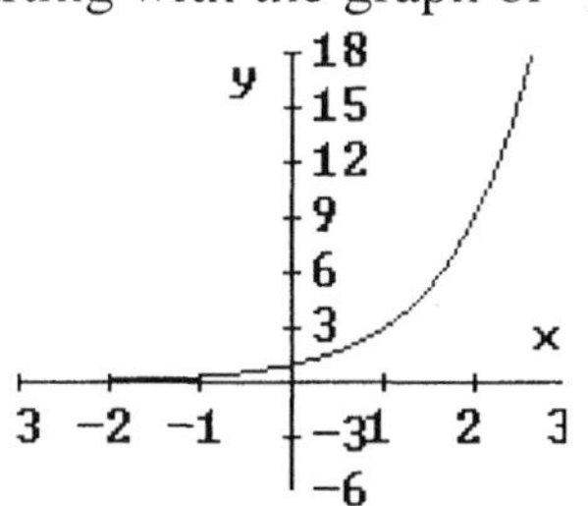

(a) The graph of $g(x) = -3^x$ is a reflection of the graph of $f(x) = 3^x$ about the x-axis.

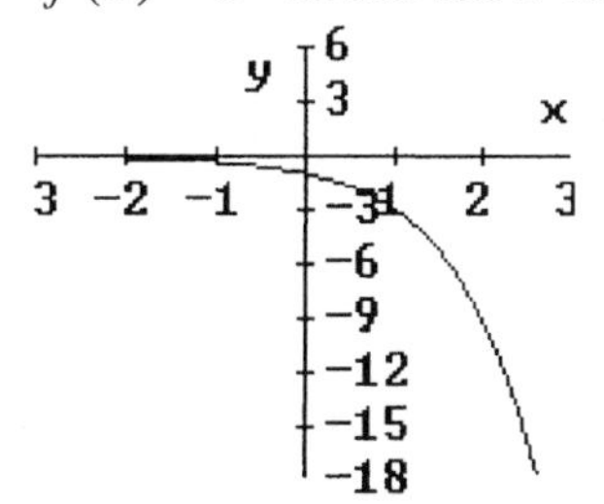

(b) The graph of $g(x) = 3^{-x}$ is a reflection of $f(x) = 3^x$ about the y- axis.

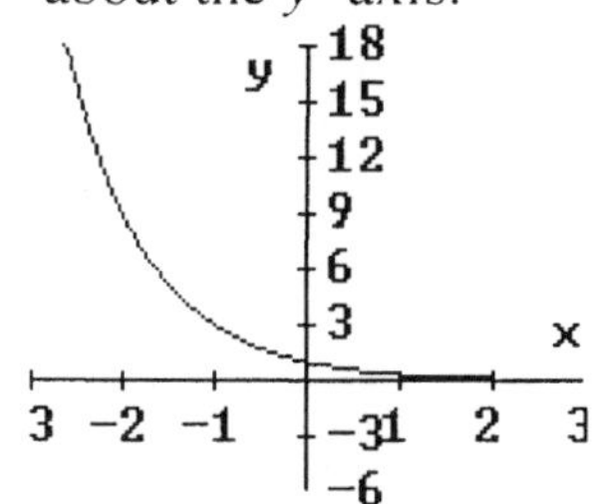

(c) The graph of $g(x) = 3^x + 2$ is a vertical shift, up 2 units, of $f(x) = 3^x$.

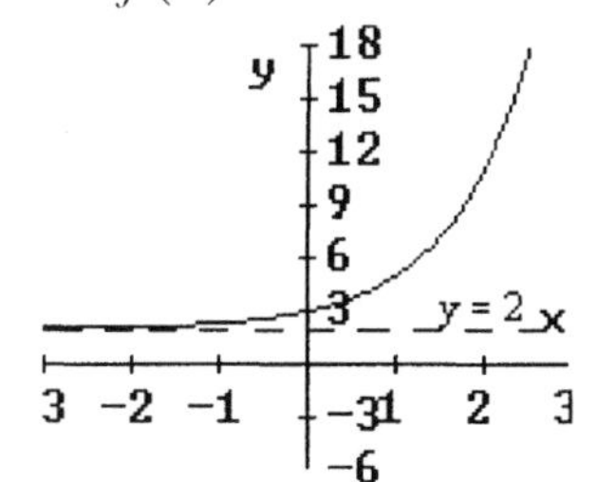

(d) The graph of $g(x) = 3^{x+2}$ is a horizontal shift, left 2 units, of $f(x) = 3^x$.

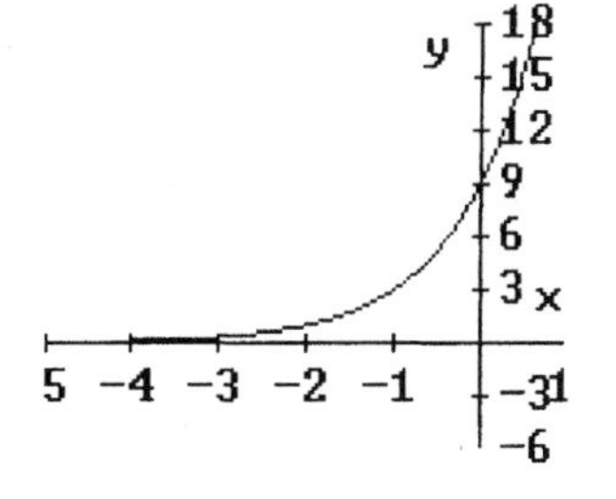

3. Given the graph of $f(x) = a^x$,

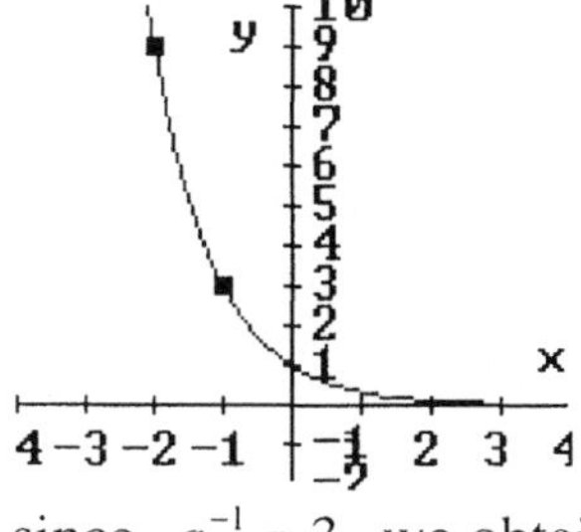

since $a^{-1} = 3$, we obtain $a = \frac{1}{3}$, and $f(x) = (\frac{1}{3})^x$

5. Given the graph of $f(x) = a^x$,

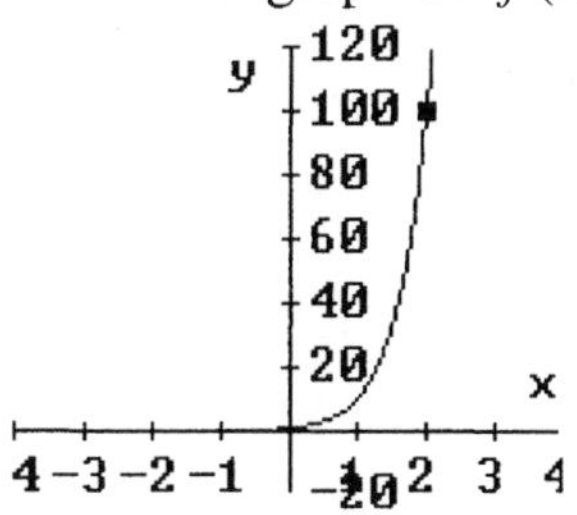

since $a^2 = 100$, we obtain $a = 10$, and $f(x) = 10^x$.

Starting with the graph of $f(x) = e^x$:

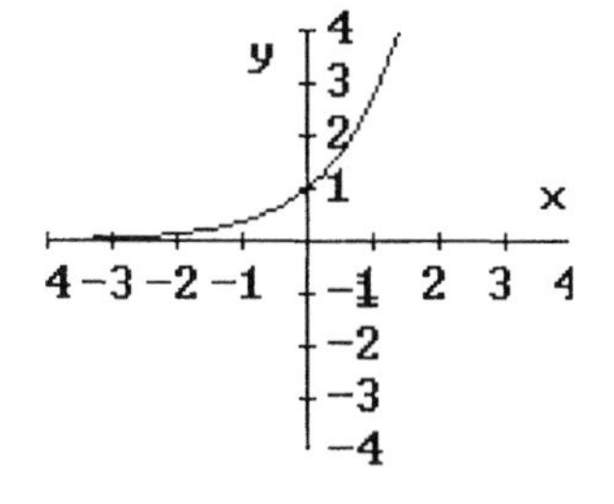

7. The graph of $g(x) = e^{-x}$ is a reflection of the graph of $f(x) = e^x$ about the y-axis.

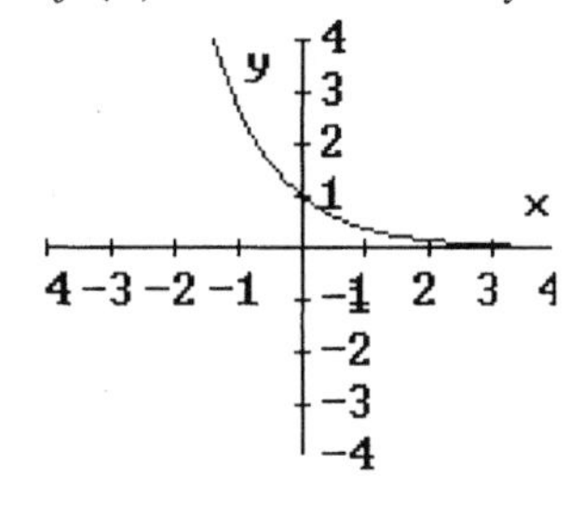

9. $g(x) = e^{x-2}$ is a horizontal shift, right 2 units, of $y = f(x)$.

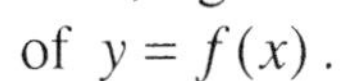

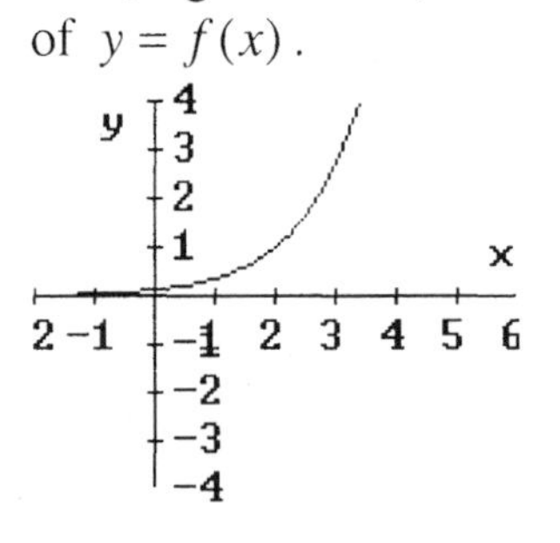

11. $4e^{0.1t} = 13$, therefore $e^{0.1t} = \frac{13}{4}$, and from the graph or table of values feature, $t \approx 11.79$.

13. $1200e^{-0.05t} = 100$, so $t \approx 13.86$ (from the graph or using the table of values feature.)

15. $3^x = 81$, that is, $3^x = 3^4$; then $x = 4$.

17. $2^{3x-4} = 32$, that is, $2^{3x-4} = 2^5$; then $3x - 4 = 5$, $3x = 9$, and $x = 3$.

19. $\frac{1}{3^x} = 27$, that is, $3^{-x} = 3^3$; then $-x = 3$, and $x = -3$.

21. $4^{x-3} = 16 \cdot 4^{2x}$, that is $4^{x-3} = 4^2 \cdot 4^{2x}$, or $4^{x-3} = 4^{2+2x}$ and $x - 3 = 2 + 2x$; therefore, $x = -5$.

23. $2^{x^2+x} = 4$, that is $2^{x^2+x} = 2^2$, therefore $x^2 + x = 2$, which is equivalent to $x^2 + x - 2 = 0$. This last equation can be factored as $(x+2)(x-1) = 0$. Then, the solutions are $x = -2,\ 1$.

To determine the balance B in an account which had an initial deposit of P dollars, and pays interest at an annual rate of interest r compounded n times per year, after t years, we use the the formula $B = P\left(1+\frac{r}{n}\right)^{nt}$. We use $B = Pe^{rt}$ for the case of interest compounded continuously.

25. $P = \$1000$, $t = 10$ years, $r = 6\%$

(a) $n = 1$; $B = 1000(1+0.06)^{10}$, or $B = 1000(1+0.06)^{10}$, $B \approx \$1,790.85$.

(b) $n = 4$; $B = 1000\left(1+\frac{0.06}{4}\right)^{40}$, or $B = 1000(1.015)^{40}$, $B \approx \$1,814.02$.

(c) $n = 12$; $B = 1000\left(1+\frac{0.06}{12}\right)^{120}$, or $B = 1000(1.005)^{120}$, $B \approx \$1,819.40$.

(d) $n = 365$; $B = 1000\left(1+\frac{0.06}{365}\right)^{3650}$, $B \approx \$1,822.03$.

(e) Compounded continuously, $B = 1000e^{0.06(10)}$

Applications

27. If a gerbil population over a period of 4 months is estimated by the exponential function $f(t) = 20\left(\frac{3}{2}\right)^t$, where t is the number of months, $0 \le t \le 4$, then the estimates are:

t (months)	$f(t)$		t (months)	$f(t)$
0	20		3	67.5
1	30		4	101.25
2	45			

If the pattern were to continue, after 6 months the estimate is 227.8125, or about 228 gerbils, and after 5 years, the estimate is $f(6) \approx 7.35\times10^{11}$ (that is, 735 billion gerbils.)

29. If \$5000 is deposited in a CD that pays 8% annual interest, compounded quarterly, after 18 years the accumulated amount is $5000\left(1+\frac{0.08}{4}\right)^{4(18)}$, approximately \$20,805.70. If the interest is compounded continuously, the accumulated amount is $5000e^{0.08(18)} \approx \$21,103.48$.

31. If $Q(t) = 23.17 \cdot 1.023^t$ (in quadrillions of BTUs) estimates the U.S. energy consumption between 1935 and 2000 ($t = 0$ corresponding to 1935), in 1940 the estimated consumption was $Q(5) = 23.17 \cdot 1.023^5$, approximately 25.96 quadrillions of BTUs. In 2000, the estimated consumption was $Q(65) = 23.17 \cdot 1.023^{65}$, approximately 101.59 quadrillions BTUs. The percent increase in consumption between 1940 and 2000 is $\left(\frac{101.59-25.96}{25.96}\right)\%$, that is a 291% increase. This model predicts an energy consumption $Q(75) \approx 127.53$ quadrillions of BTUs.

33. If a bacteria population (in thousands) after t hours is estimated by $f(t) = 2^t$, where $t = 0$ corresponds to 12:00 noon, then there will be 25,000 bacteria when $f(t) = 25$, that is $2^t = 25$. Then t is approximately 4.64 hours, using the graph or the table of values feature of a graphing calculator. We expect 25,000 bacteria at 4:38 pm.

35. The standard normal probability density function has a maximum value $1/\sqrt{2\pi}$, approximately 0.399, when $x = 0$. As x approaches infinity or minus infinity, the value of $f(x)$ approaches 0.

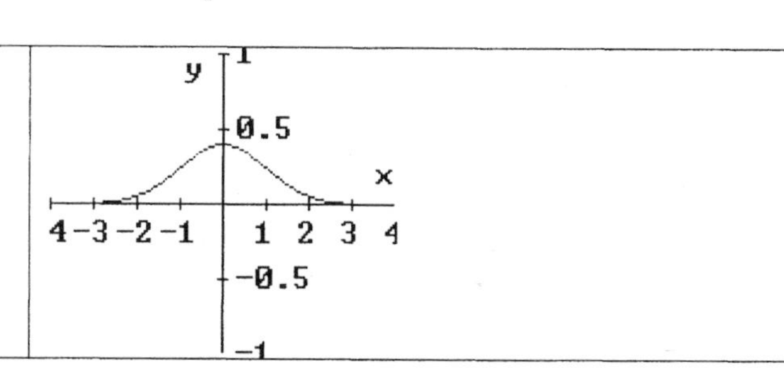

37. $N(t)=\dfrac{100{,}000}{100+900e^{-0.15t}}$ models the spread of the virus among 1000 students after t days of the return of 100 exposed students.

(a) $N(2)\approx 103.8$ and $N(5)\approx 190.4$

(b)

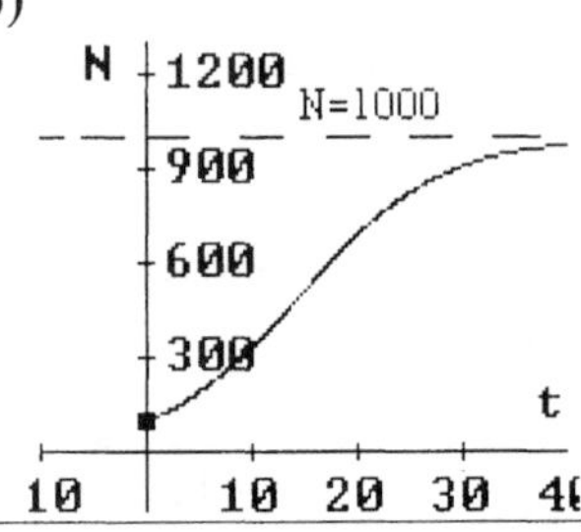

(c) There will be 800 exposed students when $\dfrac{100{,}000}{100+900e^{-0.15t}}=800$, that is, $t\approx 23.89$, about 24 days.

(d) As $t\to\infty$, $N(t)\to 1000$.

Concepts and Critical Thinking

39. The statement: "The graphs of $f(x)=2^x$ and $g(x)=\left(\frac{1}{2}\right)^x$ are reflections of one another about the y-axis" is true. In fact, $g(x)=2^{-x}$, that is $g(x)=f(-x)$.

41. The statement: "All exponential functions are 1-1" is false. If the base is 1, the function $f(x)=1^x$ is the constant function $f(x)=1$. The statement is true for all positive bases $a\neq 1$.

43. Answers may vary. An example of an exponential function that is decreasing over its entire domain is $f(x)=\left(\frac{1}{2}\right)^x$.

45. The only point that belongs to every exponential function of the form $f(x)=a^x$ is (0, 1).

Chapter 4. Section 4.2 Logarithmic Functions

1. $\ln 15.2 \approx 2.72129$

3. $\log\left(\frac{2}{3}\right) \approx -0.17609$

5. $\log\left(2\sqrt{3}\right) \approx 0.53959$

7. $(\ln 0.41)^3 \approx -0.70877$

9. $\log_2(16) = 4$

11. $\log_3 0$ does not exist

13. $\ln e^2 = 2$

15. $\log_6 216 = 3$

17. $\log_{1/5} 25 = -2$

19. $\log_{0.1} 1000 = -3$

21. $\log 10^{100} = 100$

23. $\ln\left(e^4 \cdot e^3\right) = 7$

25. $\log_5 37 \approx 2.24$

27. $\log_2 50 \approx 5.64$

29. $\log_{1/2} 108 \approx -6.75$

31. $\log_2 \frac{1}{4} = -2$ is equivalent to $2^{-2} = \frac{1}{4}$

33. $\log 1000 = 3$ is equivalent to $10^3 = 1000$

35. $\ln(x+1) = 2$ corresponds to $e^2 = x+1$

37. $\log_x 10 = 3$ is equivalent to $x^3 = 10$

39. $3^2 = 9$ is equivalent to $\log_3 9 = 2$

41. $\left(\frac{1}{2}\right)^{-3} = 8$ is equivalent to $\log_{1/2} 8 = -3$

43. $e^x = 5$ is equivalent to $\ln 5 = x$

45. $e^{-0.013t} = \frac{1}{2}$ is equivalent to $\ln \frac{1}{2} = -0.013t$

In Exercises 47-51, use the graph of $f(x) = \ln x$, and adequate transformations, obtain the graph of $g(x)$. Determine the domain of g.

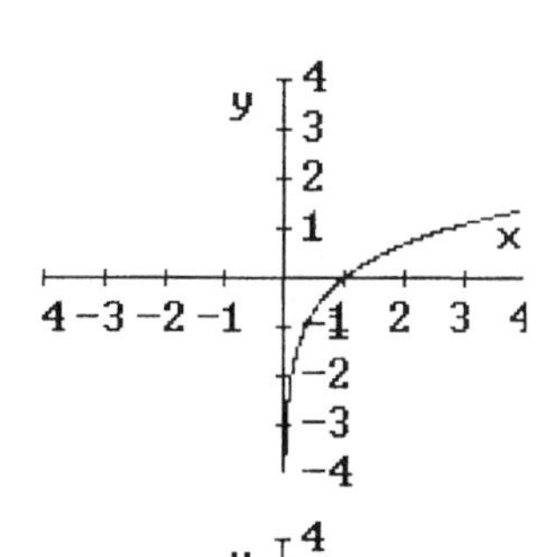

47. The graph of $g(x) = -\ln x$ is a reflection of the graph of $y = f(x)$ about the x axis. The domain of g is $(0, \infty)$.

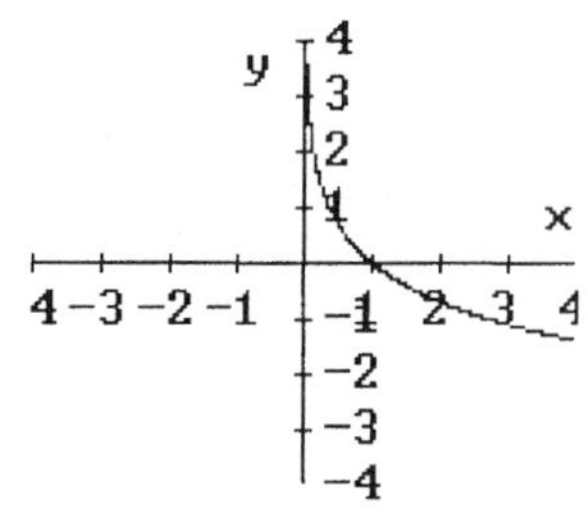

49. The graph of $g(x) = \ln(x-3)$ is a horizontal shift, to the right 3 units, of the graph of $y = f(x)$.
The domain of g is $(3, \infty)$.

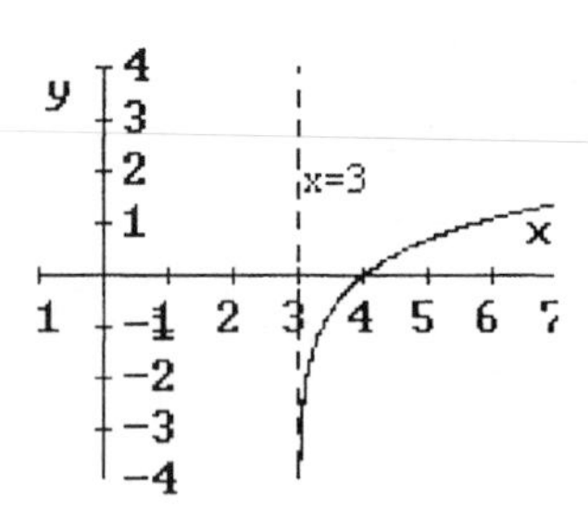

51. The graph of $g(x) = \ln(-x) - 1$ is a reflection about the y axis of the graph of $y = f(x)$, followed by a vertical shift, down 1 unit. The domain of g is $(-\infty, 0)$.

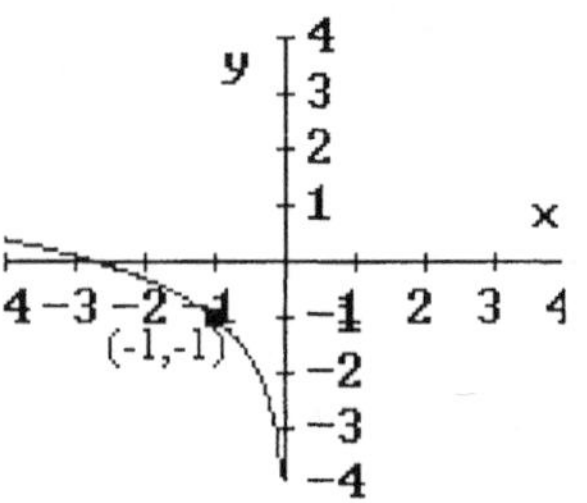

53. The inverse of $f(x) = \log_3 x$ is the function $f^{-1}(x) = 3^x$.

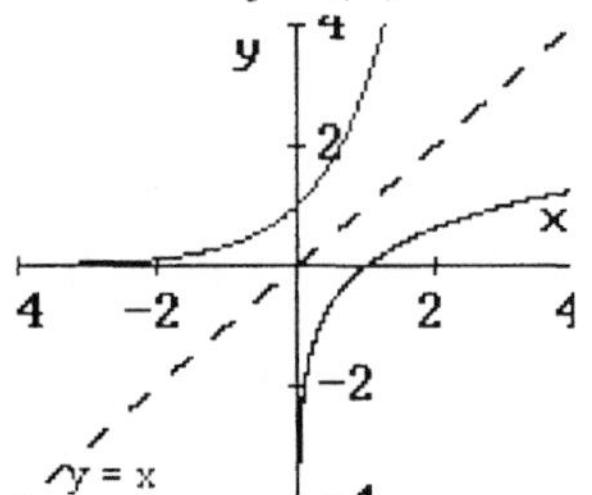

55. The inverse of $f(x) = e^x - 5$ is the function $f^{-1}(x) = \ln(x+5)$.

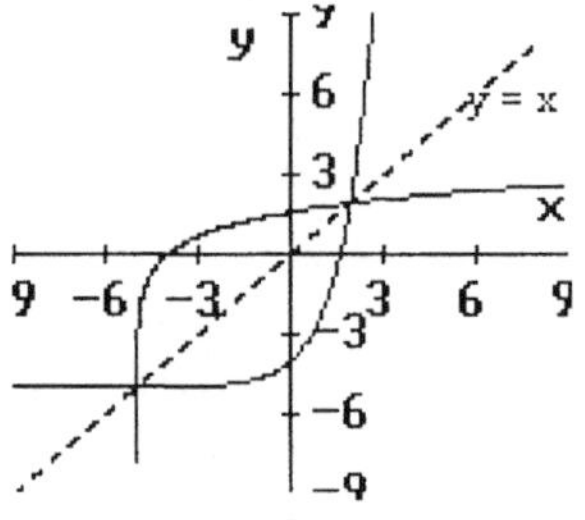

57. The inverse of $f(x) = \log(1000x)$ is the function $f^{-1}(x) = 10^{x-3}$.

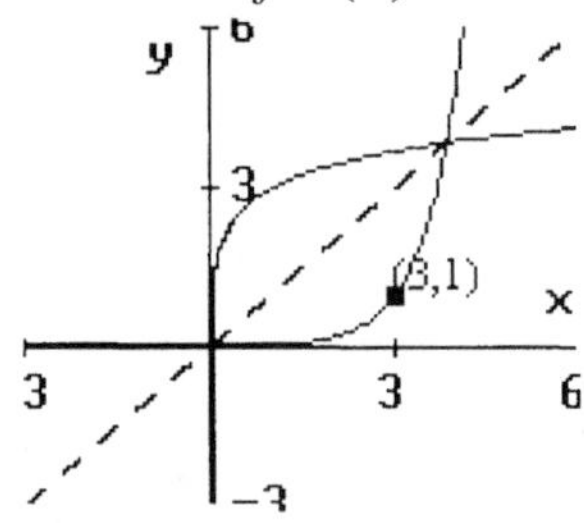

59. The domain of the function $f(x) = x \ln x$ is the set of all positive real numbers.

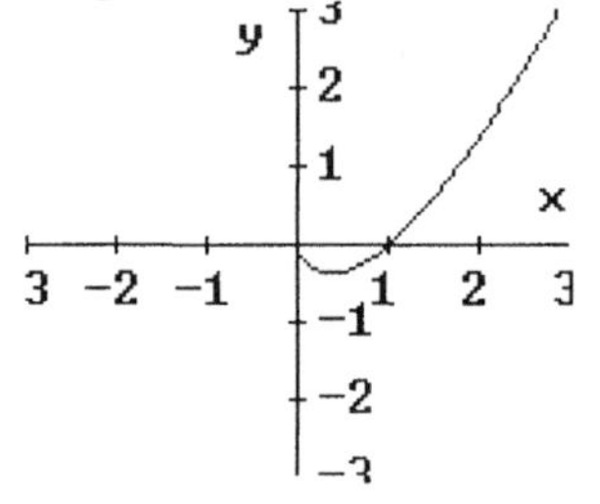

61. The domain of the function $f(x) = x/\ln(x+1)$ is the set of real numbers $x > -1$, $x \neq 0$, that is, $(-1, 0) \cup (0, \infty)$.

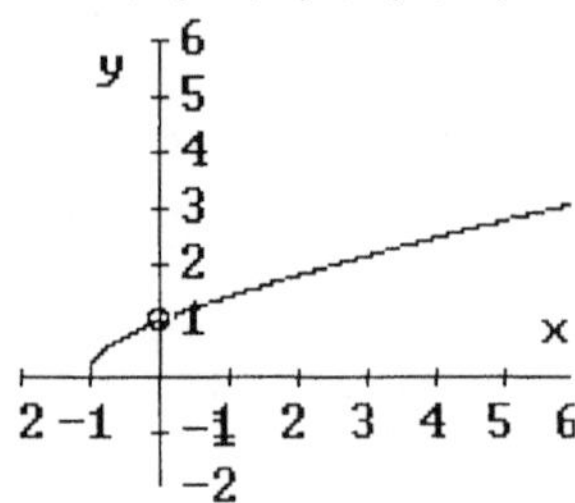

Applications

In Exercises 63-65, we apply Newton's Law of Cooling, which gives the time it takes for an object to cool from a temperature To to a temperature T, if the temperature of the surrounding air is C: $t = k\ln\left(\frac{T_0 - C}{T - C}\right)$, *where k is a constant (which will change depending on the units used for measuring the time t.)*

63. If a coffee cup has an initial temperature of 190° F, the temperature of the surrounding air is 70° F, and $k = 20$ (for time measured in minutes), then the time required for the temperature to cool to a temperature of 80° F is $t = 20\ln\left(\frac{190-70}{80-70}\right)$, or $t \approx 49.7$ min.

65. If a room has a constant temperature of 75° F and a body lying in it has a temperature of 81.2° F, down from the normal body temperature of 98.6° F, using $k = 10$ (for time measured in hours, we estimate the time since death by $t = 10\ln\left(\frac{98.6-75}{81.2-75}\right)$, or $t \approx 13.35$ hours.

In Exercises 67-69, we use the decibel rating of a sound of intensity I, $D = 10\log\frac{I}{I_0}$, *where* I_0 *is the faintest intensity perceptible to most individuals.*

67. The decibel rating of a quiet conversation with intensity $I = 9000I_0$ is $D = 10\log\frac{9000I_0}{I_0}$, or $D \approx 39.54$ decibels.

69. The decibel rating of an auto horn with intensity 100 billion times that of I_0 is $D = 10\log\frac{10^{11}I_0}{I_0}$, or $D = 110$ decibels.

In Exercises 71-73, we use the equation $\mathrm{pH} = -\log\left[\mathrm{H}^+\right]$, *to estimate the acidity of an aqueous solution, where* $\left[\mathrm{H}^+\right]$ *is the hydrogen ion concentration of the solution. The smaller the* pH, *the more acidic the solution.*

71. The acidity for orange juice, with hydrogen ion concentration $\left[\mathrm{H}^+\right] = 2.8\times10^{-4}$, is estimated by $\mathrm{pH} = -\log(2.8\times10^{-4})$, approximately $\mathrm{pH} \approx 3.55$.

73. The acidity for beer, with hydrogen ion concentration $\left[\mathrm{H}^+\right] = 3.16\times10^{-5}$, is estimated by $\mathrm{pH} = -\log(3.16\times10^{-5})$, approximately $\mathrm{pH} \approx 4.5$.

75. After each fold, the resulting thickness doubles, so the thickness is $T = 0.003\times2^t$ inches. For the thickness to be more than one mile, that is, more than 63,360 inches, we need $T \geq 63{,}360$, so $0.003\times2^t \geq 63{,}360$, or $2^t \geq \frac{63{,}360}{0.003}$, and $t \geq 24.33$.
Then, 25 folds will achieve a thickness of more than 1 mile.

77. If the time in weeks to learn to type w words per minute is estimated by $t = 5\ln\left(\frac{100}{100-w}\right)$

(a) The time to learn to type 60 words per minute is $t = 5\ln\left(\frac{100}{100-60}\right)$, or $t \approx 4.58$ weeks.

(b) As w approaches 100, the time t increases out of bounds.

(c) The relevant domain of $t = 5\ln\left(\frac{100}{100-w}\right)$ is $0 \le w < 100$.

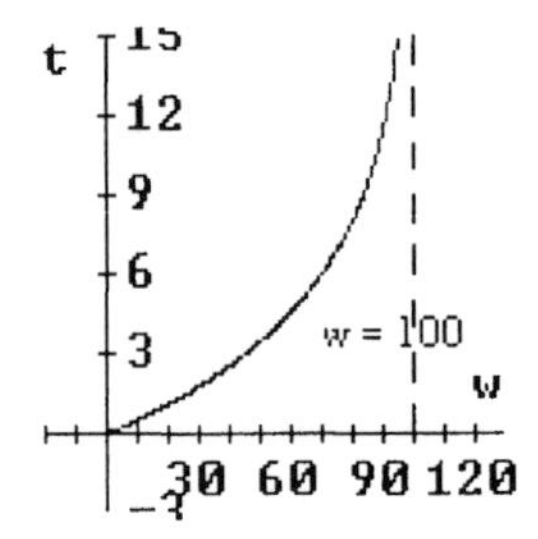

Concepts and Critical Thinking

79. The statement: "The natural logarithm and the exponential function with base e are inverses of one another" is true.

81. The statement: "The logarithm of a negative number is negative" is false. The logarithm of a negative number is undefined.

83. The statement: " $y = \log_a x$ is equivalent to $a^y = x$ " is true.

85. Answers may vary. An example of a number having a negative common logarithm is 0.1.

87. Answers may vary. Common logarithms are employed to obtain the decibel rating of sounds.

89. Answers may vary. $f(x) = x$ is an example satisfying $\ln x < f(x) < e^x$ for all $x > 0$.

91. The domain of the function $f(x) = \log\left(\frac{x+1}{2x-8}\right)$ is $(-\infty,-1)\cup(4,\infty)$. There are vertical asymptotes at $x = -1$, $x = 4$ and f has horizontal asymptote $y = \log \frac{1}{2}$.

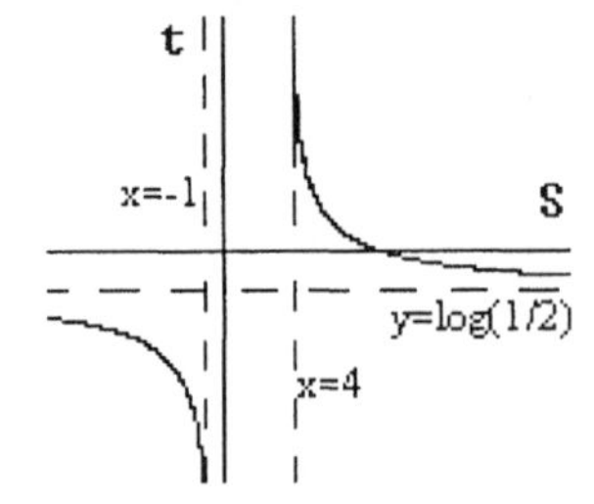

Chapter 4. Section 4.3 Logarithmic Identities and Equations

1. $\ln(x^2y) = \ln x^2 + \ln y$

$= 2\ln x + \ln y$

3. $\log_5\left(\frac{x^3y^4}{z^{-3}}\right) = \log_5 x^3 + \log_5 y^4 - \log_5 z^{-3}$

$= 3\log_5 x + 4\log_5 y + 3\log_5 z$

5. $\ln\left(\frac{e}{7x^3}\right) = \ln(e) - \ln(7x^3)$

$= 1 - \ln 7 - 3\ln x$

7. $\log\sqrt{\frac{y\sqrt{x}}{z^2}} = \frac{1}{2}\log\left(\frac{y\sqrt{x}}{z^2}\right)$

$= \frac{1}{2}\log y + \frac{1}{4}\log x - \log z$

9. $\log x - 2\log y = \log x - \log y^2$

$= \log\left(\frac{x}{y^2}\right)$

11. $3\ln x + 4\ln y - \ln z = \ln x^3 + \ln y^4 - \ln z$

$= \ln\left(\frac{x^3y^4}{z}\right)$

13. $\frac{1}{3}\ln(x-z) - \frac{2}{3}\ln(x+z) = \ln\sqrt[3]{x-z} - \ln\sqrt[3]{(x+z)^2}$

$= \ln\left(\frac{\sqrt[3]{x-z}}{\sqrt[3]{(x+z)^2}}\right)$

$= \ln\sqrt[3]{\frac{x-z}{(x+z)^2}}$

15. $2\log 4 - \log 6 + 3\log 2 = \log 16 - \log 6 + \log 8$

$= \log\left(\frac{(16)(8)}{6}\right)$

$= \log\left(\frac{64}{3}\right)$

17. $2\ln(x+3) - \ln(x^2+5x+6) = \ln(x+3)^2 - \ln[(x+2)(x+3)]$

$= \ln\left(\frac{(x+3)^2}{(x+2)(x+3)}\right)$

$= \ln\left(\frac{x+3}{x+2}\right).$

19. The equation $\frac{1}{3}\log_3 x = 1$ is equivalent to $\log_3 x^{1/3} = 1$, so $x^{1/3} = 3$, and $x = 27$.

21. $\log_x 4 = 2$ in exponential form is $x^2 = 4$. Then $x = -2$, or $x = 2$. But we only consider positive bases, so we rule out -2 as a solution. The solution is $x = 2$.

23. $3\log_5 x = 2$ is equivalent to $\log_5 x = \frac{2}{3}$, which in exponential form is $x = 5^{2/3}$, or approximately, $x \approx 2.92$.

25. $\log_4(2x+1) = -2$ in exponential form is $4^{-2} = 2x+1$, that is $2x = -\frac{15}{16}$, or $x = -\frac{15}{32}$.

27. $\log_3(27^x) = -1$ is equivalent to $x\log_3 27 = -1$, that is, $3x = -1$ and $x = -\frac{1}{3}$.

29. $2\log_3(x-1) + \log_3 9 = 2$

$$2\log_3(x-1) + 2 = 2$$

$$2\log_3(x-1) = 0$$

Then, $\log_3(x-1) = 0$, or $x - 1 = 1$ and $x = 2$.

31. $\ln(x+7) - \ln(x+2) = \ln(x+1)$

$$\ln\left(\frac{x+7}{x+2}\right) = \ln(x+1)$$

$$\frac{x+7}{x+2} = x+1$$

$x + 7 = x^2 + 3x + 2$, or $x^2 + 2x - 5 = 0$, which has two solutions, $x = -1 \pm \sqrt{6}$.

The only relevant solution in our case is $x = -1 + \sqrt{6}$.

33. The equation $\log(2x^2 + 11x + 5) - \log(2x+1) = 2$ is equivalent to $\log\left(\frac{2x^2 + 11x + 5}{2x+1}\right) = 2$, or

$\frac{2x^2 + 11x + 5}{2x+1} = 100$, or $2x^2 - 189x - 95 = 0$, which has solutions $x = -\frac{1}{2}$ and $x = 95$. As we check back in the original equation, we conclude that only $x = 95$ is a solution.

35. Since $27x^3 - 1 = (3x-1)(9x^2 + 3x + 1)$, the equation $\log_2(27x^3 - 1) - \log_2(9x^2 + 3x + 1) = 1$ is equivalent to $\log_2((3x-1)(9x^2 + 3x + 1)) - \log_2(9x^2 + 3x + 1) = 1$, or

$\log_2(3x-1) + \log(9x^2 + 3x + 1) - \log_2(9x^2 + 3x + 1) = 1$; that is, $\log_2(3x-1) = 1$, which in turn implies that $3x - 1 = 2$, and $x = 1$.

37. $\log_a(x+7) = \log_a(x+1)$ implies $x + 7 = x + 1$. No solution exists.

39. $\log\sqrt{x^2 + 1999} = 3$ is equivalent to $\sqrt{x^2 + 1999} = 1000$, then $x^2 + 1999 = 1{,}000{,}000$, so $x^2 + 1999 = 1{,}000{,}000$, $x^2 = 998{,}001$. The solutions to the last equation are $x = \pm 999$. Both solutions are legitimate solutions of the original equation $\log\sqrt{x^2 + 1999} = 3$.

41. The equation $\ln x^2 = 2\ln x$ is an identity in the domain of $\ln x$. That is, all positive real numbers are solutions of this equation.

43. If $\ln(\ln x) = 1$, then $\ln x = e$, therefore $x = e^e$.

45. $\log_3 16 = \dfrac{\ln 16}{\ln 3}$, approximately 2.52372.

47. $\log_{12} 0.341 = \dfrac{\ln 0.341}{\ln 12}$, approximately -0.43296.

49. $\log_{\pi} 10 = \dfrac{\ln 10}{\ln \pi}$, approximately 2.01147.

51. There are two solutions of $x - 2 = \ln x$, $x \approx 0.16$ and $x \approx 3.15$.

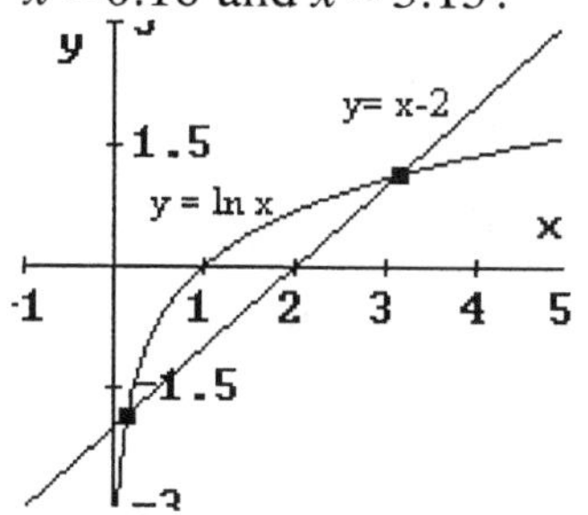

53. There are two solutions of $x - 2 = \ln x$, $x \approx 0.04$ and $x \approx 1.5$.

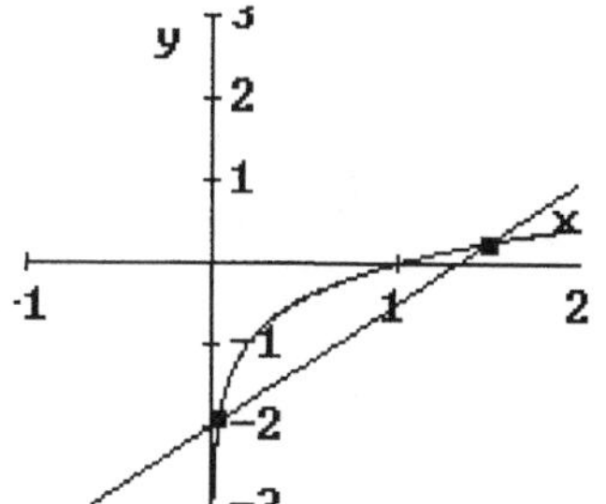

55. There are three solutions of $0.288x^2 = \ln(1 - x) + 0.002x^4 + 1$, $x \approx -11.42,\ -2.97$, and $x \approx 0.59$.

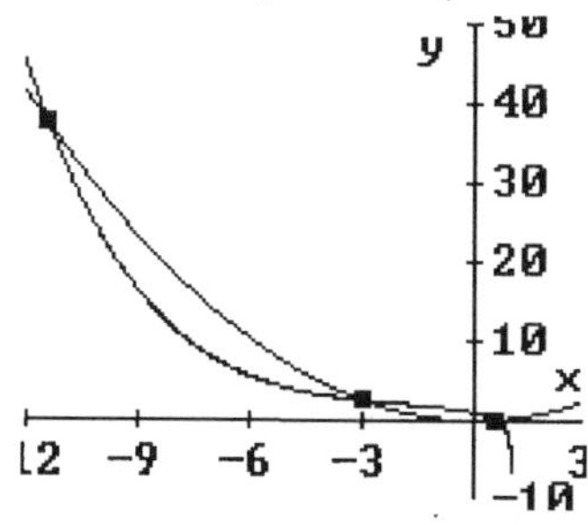

In Exercises 57-59, we use the decibel rating D of a sound with intensity D as given by the equation $D = 10\log\left(I/I_0 \right)$, *where* I_0 *is the faintest sound perceptible to the human ear.*

57. A sound of 160 decibels has an intensity I that satisfies the equation $160 = 10\log\left(\frac{I}{I_0}\right)$, so $\frac{I}{I_0} = 10^{16}$, and $I = 10^{16} I_0$. Similarly, for a 120-decibel the intensity is $I = 10^{12} I_0$, then the ratio between the two intensities is $\dfrac{10^{16} I_0}{10^{12} I_0}$, or 10,000. A 160-decibels sound is 10,000 times as intense as a 120-decibels sound.

59. If 1000 people sing a note at precisely the same volume and pitch, and I is the intensity that corresponds to an individual singer, then the intensity for the 1000 people is $1000I$.

The sound level for these 1000 people is, then, $D = 10\log\left(\dfrac{1000I}{I_0}\right)$, or $D = 10\left(3 + \log\left(\frac{I}{I_0}\right)\right)$, so $D = 30 + 10\log\left(\frac{I}{I_0}\right)$. The sound level is 30 decibels more for 1000 people than for one.

In Exercises 61-63, we use the equation that relates the released energy E (in ergs) to the Richter scale reading R of an earthquake: $\log E = 11.8 + 1.5R$.

61. If the Richter scale reading is 5, the corresponding energy E satisfies: $\log E = 11.8 + 1.5(5)$, that is, $\log E = 19.3$, so $E = 10^{19.3}$ ergs. The energy released by 1 ton of TNT is 3.4×10^{16} ergs so to determine how many tons of TNT would carry the energy of an earthquake registered 5 on the Richter scale, we divide $\frac{10^{19.3}}{3.4 \times 10^{16}}$, to obtain 586.8 tons of TNT.

63. If the Armenia earthquake in 1988 and the San Francisco earthquake in 1906 registered 6.8 and 8.3 in the Richter scale, respectively. Then $\log(E_1) = 11.8 + 1.5(6.8)$, or $E_1 = 10^{22}$ ergs for the Armenia earthquake. $\log(E_2) = 11.8 + 1.5(8.3)$, or $E_2 = 10^{24.25}$ ergs. The San Francisco earthquake in 1906 was 178 times as powerful as the Armenia's earthquake.

Concepts and Critical Thinking

65. The statement: "$\log xy = (\log x)(\log y)$" is false. For example, $\log(10 \cdot 1) = 1$, but $(\log 10)(\log 1) = 0$

67. The statement: "$\log xy = \log x + \log y$" is true for all positive x and y.

69. The statement: "$y \ln x = \ln x^y$" is true for all positive x and y.

71. The statement: "$\log_2 \frac{x}{y} = \log_2 x - \log_2 y$" is true.

73. Answers may vary. A reason why solution candidates to logarithmic equations must be checked is that as we apply logarithmic identities, the identities assume that the argument of each term is in the domain of the term, and as we solve a subsequent equation, we may come up with a value that is not. For example, to solve the equation $2\log x = 2$, we may proceed saying that the left side is equivalent to $\log x^2 = 2$, and further, to $x^2 = 100$. The solutions to the last equation are $x = \pm 10$, but $x = -10$ is not in the domain of the original.

75. Answers may vary. Examples of two pairs of numbers x and a such that $\log_a x = 2$ are $(x = 9,\ a = 3)$ and $(x = 16,\ a = 4)$, since $\log_3 9 = 2$ and $\log_4 16 = 2$.

77. If $f(x) = \log x$, $g(x) = \log(x \cdot 10)$, and $h(x) = \log(x \cdot 0.1)$, then $g(x) = \log x + 1$ and $h(x) = \log x - 1$. The graphs of $y = g(x)$ and $y = h(x)$ are vertical shifts of $y = f(x)$.

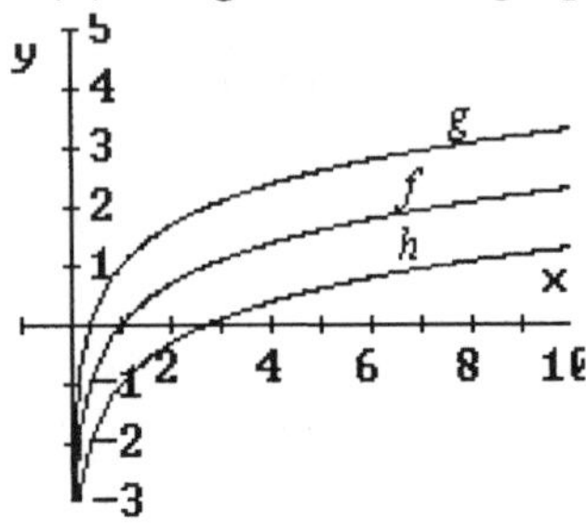

Chapter 4. Section 4.4 Exponential Equations and Applications

In Exercises 1-17, an algebraic solution of each exponential equation is provided.

1. $2^x = 16$

$x = 4$

3. $25^{3x} = 125^{x+1}$

$(5^2)^{3x} = 5^{3(x+1)}$

$6x = 3x + 3$

$3x = 3$

$x = 1$

5. $7^x = 10$, so $x\ln(7) = \ln(10)$

$x = \frac{\ln 10}{\ln 7}$

$x \approx 1.18329$

7. $3^{-x} = 11$, so $-x\ln(3) = \ln(11)$

$x = -\frac{\ln 11}{\ln 3}$

$x \approx -2.18265$

9. $\left(\frac{3}{5}\right)^x = 6^{2-x}$, so $x\ln\left(\frac{3}{5}\right) = (2-x)\ln(6)$

$x(\ln(3) - \ln(5) + \ln(6)) = 2\ln(6)$

$x = \frac{2\ln 6}{\ln 3 - \ln 5 + \ln 6}$

$x \approx 2.79758$

11. $e^x = 2^{x+1}$, so $x = (x+1)\ln(2)$

$x(1 - \ln(2)) = \ln(2)$

$x = \frac{\ln 2}{1 - \ln 2}$

$x \approx 2.25889$

13. $e^{2x-1} = \pi^{2x}$ so $2x - 1 = 2x\ln\pi$

$2x(1 - \ln\pi) = 1$

$x = \frac{1}{2(1-\ln\pi)}$

$x \approx -3.45471$

15. $\frac{1}{2^{x+2}} = \frac{1}{8}$

$x + 2 = 3$

$x = 1$

17. $1.06^t = 1000$

$t\ln(1.06) = \ln(1000)$

$t = \frac{\ln(1000)}{\ln(1.06)}$

$t \approx 118.54959$

19. The equation $e^{2x} - 5e^x + 6 = 0$ has two solutions, $x \approx 0.69$ and $x \approx 1.10$.

21. The equation $e^{-x} = x$ has one solution, $x \approx 0.57$.

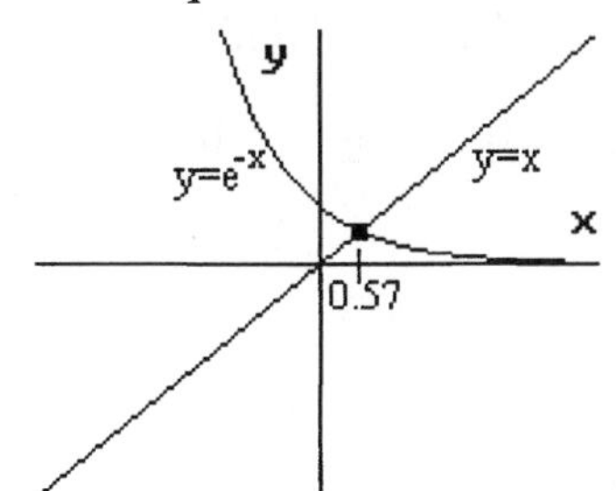

23. The equation $\frac{1}{25}x^4 + 4 = x^2 + e^{x+1}$ has three solutions: $x \approx -4.48,\ -2.11,\ 0.35$.

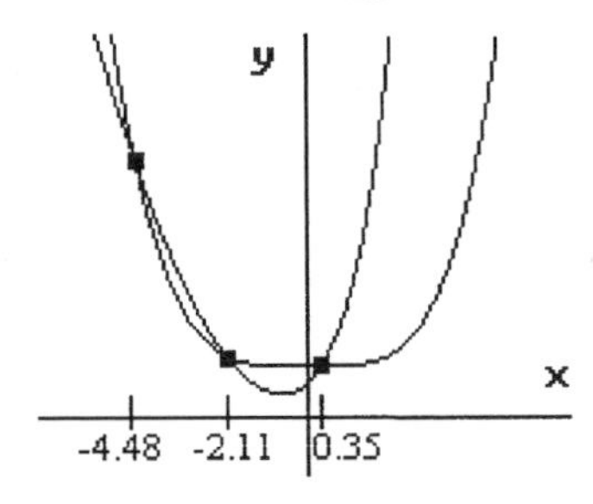

Applications

25. If a principal of \$200 is invested in an account that pays interest at an annual rate of 4% compounded semiannually, it will take 10.24 for the balance to reach \$300.

To solve for t, $200\left(1+\frac{0.04}{2}\right)^{2t} = 300$

$$\left(1+\frac{0.04}{2}\right)^{2t} = 1.5$$

$$2t\ln(1.02) = \ln(1.5)$$

$$t = \frac{\ln(1.5)}{2\ln(1.02)}$$

so, $t \approx 10.24$ years.

27. If the balance in an account that pays interest at the annual rate r compounded continuously doubles in size every 12 years, then we have: $Pe^{r(12)} = 2P$, so $e^{12r} = 2$ and $12r = \ln(2)$; then $r = \frac{\ln 2}{12}$, or $r \approx 0.05776$, or an annual rate of approximately 5.78%.

29. Account 1 has an initial deposit of \$200 on January 1, 2002, and pays interest at an annual rate of 5% compounded annually. The balance in Account 1 is given by $B_1 = 200(1.05)^t$ (here t is the time in years, with $t = 0$ representing January 1, 2002.) Account 2 has an initial deposit of \$200 on January 1, 2003, but pays interest at an annual rate of 6% compounded annually. The balance in Account 2 is given by $B_2 = 200(1.06)^{t-1}$ (t is the time in years, with $t = 0$ representing January 1, 2002 as well.) For the two accounts to have the same balance, we need $200(1.05)^t = 200(1.06)^{t-1}$. Solving for t: $t\ln(1.05) = (t-1)\ln(1.06)$; then, $t\left(\ln(1.05) - \ln(1.06)\right) = -\ln(1.06)$, and $t = \frac{\ln(1.06)}{\ln(1.06)-\ln(1.05)}$, approximately 6.15 years, which corresponds to February of 2008.

31. Starting with a price per compact disk of \$18 in January 1, 2003, and assuming an increase of 3% per year, the price is given by $P = 18(1.03)^t$. At this rate, the price of a compact disk will reach \$100 when $18(1.03)^t = 100$, or $1.03^t = \frac{50}{9}$. Solving for t: $t\ln(1.03) = \ln(\frac{50}{9})$, so $t = \frac{\ln\left(\frac{50}{9}\right)}{\ln(1.03)}$, approximately 58.01 years, which corresponds to January of 2060.

33. If the Consumer Price Index increases exponentially (with rate unchanged), and in 1983 it was \$99.6 and in 2000 it had increased to \$172.20, then $172.20 = 99.6e^{17r}$. So, $e^{17r} = \frac{172.20}{99.6}$, or $e^{17r} \approx 1.728915663$. Then, $17r \approx \ln(1.728915663)$, and $r \approx 0.0322$. Under this model, in 2010 the CPI is predicted to be $99.6e^{0.0322(27)}$, or approximately \$237.59. The level \$300 would be reached when $99.6e^{0.0322t} = 300$, that is, when $e^{0.0322t} \approx 3.012$; solving for *t*, we get $0.0322t \approx \ln(3.012)$, $t \approx \frac{\ln(3.012)}{0.0322}$, approximately 34.24 years since 1983, or in 2017.

35. An initial colony of 1000 *Mycoplasma laidlawii*, with an average weight of 10^{-16} grams has a mass of $1000 \times 10^{-16} = 10^{-13}$ grams at 8:00 A.M., July 4, 2003. Since 1 ton = 2000 lbs, Since 1 lb = 454 grams: 190 tons = 190(2000)(454) = 172,520,000 grams. The colony doubles every hour, that is $P = P_o \bullet 2^t$, where *t* is in hours. The mass of the colony will be 190 tons when $10^{-13} \bullet 2^t = 172{,}500{,}000$, or $2^t = 172{,}500{,}000\left(10^{13}\right)$. Then $t = \frac{\log\left(1.7252 \times 10^{21}\right)}{\log 2}$, that is $t \approx 70.54$ hours. This corresponds to approximately 6:30 am on July 7, 2003.

37. Assuming exponential decrease for the population of Steubenvile-Weirton, we use the census information of 1990 (pop. 142,523) and 2000 (pop. 132,008), to obtain: $132{,}008 = 142{,}523e^{10r}$, or $e^{10r} = \frac{132{,}008}{142{,}523}$, $r = 0.1\ln\left(\frac{132{,}008}{142{,}523}\right) \approx -0.0077$. In the year 2010, at this rate of decrease, the population in this area will be $142{,}523e^{-0.0077(20)}$, or 122,181 people.

39. If the half-life of Carbon-14 is assumed to be 5760 years, then we have $\left(\frac{1}{2}\right)Q_o = Q_o e^{5760k}$, so $5760k = \ln\left(\frac{1}{2}\right)$, and the decay constant for Carbon-14 is, then, $k = \frac{\ln\left(\frac{1}{2}\right)}{5760}$ ≈ -0.00012034. If an object contains 25% of its original Carbon-14, then $0.25 = e^{-0.00012034t}$. Solving for *t*, $-0.00012034t = \ln(0.25)$, and $t = \frac{\ln(0.25)}{-0.00012034}$, or approximately 11,520 years old.

41. Assuming that the concentration of pollution decreases from a level *L* to 80% of *L* in six months or 0.5 years, then we have: $0.8L = Le^{0.5r}$. Therefore, $0.5r = \ln(0.8)$, and $r = \frac{\ln(0.8)}{0.5}$, or $r \approx -0.446287$ (a decline rate of approximately 44.63% per year.) For the concentration of pollution to decrease to 10% of the current level, assuming there is no new incoming pollution, we solve the equation $0.1L = Le^{-0.446287t}$: $-0.446287t = \ln(0.1)$, or $t = \frac{\ln(0.1)}{-0.446287}$, approximately 5.16 years.

Concepts and Critical Thinking

43. The statement: "The half-life of a substance is equal to one-half the age of the substance" is false.

45. The statement: "A quantity will grow exponentially if it grows at a constant rate" is false.

47. Answers may vary. An example of an exponential equation with no solution is $2^{x+3} = 0$.

49. Answers may vary. A real-world context in which exponential growth occurs is the uninhibited growth of a species, as is the case when a new species is introduced in a region where there are no natural predators. The growth is exponential at least for a period of time.

51. If $Q = Q_o e^{kt}$, then $Q = Q_o \left(e^k\right)^t$, so if $a = e^k$, then $Q = Ca^t$, where $C = Q_o$. Now, if $Q = Ca^t$, where $a > 0$, there exists k such that $e^k = a$, since the range of $f(x) = e^x$ is the set of all positive real numbers. If we make $Q_o = C$, then we have $Q = Q_o e^{kt}$. So the equations $Q = Q_o e^{kt}$ and $Q = Ca^t$ are equivalent.

Chapter 4. Section 4.5 Modeling with Exponential and Logarithmic Functions

1. **(a)-(b)** A plot of the collected data on rabbit population over a 5-month period, with an empirical visual estimate of the curve that fits the data follows:

Month	0	1	2	3	4	5
Rabbits	20	24	30	36	45	54

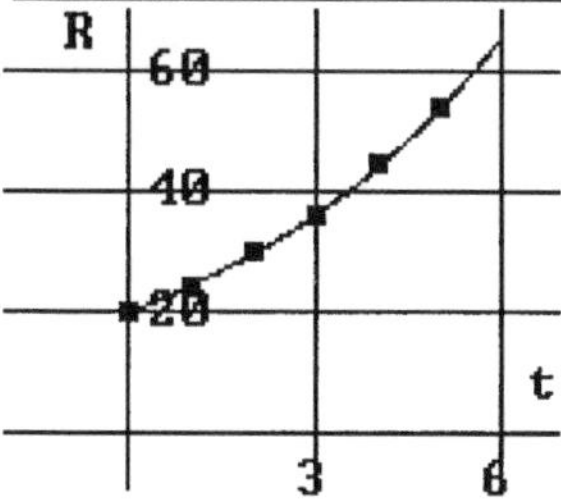

(c) A rough estimate of the rabbit population by the 6th month is 65.

3. If the rabbit population in Exercise 1 is known to be growing exponentially, that is $R(t) = ae^{bt}$, with 20 rabbits initially, and 24 rabbits after 1 month, then we have $ae^0 = 20$, so $a = 20$, $20e^{b(1)} = 24$, and $e^b = 1.2$, or $b = \ln 1.2$, approximately $b \approx 0.182$. This predicts:

Month	0	1	2	3	4	5
Rabbits	20	24	28.8	34.56	41.47	49.77
AbsoluteError	0	0	1.2	1.44	3.53	4.23

The largest absolute error occurs the 5th year. The sum of the absolute errors is 10.4.

5. If the rabbit population in Exercises 1 and 3 is known to be growing exponentially, that is $R(t) = ae^{bt}$, with 20 rabbits initially, and 54 rabbits after 5 months, then we have $ae^0 = 20$, so $a = 20$, $20e^{b(5)} = 54$, and $e^{5b} = 2.7$, or $b = \frac{\ln(2.7)}{5}$, approximately $b \approx 0.198$. With this:

Month	0	1	2	3	4	5
Rabbits	20	24.4	29.76	36.3	44.27	54
AbsoluteError	0	0.4	0.24	0.3	0.73	0

The largest absolute error occurs the 4th month. The sum of the absolute errors is 1.67. This model seems to have a better fit to the data.

7. The exponential regression model for the rabbit data in Exercise 1 is $R(t) = 19.88e^{0.201t}$, which predicts the following values for the rabbit population:

Month	0	1	2	3	4	5
Rabbits	19.88	24.31	29.72	36.33	44.42	54.31
AbsoluteError	0.12	0.31	0.28	0.33	0.58	0.31

The sum of the absolute errors is 1.93. This fits the data better than the model in Exercise 3.

9. Referring to the Alaska population data:

Year	1960	1970	1980	1990	2000
Population	226,000	303,000	402,000	550,000	627,000

we obtain the model $P(t) = 232.505814e^{0.02637t}$ (P in thousands, t in years since 1960). The population in Alaska in 2010 is predicted to be $\approx$869,000, and to be at least 1,500,000, or $232.505814e^{0.02637t} \geq 1500$, to obtain $0.02637t \geq \ln 6.45145$, $t \geq 70.7$, by the year 2030.

11. For the data on sales for a 4-month period, in thousands:

Month (t)	0	1	2	3	4
Sales (in thousands) (S)	80	72	66	61	58

we obtain the exponential model $S(t) = 78.71e^{-0.0809t}$, t is in months and S in thousands. For the sixth month, the predicted sales are $S(6) = 78.71e^{-0.0809(6)} \approx 48{,}440$. According to this, sales will reach 20,000 at $78.71e^{-0.0809t} = 20$, $e^{-0.0809t} = 20/78.71$, and $t \approx 16.93$, that is, by the 17th month sales will be under 20,000.

13. For the data on the percent of American companies which encountered viruses in the space of one year (fourth quarter of 1990 through the third quarter of 1991)

Quarter	0 (Oct-Dec 1990)	1 (Jan-Mar1991)	2 (Apr-Jun 1991)	3 (Jul-Sep 1991)
Percent	8	19	26	40

we obtain the exponential model $p(t) = 9.22e^{0.5142t}$, where t is in quarters. The percent will reach 90 when $9.22e^{0.5142t} = 90$, so $e^{0.5142t} = 90/9.22$, and $t \approx 4.43$, that is, by the 5th quarter (counted from the first quarter of 1991), the percent of American companies that encountered a virus would have been more than 90%. This predicts percents higher than 100 beginning at $t \approx 4.63$, or the first quarter of 1992. For 200 larger values of t, then, the model is void. An exponential model is not adequate for long periods or beyond $t = 5$. A logistic model is more appropriate in this situation.

15. For the data on the number of transistors in Intel microprocessors introduced since 1971,

Year	1971	1974	1978	1982	1985	1989	1993	1997	1999	2001
Processor	4004	8080	8086	286	386	486	Pentium	P II	P III	P 4
Transistors (thousands)	2	6	29	134	275	1200	3100	7500	24,000	42,000

we get $N(t) = 2.6794e^{0.322t}$, t in years since 1971, N is in thousands. For 2005, $t = 34$, the number of transistors in Intel processors would be $N(34) = 2.6794e^{0.322(34)} \approx 152{,}2978{,}000$ This number would be 100 billion if $100{,}000{,}000 = 2.6794e^{0.322t}$, $t \approx 54.15$ years, or by 2025.

17. If $N = \frac{2000}{1+499e^{-0.3t}}$ estimates the number of people who have heard a rumor after t hours,

(a) the number of people who start the rumor is $N(0) = 4$;

(b) after 10 hours, the number of people who have heard the rumor is $N(10) \approx 77.39$, or about 77 individuals;

(c) for 500 people to hear the rumor: $\frac{2000}{1+499e^{-0.3t}} = 500$, so $t \approx 17.05$ hours;

(d) the limiting number of people who will hear the rumor is 2000,

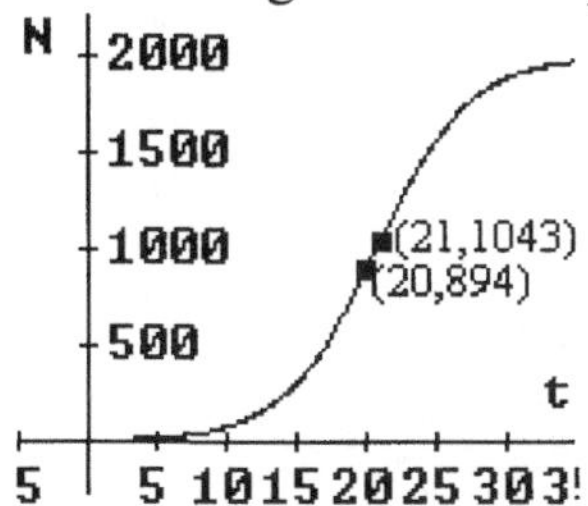

(e) the hours during which the greatest number of people hear the rumor is between $t = 20$ and $t = 21$.

19. If $S = \frac{700}{1+6.8e^{-0.057t}}$ estimates the 1-mile automobile speed record, where t is in years ($t = 0$ corresponding to 1906), and S in miles per hour,

(a) the speed record exceeded 400 mph when $\frac{700}{1+6.8e^{-0.057t}} > 400$, or for $t \approx 38.68$, in 1944;

(b) this model has an upper bound of 700 mph on the speed record;

(c) for 1997, this predicts $S(91) \approx 674.4$ mph, an underestimate of the actual 763 mph.

21. For the data on average walking speed of pedestrian and population of a city,

Population	5,500	14,000	71,000	138,000	342,000
Velocity (ft/sec)	3.3	3.7	4.3	4.4	4.8

we obtain the logarithmic model $v = 0.812 \log P + 0.29978$. This predicts: for Little Rock, with a population of 183,000, the predicted average walking speed is 4.57 *ft/sec*; for New York, with a population of 8,008,000 the predicted average walking speed is 5.9 *ft/sec*, and for Mexico City, pop. 18,330,000, the predicted average walking speed is 6.2 *ft/sec*.

Concepts and Critical Thinking

23. The statement: "If a mathematical model fits known data exactly, then we can have 100% confidence in any predictions made that are based on the model" is false. One problem may be extrapolation of the predictions to ranges of values out of the range of observations.

25. The statement: "Exponential regression is used to fit a model of the form $P(t) = ae^{bt}$ to a collection of data" is true.

27. Answers may vary. A reason why populations tend not to continue to grow exponentially over long periods of time is that most populations face curbing factors in their growth, like the presence of predators, or the carrying capacity of their environment.

29. Answers may vary. A graphical feature of a logistic model that distinguishes it from an exponential model is that a logistic model is bounded, as opposed to the fact that an exponential model is unbounded.

Chapter 4. Review

1. Sketch of the graph of $f(x)=2^x$

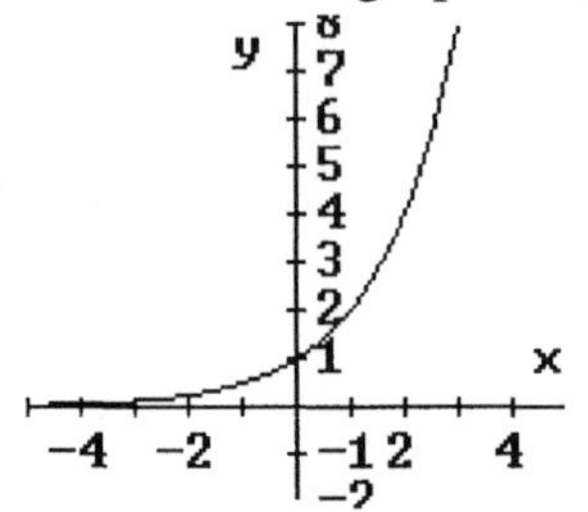

(a) Sketch of the graph of $g(x)=2^{-x}$

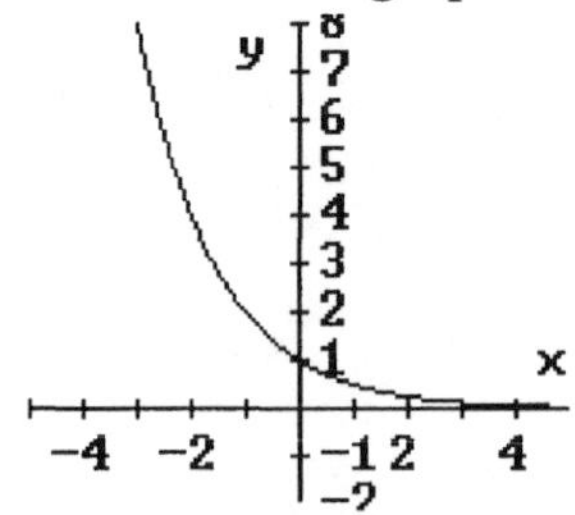

(b) Sketch of the graph of $g(x)=2^x+2$

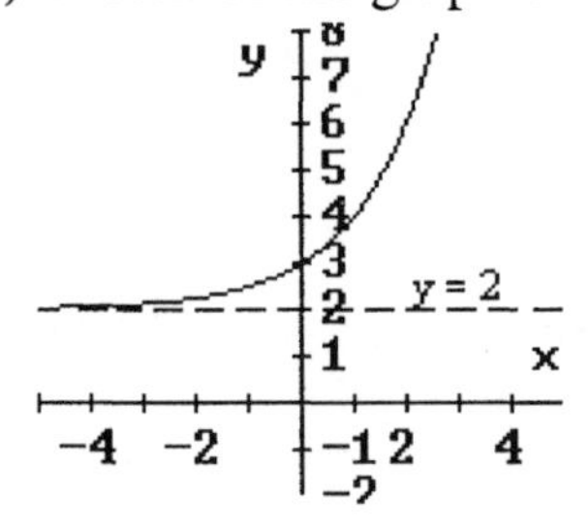

(c) Sketch of the graph of $g(x)=2^{x+1}$

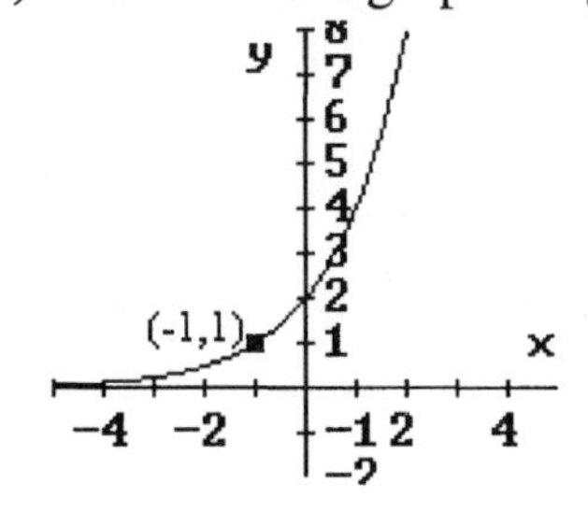

3. If $3^x=27$, we have $3^x=3^3$, therefore $x=3$.

5. The equation $5^{1-2x}=25$ is equivalent to $5^{1-2x}=5^2$, therefore $1-2x=2$, and $x=-\frac{1}{2}$.

7. $\log_2\frac{1}{4}=-2$, since $2^{-2}=\frac{1}{4}$.

9. $\log 0.01=-2$, since $10^{-2}=0.01$.

11. $\log_5\sqrt{5}=\frac{1}{2}$, since $5^{\frac{1}{2}}=\sqrt{5}$.

13. The logarithmic equation $\log_2\frac{1}{16}=-4$ is equivalent to the exponential equation $2^{-4}=\frac{1}{16}$.

15. The logarithmic equation $\log(2x+1)=4$ is equivalent to the equation $2x+1=10{,}000$.

17. The exponential equation $e^5=x$ is equivalent to the logarithmic equation $\ln x=5$.

19. The exponential equation $2^{x+3}=5$ is equivalent to the equation $x+3=\log_2 5$.

21. Sketch of the graph of $f(x)=\log_5 x$.

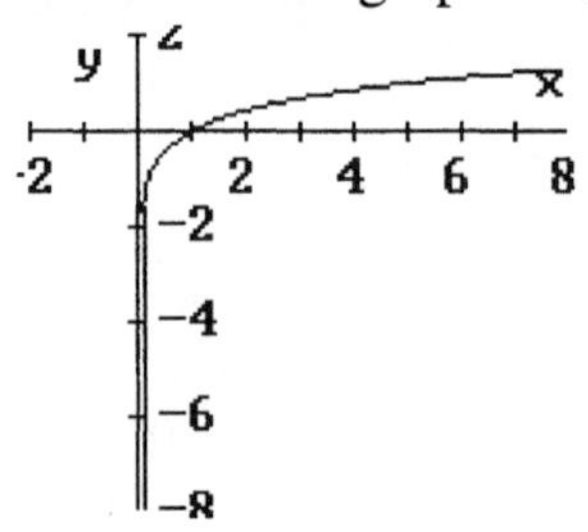

23. Sketch of the graph of $f(x) = \ln(x-2)$.

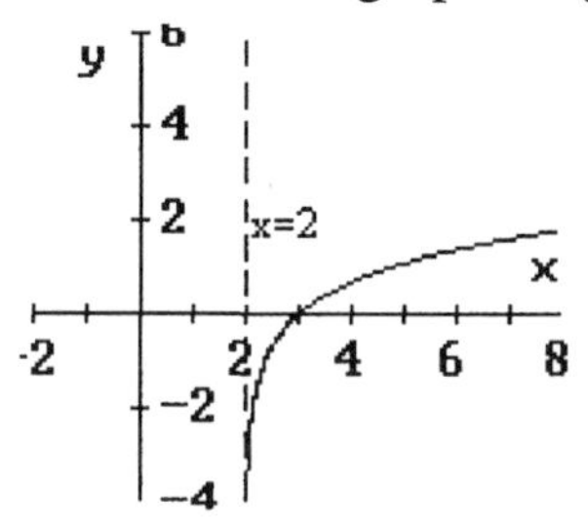

25. $\log_2(xy^2) = \log_2 x + 2\log_2 y$

27. $\log\sqrt[3]{x^2 y} = \frac{2}{3}\log x + \frac{1}{3}\log y$

29. $2\log x + 3\log y = \log(x^2 y^3)$

31. $\frac{1}{3}\ln x + \frac{2}{3}\ln y - \ln z = \ln\left(\frac{\sqrt[3]{xy^2}}{z}\right)$

33. $\log_4 10 = \frac{\log 10}{\log 4} \approx 1.66096$

35. If $\log_2 x = -1$, then $x = \frac{1}{2}$.

37. The equation $\log_5(2x+4) = \log_5 10$ is equivalent to $2x+4=10$ since the function $f(x) = \log_5 x$ is one-to-one; so, the solution of this equation is $x = 3$.

39. The equation $\log(5y-2) = 1$ is equivalent to $5y-2=10$, so the solution is $y = \frac{12}{5}$.

41. To solve: $\ln(x-1) + \ln(x+2) = \ln 4$, we note that: $\ln[(x-1)(x+2)] = \ln 4$, that is $\ln(x^2+x-2) = \ln 4$, so $x^2+x-2=4$, or $x^2+x-6=0$, and $(x+3)(x-2)=0$
This equation has solutions $x=-3$, $x=2$; $x=-3$ is not a solution of the original equation. The value $x=2$ is the only solution of $\ln(x-1)+\ln(x+2)=\ln 4$.

43. The solution of the equation $5^x = 10$ is $x = \log_5 10$, or $x = \frac{\log 10}{\log 5}$, that is $x \approx 1.43067$.

45. To solve the equation $1.05^t = 10$, we take logarithms on both sides of the equation: $\log 1.05^t = \log 10$, then $t\log 1.05 = 1$, and $t = \frac{1}{\log 1.05}$, approximately $x \approx 47.19363$.

47. To solve the equation $4e^{0.1x} = 20$, we divide both sides by 4, to $e^{0.1x} = 5$, and $x \approx 16.09437$.

49. To solve the equation $3^{x+2} = 2^{2x}$, we take logarithms of both sides, to obtain $(x+2)\ln 3 = 2x\ln 2$. Then, $x\ln 3 - 2x\ln 2 = -2\ln 3$, so $x = \frac{-2\ln 3}{(\ln 3 - 2\ln 2)}$, and $x \approx 7.63768$.

51. The approximate solutions of $e^x = 8 - x^2$ are $x \approx -2.82$ and $x \approx 1.66$.

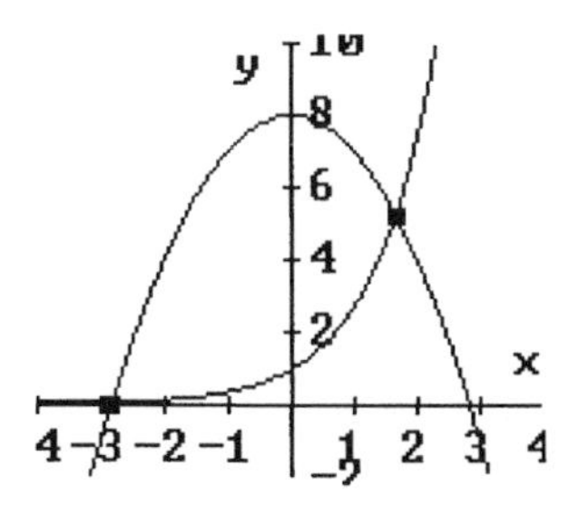

53. The approximate solutions of $\ln(3x)+0.02x^3+4x=x^2$ are $x\approx 0.17$, $x\approx 5.05$ and $x\approx 45.48$.

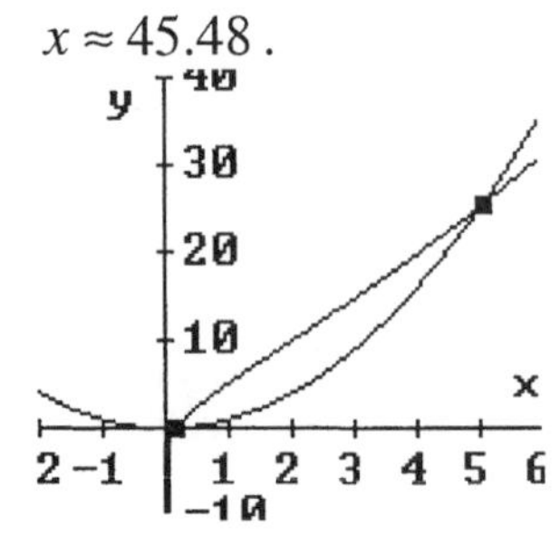

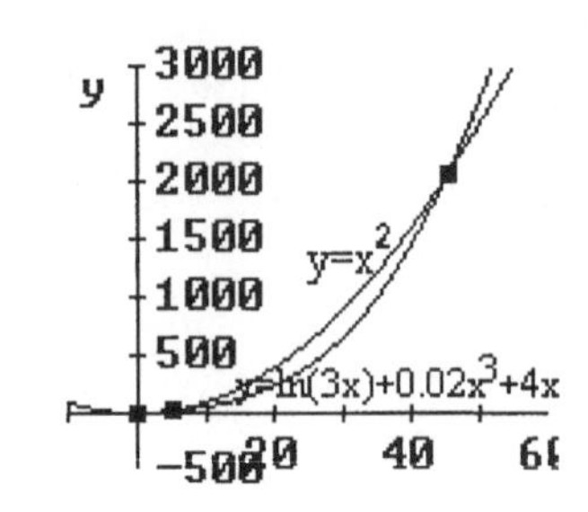

55. **(a)** $\frac{2^x}{3^x}=\left(\frac{2}{3}\right)^x$

(b) The solution of $\frac{2^x}{3^x}=2$ is the solution of $\left(\frac{2}{3}\right)^x=2$, that is $x=\frac{\ln 2}{\ln 2-\ln 3}$, $x\approx -1.71$.

57. **(a)** $x^2+x-2=0$ has solutions $x=-2$ and $x=1$, since $x^2+x-2=(x+2)(x-1)$.

(b) If $3^{x^2+x}=9$, then $3^{x^2+x}=3^2$, so $x^2+x=2$, or $x^2+x-2=0$. The solutions are $x=-2$, 1.

59. **(a)** Sketch of the graph of $y=e^{-x}$ and $y=x$ on the same set of axes.

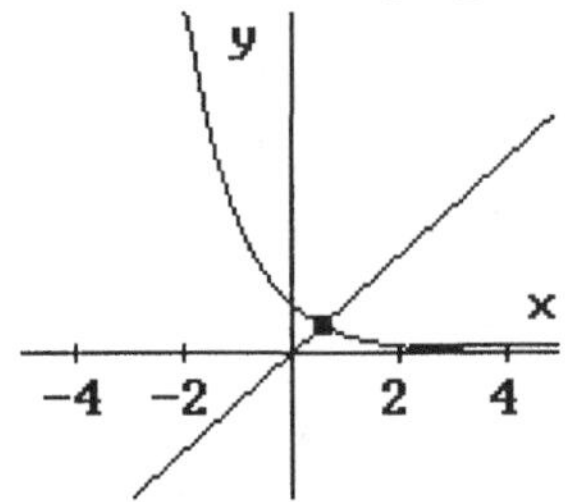

(b) There is only one solution of the equation $e^{-x}-x=0$, approximately $x\approx 0.57$.

61. **(a)** The solution set of the inequality $2x-3\le 0$ is the set of real numbers $x\le\frac{3}{2}$.

(b) The domain of $f(x)=\ln(2x-3)$ is the set of real x such that $2x-3>0$, or $x>\frac{3}{2}$.

63. **(a)** To solve the equation $u+1/u=2$, we perform the addition on the left hand side, to obtain $\frac{u^2+1}{u}=2$, which is equivalent to $u^2-2u+1=0$, or $(u-1)^2=0$. Then, $u=1$.

(b) To solve the equation $e^x-e^{-x}=2$, we make $u=e^x$, so the original equation becomes $u-\frac{1}{u}=2$. Then $\frac{u^2-1}{u}=2$, $u^2-2u-1=0$, and $u=1\pm\sqrt{2}$. $e^x>0$ for all x, so the only solution is $u=1+\sqrt{2}$, that is $e^x=1+\sqrt{2}$, so $x=\ln\left(1+\sqrt{2}\right)$, approximately $x\approx 0.88137$.

65. **(a)** Sketch of the graph of $y=4x-x^2$:

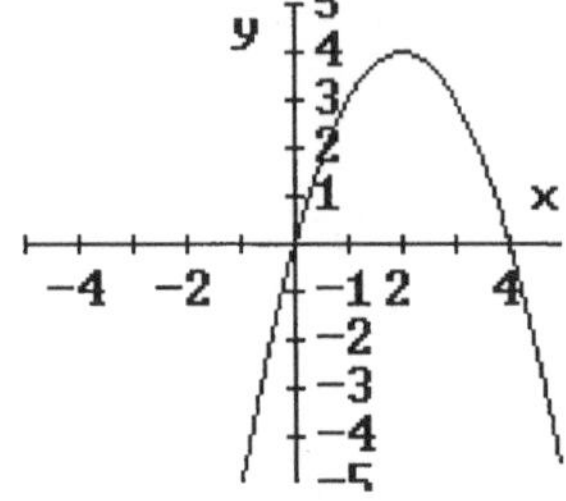

(b) The domain of $g(x)=\ln(4x-x^2)$ is $0<x<4$ (because we need $4x-x^2>0$.)

67. If \$50 is deposited in an account that pays 4% interest compounded quarterly, then the balance in the account after 3 years is $50\left(1+{}^{0.04}\!/_{4}\right)^{4(3)}$, or \$56.34.

69. In an account that pays 3% interest compounded monthly, the time that it takes to double the deposit is obtained solving for t in the equation $P\left(1+{}^{0.03}\!/_{12}\right)^{12t}=2P$, which is equivalent to $12t=\frac{\ln 2}{\ln 1.0025}$, that is $t\approx 23.13$ years.

71. If a bacteria colony is growing exponentially, beginning with 10, increasing to 25 after 1 hour, then we have $P(t)=ae^{bt}$, with $P(0)=10$ and $P(1)=25$. So, $ae^0=10$ and $ae^b=25$. From the first equation we get $a=10$, therefore $10e^b=25$ and $e^b=2.5$, from which we conclude that $b=\ln 2.5$, approximately $b\approx 0.9163$. To reach the 1 million mark, we need $10e^{0.9163t}=1{,}000{,}000$, or $0.9163t=\ln 100{,}000$, about $t\approx 12.56$ hours.

73. If $Q=\dfrac{200{,}000}{1+1999e^{-0.08t}}$ estimates the number of infected people after t days from the first exposure, then: **(a)** after 5 days, the number of infected individuals is $Q(5)=\dfrac{200{,}000}{1+1999e^{-0.08(5)}}$, or about 149; **(b)** the time it will take for 1000 people to be infected can be obtained from the $\dfrac{200{,}000}{1+1999e^{-0.08t}}=1000$, that is $1+1999e^{-0.08t}=200$, so $1999e^{-0.08t}=199$, $t=\dfrac{\ln\left({}^{199}\!/_{1999}\right)}{-0.08}$, approximately 28.84 days;
(c) the limiting number of people that will become infected is 200,000

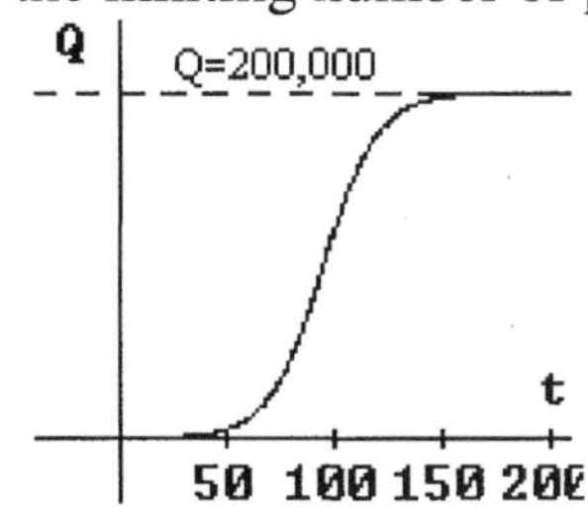

(d) among days 5, 96, and 120 after the first exposure, it was on day 96 that the greatest number of infections occurred.

74. If $y=m\log(t+1)+b$ estimates the number of names that a subject remembers after being introduced to 100 people at a party, and it is known that he remembers 50 names immediately after the party, and 36 names 2 years after the party, then $50=m\log(0+1)+b$, so $b=50$, and $36=m\log(24+1)+50$, so $m=-{}^{14}\!/_{\log(25)}$, or approximately $m\approx -10.015$. After 1 year, the number of names remembered by this subject is estimated by $y=-10.015\log(12+1)+50$, approximately 39 names.

75. For the depreciation data:

Year	2000	2001	2002	2003	2004
Value	\$28,000	\$24,000	\$20,800	\$17,800	\$15,400

we fit the exponential model $V(t)=27{,}956.82e^{-0.14945t}$. The value will be \$4000 $t\approx 13$ years after 2000, or by 2013.

Chapter 4. Test

1. The statement: "$\log_a x \cdot \log_a y = \log_a x + \log_a y$ for all $x > 0$ and $y > 0$" is false. A true statement is: "$\log_a(xy) = \log_a x + \log_a y$ for all $x > 0$ and $y > 0$."

3. The statement: "If $f(x) = a^x$, then $f^{-1}(x) = \frac{1}{a^x}$" is false.

5. The statement: "The logarithm of a negative number is not defined" is true.

7. The statement: "It is possible that a sample of a substance undergoing exponential decay will take longer to decay from 100 grams to 50 grams than from 50 grams to 25 grams" is false. If this happens, the decay is not exponential.

9. Answers may vary. The function $f(x) = \left(\frac{2}{3}\right)^x$ is always decreasing.

11. Answers may vary. A real number with negative common logarithm is 0.1, as $\log(0.1) = -1$.

13. To solve the equation $2^{x^2+x} = 4$, we note that $2^{x^2+x} = 2^2$, therefore $x^2 + x = 2$, or $x^2 + x - 2 = 0$. The solutions are $x = -2$ and $x = 1$.

15. To solve the equation $3^{x-1} = 2^x$, we take logarithms on both sides of the equation, to obtain $(x-1)\log 3 = x \log 2$, $x \log 3 - x \log 2 = \log 3$, and $x = \frac{\log 3}{\log 3 - \log 2} \approx 2.70951$.

17. $\log_2 32 = 5$

19. Sketch of the graph of $f(x) = e^{x-2}$

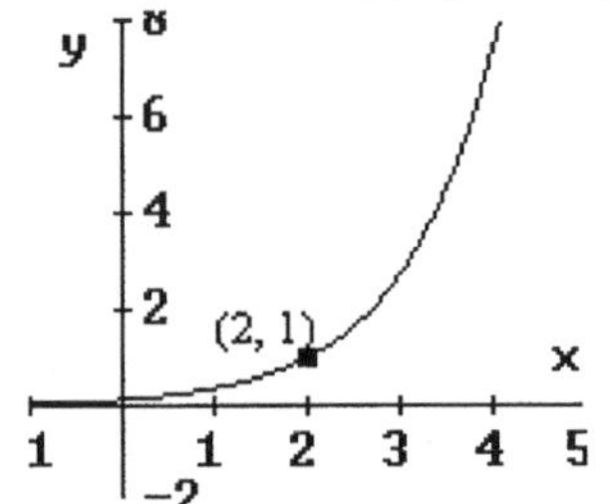

21. Sketch of the graph of $f(x) = \ln(x+3)$

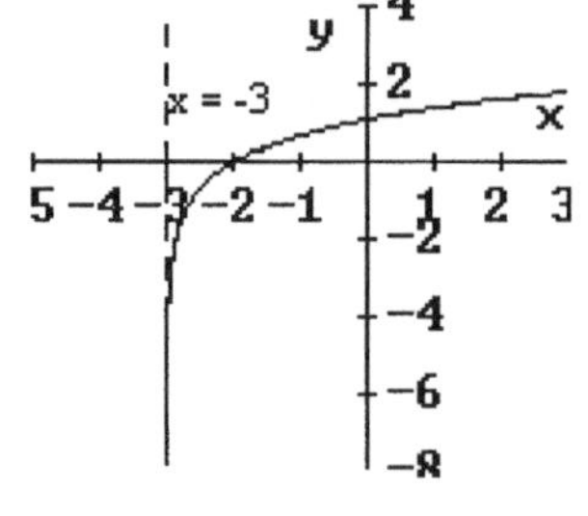

23. $3\ln x - 4\ln y + 5\ln z^2 = \ln\left(\frac{x^3 z^{10}}{y^4}\right)$

25. If a radioactive sample decays exponentially from 10 grams on January 1, 1999, to 7 grams on January 1, 2004, then $7 = 10e^{k(5)}$, so $5k = \ln\left(\frac{7}{10}\right)$, and $k \approx -0.071335$. For the sample to decay to 1 gram, we solve for t in the equation $1 = 10e^{-0.071335t}$, that is $t \approx \frac{\ln(1/10)}{-0.071335}$, approximately $t \approx 32.28$ years, or during the year 2031.

Chapter 5. TRIGONOMETRIC FUNCTIONS

Section 5.1 Angles and Their Measurement

1. 30° corresponds to sketch (v).

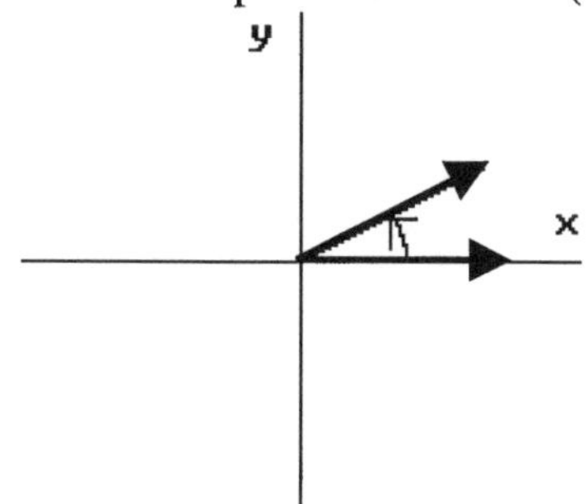

3. -135° corresponds to sketch (viii).

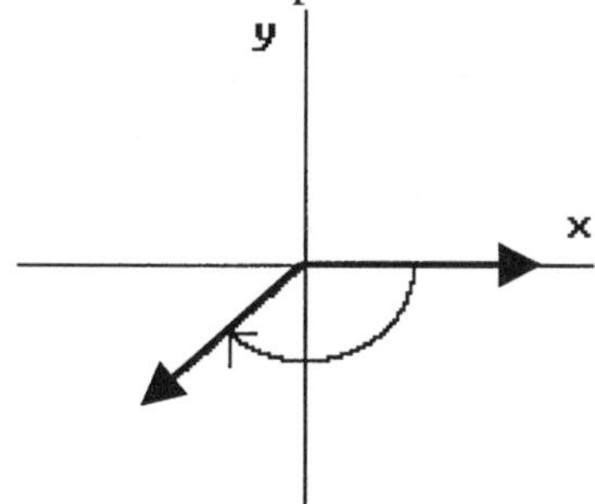

5. 540° corresponds to sketch (x).

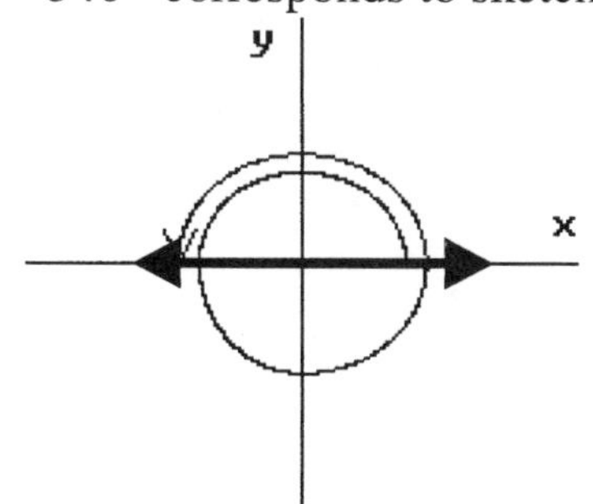

7. $\pi/3$ corresponds to sketch (iii).

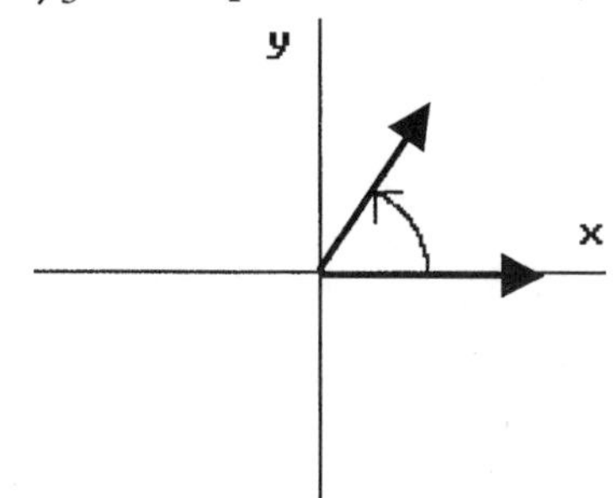

9. $-3\pi/2$ corresponds to sketch (iv).

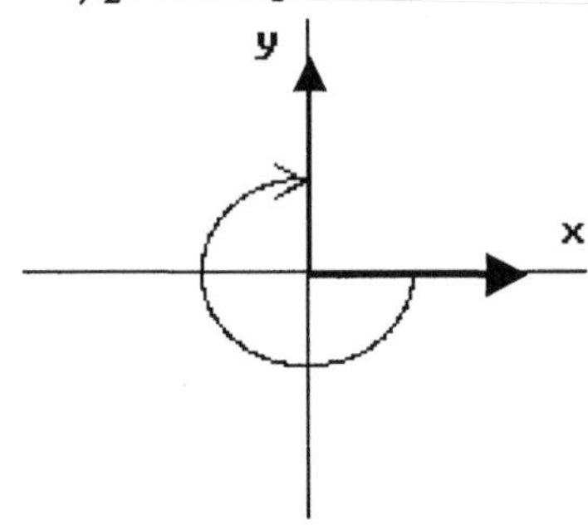

In Exercises 11-17, answers may vary from the ones given here.

	Angle	Positive Coterminal Angle	Negative Coterminal Angle
11.	30°	390°	-330°
13.	-135°	225°	-495°
15.	$\pi/3$	$7\pi/3$	$-5\pi/3$
17.	$-3\pi/2$	$\pi/2$	$-7\pi/2$

19. The complementary angle of 20° is 70°; the supplementary angle of 20° is 160°.

21. The complementary angle of $\pi/6$ is $\pi/3$; the supplementary angle of $\pi/6$ is $5\pi/6$.

23. $25^\circ\ 16' = \left(25 + 16/60\right)^\circ$; that is, rounded to the nearest hundredth, $25^\circ\ 16' \approx 25.27^\circ$.

25. $173^\circ\ 20'\ 35'' = \left(173 + 20/60 + 35/3600\right)^\circ$; that is, rounded to the nearest hundredth, $173^\circ\ 20'\ 35'' \approx 173.34^\circ$.

27. The angle measure 132.4°, in degree-minute-second form is $132^\circ + 0.4(60)'$, or $132^\circ\ 24'$.

29. The angle measure 15.625°, in degree-minute-second form is $15^\circ + 0.625(60)'$, or $15^\circ\ 37.5'$, which is equivalent to $15^\circ\ 37' + 0.5(60)''$, or $15^\circ\ 37'\ 30''$.

	Angle Measure (Degrees)	Angle Measure (Radians)		Angle Measure (Degrees)	Angle Measure (Radians)
31.	30°	$\pi/6$	**33.**	225°	$5\pi/4$
35.	-150°	$-5\pi/6$	**37.**	15°	0.262
39.	-117.4°	-2.049	**40.**	326.7°	5. 702

	Angle Measure (Radians)	Angle Measure (Degrees)		Angle Measure (Radians)	Angle Measure (Degrees)
41.	$3\pi/2$	270°	**43.**	$-4\pi/3$	-240°
45.	$11\pi/12$	165°	**47.**	$-2\pi/7$	-51.429°
49.	2	114.592°	**50.**	-3	-171.887°

	Central Angle θ	Radius	Length of the Arc Subtended by θ ($s = r\,\theta$)	Area of the Sector Determined by θ ($A = \frac{1}{2}r\,\theta$)
51.	$\theta = 5\pi/3$	$r = 20$ meters	$s = 100\pi/3$, $s \approx 104.72$ m	$A = 1000\pi/3$, $A \approx 1047.2\ \text{m}^2$
53.	$\theta = 75^\circ$	$r = 4$ inches	$s = 10\pi/3$, $s \approx 10.47$ in	$A = 20\pi/3$, $A \approx 20.94\ \text{in}^2$

Applications

55. If a bucket is raised turning a crank attached to an 8-inch pulley, then:
(a) if the crank is turned through an angle of 10π (5 revolutions), then the bucket is raised a length $s = r\theta$, or $s = 8(10\pi)$, approximately $s \approx 251.33$ inches, or $s \approx 20.94$ feet;
(b) to raise the bucket 9 feet, or 108 inches, the angle θ must satisfy $8\theta = 108$, so $\theta = 13.5$, or $\theta \approx 773.49^\circ$, approximately 2.15 revolutions.

57. If a wind turbine rotates at a rate of 40 revolutions per minute, the linear speed of the tip of a 100-foot blade is $v = \dfrac{r\theta}{t}$, or $v = \dfrac{100(40(2\pi))\ \text{feet}}{1\ \text{min}}$, that is $v = 8000\pi\ ft/min$, approximately $v \approx 25{,}132.74\ ft/min$.

59. If the tip on the minute hand on the four faces of the Big Ben is approximately 11 feet from the center of the face, its linear speed is 1 revolution per hour, $v = 11(2\pi)\ ft/hr$, approximately $v = 69.12\ ft/hr$; in miles per hour, $v = 0.013$ mph.

61. If the tires on a certain car have a 14-inch radius, in order to attain a rate of $90\ ft/sec$, the number n of revolutions per second satisfies $90\ ft/sec = \dfrac{(14\ \text{in})(2\pi n)}{1\text{sec}}\dfrac{1\,ft}{12\ \text{in}}$, so $n = \dfrac{6\times 90}{14\pi}$, or approximately $n = 12.277$ revolutions per second, 736.67 revolutions per minute.

63. If a piece of stained glass is cut in the shape of a circular sector with a radius of 12 inches and a central angle of 120°, then:
(a) the area of the piece of glass is $A = \frac{1}{2}(12^2)(2\pi/3)$, approximately $A \approx 150.8\ in^2$
(b) the length of a lead strip to be placed around the edges of this piece of glass is $s = 12 + \dfrac{12(2\pi)}{3} + 12$ inches, that is $s = 24 + 8\pi$ inches, approximately 49.13 inches.

65. When a ray of light passes by a massive heavenly body, the ray is deflected by an angle α, $\alpha = 4.05\dfrac{Gm}{rc^2}$, where *m* is in kilograms, *r* is in meters, $G \approx 6.67\times10^{-11}\ {m^3}/{(kg.sec^2)}$ is Newton's gravitational constant, and $c \approx 3\times10^8\ {m}/{sec}$. If a light ray passes close to the surface of the sun, in which case $m \approx 1.989\times10^{30}$ kilograms and $r \approx 6.95\times10^8$ meters, then it gets deflected by an angle $\alpha \approx 0.000086$ radians, or $\alpha \approx 0.005^\circ$.

	Cities and Latitudes	Central Angle θ Between the Two Latitudes	Length of Arc Subtended by θ (Distance Between the Cities)
67.	Phoenix $33^\circ\ 30'\ N$ Salt Lake City $40^\circ\ 45'\ N$	$\theta = 40^\circ\ 45' - 33^\circ\ 30'$ $= 7^\circ\ 15' = 7.25^\circ$	$s = 3963\left(7.25\times{\pi}/{180}\right)$ ≈ 501.46 miles
69.	Montreal, Canada $45^\circ\ 30'\ N$ Bogotá, Colombia $4^\circ\ 38'\ N$	$\theta = 45^\circ\ 30' - 4^\circ\ 38'$ $= 40^\circ\ 52' \approx 40.87^\circ$	$s = 3963\left(40.87\times{\pi}/{180}\right)$ ≈ 2826.9 miles

The sum of the angles on a spherical triangle is related to the area of this triangle by the equation $A+B+C = 180^\circ + \left({T}/{S}\right)720^\circ$, *where T is the area of the triangle and S the surface area of the sphere.*

71. For the Bermuda Triangle, the region in the Atlantic Ocean bounded by the imaginary lines connecting Bermuda, Puerto Rico, and Melbourne, Florida, the area is $T \approx 490{,}000$ square miles. The surface area of the Earth is $S = 4\pi r^2$, approximately $S = 4\pi\left(3963^2\right)$, or $S \approx 197{,}359{,}488$ square miles. The sum of the angles of the Bermuda Triangle is $A+B+C = 180^\circ + \left({490{,}000}/{197{,}359{,}488}\right)720^\circ$, that is $A+B+C = 181.79^\circ$ or ≈ 3.173 radians.

Concepts and Critical Thinking

73. The statement: "If two angles are coterminal, then their measures must differ by 360 degrees" is false. For example, 60° and 780° are coterminal, and they differ by 720°.

75. The statement: "There are 3600 seconds in 1°" is true.

77. Answers may vary. An example of a pair of complementary angles is $\left(32^\circ,\ 58^\circ\right)$.

79. Answers may vary. An example of a quadrant angle is 270°.

81. If α and β are coterminal central angles of a circle with radius *r*, the arc lengths subtended by α and β don't need to be the same. For example, the angles $\alpha = {\pi}/{2}$ and $\beta = {5\pi}/{2}$ are coterminal, yet the arc lengths in the circle of radius 2 subtended by α and β are π and 5π, respectively.

Chapter 5. Section 5.2 Trigonometric Functions of Acute Angles

1.

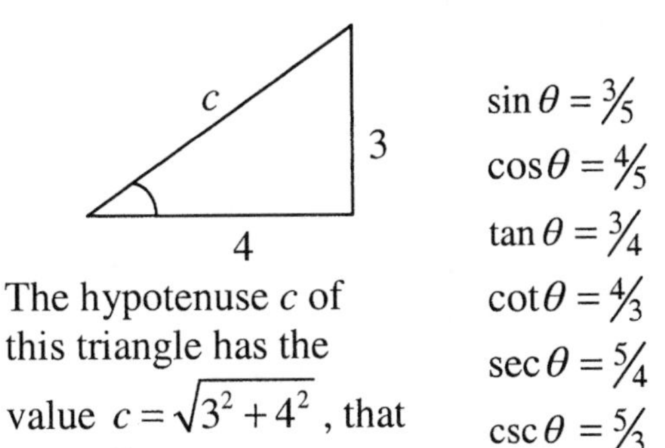

The hypotenuse c of this triangle has the value $c=\sqrt{3^2+4^2}$, that is $c=5$.

$\sin\theta = 3/5$
$\cos\theta = 4/5$
$\tan\theta = 3/4$
$\cot\theta = 4/3$
$\sec\theta = 5/4$
$\csc\theta = 5/3$

3.

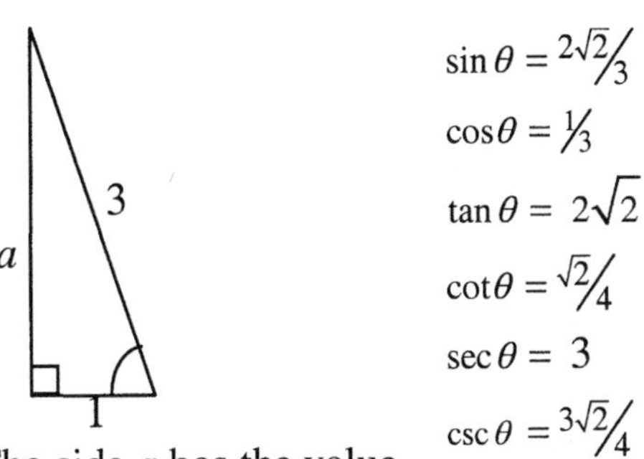

The side a has the value $a=\sqrt{3^2-1^2}$, $a=2\sqrt{2}$.

$\sin\theta = 2\sqrt{2}/3$
$\cos\theta = 1/3$
$\tan\theta = 2\sqrt{2}$
$\cot\theta = \sqrt{2}/4$
$\sec\theta = 3$
$\csc\theta = 3\sqrt{2}/4$

5.

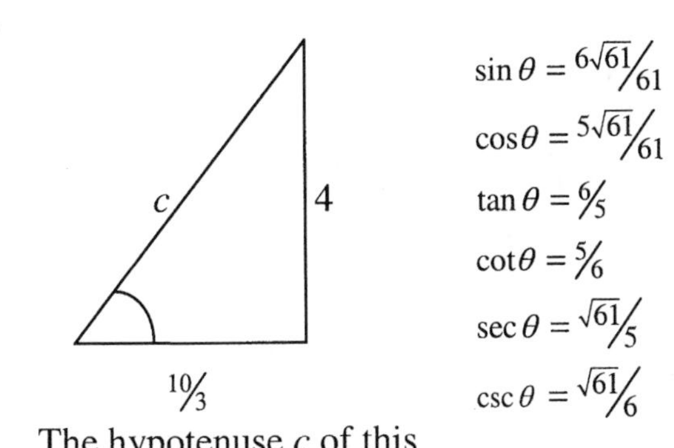

The hypotenuse c of this triangle has the value $c=\sqrt{(10/3)^2+4^2} = 2\sqrt{61}/3$.

$\sin\theta = 6\sqrt{61}/61$
$\cos\theta = 5\sqrt{61}/61$
$\tan\theta = 6/5$
$\cot\theta = 5/6$
$\sec\theta = \sqrt{61}/5$
$\csc\theta = \sqrt{61}/6$

7. If θ is an acute angle and $\sin\theta = \frac{5}{13}$, we can refer to the right triangle:

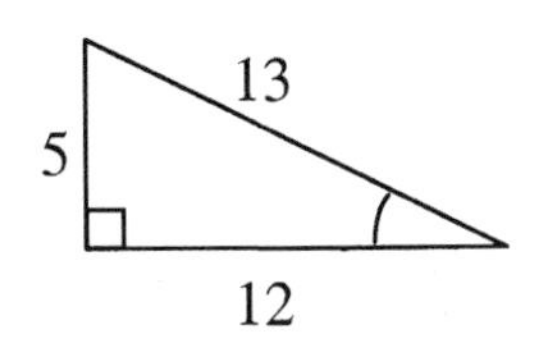

and we have: $\cos\theta = \dfrac{12}{13}$, $\sec\theta = \dfrac{13}{12}$,

$\tan\theta = \dfrac{5}{12}$, $\cot\theta = \dfrac{12}{5}$, $\csc\theta = \dfrac{13}{5}$.

9. If θ is an acute angle and $\sec\theta = 4$, we can refer to the right triangle:

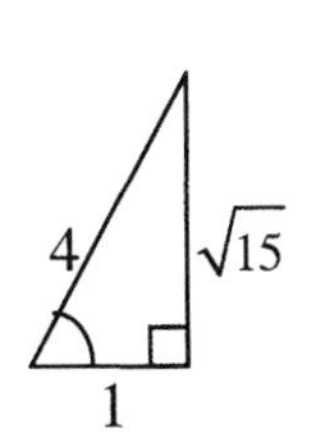

and we have: $\sin\theta = \dfrac{\sqrt{15}}{4}$, $\cos\theta = \dfrac{1}{4}$,

$\tan\theta = \sqrt{15}$, $\cot\theta = \dfrac{\sqrt{15}}{15}$, $\csc\theta = \dfrac{4\sqrt{15}}{15}$.

11. If θ is an acute angle and $\tan\theta = \frac{2}{3}$, we can refer to the right triangle:

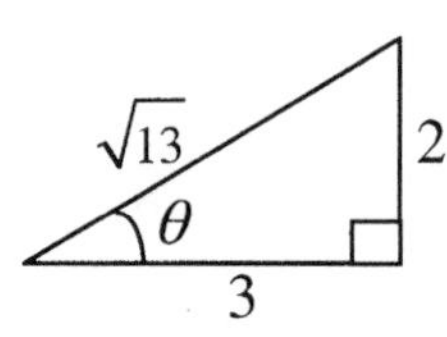

and we have: $\sin\theta = \frac{2\sqrt{13}}{13}$, $\cos\theta = \frac{3\sqrt{13}}{13}$,

$\cot\theta = \frac{3}{2}$, $\sec\theta = \frac{\sqrt{13}}{3}$, $\csc\theta = \frac{\sqrt{13}}{2}$.

13. If $\sec\theta = 2$, then $\cos\theta = \frac{1}{\sec\theta}$, or $\cos\theta = \frac{1}{2}$.

15. If $\cos\theta = \frac{1}{3}$, since $\sin\theta = \pm\sqrt{1-\cos^2\theta}$ and θ is acute, we have $\sin\theta = \sqrt{1-\frac{1}{9}}$,

15. If $\cos\theta = \frac{1}{3}$, since $\sin\theta = \pm\sqrt{1-\cos^2\theta}$ and θ is acute, we have $\sin\theta = \sqrt{1-\frac{1}{9}}$, that is $\sin\theta = \frac{2\sqrt{2}}{3}$.

17. If $\sec\theta = \frac{25}{24}$, since $\tan\theta = \pm\sqrt{\sec^2\theta - 1}$ and θ is acute, we get $\tan\theta = \sqrt{\frac{625}{576}-1}$, or $\tan\theta = \frac{7}{24}$ and $\cot\theta = \frac{24}{7}$.

19. If $\cot(90^\circ - \theta) = 8$, then $\tan\theta = 8$, since $\cot(90^\circ - \theta) = \tan\theta$ is an identity.

21. $\sin 30^\circ = \frac{1}{2}$

23. $\tan\frac{\pi}{3} = \sqrt{3}$

25. $\sec 30^\circ = \frac{2\sqrt{3}}{3}$

27. $\cot\frac{\pi}{6} = \sqrt{3}$

29. Given the value $\cos\frac{5\pi}{12} = \frac{1}{4}\left(\sqrt{6}-\sqrt{2}\right)$, we use the cofunction identity $\sin(\theta) = \cos\left(\frac{\pi}{2}-\theta\right)$, to obtain $\sin\left(\frac{\pi}{12}\right) = \cos\left(\frac{\pi}{2}-\frac{\pi}{12}\right)$, or $\sin\left(\frac{\pi}{12}\right) = \cos\left(\frac{5\pi}{12}\right)$, that is $\sin\left(\frac{\pi}{12}\right) = \frac{1}{4}\left(\sqrt{6}-\sqrt{2}\right)$.

33. $\sin 20^\circ \approx 0.3420$

35. $\cot 88.4^0 \approx 0.0279$

37. $\sec(20^\circ\ 42') \approx \sec(20.7^\circ)$

≈ 1.0690

39. $\cos\frac{\pi}{8} \approx 0.9238$

41. $\csc 1 \approx 1.1883$

43. The acute angle θ such that $\cos\theta = 0.2$, correct to four decimal places, is $\theta \approx 78.4630^{\circ}$.

45. The acute angle θ such that $\tan\theta = 4.36$, correct to four decimal places, is $\theta \approx 77.0821^{\circ}$.

47. The acute angle θ such that $\csc\theta = 5.6$ (or, equivalently, such that $\sin\theta = \frac{1}{5.6}$), correct to four decimal places, is $\theta \approx 10.2865^{\circ}$.

49.

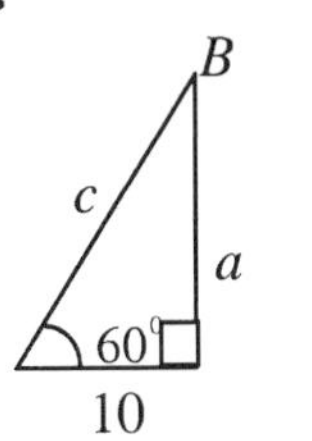

We have $\tan 60^{\circ} = a/10$, then $a = 10\sqrt{3}$.

$B = 30^{0}$; since $\sin B = \frac{10}{c}$, we get $\frac{1}{2} = \frac{10}{c}$, so $c = 20$.

51.

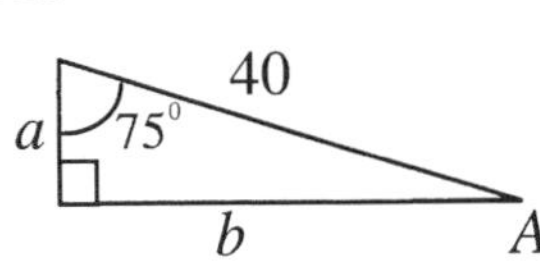

We have $\sin 75^{\circ} = \frac{b}{40}$, therefore $b = 40\sin 75^{\circ}$, or $b \approx 38.64$. Also, $\cos 75^{\circ} = \frac{a}{40}$, so $a = 40\cos 75^{\circ}$, or $a \approx 10.35$. A measures 15°.

53.

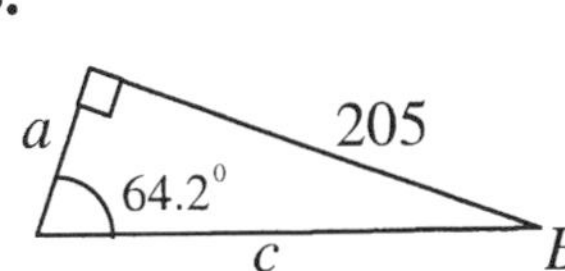

We have $\sin 64.2^{\circ} = \frac{205}{c}$, then $c = \frac{205}{\sin 64.2^{\circ}}$, or $c \approx 227.7$. Also, $\cos 64.2^{\circ} = \frac{a}{205}$, so $a = 205\cos 64.2^{\circ}$, or $a \approx 89.22$.

B measures 25.8°.

55.

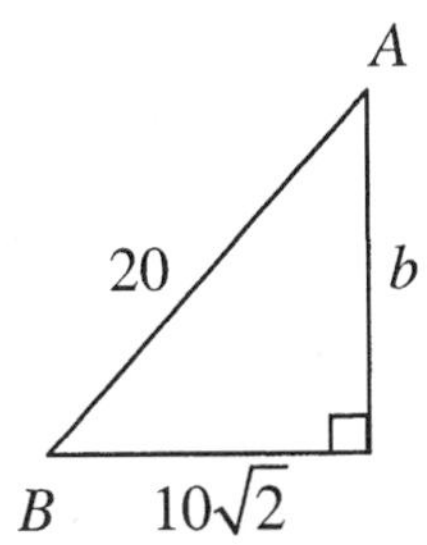

According to the Pythagorean Theorem, $\left(10\sqrt{2}\right)^{2} + b^{2} = 20^{2}$, that is $200 + b^{2} = 400$. Then $b^{2} = 200$, so $b = 10\sqrt{2}$ (the negative square root of 200 is not relevant here.)

The angles A and B have the same measure, 45°.

57.

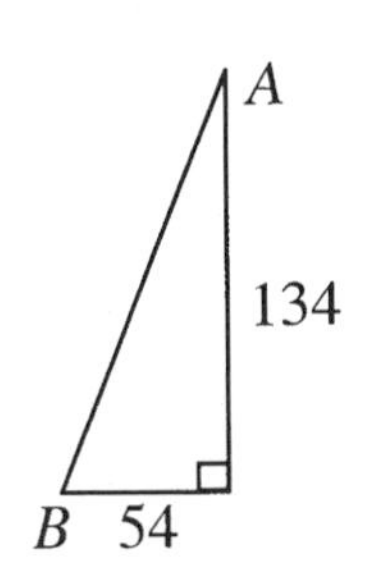

According to the Pythagorean Theorem, $54^{2} + 134^{2} = c^{2}$, that is $c^{2} = 20872$, so $c = \sqrt{20872}$, approximately 144.47 (the negative square root of 20872 is not relevant here.)

We have $\tan B = \frac{134}{54}$, and B is an acute angle, so $B \approx 68.05^{\circ}$, and $A \approx 21.95^{\circ}$

Applications

59. At a distance of 200 *ft* from the base of the tree, the angle of elevation to the top of the Montgomery State Reserve redwood tree, recorded as the tallest in the world, is 61.5°. Then, the height of the tree is $h = 200\tan 61.5^{\circ}$, approximately $h \approx 368.35$ *ft*.

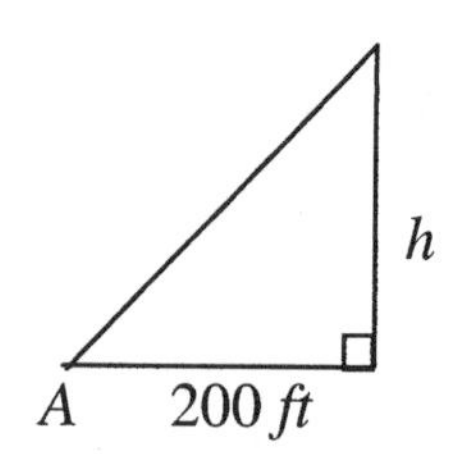

61. The horizontal distance x from a plane flying at a height of 500 *ft* to a fire in the distance if the angle of depression A is 12° satisfies $\tan 12^{\circ} = 500/x$. Then $x = 500/\tan 12^{\circ}$, approximately $x \approx 2{,}352.32$ *ft*.

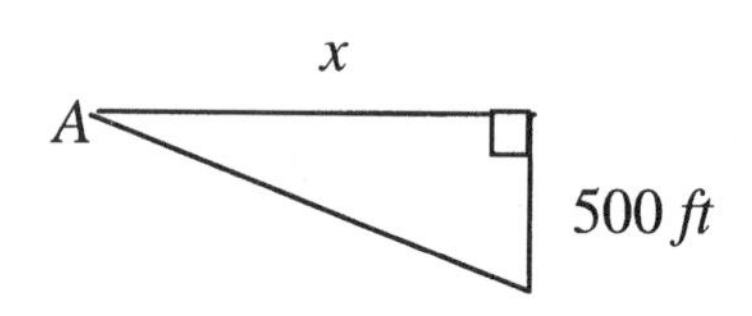

63. If the diagonal of a rectangular plot of land measures 225 *ft*, and the angle that it makes with one of the sides b is 20°, then b satisfies $\cos 20^{\circ} = b/225$, so $b = 225\cos 20^{\circ}$, or $b \approx 211.43$ *ft*. The other side, a, measures $a = 225\sin 20^{\circ}$, approximately $a \approx 76.95$ *ft*. The area of this rectangular plot is, approximately, $A = 16{,}269.54\ ft^2$.

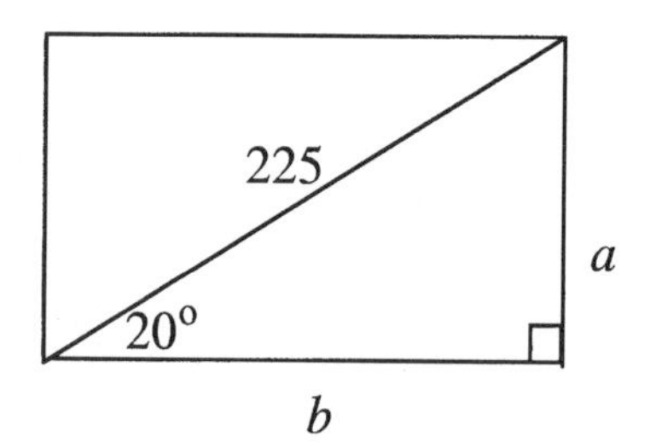

65. If the angle of elevation A from a point 5 feet off the ground to the top of a 7-foot golf flag is 1.2°, then the distance x to the flag is $x = 2/\tan 1.2^{\circ}$, approximately $x \approx 95.48$ *ft*.
$x \approx 95.48$ *ft*.

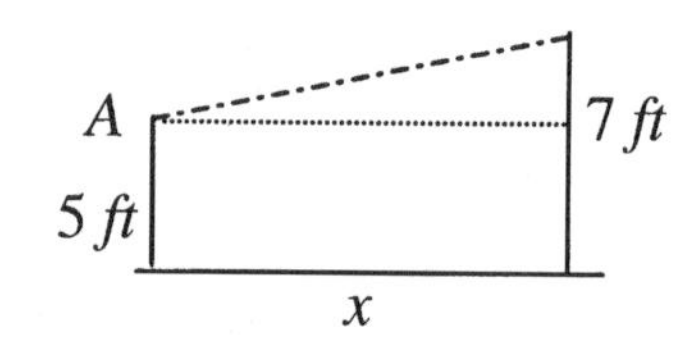

67. The base of a conical pile of grain has a circumference of 160 feet, and the angle made between the ground and the side of the pile is 35°. The radius of the base is $r = 160/(2\pi)$, so the height of the pile h is $h = r\tan 35^{\circ}$, or $h = 160\tan 35^{\circ}/(2\pi)$. The volume of the pile is given by $V = \frac{1}{3}\pi r^2 h$, so in our case $V = \frac{1}{3}\pi\left(\frac{160}{2\pi}\right)^2\left(\frac{160\tan 35^{\circ}}{2\pi}\right)$, approximately $V \approx 12{,}108\ ft^3$.

69. A hiker leaves a north-south highway, on a course 153° clockwise from north. After 3 miles, his distance x from the highway satisfies $\sin 27^{\circ} = x/3$. So, $x = 3\sin 27^{\circ}$, approximately $x \approx 1.36$ miles.

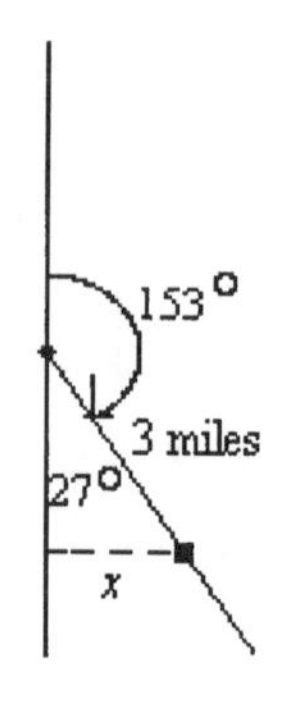

71. A plane is traveling on a course of $N\ 35^{\circ}\ E$, at a speed of Mach 2, or twice the speed of sound (1 Mach is 742 mph.) At 7:00 pm, 100 miles south from an east-west line, the distance l to this line along the course that the plane is flying satisfies $\cos 35^{\circ} = {}^{100}\!/_{l}$, so $l = {}^{100}\!/_{\cos 35^{\circ}}$, approximately $l \approx 122.08$ miles.

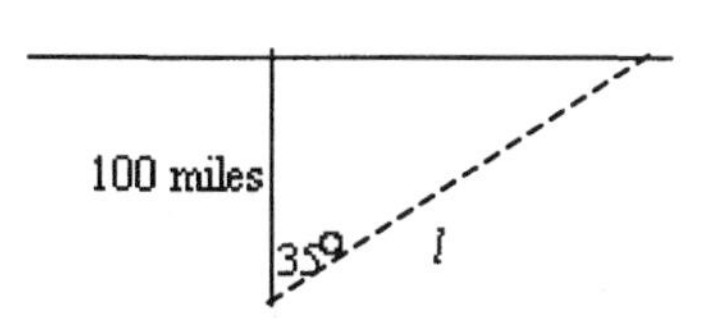

The time to travel this distance is $t \approx \dfrac{122.08}{2 \times 742}$, approximately $t \approx 0.0823$ hrs, or $t \approx 4.94$ min.

73. With an angle of elevation of the sun of $45^{\circ}\ 59'$, the length x of the shadow of a 6 feet tall individual would satisfy $\tan 45^{\circ}\ 59' = {}^{6}\!/_{x}$, so $x = {}^{6}\!/_{\tan 45^{\circ}\ 59'}$, approximately $x \approx 5.04$ feet. This is, of course, much shorter than the 8.7 feet long shadow on the photograph. For a shadow to be 8.7 feet long, the height of the individual should have been about 9 feet.

Concepts and Critical Thinkin

75. The statement: "The equation $\tan^2 x - \sec^2 x = 1$ is true for all acute angles x" is false. For example, if $x = 45^{\circ}$, we have $\tan 45^{\circ} = 1$ and $\sec 45^{\circ} = \sqrt{2}$, so $\tan^2 45^{\circ} - \sec^2 45^{\circ} = -1$.

77. The statement: "If both acute angle measures of a right triangle are given, then all side lengths can be found" is false. Any pair of similar right triangles have equal angle measures.

79. Answers may vary. An example of right triangle that has an angle of ${}^{\pi}\!/_{3}$ has sides ${}^{1}\!/_{2}$, ${}^{\sqrt{3}}\!/_{2}$ and hypotenuse 1. The other angle has measure ${}^{\pi}\!/_{6}$.

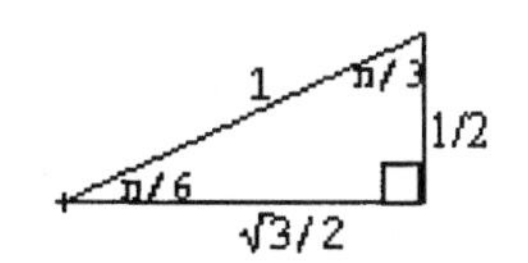

81. Answers may wildly vary. A navigational course for a hot-air balloon flying directly from New York to Cairo is to go from (New York: $40^{\circ} N$, $74^{\circ} W$) to (Cairo: $30^{\circ} N$, $27^{\circ} E$). The angle of depression from the East direction satisfies $\tan \alpha = {}^{(40-30)}\!/_{(74+27)}$, or $\alpha \approx 5.65^{\circ}$. A course that would take directly from New York to Africa is $N\ 96^{\circ}\ E$.

83. The value of $\tan \theta$ is obtained by dividing the length of the opposite side by the length of the adjacent side. If the opposite side is longer than the adjacent side, then $\tan \theta > 1$. The values of $\sin \theta$ and $\cos \theta$ may never exceed 1 because both sides are shorter than the hypotenuse.

Chapter 5. Section 5.3 Trigonometric Functions of Real Numbers

1. The point $(-1,0)$ is associated with the angle π in $[0,2\pi)$

$\sin\pi=0$ $\cos\pi=-1$ $\tan\pi=0$

$\csc\pi$ is undefined $\sec\pi=-1$ $\cot\pi$ is undefined

3. The point $\left(\frac{1}{2},-\frac{\sqrt{3}}{2}\right)$ is associated with the angle $\frac{5\pi}{3}$ in $[0,2\pi)$

$\sin\frac{5\pi}{3}=-\frac{\sqrt{3}}{2}$ $\cos\frac{5\pi}{3}=\frac{1}{2}$ $\tan\frac{5\pi}{3}=-\sqrt{3}$

$\csc\frac{5\pi}{3}=-\frac{2\sqrt{3}}{3}$ $\sec\frac{5\pi}{3}=2$ $\cot\frac{5\pi}{3}=-\frac{\sqrt{3}}{3}$

5. The point $\left(-\frac{\sqrt{2}}{2},\frac{\sqrt{2}}{2}\right)$ is associated with the angle $\frac{3\pi}{4}$ in $[0,2\pi)$

$\sin\frac{3\pi}{4}=\frac{\sqrt{2}}{2}$ $\cos\frac{3\pi}{4}=-\frac{\sqrt{2}}{2}$ $\tan\frac{3\pi}{4}=-1$

$\csc\frac{3\pi}{4}=\sqrt{2}$ $\sec\frac{3\pi}{4}=-\sqrt{2}$ $\cot\frac{3\pi}{4}=-1$

7. The point on the unit circle associated with the angle $\theta=-90^{\circ}$ is $(0,-1)$

$\sin\left(-90^{\circ}\right)=-1$ $\cos\left(-90^{\circ}\right)=0$ $\tan\left(-90^{\circ}\right)$ is undefined

$\csc\left(-90^{\circ}\right)=-1$ $\sec\left(-90^{\circ}\right)$ is undefined $\cot\left(-90^{\circ}\right)=0$

9. The point on the unit circle associated with the angle $\theta=\frac{5\pi}{4}$ is $\left(-\frac{\sqrt{2}}{2},-\frac{\sqrt{2}}{2}\right)$

$\sin\frac{5\pi}{4}=-\frac{\sqrt{2}}{2}$ $\cos\frac{5\pi}{4}=-\frac{\sqrt{2}}{2}$ $\tan\frac{5\pi}{4}=1$

$\csc\frac{5\pi}{4}=-\sqrt{2}$ $\sec\frac{5\pi}{4}=-\sqrt{2}$ $\cot\frac{5\pi}{4}=1$

11. If $\sin\theta<0$ and $\cos\theta>0$, then θ lies in the fourth quadrant.

13. If $\sec\theta<0$ and $\tan\theta>0$, then θ lies in the third quadrant.

15. If $\cot\theta<0$ and $\csc\theta>0$, then θ lies in the second quadrant.

17. If $\sin\theta=\frac{1}{3}$ and $\tan\theta<0$, then $\cos\theta<0$, so $\cos\theta=-\sqrt{1-\frac{1}{9}}$, that is $\cos\theta=-\frac{2\sqrt{2}}{3}$.

19. If $\sec\theta=-3$ and $\sin\theta<0$, then $\cos\theta=-\frac{1}{3}$, $\sin\theta=-\sqrt{1-\frac{1}{9}}$, that is $\sin\theta=-\frac{2\sqrt{2}}{3}$ and

$\cot\theta=\dfrac{1}{2\sqrt{2}}$, or $\cot\theta=\dfrac{\sqrt{2}}{4}$.

21. If $\tan\theta=2$ and $\sec\theta<0$, then $\sec\theta=-\sqrt{1+4}$, that is $\sec\theta=-\sqrt{5}$; therefore $\cos\theta=-\frac{\sqrt{5}}{5}$.

	Angle θ	Reference Angle		Angle θ	Reference Angle
23.	$\theta = 120^\circ$	60°	**25.**	$\theta = -50^\circ$	50°
27.	$\theta = 5\pi/4$	$\pi/4$	**29.**	$\theta = -2$	$\pi - 2 \approx 1.14159$

31. The reference angle of $\theta = 210^\circ$ is 30°; $\sin 210^\circ = -1/2$.

33. The reference angle of $\theta = -30^\circ$ is 30°; $\cos(-30^\circ) = \sqrt{3}/2$.

35. The reference angle of $\theta = 5\pi/6$ is $\pi/6$; $\tan 5\pi/6 = -\sqrt{3}/3$.

37. The reference angle of $\theta = -3\pi/4$ is $\pi/4$; $\cot(-3\pi/4) = -1$.

39. The reference angle of $\theta = 765^\circ$ is 45°; $\csc 765^\circ = \sqrt{2}$.

41. $\sin 110^\circ \approx 0.9396$

43. $\sec 221.4^\circ \approx -1.3331$

45. $\cot(-2.3) \approx 0.8934$

47. If $f(t) = \cos t$

(a) $f(0) = 1$

(b) $f(-\pi/4) = \sqrt{2}/2$

(c) $f(2) \approx -0.4161$

49. If $f(t) = \csc t$

(a) $f(-2\pi/3) = -2\sqrt{3}/3$

(b) $f(\pi)$ is undefined

(c) $f(5.3) \approx -1.2015$

51. The domain of $f(x) = \cos x$ is the set of all real numbers; the range of f is $[-1, 1]$.

53. The domain of $f(x) = \csc x$ is the set of real numbers $x \neq \pi k$ for any integer k; the range of f is $(-\infty, -1] \cup [1, \infty)$.

Applications

55. For a javelin throw, ignoring air resistance, if the initial speed is v, the initial height is h, and the initial angle is θ, then the range in feet is $R = \frac{v\cos\theta}{32}\left(\sqrt{v^2\sin^2\theta + 64h} + v\sin\theta\right)$.

(a) If $\theta = 0^\circ$, then $R = 61.24$ *ft*;
if $\theta = 30^\circ$, then $R = 280.65$;
if $\theta = 90^\circ$, then $R = 0$ *ft*.

(b) For an initial speed of 100 $ft/\sec$, the maximum range is 318.44 feet, so the world record of 323.1 feet could not have been achieved with this initial speed.

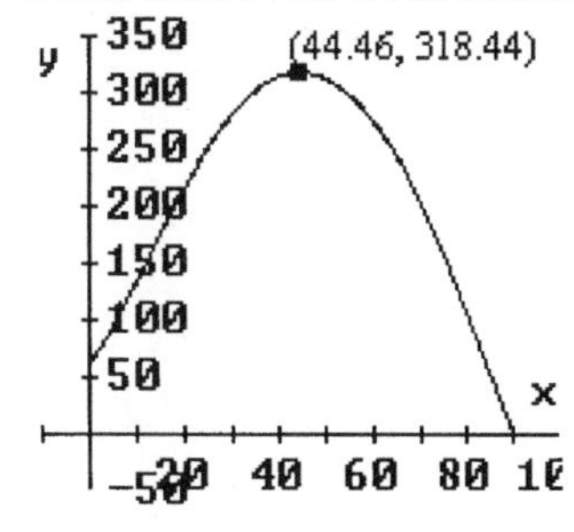

57. If the average monthly temperature for Orlando, Florida, is modeled by $f(t)=73-5.5\cos\left(\frac{\pi t}{6}\right)-13\sin\left(\frac{\pi t}{6}\right)$, where $t=1$ corresponding to January 2003, $t=2$ corresponding to February 2003, and so on, then:

(a) $f(3)=60$, and $f(15)=60$; that is, the average monthly temperature in March is 60°.

(b) The highest average monthly temperature is 87.11°, when $t\approx 8.23$ (August); the lowest average monthly temperature is 58.9° for $t\approx 2.24$ (February).

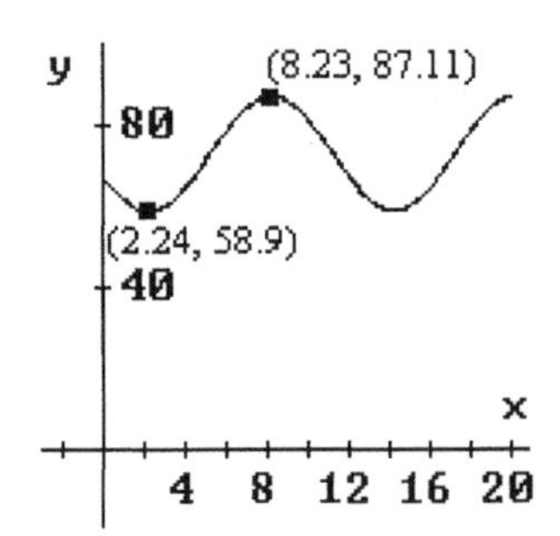

Concepts and Critical Thinking

59. The statement: "An angle whose sine and cosine are both negative is in the third quadrant" is true.

61. The statement: "The tangent function is undefined for all integer multiples of π" is false. At integer multiples of π, the value of the tangent function is 0.

63. Answers may vary. An example of a nonacute angle with reference angle 10° is 170°.

65. Answers may vary. An example of an angle whose secant is undefined is $\frac{\pi}{2}$.

67. The value $\sec\frac{\pi}{2}$ is undefined because $\sec x=\dfrac{1}{\cos x}$, and $\cos\frac{\pi}{2}=0$.

69. If α and β are two angles that have the same reference angle, then $\sin\alpha=\sin\beta$. The converse statement is also true, if $\sin\alpha=\sin\beta$, then α and β have the same reference angle.

Chapter 5. Section 5.4 Graphs of Sine and Cosine

1. Given the graph of a function:

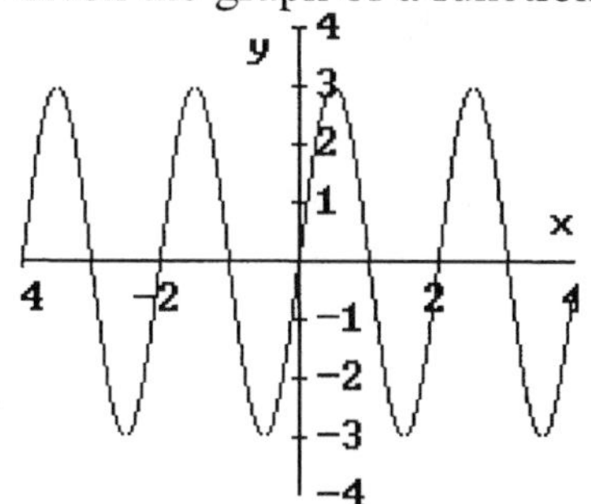

the period of the function is 2; the amplitude of the function is $\frac{3-(-3)}{2}=3$.

3. Given the graph of a function:

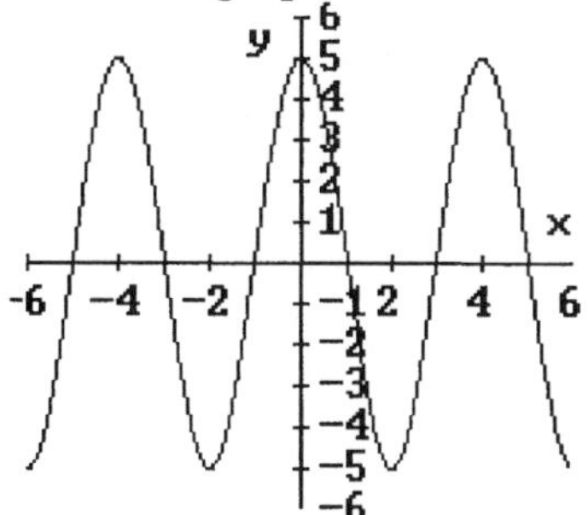

the period of the function is 4; the amplitude of the function is $\frac{5-(-5)}{2}=5$.

5. Given the graph of a function:

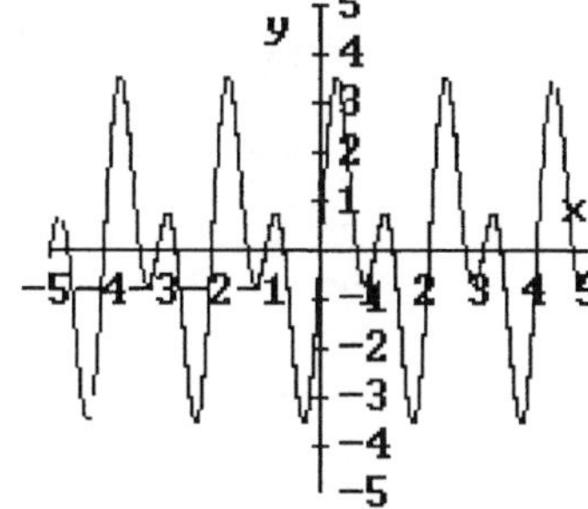

the period of the function is 2; the amplitude of the function is $\frac{3.5-(-3.5)}{2}=3.5$.

7. Given the graph of a function:

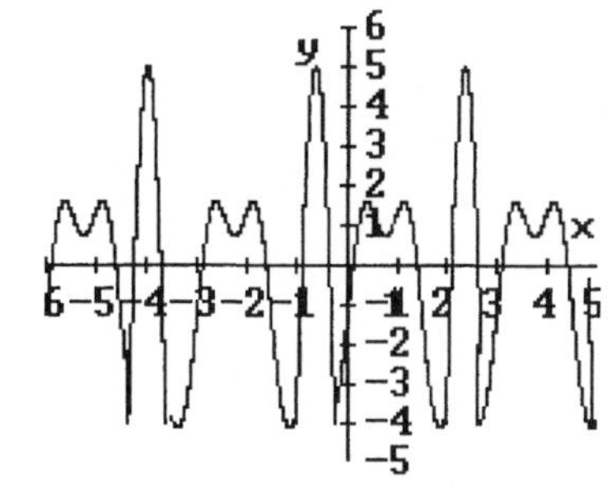

the period of the function is 3; the amplitude of the function is $\frac{5-(-4)}{2}=4.5$.

9. For the function $f(x)=5\sin 4x$

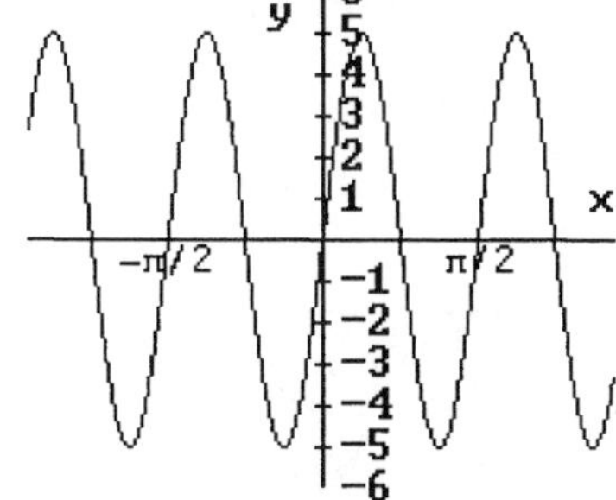

the period of f is ${}^{2\pi}/_{4}={}^{\pi}/_{2}$; the amplitude of the function is 5.

11. For the function $h(x)=\sin 3x-2\cos x$

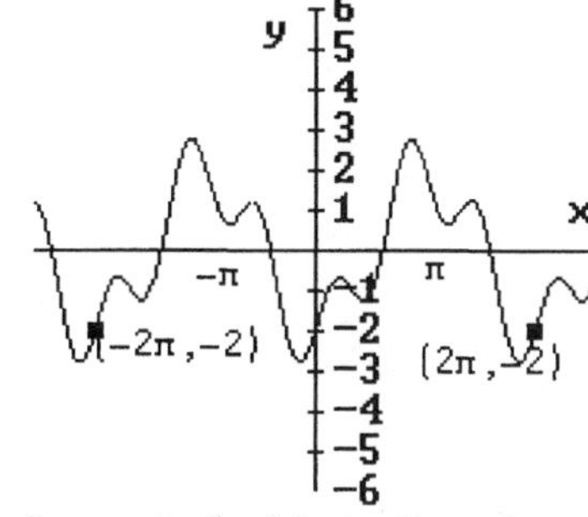

the period of h is 2π; the amplitude of the function is 2.8.

13. For the function $g(x) = 2^{\sin x} - 3^{\cos x}$

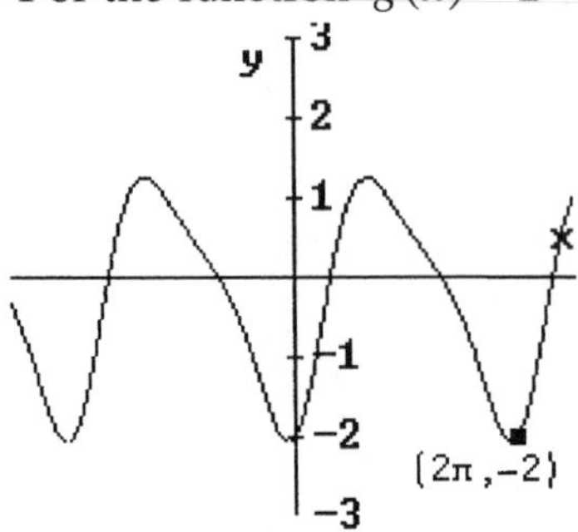

the period of g is 2π ; the amplitude of the function is $\approx \dfrac{1.247-(-2.064)}{2}$, or ≈ 1.66.

15. For the function $h(x) = \cos 2x$

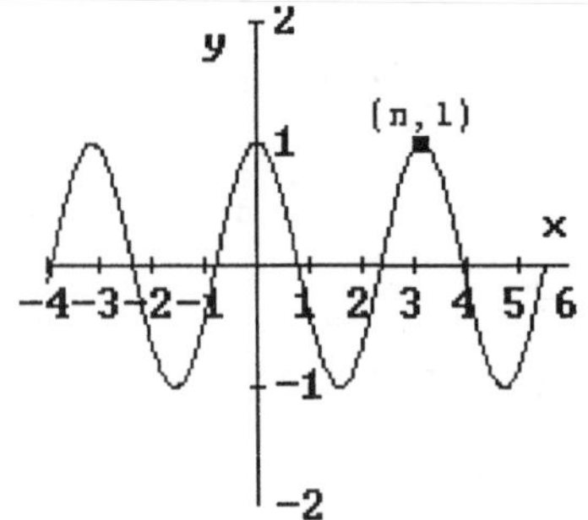

the period of g is π, the amplitude of the function is 1, and the phase shift is 0.

17. For the function $f(y) = 2\sin 2\pi y$

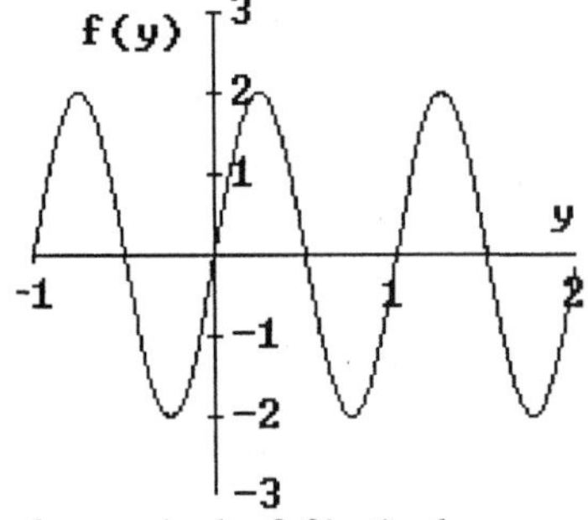

the period of f is 1, the amplitude of the function is 2, and the phase shift is 0.

19. For the function $p(t) = 3\cos(2t+1)$

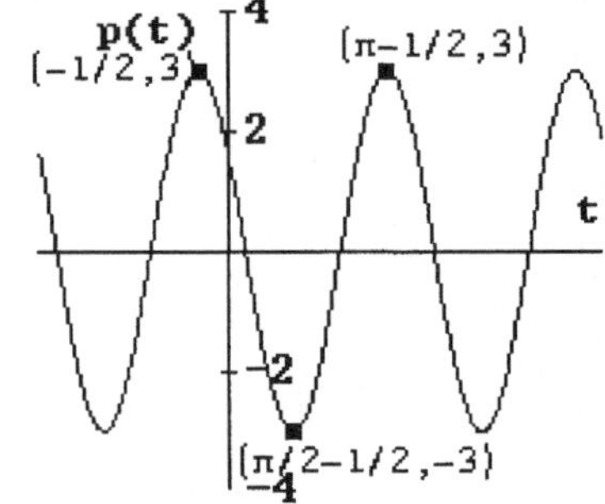

the period of p is π, the amplitude of the function is 3, and the phase shift is $-\frac{1}{2}$ ($\frac{1}{2}$ to the left) since $p(t) = 3\cos 2\left(t-\left(-\frac{1}{2}\right)\right)$

23. For the function $f(x) = 3\sin \pi x$

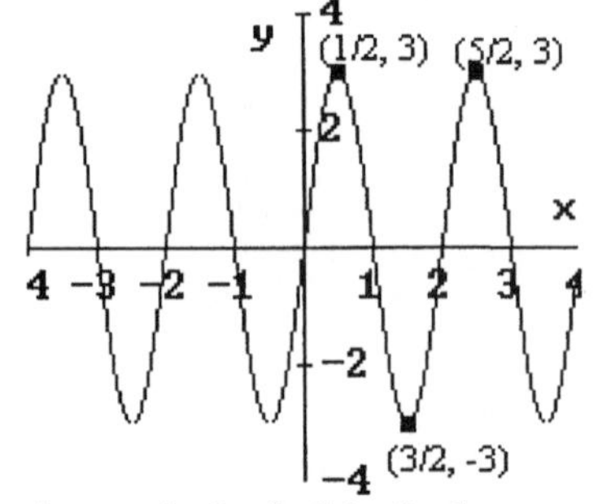

the period of f is 2, the amplitude of the function is 3, and the phase shift is 0. The maximum value of f is 3, and the minimum value of f is −3.

24. For the function $g(x) = 4\cos 2\pi x$

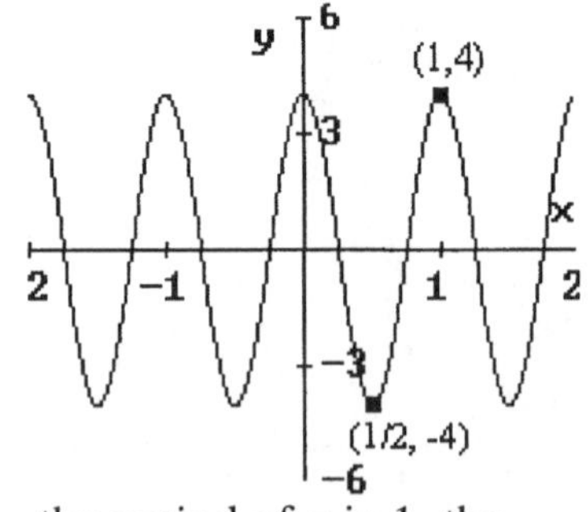

the period of g is 1, the amplitude of the function is 4, and the phase shift is 0. The maximum value of g is 4, and the minimum value of g is −4.

25. For the function $p(t) = -4\sin(2t+1)$

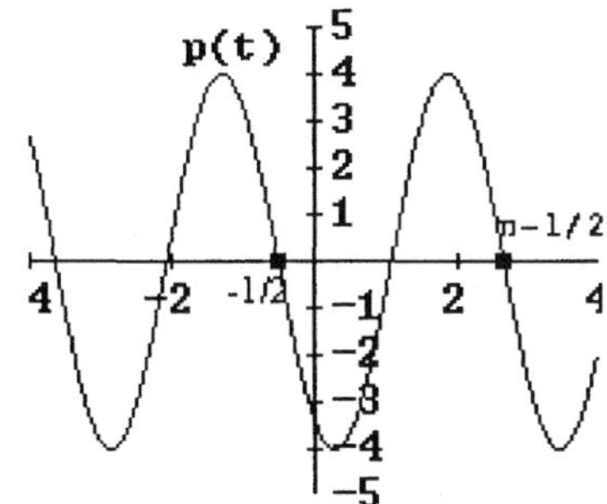

the period of f is π, the amplitude of the function is 4, and the phase shift is $-\frac{1}{2}$ ($\frac{1}{2}$ to the left) since $p(t) = -4\sin 2\left(t - \left(-\frac{1}{2}\right)\right)$. f has maximum value $= 4$, and minimum value $= -4$.

27. For the function $h(t) = 6\sin\left(2t + \frac{\pi}{2}\right)$

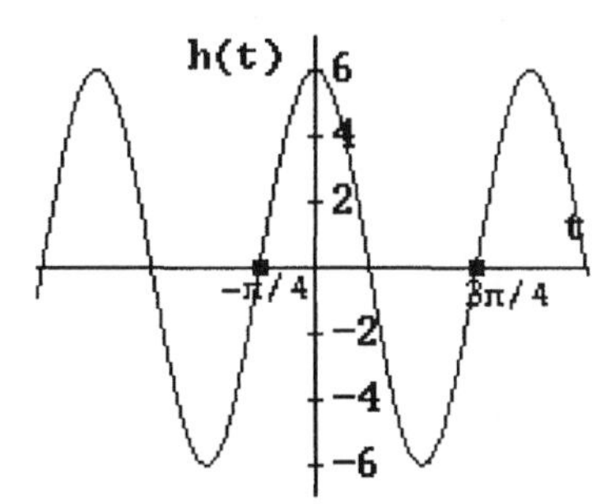

the period of h is π, the amplitude is 6; the phase shift is $-\frac{\pi}{4}$ ($\frac{\pi}{4}$ to the left) $\left(h(t) = 6\sin 2\left(t - \left(-\frac{\pi}{4}\right)\right)\right)$.
The maximum value of f is 6, and its minimum value -6.

29. For $g(y) = 5 - 2\cos(3y-2)$:

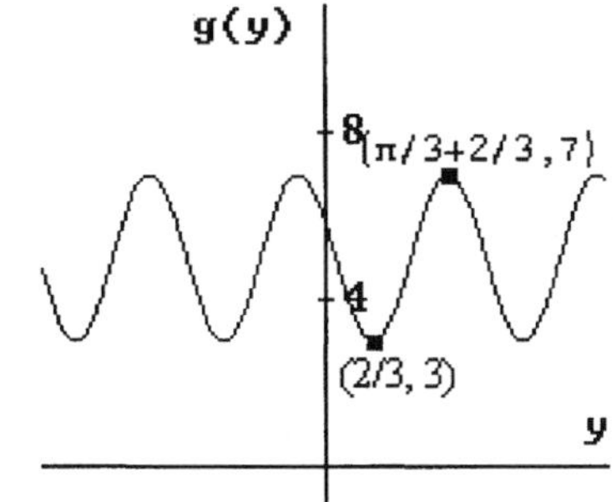

period $= \frac{2\pi}{3}$, amplitude $= 2$, phase shift $= \frac{2}{3}$; maximum value $= 7$, minimum value $= 3$.

31. The function $f(x) = \sin 2x$ is decreasing on the interval $\left(\frac{\pi}{4}, \frac{3\pi}{4}\right)$.

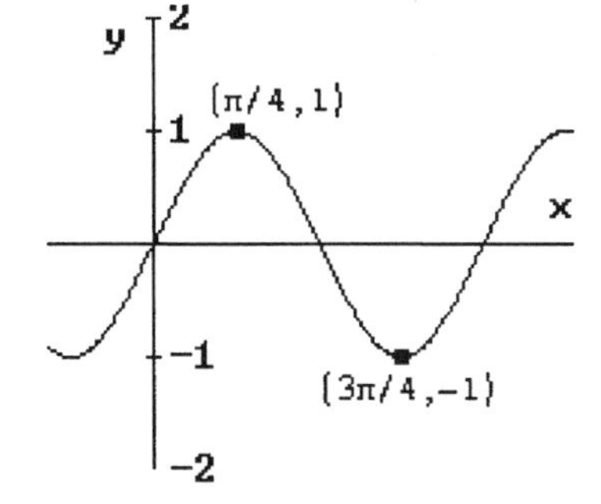

33. The function $h(x) = 2\cos(\pi x - \pi)$ changes direction at $x = 0$ on $\left(-\frac{1}{10}, \frac{1}{10}\right)$.

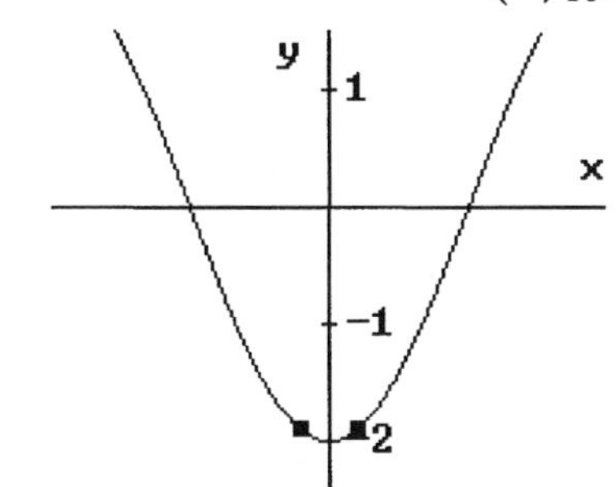

35. The function $g(x) = e^{\sin x}$ is decreasing on the interval $\left(0, \frac{\pi}{2}\right)$.

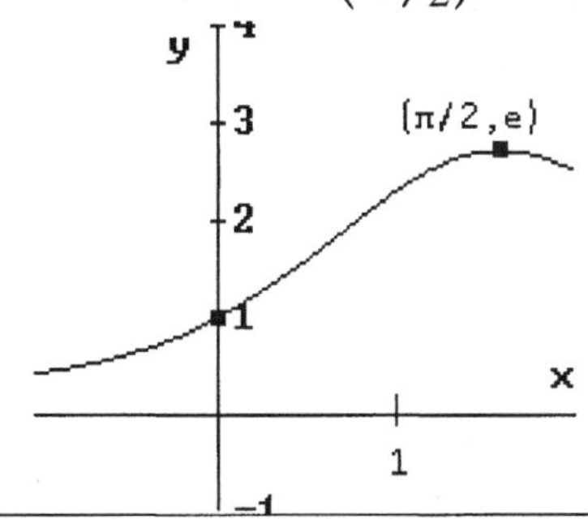

37. This graph corresponds to $f(x) = 2\sin \pi x$.

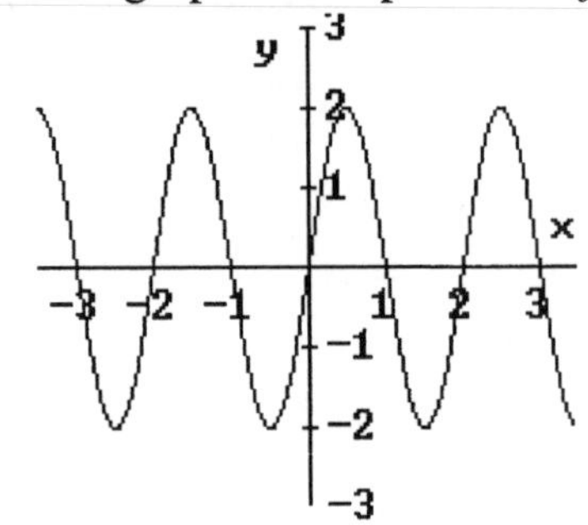

39. This graph corresponds to $f(x) = 2\sin\left(\frac{\pi}{2}x - \frac{\pi}{4}\right)$

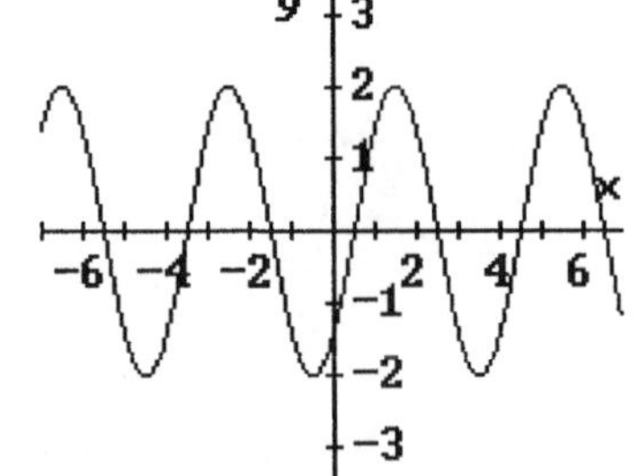

41. This graph corresponds to $f(x) = \frac{9}{4}\sin(\pi x - \pi)$

43. The equation $\sin(-x) = \sin x$ is not an identity. For example, $\sin\left(-\frac{\pi}{2}\right) = -1$ and $\sin\frac{\pi}{2} = 1$.

45. The equation $\sin 3x - 3\sin x = 4\sin^3 x$ is not an identity. For example, $\sin 3\left(\frac{\pi}{2}\right) - 3\sin\frac{\pi}{2} = -4$ while $4\sin^3\frac{\pi}{2} = 4$.

47. The equation $\sin\frac{x}{2}\cos\frac{x}{2} - \frac{\sin x}{2} = 0$ is an identity.

49. A function of the form $f(x) = a\sin(bx + c)$ with amplitude 3, period π, and phase shift $\frac{\pi}{3}$ is $f(x) = 3\sin(2x - \frac{2\pi}{3})$.

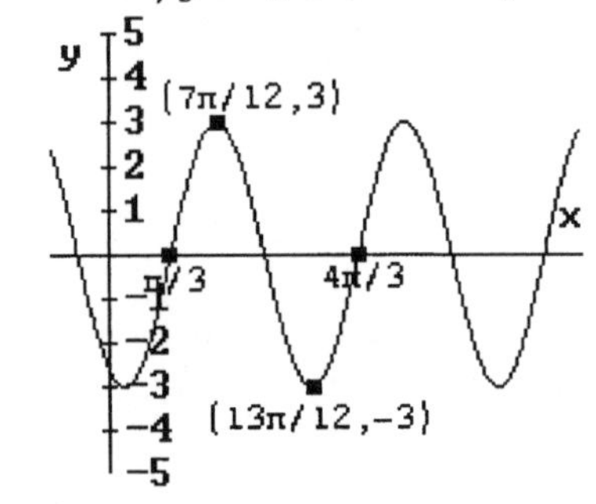

51. A function of the form $f(x) = a\sin(bx + c)$ with amplitude $\sqrt{2}$, period $\frac{\pi}{4}$, and phase shift $\frac{\pi}{8}$ is $f(x) = \sqrt{2}\sin(8x - \pi)$.

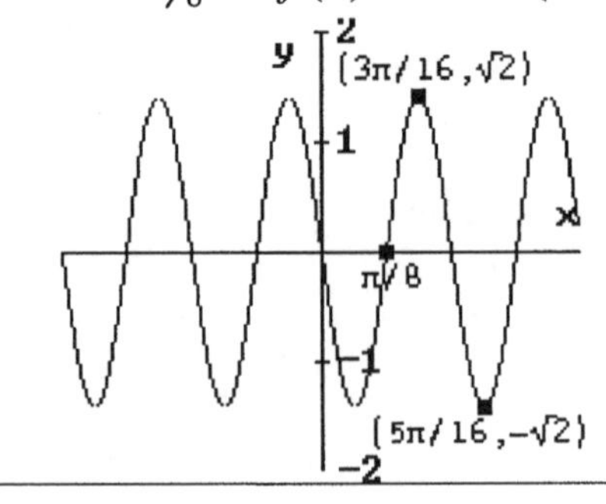

53. The zeros of the function $f(x) = 4\sin(3x+1)+2$ are the values of x for which $3x+1 = {}^{7\pi}\!/_6 + 2\pi k$, and $3x+1 = {}^{11\pi}\!/_6 + 2\pi k$. The smallest positive zeros of this function are $x = \dfrac{7\pi - 6}{18}$ and $x = \dfrac{11\pi - 6}{18}$ approximately $x \approx 0.89$ and $x \approx 1.59$.

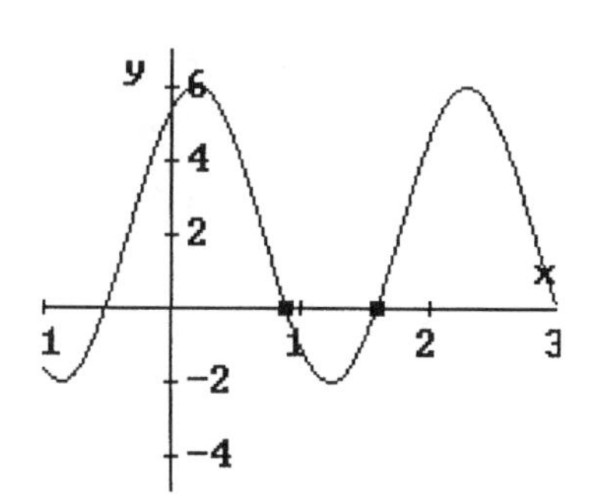

55. The smallest positive zeros of the function $f(x) = {}^4\!/_5 \cos 20x - {}^1\!/_2$ are, approximately $x \approx 0.1$ and $x \approx 0.26$.

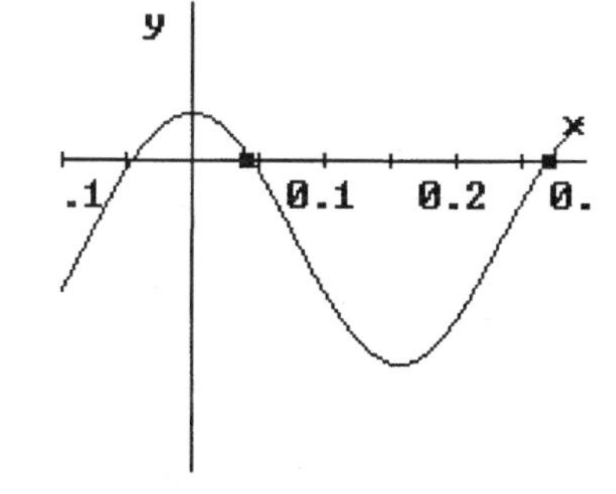

Applications

57. The wave created by a tuning fork producing a C on the musical scale, modeled by the function $f(x) = 0.002\sin 528\pi x$, has a period of $\dfrac{2\pi}{528\pi} = \dfrac{1}{264}$. The amplitude is 0.002.

59. If the displacement of a spring in centimeters from the equilibrium (resting) position t seconds after releasing the spring from a stretched position is approximated by $y(t) = 5\sin\left(4\pi t - {}^{\pi}\!/_2\right)$, then:

(a) the period of y is ${}^1\!/_2$ second

(b) the maximum displacement (from the rest position) is 5 cm (10 cm measured from the stretched position.

61. If the voltage of a 110-volt outlet varies according to the function $V(t) = 110\sqrt{2}\sin 120\pi t$, where t is in seconds (measured from the time the circuit is closed), then:

(a) the amplitude of V is $110\sqrt{2}$, or approximately 155.56 volts; the period of V is $\dfrac{2\pi}{120\pi}$ or ${}^1\!/_{60}$ seconds

(b) the first positive time when V is 110 volts is when $\sin 120\pi t = \dfrac{\sqrt{2}}{2}$, or $120\pi t = \dfrac{\pi}{4}$, that is $t = {}^1\!/_{480}$ seconds.

63. If the height in feet of a passenger on a Ferris wheel is approximated by the function $y(t) = 55 + 50\sin\left(\pi t/15 - \pi/2\right)$, where t is the time in seconds from the moment that the wheel is set in motion, then:

(a) The initial height of the passenger is $y(0) = 5$ feet

(b) The maximum height is 105 feet, and the minimum height is 5 feet. The maximum is attained when $\pi t/15 - \pi/2 = \pi/2$, or $t = 15$ seconds and the minimum when $t = 0,\ 30$ seconds

(c) One complete revolution corresponds to one period of y, $\dfrac{2\pi}{\left(\pi/15\right)} = 30$ seconds.

65. If the body temperature of a healthy individual is modeled by $f(t) = 0.8\cos\left(0.2618t + 1.5708\right) + 98.4$, where t is the time in hours after midnight, then the period of f is $\dfrac{2\pi}{0.2618} \approx 24$ hours. The minimum temperature is $98.4 - 0.8 = 97.6$ degrees Fahrenheit and it occurs when $0.2618t + 1.5708 = \pi$, that is $t \approx 6$ (this means 6 am.) The maximum temperature is $98.4 + 0.8 = 99.2$ degrees Fahrenheit, which occurs when $0.2618t + 1.5708 = 2\pi$, that is $t \approx 18$ (this means 6 pm.)

Concepts and Critical Thinking

67. The statement: "The graph of the cosine function is a translation of the graph of the sine function" is true.

69. The statement: "The amplitude of a function of the form $f(x) = a\sin(bx + c)$ equals the maximum value of $f(x)$" is true. The amplitude of such a function is $|a|$.

71. Answers may vary. An example of a function of the form $f(x) = a\sin\left(bx + c\right)$ whose period is the same as its amplitude is $f(x) = 2\sin\left(\pi x + 1\right)$.

73. Answers may vary. $f(x) = \sin x$ is a function that has π as one of its zeros.

75. The phase shift of the function $f(x) = \sin\left(2x - \pi/3\right)$ is $\pi/6$. $\left(\sin\left(2x - \pi/3\right) = \sin 2\left(x - \pi/6\right)\right)$.

Chapter 5. Section 5.5 Graphs of Other Trigonometric Functions

1. The graph of $y = \tan 2x$ is (vii):

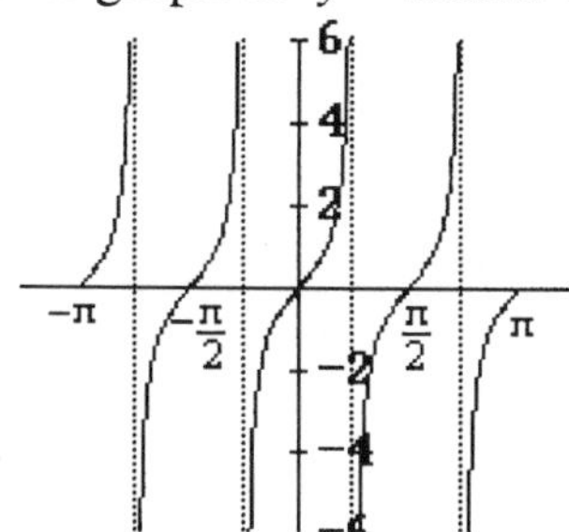

3. The graph of $y = \sec \frac{\pi x}{2}$ is (ii):

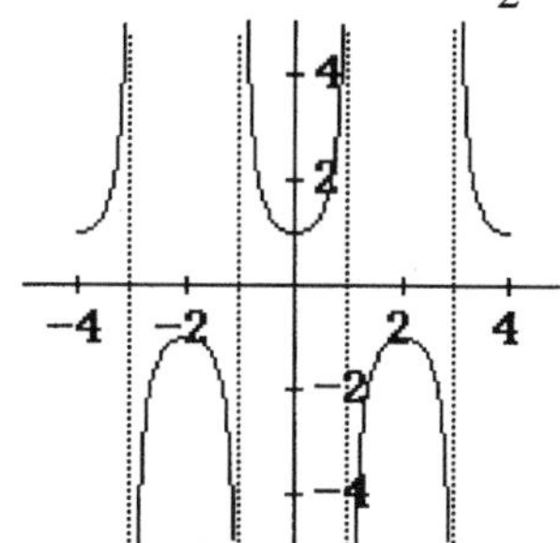

5. The graph of $y = \cot\left(\pi x - \frac{\pi}{2}\right)$ is (vi):

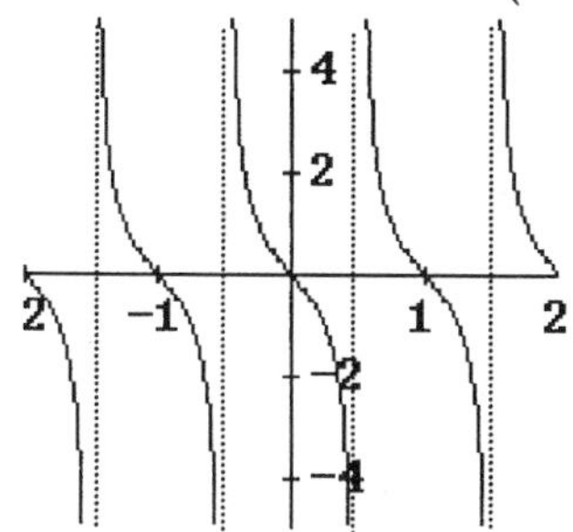

7. The graph of $y = \tan\left(2\pi x - \pi\right)$ is (v):

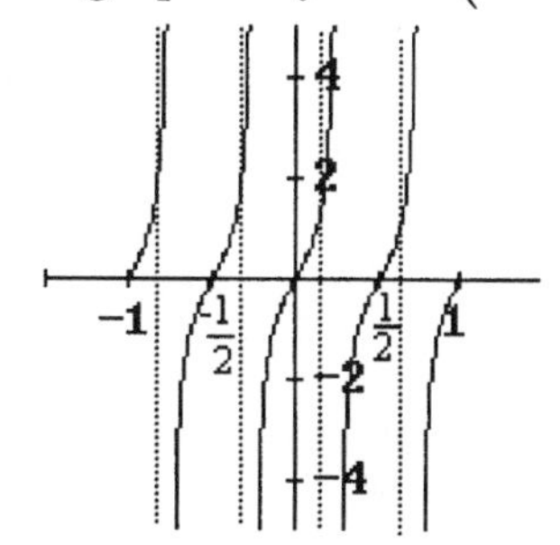

In the following exercises, k represents any integer value.

9. The graph of $f(x) = \tan 3x$ is:

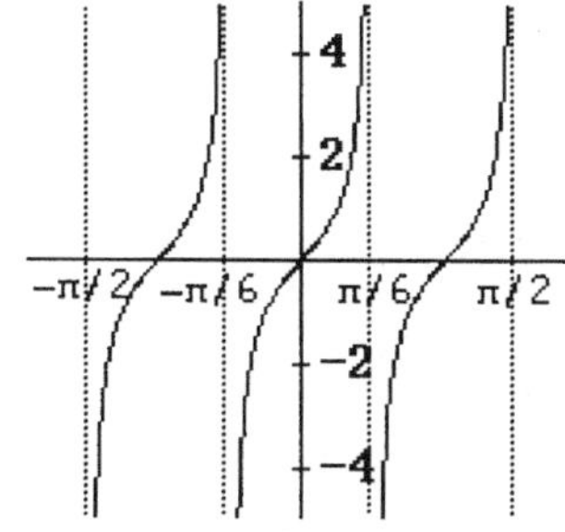

The period of this function is $\frac{\pi}{3}$, and its asymptotes are $x = \frac{\pi}{6} + \frac{\pi}{3}k$.

11. The graph of $f(x) = \cot \frac{\pi x}{2}$ is:

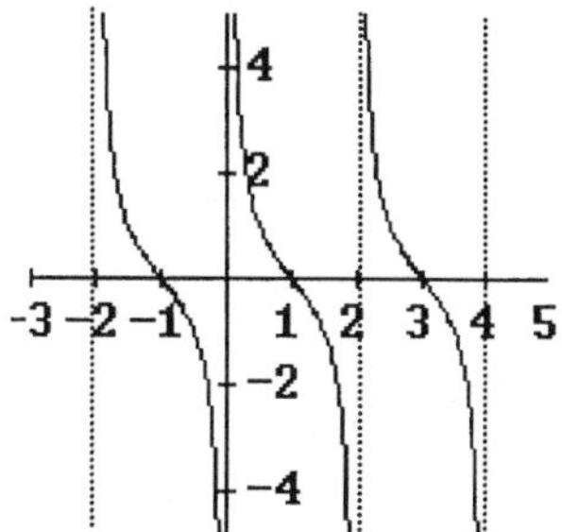

The period of this function is 2, and its asymptotes are $x = 2 + 2k$.

13. The graph of $f(x)=\tan\left(x-\frac{\pi}{2}\right)$ is:

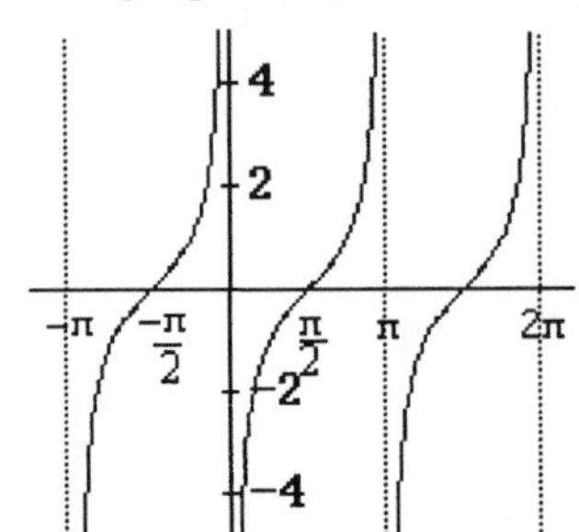

The period of this function is π, and its asymptotes are $x=\pi+\pi k$.

15. The graph of $f(x)=\cot\left(\pi x+\frac{\pi}{4}\right)$ is:

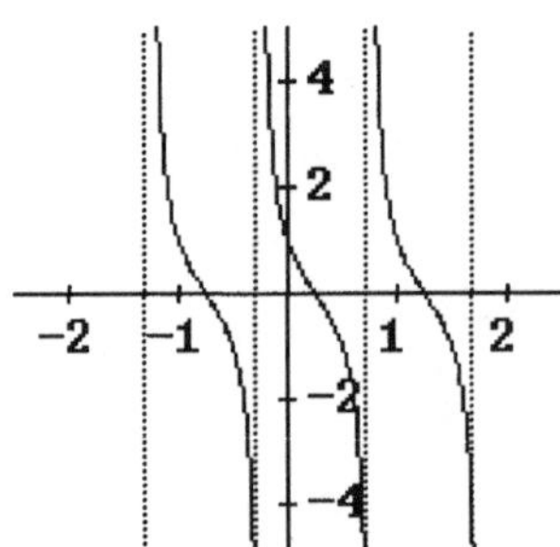

The period of this function is 1, and its asymptotes are $x=\frac{3}{4}+k$.

17. The graph of $f(x)=\sec 4x$ is:

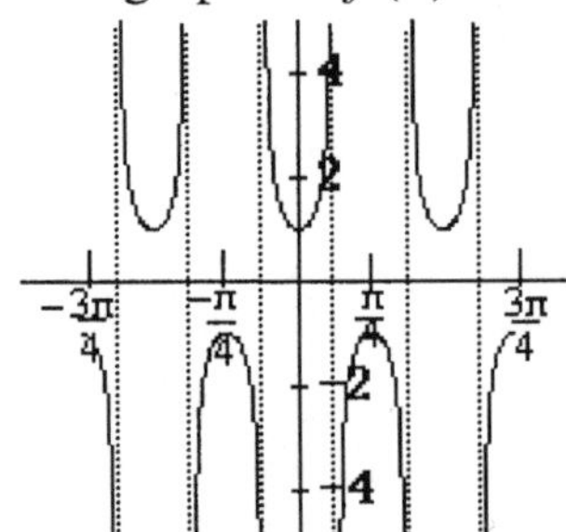

The period of this function is $\frac{\pi}{2}$, and its asymptotes are $x=\frac{\pi}{8}+\frac{\pi}{4}k$.

19. The graph of $f(x)=\sec(\pi x+\pi)$ is:

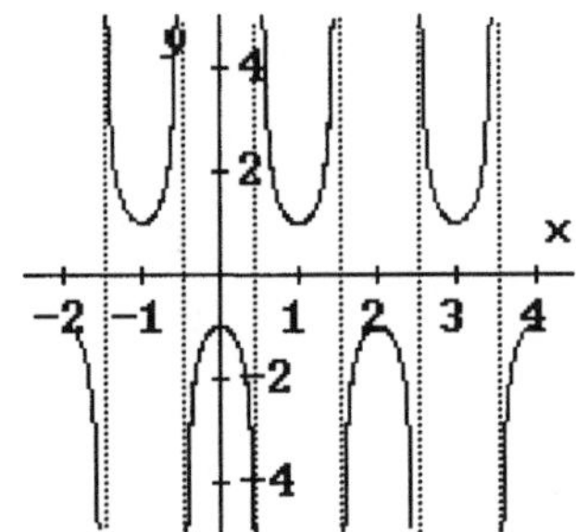

The period of this function is 2, and its asymptotes are $x=\frac{1}{2}+k$.

21. The graph of $f(x)=\csc\left(2x-\frac{\pi}{2}\right)$ is:

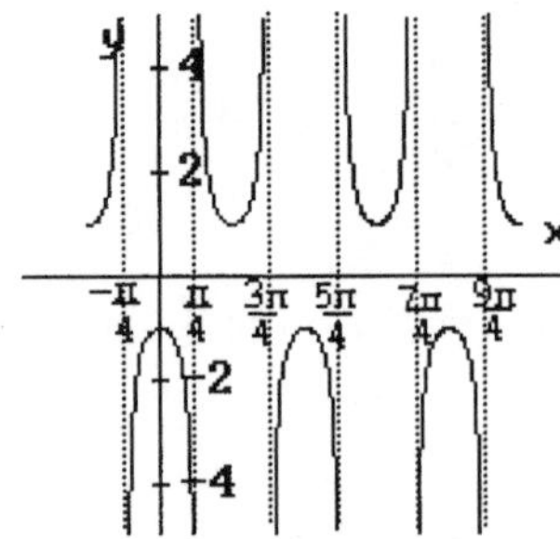

The period of this function is π, and its asymptotes are $x=\frac{\pi}{4}+\frac{\pi}{2}k$.

23. The smallest positive zero of the function $f(x)=\tan x+\tan 2x$ is, approximately, $x\approx 1.05$.

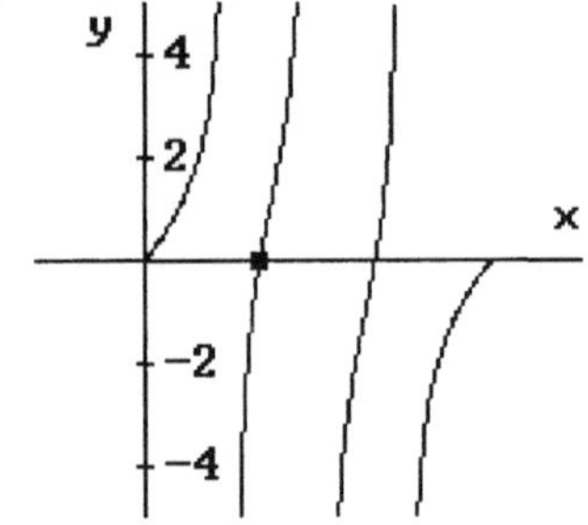

25. The function $f(x) = 1 - \tan x - \cot x$ has no zeros.

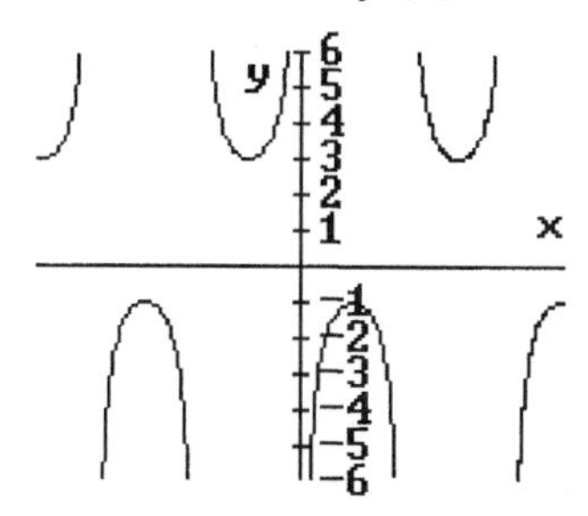

Applications

27. **(a)** If an observer is placed 5 miles away from the launch pad of a space shuttle, and θ is the angle of elevation from the observer to the shuttle, then the height h is related to θ by the equation $\tan\theta = \frac{h}{5}$, that is $h = 5\tan\theta$.

(b) The equation $h = 5\tan\theta$ is relevant only for the values $0 < \theta < \frac{\pi}{2}$

(c) The graph of h as a function of θ on the interval $0 < \theta < \frac{\pi}{2}$:

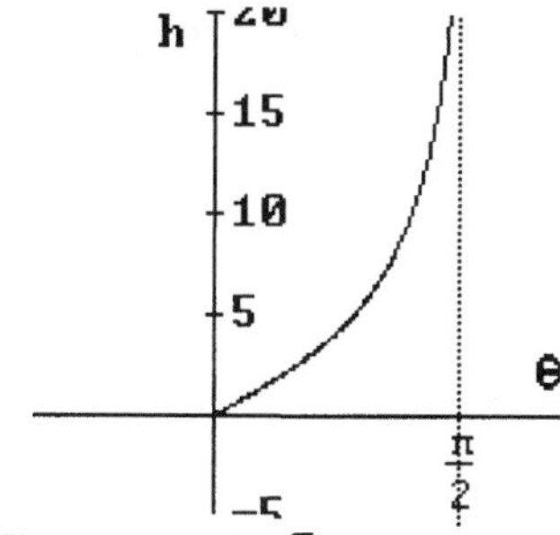

(d) Near $\theta = \frac{\pi}{2}$, the value of h increases without bounds.

Concepts and Critical Thinking

29. The statement: "The graph of $f(x) = \sec(bx + c)$ has an asymptote wherever the graph of $h(x) = \cos(bx + c)$ has an x-intercept" is true.

31. The statement: "Functions of the type $f(x) = \tan(bx + c)$ have infinitely many vertical asymptotes" is true. (Of course, assuming the function on its natural domain, the set of values x such that $bx + c \neq \pm\frac{\pi}{2} + \pi k$.)

33. Answers may vary. A trigonometric function with range $(-\infty, -1] \cup [1, \infty)$ is $f(x) = \sec x$.

35. Answers may vary. An example of a trigonometric function with vertical asymptotes at $x = 0$ and $x = 2$, but none between is $f(x) = \cot(\pi x)$.

37. The secant and cosecant function have no x-intercepts since $\sec x = \dfrac{1}{\cos x}$ is never 0, and $\csc x = \dfrac{1}{\sin x}$ is never 0 either.

Chapter 5. Section 5.6 Inverse Trigonometric Functions

1. $\arcsin 1 = \dfrac{\pi}{2}$

3. $\cos^{-1}\left(-\dfrac{1}{2}\right) = \dfrac{2\pi}{3}$

5. $\tan^{-1} 1 = \dfrac{\pi}{4}$

7. $\arctan \sqrt{3} = \dfrac{\pi}{3}$

9. $\sin^{-1}\left(-\dfrac{\sqrt{2}}{2}\right) = -\dfrac{\pi}{4}$

11. $\arccos(-1) = \pi$

13. $\tan^{-1} 0 = 0$

15. $\arcsin \pi$ is not defined

17. The solutions of $\sin x = 1/3$ in $[0, 2\pi)$ are $x = \sin^{-1}(1/3)$ and $x = \pi - \sin^{-1}(1/3)$, approximately $x \approx 0.34$, and $x \approx 2.80$.

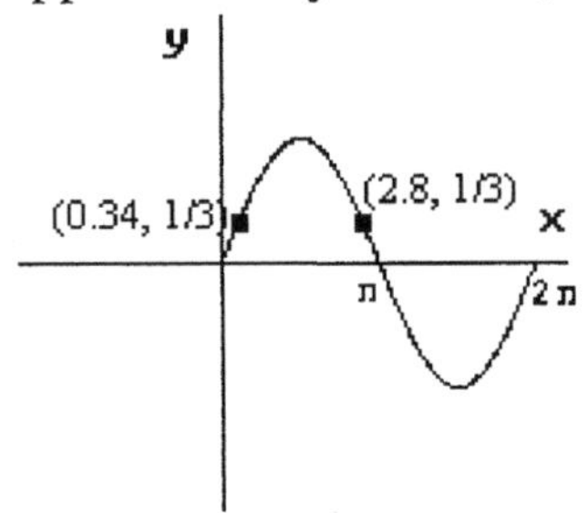

19. The solutions of $\tan x = 20$ in $[0, 2\pi)$ are $x = \tan^{-1} 20$ and $x = \tan^{-1} 20 + \pi$, approximately $x \approx 1.52$, and $x \approx 4.66$.

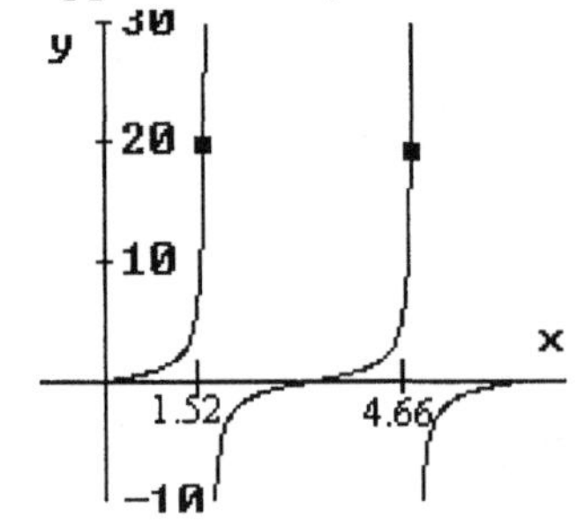

21. The solutions of $\cos x = 0.7$ in $[0, 2\pi)$ are $x = \cos^{-1} 0.7$ and $x = 2\pi - \cos^{-1} 0.7$, approximately $x \approx 0.8$, and $x \approx 5.49$.

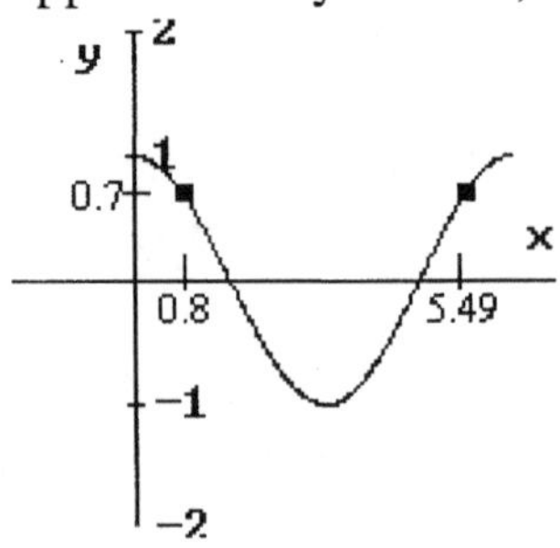

23. The equation $\sin x = \frac{7}{6}$ has no solutions.

25. $\sin\left(\arcsin \frac{1}{3}\right) = \frac{1}{3}$

27. $\arctan\left[\tan\left(-\frac{2\pi}{5}\right)\right] = -\frac{2\pi}{5}$

29. $\tan\left(\tan^{-1}(-7)\right) = -7$

31. $\arccos\left[\cos\left(-\frac{3\pi}{4}\right)\right] = \frac{3\pi}{4}$, since

$$\arccos\left[\cos\left(-\frac{3\pi}{4}\right)\right] = \arccos\left(-\frac{\sqrt{2}}{2}\right) = \frac{3\pi}{4}$$

33. $\sin^{-1}\left(\sin\frac{3\pi}{2}\right) = -\frac{\pi}{2}$, since

$$\sin^{-1}\left(\sin\frac{3\pi}{2}\right) = \sin^{-1}(-1) = -\frac{\pi}{2}$$

35. $\sin\left(\arctan\frac{4}{3}\right) = \frac{4}{5}$

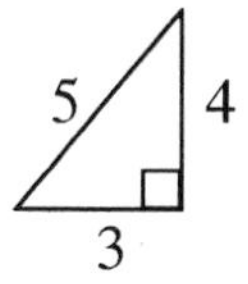

5, 4, 3

37. $\sec\left[\cos^{-1}\left(-\frac{1}{8}\right)\right] = -8$, since

$$\sec\left[\cos^{-1}\left(-\frac{1}{8}\right)\right] = \frac{1}{\cos\left[\cos^{-1}\left(-\frac{1}{8}\right)\right]}$$

39. $\cot\left(\sin^{-1}\frac{2\sqrt{5}}{5}\right) = \frac{1}{2}$

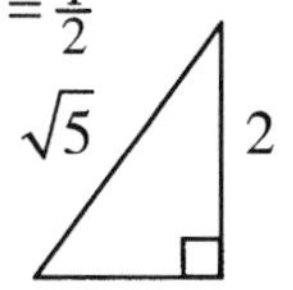

$\sqrt{5}$, 2

41. $\cos\left[\arctan\left(-\sqrt{6}\right)\right] = -\frac{\sqrt{7}}{7}$, as we now see: since $\cos\left[\arctan\left(-\sqrt{6}\right)\right] = \frac{1}{\sec\left[\arctan\left(-\sqrt{6}\right)\right]}$

Now, $\sec\left[\arctan\left(-\sqrt{6}\right)\right] = \pm\sqrt{1+\left(-\sqrt{6}\right)^2}$, where we have to choose the negative case, since $\arctan\left(-\sqrt{6}\right)$ is in the second quadrant. That is, $\cos\left[\arctan\left(-\sqrt{6}\right)\right] = \frac{1}{-\sqrt{7}}$, or $\cos\left[\arctan\left(-\sqrt{6}\right)\right] = -\frac{\sqrt{7}}{7}$.

In Exercises 43-47 it is assumed that $x > 0$.

43. $\tan(\arccos x) = \frac{\sqrt{1-x^2}}{x}$

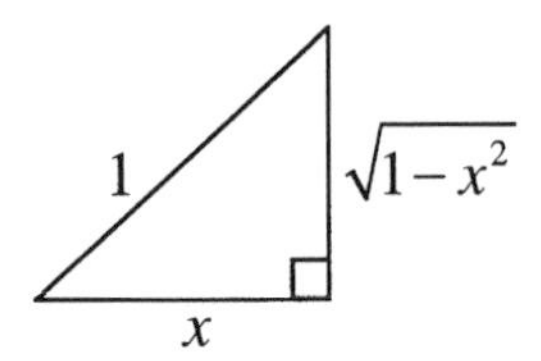

45. $\sec\left(\sin^{-1}\frac{2}{x}\right) = \frac{x}{\sqrt{x^2-4}}$

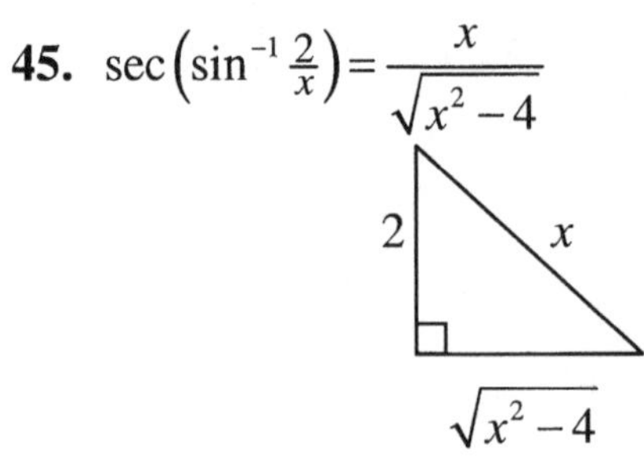

47. $\cos\left(\arctan\frac{\sqrt{x^2-4}}{2}\right) = \frac{2}{x}$

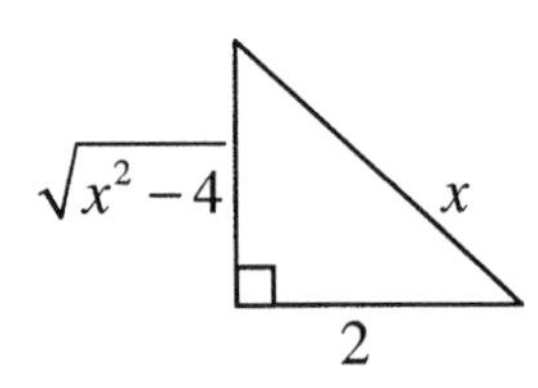

49. $y = \sin^{-1}(x-2)$

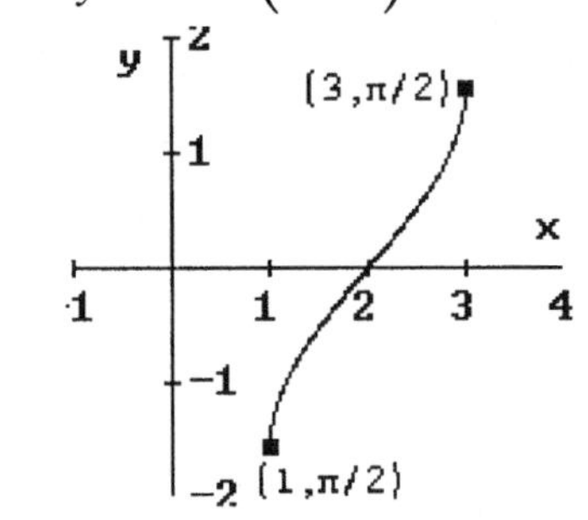

51. $y = \cos^{-1}(x) - \pi$

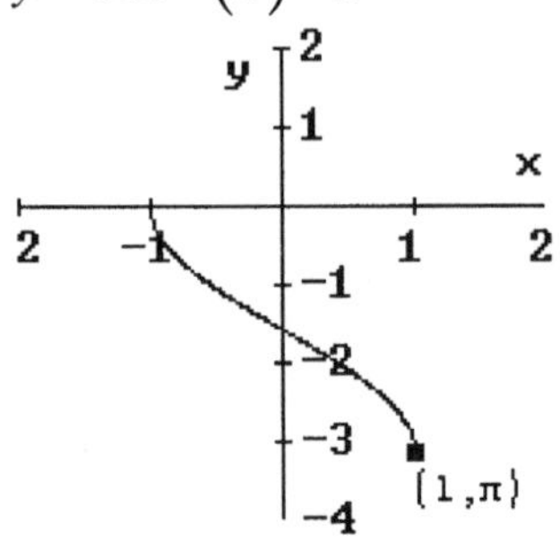

53. $y = \tan^{-1}(x-1) + \frac{\pi}{2}$

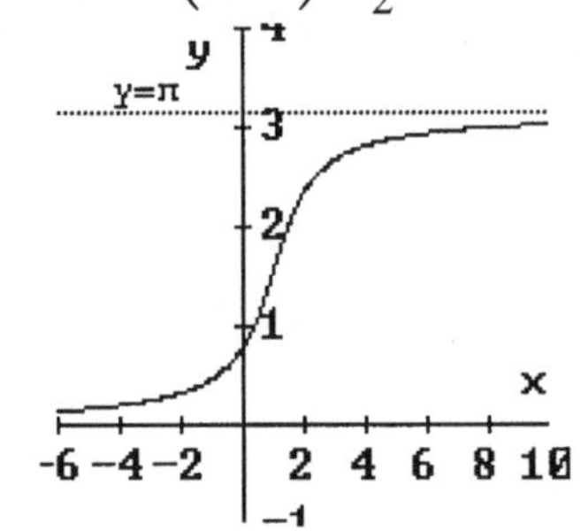

55. The solution of $\arcsin x = x + \frac{1}{2}$ is, approximately, $x \approx 1$.

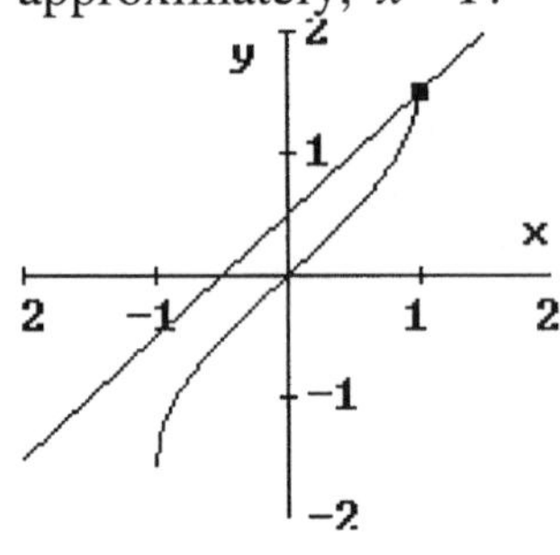

57. The solutions of $\arctan x = 1 - x^2$ are, approximately, $x \approx -1.4$ and $x \approx 0.65$.

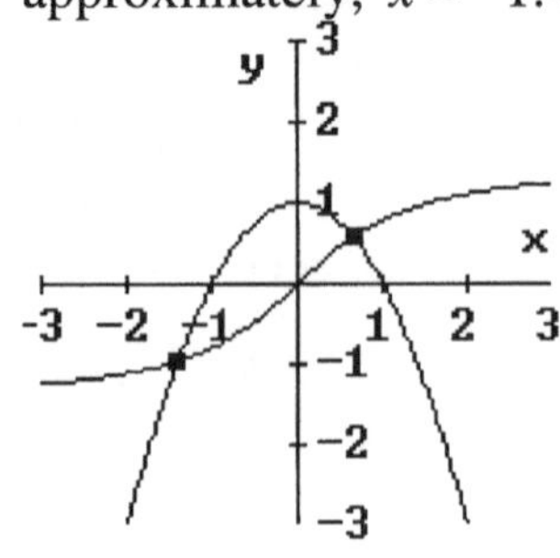

59. The equation $\arcsin^2 x + \arccos^2 x = 1$ is not an identity. It is a contradiction.

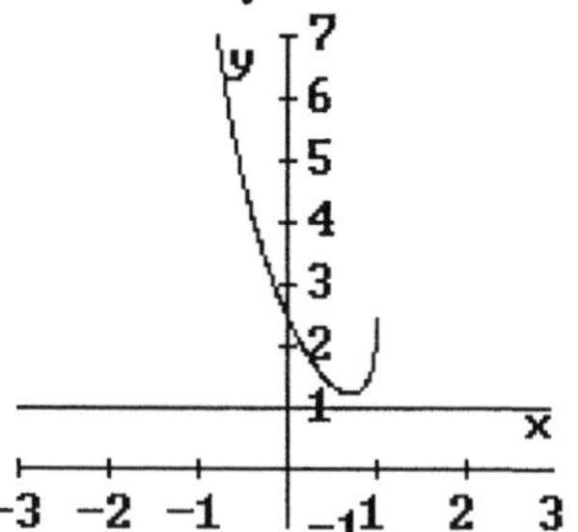

61. The equation $\cos\left(2\cos^{-1} x\right) = 2x^2 - 1$ is an identity.

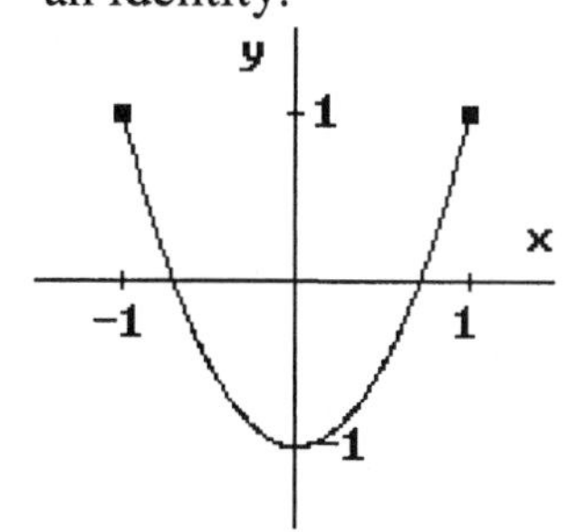

Applications

63. If a hot-air balloon rises straight into the air at a constant rate of 6 feet per second, then:
(a) the height of the balloon after 8 seconds is 48 feet;
(b) the angle of elevation of the balloon after 8 seconds of its release from an observer on the ground at a horizontal distance of 100 feet away from where the balloon is released is $\theta = \tan^{-1}\left(\frac{48}{100}\right)$, approximately $\theta \approx 25.64^0$;
(c) the angle of elevation of the balloon after t seconds of its release from an observer on the ground at a horizontal distance of 100 feet away from where the balloon is released is $\theta = \tan^{-1}\left(\frac{6t}{100}\right)$.

65. If a rescue plane is flying at a constant altitude of 2000 feet, then:
(a) the angle of depression from the plane to the boat when the horizontal distance between the plane and the boat is 1200 feet is $\theta = \tan^{-1}\frac{2000}{1200}$, approximately $\theta \approx 59.03^0$;
(b) the angle of depression from the plane to the boat when the horizontal distance between the plane and the boat is x feet is $\theta = \tan^{-1}\frac{2000}{x}$;
(c) if the pilot is scanning ahead with an angle of depression of 60^0, and x is the horizontal distance to the point being spotted, then $\tan 60^0 = \frac{2000}{x}$, so $x = \frac{2000}{\tan 60^0}$, approximately $x \approx 1154.7$ feet.

67. If a solar panel is to be placed between two buildings that have heights of 70 and 30 feet, respectively, and that are 140 feet apart, the angle θ between the lines running from the panel to the top of each building, then $\theta = \pi - \tan^{-1}\left(\frac{30}{x}\right) - \tan^{-1}\left(\frac{70}{140-x}\right)$:

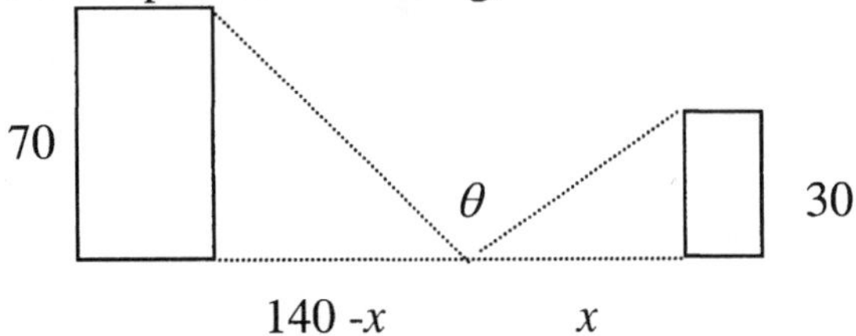

(a) for the distances $x = 20,\ 50,\ \text{and}\ 100$ feet:

x (feet)	20	50	100
θ (radians) (approximate)	1.63	1.94	1.8

(b) as the distance x increases, the angle first increases and then decreases, so it must reach a maximum;

(c) the largest value for the angle occurs when $x \approx 61.8$ feet; this is the optimal position, so the panel gets the maximum possible exposure to the sun.

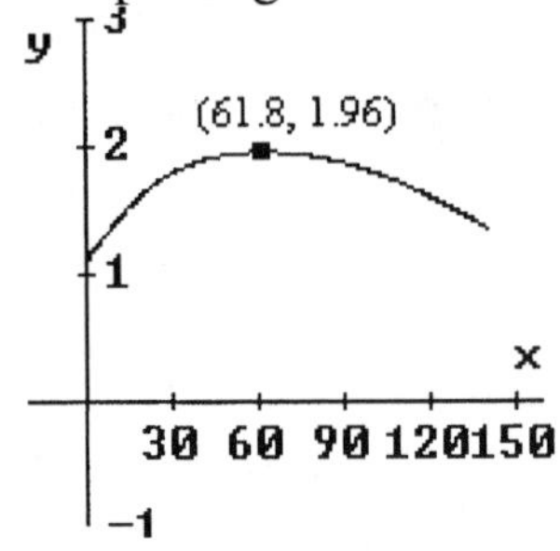

Concepts and Critical Thinking

69. The statement: "Given that $y = \arcsin x$ and $-1 \le x \le 1$, it must follow that $x = \sin y$" is true.

71. The statement: "$\cos\left(\cos^{-1} x\right) = x$ for all x" is true for all x in the domain of $\cos^{-1}$.

73. Answers may vary. If x is a value such that $|x| > 1$, then $\sin(\arcsin x)$ is undefined. However, if x is in the domain of arcsin, that is, $-1 \le x \le 1$, then $\sin(\arcsin x) = x$

75. Answers may vary. An example of an inverse trigonometric function that is increasing over its entire domain is the function defined by $f(x) = \arctan x$, for all real values x.

Chapter 5. Review

1. 60^0 corresponds to sketch (iv):

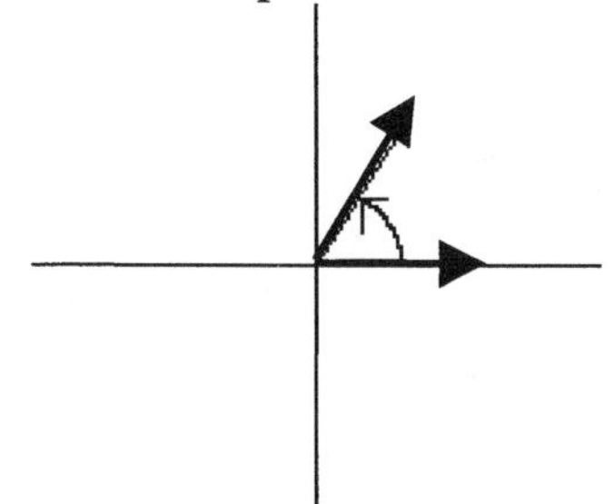

3. $-3\pi/2$ corresponds to sketch (ii):

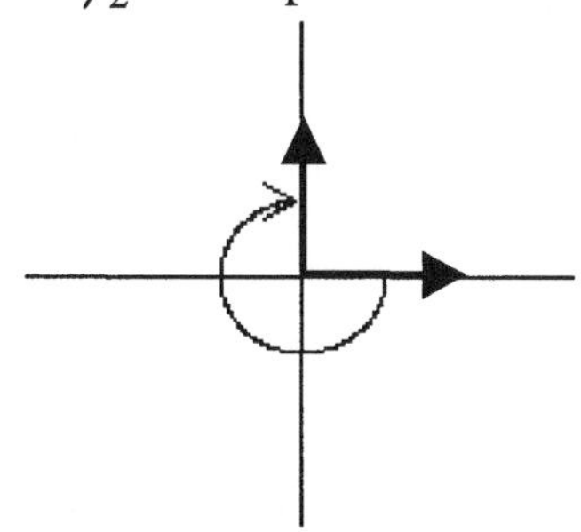

In Exercises 5-7, answers may vary.

	Angle θ	Positive Coterminal Angle for θ	Negative Coterminal Angle for θ
5.	60^0	420^0	-300^0
7.	$-3\pi/2$	$\pi/2$	$-7\pi/2$

	Angle θ	Radian Measure
9.	60^0	$\pi/3$
11.	-210^0	$-7\pi/6$

	Angle θ	Degree Measure
13.	$-5\pi/6$	-150^0
15.	$\pi/8$	22.5^0

17. The length of the arc subtended by a central angle $\theta = 2\pi/3$ in a circle of radius 6 is $s = 6(2\pi/3)$, or $s = 4\pi$, approximately $s \approx 12.57$.

In Exercises 19-21, a right triangle is given, and θ is one of its angles.

19.

$c = \sqrt{8^2 + 6^2}$, or $c = 10$;

$\sin\theta = 4/5 \quad \csc\theta = 5/4$

$\cos\theta = 3/5 \quad \sec\theta = 5/3$

$\tan\theta = 4/3 \quad \cot\theta = 3/4$

21.

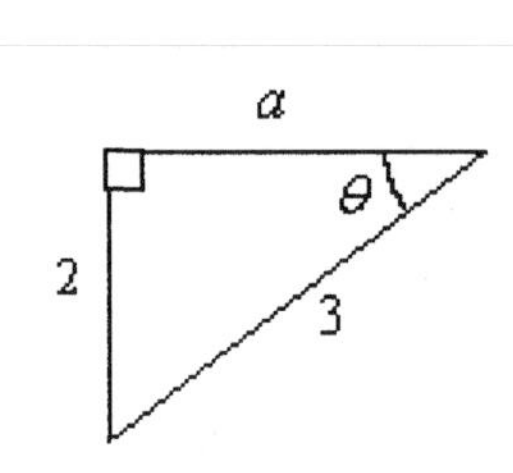

$a = \sqrt{3^2 - 2^2}$, or $a = \sqrt{5}$;

$\sin\theta = 2/3$ $\quad \csc\theta = 3/2$

$\cos\theta = \sqrt{5}/3$ $\quad \sec\theta = 3\sqrt{5}/5$

$\tan\theta = 2\sqrt{5}/5$ $\quad \cot\theta = \sqrt{5}/2$

23. If θ is an acute angle, and $\cos\theta = 15/17$, then θ may be seen as one of the angles in the right triangle of sides a, 15, 17.

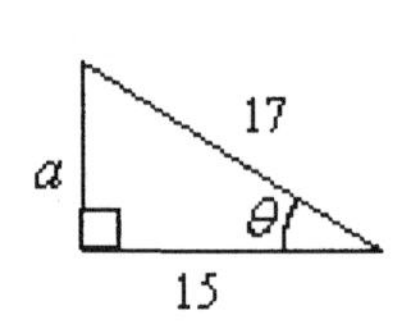

$a = \sqrt{17^2 - 15^2}$, that is $a = 8$.
Then: $\sin\theta = 8/17$, $\tan\theta = 8/15$,
$\cot\theta = 15/8$, $\sec\theta = 17/15$, $\csc\theta = 17/8$.

25. If θ is an acute angle, and $\csc\theta = 2$, then θ may be seen as one of the angles in the right triangle of sides 1, b, 2.

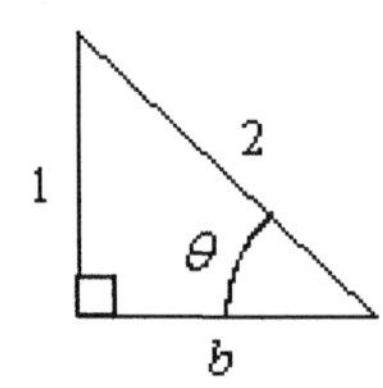

$b = \sqrt{2^2 - 1^2}$, that is $b = \sqrt{3}$.
Then: $\sin\theta = 1/2$, $\cos\theta = \sqrt{3}/2$,
$\tan\theta = \sqrt{3}/3$, $\cot\theta = \sqrt{3}$, $\sec\theta = 2\sqrt{3}/3$.

	Trigonometric Function	Approximate value of θ (if θ is an acute angle)
27.	$\tan\theta = 3.6$	$\tan^{-1} 3.6 \approx 74.4758^0$
29.	$\sin\theta = 0.32$	$\sin^{-1} 0.32 \approx 18.6629^0$

31. To solve the triangle:

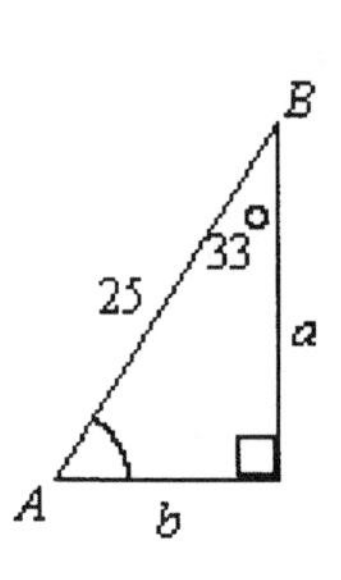

$A = 180^0 - (90^0 + 33^0)$, that is $A = 57^0$;

$a = 25\cos 33^0$, or $a \approx 20.97$;

$b = 25\sin 33^0$, that is $b \approx 13.62$.

33. To solve the triangle:

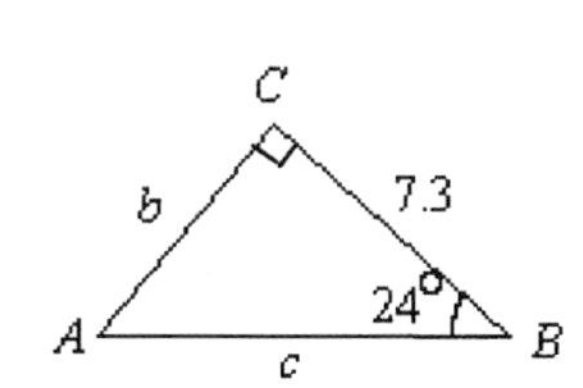

$A = 180^0 - (90^0 + 24^0)$, that is $A = 66^0$;

$\cos 24^0 = \dfrac{7.3}{c}$, so $c = \dfrac{7.3}{\cos 24^0}$, or $c \approx 7.99$;

$\tan 24^0 = \dfrac{b}{7.3}$, that is $b = 7.3 \tan 24^0$, that is $b \approx 3.25$.

	Angle	Reference Angle
35.	$\theta = 210^0$	30^0
37.	$\theta = -3\pi/4$	$\pi/4$

	Requested Value	Reference Angle	Quadrant	Answer for Requested Value
39.	$\sin 2\pi/3$	$\pi/3$	II	$\sin 2\pi/3 = \sqrt{3}/2$
41.	$\sec(-150^0)$	30^0	III	$\sec(-150^0) = -2\sqrt{3}/3$
43.	$\tan 5\pi/4$	$\pi/4$	III	$\tan 5\pi/4 = 1$

45. $\sin(-200^0) \approx 0.3420$

47. $\sec 7\pi/10 \approx -1.7013$

49. $\cot(125^0\ 10'\ 15'') \approx \dfrac{1}{\tan(125.1708)} \approx -0.7046$

51. If $\tan\theta = -2$, and $\cos\theta < 0$, then $\sin\theta > 0$. Now, $\sec^2\theta = 1 + \tan^2\theta$, therefore $\sec^2\theta = 5$ and since $\sec\theta < 0$ we have $\sec\theta = -\sqrt{5}$. Then, $\cos\theta = -\sqrt{5}/5$, and $\sin\theta = \sqrt{1 - 1/5}$, or $\sin\theta = 2\sqrt{5}/5$.

53. If $\cos\theta = -2/3$, and $\sin\theta > 0$, then $\sin\theta = \sqrt{1 - 4/9}$, that is $\sin\theta = \sqrt{5}/3$. Then, $\tan\theta = -\sqrt{5}/2$.

55. The function $f(x) = \sin \pi x$ corresponds to graph (vi).

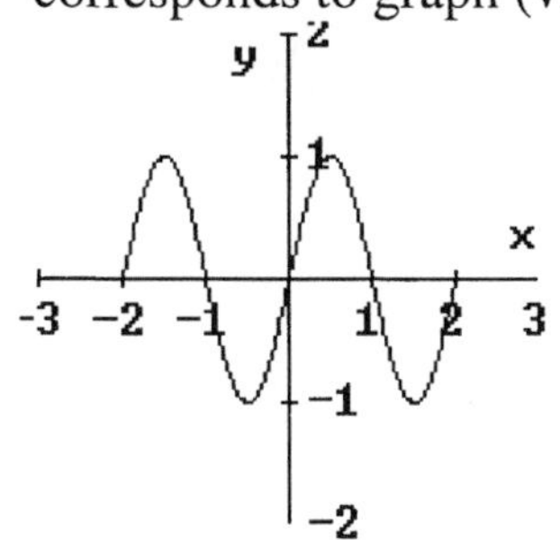

57. The function $f(x) = \cos 2x$ corresponds to graph (vii).

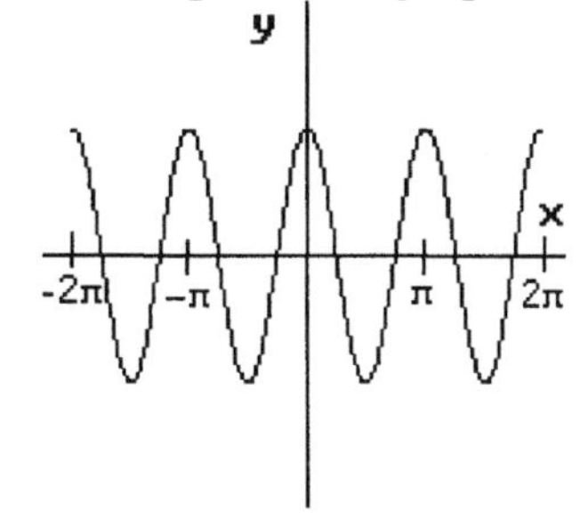

59. The function $f(x) = \tan\left(x + \pi/3\right)$ corresponds to graph (i).

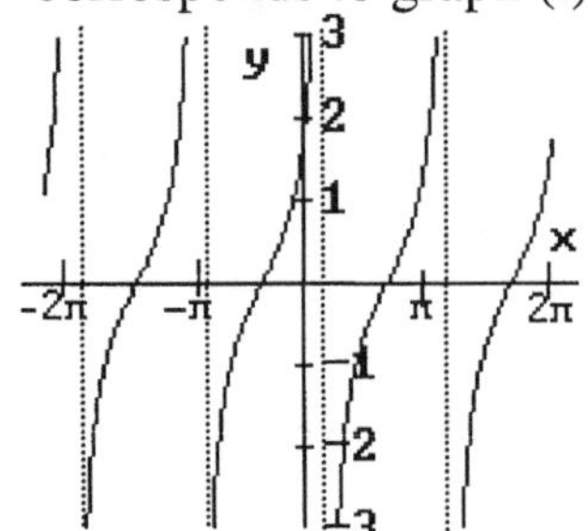

61. The function $f(x) = \sec x/2$ corresponds to graph (viii).

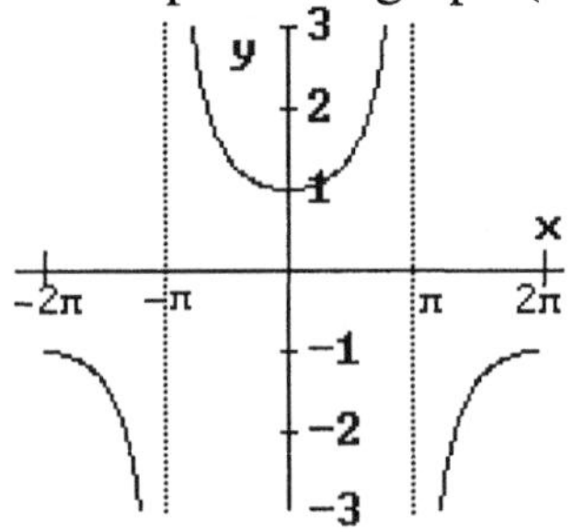

63. For the function whose graph is given below, the period is 4π and the amplitude is 3.

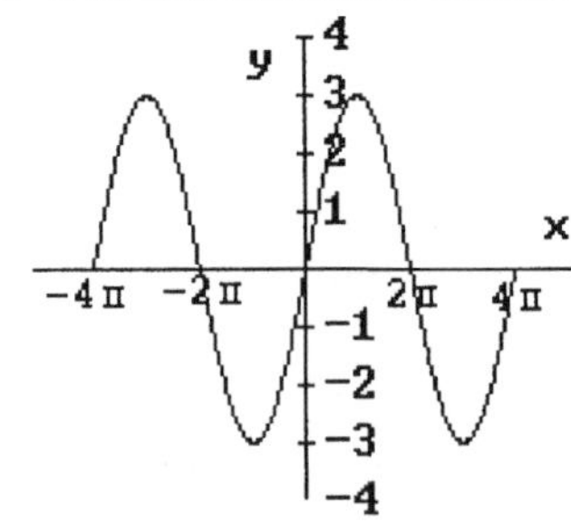

65. For the function whose graph is given below, the period is 2 and the amplitude is $\dfrac{4-(-0.5)}{2} = 2.25.$

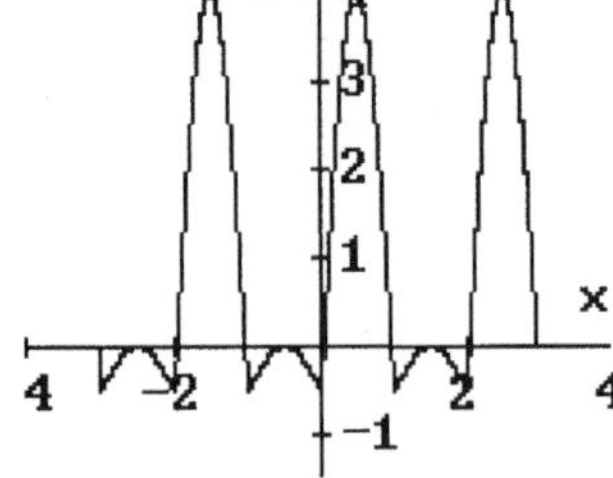

67. For the function $g(x) = 2\sin(3x + \pi)$, the period is $\dfrac{2\pi}{3}$, and the amplitude is 2.

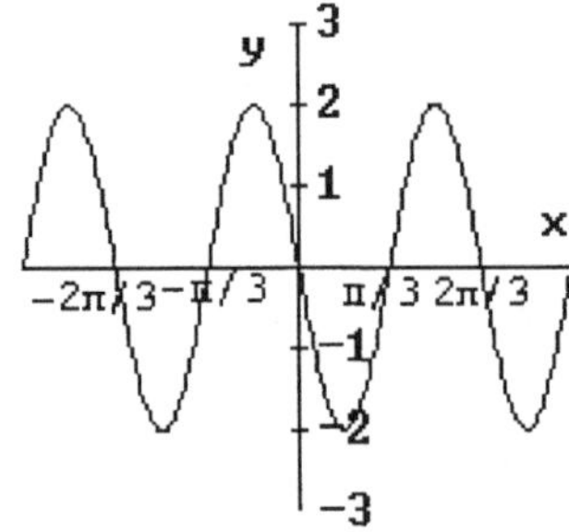

69. For the function $h(x) = 2\sin x + \cos 2x$, the period is 2π and the amplitude is $\frac{1.5-(-3)}{2} = 2.25$.

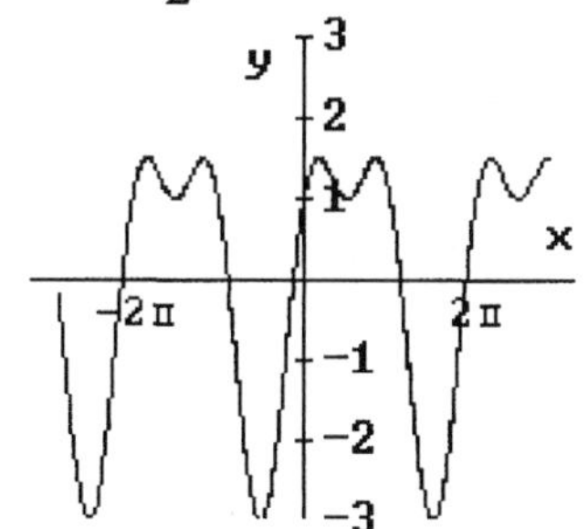

71. The function $f(x) = \sin\frac{\pi x}{2}$ has period 4, amplitude 1, phase shift 0, and no asymptotes.

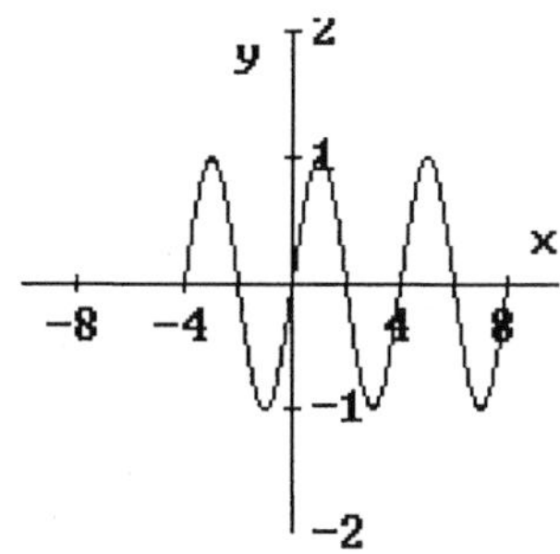

73. The function $h(t) = \cot 2t$ has period $\frac{\pi}{2}$, infinite amplitude, phase shift 0, and vertical asymptotes $\frac{\pi}{2} + \pi k$.

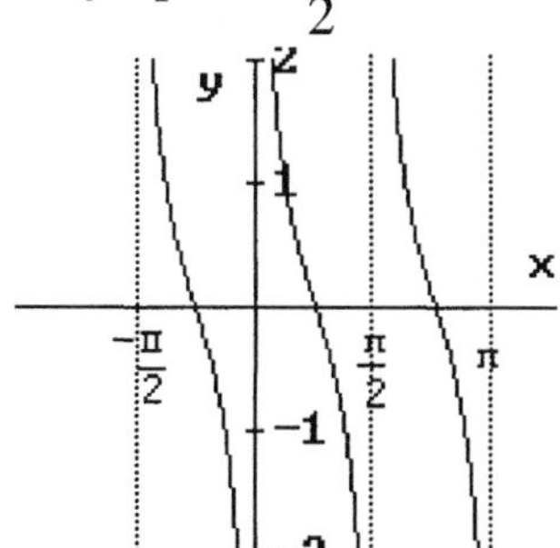

75. The function $f(t) = \sec 2\pi t$ has period 1, infinite amplitude, phase shift 0, and vertical asymptotes $t = \frac{1}{4} + \frac{1}{2}k$.

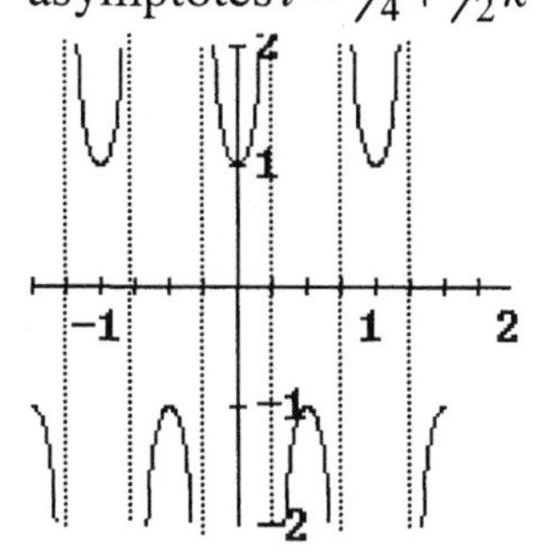

77. The function $g(x)=3\cos \frac{x}{2}-1$ has period 4π, amplitude 3, phase shift 0, no asymptotes.

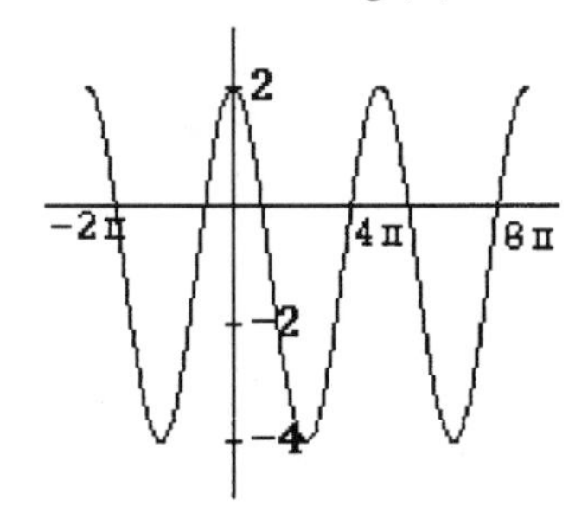

79. The function $y=\tan\left(x-\frac{\pi}{4}\right)$ has period π, infinite amplitude, phase shift $\frac{\pi}{4}$ (that is, $\frac{\pi}{4}$ units to the right) and vertical asymptotes $x=\frac{3\pi}{4}+\pi k$.

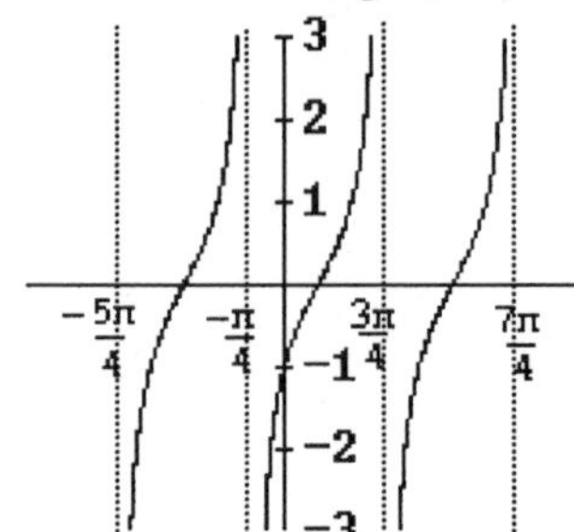

81. The function $y=\csc(\pi x-\pi)$ has period 2, infinite amplitude, phase shift 1 (that is, 1 unit to the right) and vertical asymptotes $x=1+k$.

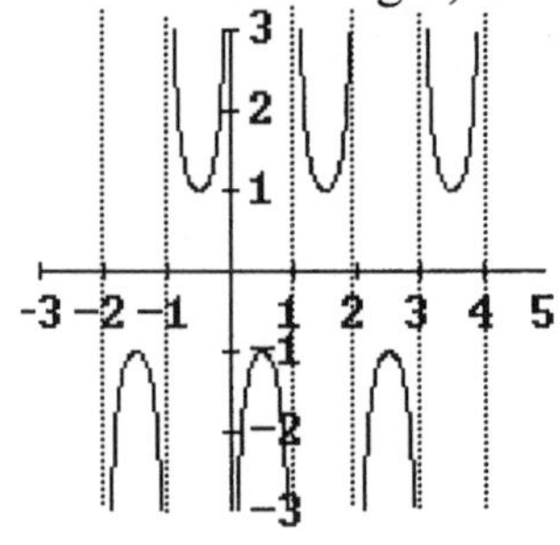

83. The function $y=-2\cos\left(\pi x+\frac{\pi}{2}\right)$ has period 2, amplitude 2, phase shift $-\frac{1}{2}$ (that is $\frac{1}{2}$ units to the left), and no asymptotes.

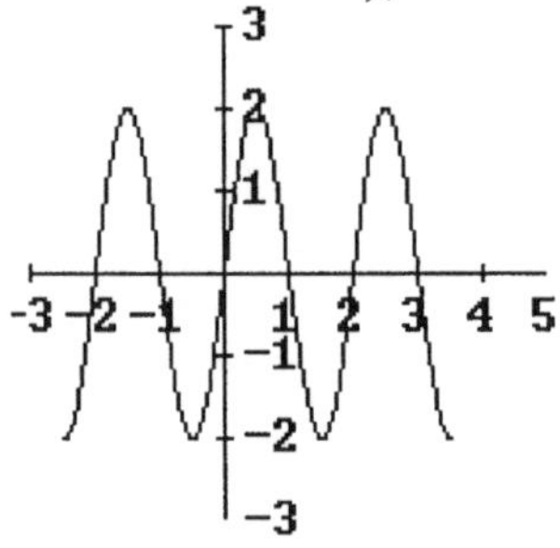

85. Answers may vary. A possible choice for a, b, c so that the graph of $f(x) = a\sin(bx+c)$ is as shown below, is $a=1,\ b=2,\ c=0$, since the period of the function is π, so $2\pi/b = \pi$. That is, $f(x) = \sin 2x$.

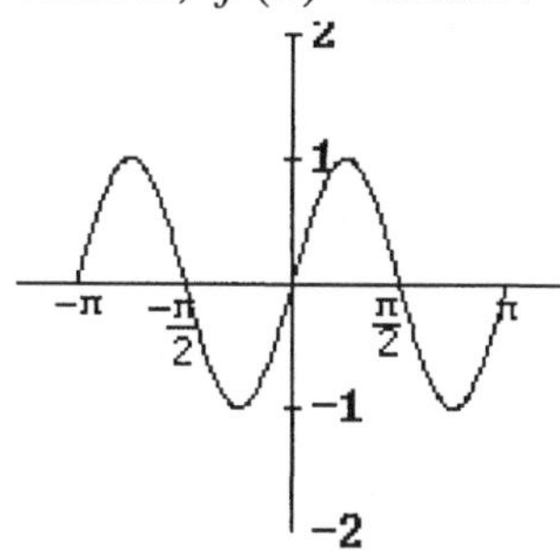

87. Answers may vary. A possible choice for a, b, c so that the graph of $f(x) = a\sin(bx+c)$ is as shown below, is $a=3,\ b=1,\ c=\pi/4$, since the period of the function is 2π, so $2\pi/b = 2\pi$ and we may consider a phase shift of $\pi/4$ to the left. That is, $f(x) = 3\sin\left(x+\pi/4\right)$.

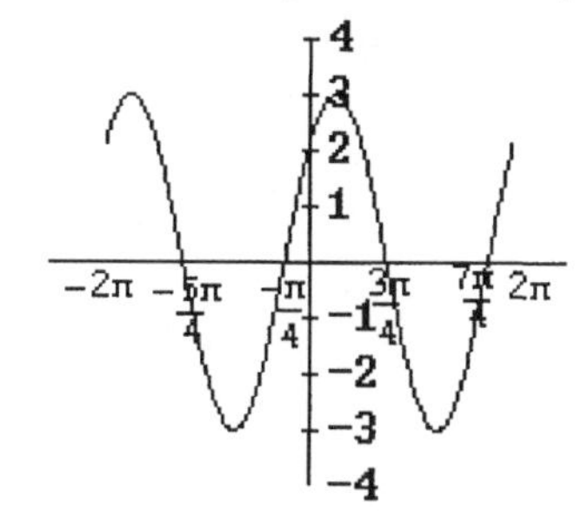

89. Answers may vary. A function of the form $f(x) = a\sin(bx+c)$ with amplitude 4, period $\pi/2$ and phase shift $\pi/4$ is $f(x) = 4\sin(4x-\pi)$.

91. The smallest positive zero of the function $f(x) = \sin x - 2\cos 3x$ is, approximately $x \approx 0.45$.

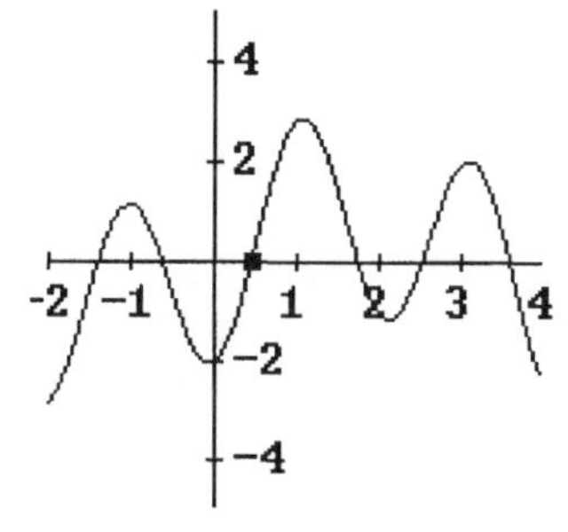

93. $\arccos 1 = 0$

95. $\sin^{-1}\dfrac{\sqrt{3}}{2} = \dfrac{\pi}{3}$

97. $\arctan\left(-\sqrt{3}/3\right) = -\dfrac{\pi}{6}$

99. The solutions of the equation $\cos x = -\frac{1}{4}$ in the interval $[0, 2\pi)$ are, $x = \cos^{-1}\left(-\frac{1}{4}\right)$ and $x = 2\pi - \cos^{-1}\left(-\frac{1}{4}\right)$, or approximately $x \approx 1.82$ and $x \approx 4.46$.

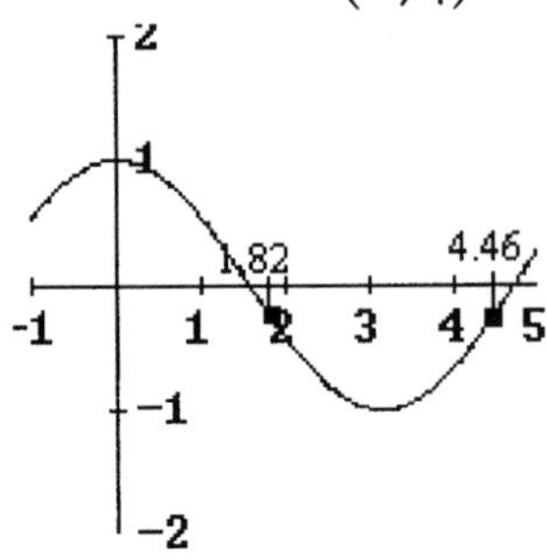

101. The solutions of the equation $\sin x = \frac{1}{\sqrt{3}}$ in the interval $[0, 2\pi)$ are, $x = \sin^{-1}\left(\frac{1}{\sqrt{3}}\right)$ and $x = \pi - \sin^{-1}\left(\frac{1}{\sqrt{3}}\right)$, or approximately $x \approx 0.62$ and $x \approx 2.53$.

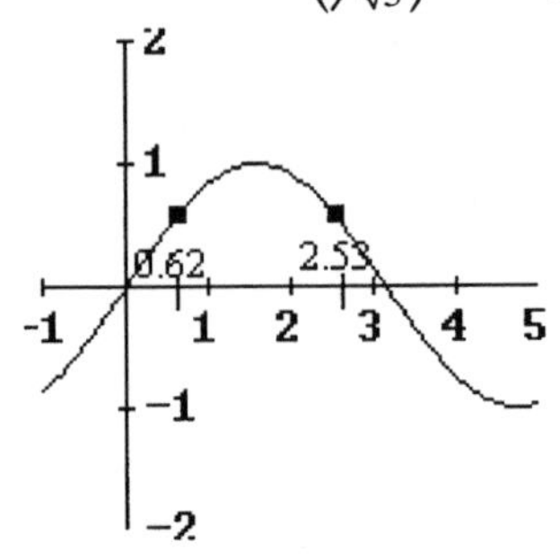

103. $\cos(\tan^{-1}\frac{3}{4}) = \frac{4}{5}$

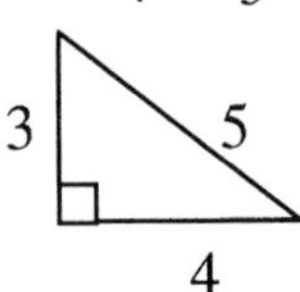

105. $\cos\left(\arccos\left(-\frac{1}{3}\right)\right) = -\frac{1}{3}$

107. $\tan\left(\tan^{-1} 16\right) = 16$

109. $\sin^{-1}(\sin 3\pi) = \sin^{-1} 0$
$= 0$

In Exercises 111-113, assume $x > 0$.

111. $\sec(\arcsin x) = \dfrac{1}{\sqrt{1 - x^2}}$

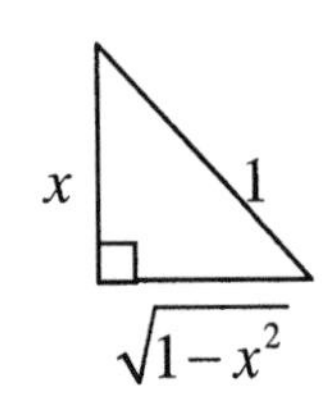

113. $\cos\left(\tan^{-1}\dfrac{x}{3}\right) = \dfrac{3}{\sqrt{x^2 + 9}}$

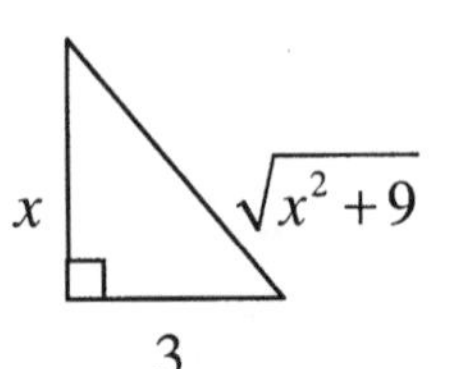

115. The graph of $y = \arctan(x+1)$ is a horizontal translation, 1 unit to the left, of the graph of $y = \arctan x$. Its x-intercept is $x = -1$, and it has horizontal asymptotes $y = \pm\dfrac{\pi}{2}$.

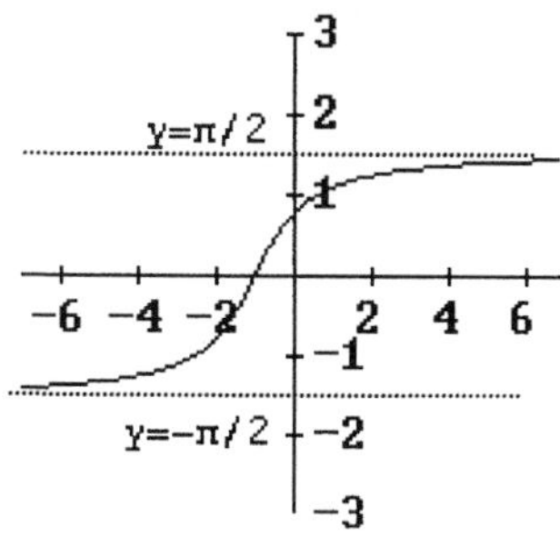

117. The graph of $y = \arccos x + \frac{\pi}{2}$ is a vertical translation, up $\frac{\pi}{2}$ units, of the graph of $y = \arccos x$. The domain of $y = \arccos x + \frac{\pi}{2}$ is $[-1,1]$ and its range is $\left[\frac{\pi}{2}, \frac{3\pi}{2}\right]$.

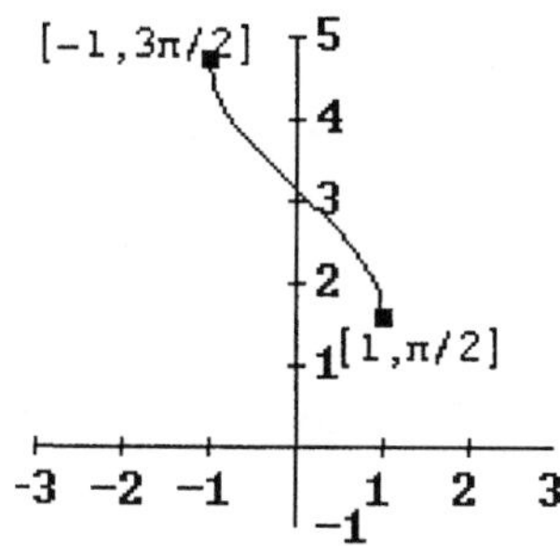

119. **(a)** The missing side length in the triangle below is $\sqrt{(x+1)^2-(x-1)^2}$, or $2\sqrt{x}$.

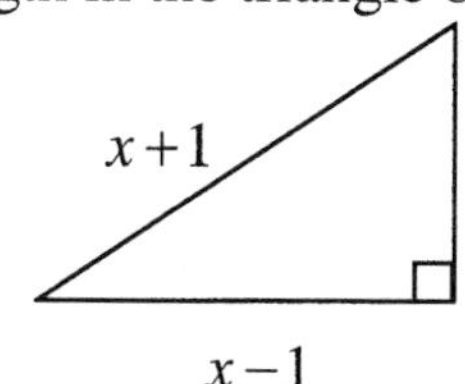

(b) $\sin\left(\arccos\left(\dfrac{x-1}{x+1}\right)\right) = \dfrac{2\sqrt{x}}{x+1}$

121. **(a)** $f\left(f^{-1}(x)\right) = x$

(b) $\tan(\arctan 2001) = 2001$

Chapter 5. Test

1. The statement: "Given any angle θ, there are exactly two angles that are coterminal with θ" is false. There are infinitely many angles that are coterminal with any given angle θ.

3. The statement: "The function $f(x) = \arcsin x$ has period 2π" is false. The function $f(x) = \arcsin x$ is not periodic.

5. The statement: "If α and β are coterminal, then $\sin\alpha = \sin\beta$" is true. The same is true for all the other trigonometric functions.

7. The statement: "The period of $f(x) = a\sin(bx + c)$ depends on the value of b but not on the values of a and c" is true. The period of this function is $\frac{2\pi}{|b|}$.

9. The statement: "A reference angle is always acute" is true.

11. Answers may vary. $f(x) = 3\cos 4x$ is a trigonometric function with amplitude, period $\frac{\pi}{2}$.

13. Answers may vary. $f(x) = \csc x$ is a trigonometric function with an asymptote at π.

15. In a circle of radius 5, the length of the arc subtended by a central angle of radius 25^0 is $s = 5\left(\frac{25\pi}{180}\right)$, approximately $s \approx 2.18$.

17. From a distance of 50 feet away from the base of a radio tower, the angle of elevation to the top of the tower is 75^0. The height of the tower, *h*, satisfies $\tan 75^0 = \frac{h}{50}$, so $h = 50\tan 75^0$, approximately $h \approx 186.6$ feet.

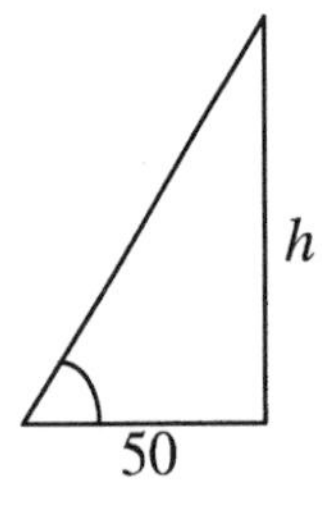

19. The function $f(x) = 2\sin(3x - \pi) + 1$ has amplitude 2, period $\frac{2\pi}{3}$, and phase shift $\frac{\pi}{3}$ (that is, $\frac{\pi}{3}$ units to the right.)

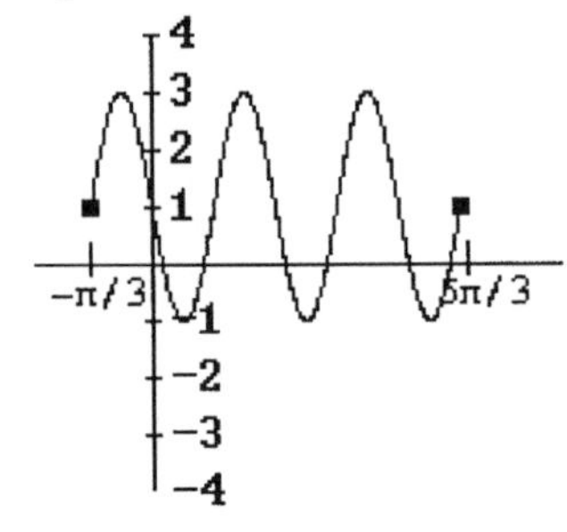

21. The function $g(x)=\tan(\pi x-\pi)$ has period 1, and its graph has asymptotes $x=\frac{1}{2}+k$, for any integer k.

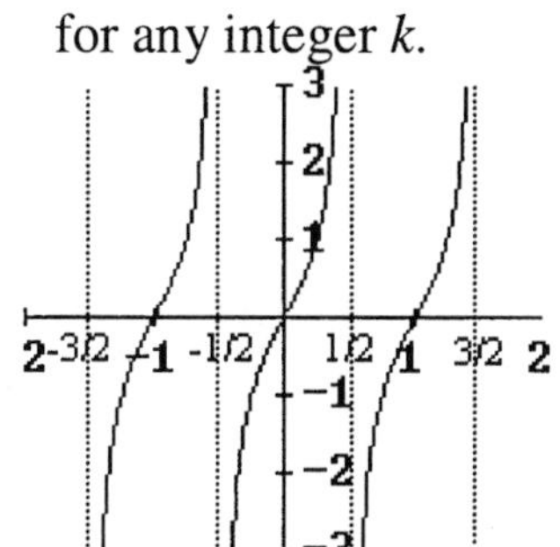

23. The solutions of $\sin x-\frac{1}{3}=0$ in the interval $[0,2\pi)$ are $x=\sin^{-1}\frac{1}{3}$ and $x=\pi-\sin^{-1}\frac{1}{3}$, approximately $x\approx 0.34$ and $x\approx 2.8$.

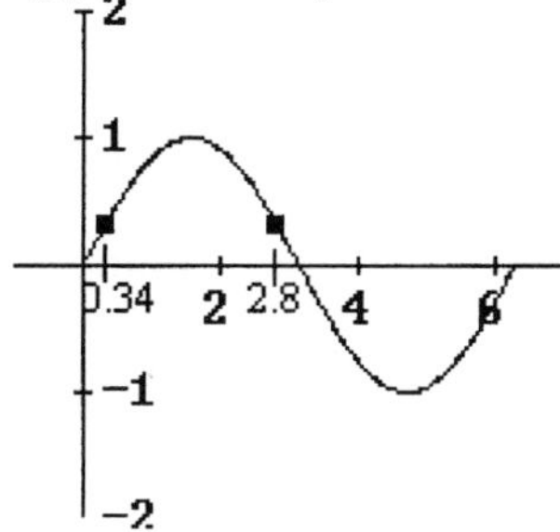

Chapter 6. TRIGONOMETRIC IDENTITIES AND EQUATIONS

Section 6.1 Fundamental Trigonometric Identities

1. $2\cos x = \sin x$ is a conditional equation.

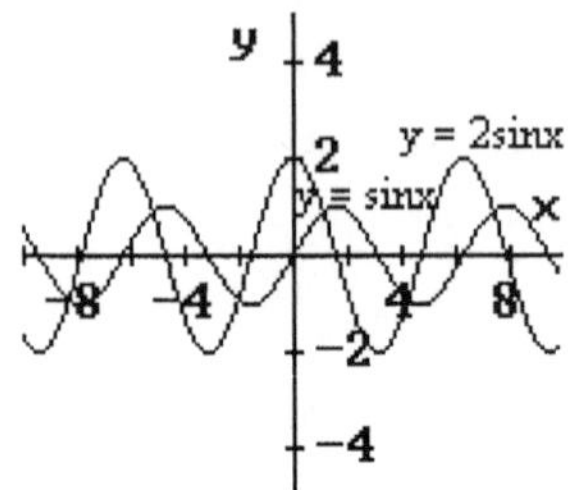

3. $\sin^2 x + \cos^2 x = 1.45$ is a contradiction, since $\sin^2 x + \cos^2 x = 1$ for all real numbers x.

5. $\sec^2 x - \csc^2 x = \tan^2 x - \cot^2 x$ is an identity: $\sec^2 x - \csc^2 x = (\tan^2 x + 1) - (\cot^2 x + 1)$

$$= \tan^2 x - \cot^2 x$$

7. $\cos 2x + \cos x = 3$ is a contradiction, since $\cos 2x \le 1$ and $\cos x \le 1$ for all real numbers x.

9. $2\cot 2x = \cot x - \tan x$ is an identity:

$$\begin{aligned}
2\cot 2x &= \frac{2\cos 2x}{\sin 2x} \\
&= \frac{2\left(\cos^2 x - \sin^2 x\right)}{2\sin x \cos x} \\
&= \frac{\cos^2 x - \sin^2 x}{\sin x \cos x} \\
&= \frac{\cos x}{\sin x} - \frac{\sin x}{\cot x} \\
&= \frac{\cos x}{\sin x} - \frac{\sin x}{\cot x} \\
&= \cot x - \tan x
\end{aligned}$$

11. $\cos x = 1 - \frac{x^2}{2} + \frac{x^4}{4}$ is a conditional equation. When $x = 0$, both sides are equal to 1.

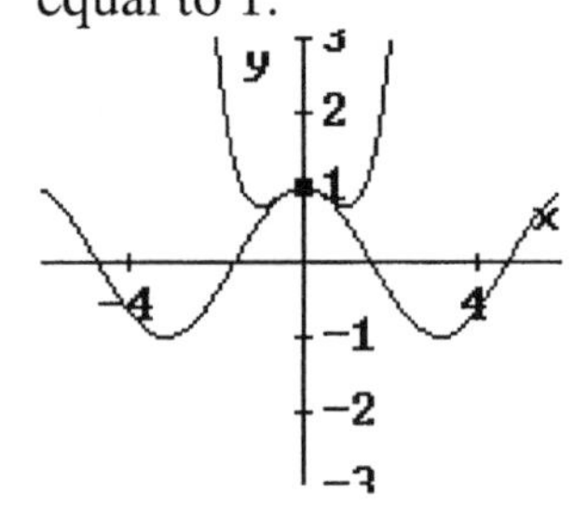

13. $2\sin^2 x - 1 = 1 - 2\cos^2 x$ is an identity: $2\sin^2 x - 1 = 2(1 - \cos^2 x) - 1$

$$= 1 - \cos^2 x$$

15. $\sin(x - \pi) = \sin x - \pi$ is a contradiction

17. $\sin x \cos x = \cos x + \sin x$ is a conditional equation, whose smallest positive solution is, approximately, $x \approx 2.65$.

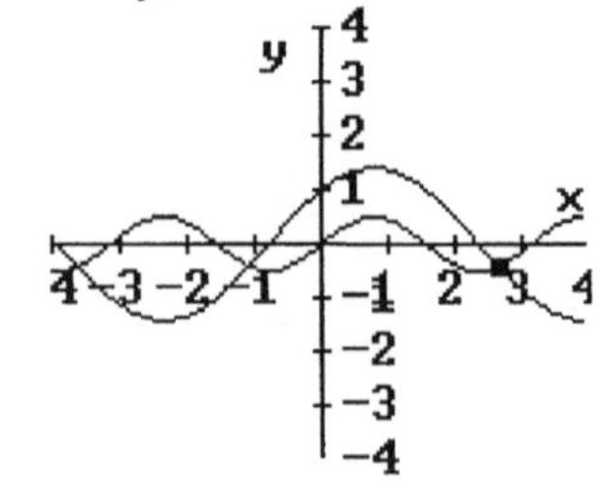

19. The equation $\tan x \cot x - 1 = 0$ is an identity, since $\tan x = 1/\cot x$ for all x in the domain of both tan x and cot x.

21. The equation $\sin 4x = 4\sin x$ is a conditional equation, and its smallest positive solution is $x = \pi$.

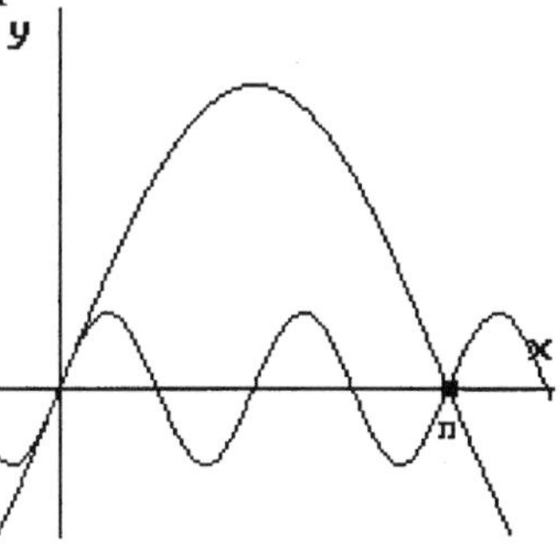

23. To verify that $\tan x/\sin x = \sec x$ is an identity:
$$\tan x/\sin x = (\sin x/\cos x)/\sin x$$
$$= 1/\cos x$$
$$= \sec x$$

25. To show that $\sec x/\csc x = \tan x$ is an identity:
$$\sec x/\csc x = (1/\cos x)/(1/\sin x)$$
$$= \sin x/\cos x$$
$$= \tan x.$$

27. The equation $\sin^2 x = 1 - \cos^2 x$ is an identity. It is simply a different version of the basic trigonometric identity $\sin^2 x + \cos^2 x = 1$.

29. The equation $\sec^2 x - \tan^2 x = 1$ is an identity:
$$\sec^2 x - \tan^2 x = 1/\cos^2 x - \sin^2 x/\cos^2 x$$
$$= (1 - \sin^2 x)/\cos^2 x$$
$$= \cos^2 x/\cos^2 x$$
$$= 1$$

31. The equation $\tan^2\beta - \sin^2\beta = \tan^2\beta \sin^2\beta$ is an identity:
$$\tan^2\beta - \sin^2\beta = \sin^2\beta/\cos^2\beta - \sin^2\beta$$
$$= \left(1/\cos^2\beta - 1\right)\sin^2\beta$$
$$= \left(\sec^2\beta - 1\right)\sin^2\beta$$
$$= \tan^2\beta\sin^2\beta$$

33. The equation $\cos t + \sin t = \dfrac{\tan t + 1}{\sec t}$ is an identity:
$$\frac{\tan t + 1}{\sec t} = \frac{\sin t/\cos t + 1}{1/\cos t}$$
$$= \frac{(\sin t + \cos t)/\cos t}{1/\cos t}$$
$$= \sin t + \cos t$$

35. The equation $\sec^2 x + \csc^2 x = \sec^2 x \csc^2 x$ is an identity: $\sec^2 x + \csc^2 x = 1/\cos^2 x + 1/\sin^2 x$

$$= (\sin^2 x + \cos^2 x)/(\sin^2 x \cos^2 x)$$
$$= 1/(\sin^2 x \cos^2 x)$$
$$= \sec^2 x \csc^2 x$$

37. The equation $\sin u + \cos u \cot u = \csc u$ is an identity: $\sin u + \cos u \cot u = \sin u + \cos^2 u/\sin u$

$$= \overbrace{(\sin^2 u + \cos^2 u)}^{=1}/\sin u$$
$$= \csc u$$

39. The equation $(\sin^2 \alpha - 1)(\cot^2 \alpha - 1) = 1 - \csc^2 \alpha$ is an identity:

$$(\sin^2 \alpha - 1)(\cot^2 \alpha - 1) = (\sin^2 \alpha - 1)(\cos^2 \alpha/\sin^2 \alpha - 1)$$
$$= \underbrace{\cos^2 \alpha + \sin^2 \alpha}_{=1} - \cos^2 \alpha/\sin^2 \alpha + 1$$
$$= -(1 - \sin^2 \alpha)/\sin^2 \alpha$$
$$= 1 - \csc^2 \alpha$$

41. The equation $\dfrac{\csc x + \cot x}{\csc x - \cot x} = (\csc x + \cot x)^2$ is an identity:

$$\frac{\csc x + \cot x}{\csc x - \cot x} = \left(\frac{\csc x + \cot x}{\csc x - \cot x}\right)\left(\frac{\csc x + \cot x}{\csc x + \cot x}\right)$$
$$= \frac{(\csc x + \cot x)^2}{\underbrace{\csc^2 x - \cot^2 x}_{=1}}$$
$$= (\csc x + \cot x)^2$$

43. The equation $\dfrac{1}{1 - \sin x} = \sec^2 x + \sec x \tan x$ is an identity:

$$\sec^2 x + \sec x \tan x = \sec x(\sec x + \tan x)$$
$$= 1/\cos x\,(1/\cos x + \sin x/\cos x)$$
$$= (1 + \sin x)/\cos^2 x$$
$$= (1 + \sin x)/(1 - \sin^2 x)$$
$$= (1 + \sin x)/[(1 - \sin x)(1 + \sin x)]$$
$$= 1/(1 - \sin x)$$

45. The equation $\sin^2 x/\cos x + \cos^2 x/\sin x = \sec x + \csc x - \cos x - \sin x$ is an identity:

$$\begin{aligned}
\sec x + \csc x - \cos x - \sin x &= 1/\cos x + 1/\sin x - \cos x - \sin x \\
&= \frac{\sin x + \cos x - \sin x \cos^2 x - \sin^2 x \cos x}{\sin x \cos x} \\
&= \frac{\sin x + \cos x - \sin x\left(1-\sin^2 x\right) - \left(1-\cos^2 x\right)\cos x}{\sin x \cos x} \\
&= \frac{\sin^3 x + \cos^3 x}{\sin x \cos x} \\
&= \sin^2 x/\cos x + \cos^2 x/\sin x
\end{aligned}$$

47. The equation $\dfrac{\sin y \tan y}{\tan y - \sin y} = \dfrac{\tan y + \sin y}{\sin y \tan y}$ is an identity. To see this, we observe that

$$\begin{aligned}
\sin^2 y \tan^2 y &= \frac{\sin^2 y \sin^2 y}{\cos^2 y} \\
&= \frac{\sin^2 y\left(1-\cos^2 y\right)}{\cos^2 y} \\
&= \frac{\sin^2 y - \sin^2 y \cos^2 y}{\cos^2 y} \\
&= \tan^2 y - \sin^2 y \\
&= (\tan y - \sin y)(\tan y + \sin y)
\end{aligned}$$

That is: $(\sin y \tan y)^2 = (\tan y - \sin y)(\tan y + \sin y)$. Therefore: $\dfrac{\sin y \tan y}{\tan y - \sin y} = \dfrac{\tan y + \sin y}{\sin y \tan y}$.

49. The equation $\sec^4 z - 2\tan^2 z \sec^2 z + \tan^4 z = 1$ is an identity:

$$\begin{aligned}
\sec^4 z - 2\tan^2 z \sec^2 z + \tan^4 z &= \left(\sec^2 z - \tan^2 z\right)^2 \\
&= 1^2 \\
&= 1
\end{aligned}$$

51. The equation $(\csc\phi-\cot\phi)^2=\dfrac{1-\cos\phi}{1+\cos\phi}$ is an identity:

$$\begin{aligned}(\csc\phi-\cot\phi)^2&=\left(\frac{1}{\sin\phi}-\frac{\cos\phi}{\sin\phi}\right)^2\\&=\frac{(1-\cos\phi)^2}{\sin^2\phi}\\&=\frac{(1-\cos\phi)^2}{1-\cos^2\phi}\\&=\frac{(1-\cos\phi)^2}{(1-\cos\phi)(1+\cos\phi)}\\&=\frac{1-\cos\phi}{1+\cos\phi}\end{aligned}$$

53. The equation $|\cos x|=\sqrt{1-\sin^2 x}$ is an identity, since $\cos^2 x=1-\sin^2 x$.

55. The equation $\ln|\cot s|=-\ln|\tan s|$ is an identity:
$$\begin{aligned}\ln|\cot s|&=\ln 1/|\tan s|\\&=\underbrace{\ln 1}_{=0}-\ln|\tan s|\\&=-\ln|\tan s|\end{aligned}$$

57. The equation $\ln|\sec w+\tan w|=-\ln|\sec w-\tan w|$ is an identity:
$|(\sec w+\tan w)(\sec w-\tan w)|=|\sec^2 w-\tan^2 w|$, so $|(\sec w+\tan w)(\sec w-\tan w)|=1$;
therefore, $\ln|(\sec w+\tan w)(\sec w-\tan w)|=0$, and $\ln|\sec w+\tan w|+\ln|\sec w-\tan w|=0$.

59. $$\begin{aligned}\frac{\sin^2 x+\cos^2 x}{\cos x}&=\frac{1}{\cos x}\\&=\sec x\end{aligned}$$

61. $$\begin{aligned}\frac{\tan x\cot x}{\sin x}&=\frac{1}{\sin x}\\&=\csc x\end{aligned}$$

63. $$\begin{aligned}&\frac{\sin^2 x-\cos^2 x}{\sin x+\cos x}+\frac{1}{\sec x}\\&=\frac{(\sin x-\cos x)(\sin x+\cos x)}{\sin x+\cos x}+\frac{1}{\sec x}\\&=\sin x-\cos x+\frac{1}{\sec x}\\&=\sin x\end{aligned}$$

65. $\tan x=\dfrac{1}{\cot x}$

67. $\cos x = \pm\sqrt{1-\sin^2 x}$; the sign depends on the quadrant corresponding to x.

69. $\cot x = \pm\sqrt{\dfrac{\cos^2 x}{1-\cos^2 x}}$; the sign depends on the quadrant corresponding to x.

71. $\sec x = \pm\sqrt{\dfrac{1}{1-\sin^2 x}}$; the sign depends on the quadrant corresponding to x.

Applications

73. If $\sin\theta = \dfrac{v^2}{32r}\cos\theta$ relates the banking angle of a curve of radius r and the maximum safe velocity of a vehicle traveling along the curve, then:

(a) $v = 4\sqrt{2r\tan\theta}$

(b) If $r = 2400\, ft$, and $\theta = 5^0$, then the maximum safe velocity of a vehicle traveling along the curve is $v = 4\sqrt{2(2400)\tan 5^0}$, that is $v \approx 82$ mph.

Concepts and Critical Thinking

75. The statement: "If the graphs of *f* and *g* are identical, then the equation $f(x) = g(x)$ is an identity" is true.

77. The statement: "If an equation is an identity, then it has infinitely many solutions" is true.

79. Answers may vary. An example of a conditional trigonometric equation is the equation $\sin x + \cos x = 0$.

81. Answers may vary. An example of an expression *E* involving only the secant and cosecant functions such that $E = \tan\theta$ is an identity is the expression $\dfrac{\sec\theta}{\csc\theta}$. (The equation $\dfrac{\sec\theta}{\csc\theta} = \tan\theta$ is an identity.)

Chapter 6. Section 6.2 Sum and Difference Identities

1. Given $\cos u = \frac{4}{5}$, $\sin u = -\frac{3}{5}$, $\cos v = \frac{5}{13}$, and $\sin v = \frac{12}{13}$, then:

(a) $\cos(u-v) = \cos u \cos v + \sin u \sin v = \left(\frac{4}{5}\right)\left(\frac{5}{13}\right) + \left(-\frac{3}{5}\right)\left(\frac{12}{13}\right) = -\frac{16}{65}$

(b) $\cos(u+v) = \cos u \cos v - \sin u \sin v = \left(\frac{4}{5}\right)\left(\frac{5}{13}\right) - \left(-\frac{3}{5}\right)\left(\frac{12}{13}\right) = \frac{56}{65}$

(c) $\sin(u-v) = \sin u \cos v - \sin v \cos u = \left(-\frac{3}{5}\right)\left(\frac{5}{13}\right) - \left(\frac{12}{13}\right)\left(\frac{4}{5}\right) = -\frac{63}{65}$

(d) $\sin(u+v) = \sin u \cos v + \sin v \cos u = \left(-\frac{3}{5}\right)\left(\frac{5}{13}\right) + \left(\frac{12}{13}\right)\left(\frac{4}{5}\right) = \frac{33}{65}$

3. Given $\cos u = \frac{3}{5}$, $0 \le u < \frac{\pi}{2}$, $\sin v = \frac{12}{13}$, $0 \le v < \frac{\pi}{2}$, then $\sin u = \frac{4}{5}$, and $\cos v = \frac{5}{13}$, so:

(a) $\cos(u-v) = \cos u \cos v + \sin u \sin v = \left(\frac{3}{5}\right)\left(\frac{5}{13}\right) + \left(\frac{4}{5}\right)\left(\frac{12}{13}\right) = \frac{63}{65}$

(b) $\cos(u+v) = \cos u \cos v - \sin u \sin v = \left(\frac{3}{5}\right)\left(\frac{5}{13}\right) - \left(\frac{4}{5}\right)\left(\frac{12}{13}\right) = -\frac{33}{65}$

(c) $\sin(u-v) = \sin u \cos v - \sin v \cos u = \left(\frac{4}{5}\right)\left(\frac{5}{13}\right) - \left(\frac{12}{13}\right)\left(\frac{3}{5}\right) = -\frac{16}{65}$

(d) $\sin(u+v) = \sin u \cos v + \sin v \cos u = \left(\frac{4}{5}\right)\left(\frac{5}{13}\right) + \left(\frac{12}{13}\right)\left(\frac{3}{5}\right) = \frac{56}{65}$

5. Given $\sin r = -\frac{4}{5}$, $\pi \le r < \frac{3\pi}{2}$, $\sin s = \frac{3}{5}$, $0 \le s < \frac{\pi}{2}$, then $\cos r = -\frac{3}{5}$ and $\cos s = \frac{4}{5}$:

(a) $\cos(r-s) = \cos r \cos s + \sin r \sin s = \left(-\frac{3}{5}\right)\left(\frac{4}{5}\right) + \left(-\frac{4}{5}\right)\left(\frac{3}{5}\right) = -\frac{24}{25}$

(b) $\cos(r+s) = \cos r \cos s - \sin r \sin s = \left(-\frac{3}{5}\right)\left(\frac{4}{5}\right) - \left(-\frac{4}{5}\right)\left(\frac{3}{5}\right) = 0$

(c) $\sin(r-s) = \sin r \cos s - \sin s \cos r = \left(-\frac{4}{5}\right)\left(\frac{4}{5}\right) - \left(\frac{3}{5}\right)\left(-\frac{3}{5}\right) = -\frac{7}{25}$

(d) $\sin(r+s) = \sin r \cos s + \sin s \cos r = \left(-\frac{4}{5}\right)\left(\frac{4}{5}\right) + \left(\frac{3}{5}\right)\left(-\frac{3}{5}\right) = -1$

7. Given $\sec v = \frac{5}{4}$, $0 \le v < \frac{\pi}{2}$, $\csc w = -\frac{13}{12}$, $\pi \le \beta < \frac{3\pi}{2}$, then $\cos v = \frac{4}{5}$, $\sin v = \frac{3}{5}$, $\sin w = -\frac{12}{13}$, and $\cos w = -\frac{5}{13}$. Therefore,

(a) $\cos(v-w) = \cos v \cos w + \sin w \sin v = \left(\frac{4}{5}\right)\left(-\frac{5}{13}\right) + \left(\frac{3}{5}\right)\left(-\frac{12}{13}\right) = -\frac{56}{65}$

(b) $\cos(v+w) = \cos v \cos w - \sin w \sin v = \left(\frac{4}{5}\right)\left(-\frac{5}{13}\right) - \left(\frac{3}{5}\right)\left(-\frac{12}{13}\right) = \frac{16}{65}$

(c) $\sin(v-w) = \sin v \cos w - \sin w \cos v = \left(\frac{3}{5}\right)\left(-\frac{5}{13}\right) + \left(-\frac{12}{13}\right)\left(\frac{3}{5}\right) = \frac{33}{65}$

(d) $\sin(v+w) = \sin v \cos w + \sin w \cos v = \left(\frac{3}{5}\right)\left(-\frac{5}{13}\right) + \left(-\frac{12}{13}\right)\left(\frac{4}{5}\right) = -\frac{63}{65}$

9. $\sin 62^0 \cos 32^0 - \sin 32^0 \cos 62^0$
$$= \sin\left(62^0 - 32^0\right) = \sin 30^0 = \tfrac{1}{2}$$

11. $\cos \frac{\pi}{8} \cos \frac{5\pi}{8} - \sin \frac{\pi}{8} \sin \frac{5\pi}{8}$
$$= \cos\left(\tfrac{\pi}{8} + \tfrac{5\pi}{8}\right) = \cos \tfrac{3\pi}{4} = -\tfrac{\sqrt{2}}{2}$$

13. $\sin \frac{\pi}{5} \cos \frac{\pi}{15} + \sin \frac{\pi}{15} \cos \frac{\pi}{5}$
$$= \sin\left(\tfrac{\pi}{5} + \tfrac{\pi}{15}\right) = \sin \tfrac{4\pi}{15} \approx 0.743144$$

15. $\cos y \cos 2y - \sin y \sin 2y = \cos\left(y + 2y\right) = \cos 3y$

17. $\dfrac{\tan 4a - \tan 2a}{1 + \tan 4a \tan 2a} = \tan\left(4a - 2a\right) = \tan 2a$

19. $\cos 105^0 = \cos\left(60^0 + 45^0\right)$
$$= \cos 60^0 \cos 45^0 - \sin 60^0 \sin 45^0 = \left(\tfrac{1}{2}\right)\left(\tfrac{\sqrt{2}}{2}\right) - \left(\tfrac{\sqrt{3}}{2}\right)\left(\tfrac{\sqrt{2}}{2}\right) = \frac{\sqrt{2}\left(1 - \sqrt{3}\right)}{4}$$

21. $\tan \frac{11\pi}{12} = \tan\left(\frac{3\pi}{4} + \frac{\pi}{6}\right)$
$$= \frac{\tan \frac{3\pi}{4} + \tan \frac{\pi}{6}}{1 - \tan \frac{3\pi}{4} \tan \frac{\pi}{6}} = \frac{-1 + \frac{\sqrt{3}}{3}}{1 - (-1)\frac{\sqrt{3}}{3}} = \frac{-3 + \sqrt{3}}{3 + \sqrt{3}}$$

23. $\cos\left(x + \pi\right) = \cos x \cos \pi - \sin x \sin \pi$
$$= \cos x \cdot (-1) - \sin x \cdot 0 = -\cos x$$

25. $\sin\left(\frac{\pi}{2} + x\right) = \sin \frac{\pi}{2} \cos x + \sin x \cos \frac{\pi}{2}$
$$= 1 \cdot \cos x + \sin x \cdot 0 = \cos x$$

27. $\cos\left(\frac{\pi}{3} - x\right) = \cos \frac{\pi}{3} \cos x + \sin \frac{\pi}{3} \sin x$
$$= \tfrac{1}{2} \cdot \cos x + \tfrac{\sqrt{3}}{2} \sin x$$

29. $\cos\left(x + \frac{2\pi}{3}\right) = \cos x \cos \frac{2\pi}{3} + \sin x \sin \frac{2\pi}{3}$
$$= -\tfrac{1}{2} \cdot \cos x + \tfrac{\sqrt{3}}{2} \sin x$$

31. $\tan\left(x - \frac{\pi}{4}\right) = \dfrac{\tan x - \tan \frac{\pi}{4}}{1 + \tan x \tan \frac{\pi}{4}}$

31. $\tan\left(x - \frac{\pi}{4}\right) = \dfrac{\tan x - \tan \frac{\pi}{4}}{1 + \tan x \tan \frac{\pi}{4}}$
$$= \frac{\tan x - 1}{1 + \tan x \cdot 1} = \frac{\tan x - 1}{\tan x + 1}$$

33. $\sin\left(\arcsin \tfrac{3}{5} + \arccos \tfrac{3}{5}\right) = \sin\left(\arcsin \tfrac{3}{5}\right)\cos\left(\arccos \tfrac{3}{5}\right) + \sin\left(\arccos \tfrac{3}{5}\right)\cos\left(\arcsin \tfrac{3}{5}\right)$

$$= \left(\tfrac{3}{5}\right)\left(\tfrac{3}{5}\right) + \sqrt{1-\left(\tfrac{3}{5}\right)^2}\ \sqrt{1-\left(\tfrac{3}{5}\right)^2}$$
$$= \tfrac{9}{25} + \tfrac{16}{25}$$
$$= 1$$

35. $\tan\left(\tfrac{\pi}{4} - \arctan \tfrac{25}{24}\right) = \dfrac{\tan \tfrac{\pi}{4} - \tan\left(\arctan \tfrac{25}{24}\right)}{1 + \tan \tfrac{\pi}{4} \tan\left(\arctan \tfrac{25}{24}\right)}$

$$= \frac{1 - \tfrac{25}{24}}{1 + \tfrac{25}{24}}$$
$$= -\tfrac{1}{49}$$

37. To verify that $\cos\left(\tfrac{\pi}{2} - x\right) = \sin x$ is an identity: $\cos\left(\tfrac{\pi}{2} - x\right) = \cos \tfrac{\pi}{2} \cos x + \sin \tfrac{\pi}{2} \sin x$

$$= 0 \cdot \cos x + 1 \cdot \sin x$$
$$= \sin x$$

39. To verify that $\sec(u+v) = \dfrac{\csc u \sec v}{\cot u - \tan v}$ is an identity:

$$\sec(u+v) = \tfrac{1}{\cos(u+v)}$$
$$= \frac{1}{\cos u \cos v - \sin u \sin v}$$
$$= \frac{1/(\sin u \cos v)}{(\cos u \cos v - \sin u \sin v)/(\sin u \cos v)}$$
$$= \frac{\csc u \sec v}{\cot u - \tan v}$$

41. To verify that $\cot(u-v) = \dfrac{\cot u \cot v + 1}{\cot v - \cot u}$ is an identity:

$$\cot(u-v) = \frac{\cos(u-v)}{\sin(u-v)}$$
$$= \frac{\cos u \cos v + \sin u \sin v}{\sin u \cos v - \sin v \cos u}$$
$$= \frac{\dfrac{\cos u \cos v - \sin u \sin v}{\sin u \sin v}}{\dfrac{\sin u \cos v - \sin v \cos u}{\sin u \sin v}}$$
$$= \frac{\cot u \cot v + 1}{\cot v - \cot u}$$

43. To verify that $\cos(-x)=\cos x$ is an identity:

$$\begin{aligned}\cos(-x)&=\cos(0-x)\\&=\cos 0\cos x+\sin 0\sin x\\&=1\cdot\cos x+0\cdot\sin x\\&=\cos x\end{aligned}$$

45. To verify that $\sin u\sin v=\frac{1}{2}[\cos(u-v)-\cos(u+v)]$ is an identity:

$$\begin{aligned}\tfrac{1}{2}[\cos(u-v)-\cos(u+v)]&=\tfrac{1}{2}[(\cos u\cos v+\sin u\sin v)-(\cos u\cos v-\sin u\sin v)]\\&=\tfrac{1}{2}(2\sin u\sin v)\\&=\sin u\sin v\end{aligned}$$

47. To verify that $\sin u\cos v=\frac{1}{2}[\sin(u+v)+\sin(u-v)]$ is an identity:

$$\begin{aligned}\tfrac{1}{2}[\sin(u+v)+\sin(u-v)]&=\tfrac{1}{2}[(\sin u\cos v+\sin v\cos u)+(\sin u\cos v-\sin v\cos u)]\\&=\tfrac{1}{2}(2\sin u\cos v)\\&=\sin u\cos v\end{aligned}$$

49. To verify that $\sin 2x=2\sin x\cos x$ is an identity:

$$\begin{aligned}\sin 2x&=\sin(x+x)\\&=\sin x\cos x+\sin x\cos x\\&=2\sin x\cos x\end{aligned}$$

Concepts and Critical Thinking

51. The statement: "$\sin(\pi/2+x)=\cos x$ for all x" is true; $\sin(\pi/2+x)=\underbrace{\sin\pi/2}_{=1}\cos x+\sin x\underbrace{\cos\pi/2}_{=0}$

$$=\cos x$$

53. The statement: "$\cos(\pi/2+x)=\sin x$ for all x" is false; $\cos(\pi/2+x)=\underbrace{\cos\pi/2}_{=0}\cos x-\underbrace{\sin\pi/2}_{=1}\sin x$

$$=-\sin x$$

55. Answers may vary. An example of an angle θ that is not a multiple of 30^0 for which a sum identity can be used to determine $\sin\theta$ is $\theta=75^0$; $\sin 75^0=\sin(45^0+30^0)$, therefore

$\sin 75^0=\sin 45^0\cos 30^0+\sin 30^0\cos 45^0$, that is $\sin 75^0=\dfrac{\sqrt{6}+\sqrt{2}}{4}$.

57. Answers may vary. An example of pair of angles x and y for which $\cos(x+y)=\cos x+\cos y$ is $x=\pi/4$, $y=5\pi/4$. We have $\cos(\pi/4+5\pi/4)=\cos(3\pi/2)$, that is, $\cos(\pi/4+5\pi/4)=0$. On the other hand, $\cos\pi/4+\cos 5\pi/4=\sqrt{2}/2-\sqrt{2}/2$, or $\cos\pi/4+\cos 5\pi/4=0$ also.

59. To prove that if u and v are angles in the first quadrant, then $u+v$ is also in the first quadrant if and only if $\tan u \tan v < 1$. If $u+v$ is in the first quadrant, then $\tan(u+v) > 0$, that is $\tan(u+v) = \dfrac{\tan u + \tan v}{1-\tan u \tan v} > 0$. Since $\tan u > 0$ and $\tan v > 0$, then $1-\tan u \tan v > 0$, so $\tan u \tan v < 1$. For the converse, first observe that if u and v are angles in the first quadrant, say $2a\pi < u < 2a\pi + \pi/2$ and $2c\pi < v < 2d\pi + \pi/2$ for some integers a, b, c, d, then $2(a+c)\pi < u+v < 2(c+d)\pi + \pi$, so $u+v$ is in the first or the second quadrant. So, if $\tan u \tan v < 1$, we have $\tan(u+v) = \dfrac{\tan u + \tan v}{1-\tan u \tan v} > 0$, which makes us conclude that $u+v$ is in the first quadrant.

61.
$$\begin{aligned}\cos(x+y+z) &= \cos(x+y)\cos z - \sin(x+y)\sin z \\ &= (\cos x \cos y - \sin x \sin y)\cos z - (\sin x \cos y + \sin y \cos x)\sin z \\ &= \cos x \cos y \cos z - \sin x \sin y \cos z - \sin x \cos y \sin z - \sin y \cos x \sin z\end{aligned}$$

$$\begin{aligned}\sin(x+y+z) &= \sin(x+y)\cos z + \sin z \cos(x+y) \\ &= (\sin x \cos y + \sin y \cos x)\cos z + \sin z(\cos x \cos y - \sin x \sin y) \\ &= \sin x \cos y \cos z + \sin y \cos x \cos z + \sin z \cos x \cos y - \sin z \sin x \sin y\end{aligned}$$

Chapter 6. Section 6.3 Double-Angle and Half-Angle Identities

1. Given $\sin\theta = 3/5$, $\cos\theta = -4/5$, we have:

(a) $\sin 2\theta = 2\left(\frac{3}{5}\right)\left(-\frac{4}{5}\right)$, or $\sin 2\theta = -\frac{24}{25}$

(b) $\cos 2\theta = \left(-\frac{4}{5}\right)^2 - \left(\frac{3}{5}\right)^2$, or $\cos 2\theta = \frac{7}{25}$

(c) $\tan 2\theta = -\frac{24}{7}$

3. Given $\sin\theta = -3/5$, $\pi \le \theta < \frac{3\pi}{2}$, then $\cos\theta = -\sqrt{1-\left(-\frac{3}{5}\right)^2}$, that is, $\cos\theta = -\frac{4}{5}$; therefore:

(a) $\sin 2\theta = 2\left(-\frac{3}{5}\right)\left(-\frac{4}{5}\right)$, or $\sin 2\theta = \frac{24}{25}$

(b) $\cos 2\theta = \left(-\frac{4}{5}\right)^2 - \left(-\frac{3}{5}\right)^2$, or $\cos 2\theta = \frac{7}{25}$

(c) $\tan 2\theta = \frac{24}{7}$

5. Given $\tan\theta = 8/15$, $0 \le \theta < \pi/2$, then $\sec\theta = \sqrt{1+(8/15)^2}$, that is, $\sec\theta = 17/15$; therefore, $\sin\theta = 8/17$ and $\sin\theta = 15/17$.

(a) $\sin 2\theta = 2\left(\frac{12}{13}\right)\left(\frac{5}{13}\right)$, or $\sin 2\theta = \frac{240}{289}$

(b) $\cos 2\theta = \left(\frac{15}{17}\right)^2 - \left(\frac{8}{17}\right)^2$, or $\cos 2\theta = \frac{161}{289}$

(c) $\tan 2\theta = \frac{240}{161}$

7. If $\cot\theta = -12/5$, $\pi/2 \le \theta < \pi$, then $\tan\theta = -5/12$, and $\sec\theta = -\sqrt{1+(-5/12)^2}$, or $\sec\theta = -13/12$; then, $\cos\theta = -12/13$ and $\sin\theta = 5/13$.

(a) $\sin 2\theta = 2\left(\frac{5}{13}\right)\left(-\frac{12}{13}\right)$, or $\sin 2\theta = -\frac{120}{169}$

(b) $\cos 2\theta = \left(-\frac{12}{13}\right)^2 - \left(\frac{5}{13}\right)^2$, or $\cos 2\theta = \frac{119}{169}$

(c) $\tan 2\theta = -\frac{120}{119}$

9. If $\cos x = 7/25$, where $0 \le x < \pi/2$, then:

(a) $\sin\frac{x}{2} = \sqrt{\frac{1-7/25}{2}}$, that is $\sin\frac{x}{2} = \frac{3}{5}$

(b) $\cos\frac{x}{2} = \frac{4}{5}$

(c) $\tan\frac{x}{2} = \frac{3}{4}$

11. If $\cos x = -1/9$, where $\pi/2 \le x < \pi$, then $\pi/4 \le x/2 < \pi/2$;

(a) $\sin\frac{x}{2} = \sqrt{\frac{1+1/9}{2}}$, that is $\sin\frac{x}{2} = \frac{\sqrt{5}}{3}$

(b) $\cos\frac{x}{2} = \sqrt{\frac{1-1/9}{2}}$, or $\cos\frac{x}{2} = \frac{2}{3}$

(c) $\tan\frac{x}{2} = \frac{\sqrt{5}}{2}$

13. Given $\sec x = 9/7$, where $3\pi/2 \le x < 2\pi$, then $\cos x = 7/9$ and $3\pi/4 \le x/2 < \pi$;

(a) $\sin\frac{x}{2} = \sqrt{\frac{1-7/9}{2}}$, that is $\sin\frac{x}{2} = \frac{1}{3}$

(b) $\cos\frac{x}{2} = -\sqrt{\frac{1+7/9}{2}}$, so $\cos\frac{x}{2} = -2\sqrt{2}/3$

(c) $\tan\frac{x}{2} = -\sqrt{2}/4$

15. If $\sin x = -\sqrt{15}/8$, where $3\pi/2 \le x < 2\pi$, then $\cos x = \sqrt{1-15/64}$, or $\cos x = 7/8$, and $3\pi/4 \le x/2 < \pi$;

(a) $\sin\frac{x}{2} = \sqrt{\frac{1-7/8}{2}}$, that is $\sin\frac{x}{2} = \frac{1}{4}$

(b) $\cos\frac{x}{2} = -\sqrt{\frac{1+7/8}{2}}$, so $\cos\frac{x}{2} = -\frac{\sqrt{15}}{2}$

(c) $\tan x/2 = -\sqrt{15}/30$

(c) $\tan x/2 = -\sqrt{5}/2$

17. $\sin 15^0 = \sin\frac{30^0}{2}$, so $\sin 15^0 = \sqrt{\frac{1-\cos 30^0}{2}}$, therefore $\sin 15^0 = \frac{\sqrt{2-\sqrt{3}}}{2}$.

19. $\sec \frac{\pi}{8} = \frac{1}{\cos \frac{\pi}{8}}$, so $\sec \frac{\pi}{8} = \frac{2}{\sqrt{2+\sqrt{2}}}$, that is $\sec \frac{\pi}{8} = \sqrt{2}\sqrt{2-\sqrt{2}}$.

21. $\tan 2x = \frac{2\cot x}{\cot^2 x - 1}$, as we now prove:

$$\frac{2\cot x}{\cot^2 x - 1} = \frac{\left(\frac{2\cos x}{\sin x}\right)}{\left(\frac{\cos x}{\sin x}\right)^2 - 1}$$

$$= \left(\frac{2\cos x}{\sin x}\right) \Big/ \left(\frac{\cos^2 x - \sin^2 x}{\sin^2 x}\right)$$

$$= \frac{2\sin x \cos x}{\cos^2 x - \sin^2 x}$$

$$= \frac{\sin 2x}{\cos 2x}$$

$$= \tan 2x$$

23. To show $\sin x \cos x \csc 2x = \frac{1}{2}$: $2\sin x \cos x \csc 2x = \sin 2x \csc 2x$, so $2\sin x \cos x \csc 2x = 1$
Then, $\sin x \cos x \csc 2x = \frac{1}{2}$, as required.

25. To show $\tan 2x = \frac{2\sin x}{2\cos x - \sec x}$:

$$\frac{2\sin x}{2\cos x - \sec x} = \frac{2\sin x}{2\cos x - \frac{1}{\cos x}}$$

$$= \frac{2\sin x \cos x}{2\cos^2 x - 1}$$

$$= \frac{2\sin x \cos x}{\cos^2 x + (\cos^2 x - 1)}$$

$$= \frac{2\sin x \cos x}{\cos^2 x - \sin^2 x}$$

$$= \sin 2x / \cos 2x$$

$$= \tan 2x$$

27. To prove that $2\tan x \cot 2x = 2 - \sec^2 x$:

$$2\tan x \cot 2x = 2\left(\frac{\sin x}{\cos x}\right)\left(\frac{\cos 2x}{\sin 2x}\right)$$

$$= 2\left(\frac{\sin x}{\cos x}\right)\left(\frac{\cos^2 x - \sin^2 x}{2\sin x \cos x}\right)$$

$$= \frac{\cos^2 x - (1 - \cos^2 x)}{\cos^2 x}$$

$$= \frac{2\cos^2 x - 1}{\cos^2 x}$$

$$= 2 - \sec^2 x$$

29. To prove the identity $2\csc 30x = \csc 15x \sec 15x$:

$$\begin{aligned} 2\csc 30x &= 2/\sin 30x \\ &= \frac{1}{\sin 15x \cos 15x} \\ &= \csc 15x \sec 15x \end{aligned}$$

31. $$\begin{aligned} 4\sin^2 x\cos^2 x &= (2\sin x\cos x)^2 \\ &= (\sin 2x)^2 \end{aligned}$$

33. $\dfrac{1-\tan^2 x}{\tan x} = \dfrac{1-\dfrac{\sin^2 x}{\cos^2 x}}{\sin x/\cos x}$, or $\dfrac{1-\tan^2 x}{\tan x} = \dfrac{\dfrac{\cos^2 x-\sin^2 x}{\cos^2 x}}{\sin x/\cos x}$, and $\dfrac{1-\tan^2 x}{\tan x} = \dfrac{\cos 2x}{\sin x\cos x}$.

The last identity is equivalent to $\dfrac{1-\tan^2 x}{\tan x} = \dfrac{\cos 2x}{\left(\sin 2x/2\right)}$, so $\dfrac{1-\tan^2 x}{\tan x} = 2\cot 2x$.

35. To show that $\sin\left(2x-\pi/3\right)+\sqrt{3}/2 = \sin x(\cos x+\sqrt{3}\sin x)$ is an identity:

$$\begin{aligned} \sin\left(2x-\pi/3\right)+\sqrt{3}/2 &= \sin 2x\cos \pi/3 - \sin \pi/3 \cos 2x + \sqrt{3}/2 \\ &= 2\sin x\cos x\left(1/2\right) - \sqrt{3}/2\left(\cos^2 x - \sin^2 x\right) + \sqrt{3}/2 \\ &= \sin x\cos x - \sqrt{3}/2\left(1-2\sin^2 x\right) + \sqrt{3}/2 \\ &= \sin x\cos x + \sqrt{3}\sin^2 x \\ &= \sin x\left(\cos x + \sqrt{3}\sin x\right) \end{aligned}$$

37. To show that $\cot x/2 = \csc x + \cot x$ is an identity:

$$\begin{aligned} \cot x/2 &= 1/\tan x/2 \\ &= \frac{1}{\left(\sin x/(1+\cos x)\right)} \\ &= \frac{1+\cos x}{\sin x} \\ &= \csc x + \cot x \end{aligned}$$

39. To verify the identity $\tan \frac{3x}{2} = \dfrac{3\sin x - 4\sin^3 x}{4\cos^3 x - 3\cos x + 1}$:

$$\begin{aligned}
\tan \tfrac{3x}{2} &= \frac{\sin 3x}{1+\cos 3x} \\
&= \frac{\sin 2x \cos x + \sin x \cos 2x}{1 + \cos 2x \cos x - \sin 2x \sin x} \\
&= \frac{2\sin x \cos^2 x + \sin x(\cos^2 x - \sin^2 x)}{1 + (\cos^2 x - \sin^2 x)\cos x - 2\sin^2 x \cos x} \\
&= \frac{3\sin x \cos^2 x - \sin^3 x}{1 + \cos^3 x - \cos x + \cos^3 x - 2\cos x + 2\cos^3 x} \\
&= \frac{3\sin x(1-\sin^2 x) - \sin^3 x}{4\cos^3 x - 3\cos x + 1} \\
&= \frac{3\sin x - 4\sin^3 x}{4\cos^3 x - 3\cos x + 1}
\end{aligned}$$

41. To verify the identity $\tan\left(\frac{x}{2} + \frac{\pi}{4}\right) = \tan x + \sec x$:

$$\begin{aligned}
\tan\left(\tfrac{x}{2} + \tfrac{\pi}{4}\right) &= \tan\left(\tfrac{(x+\pi/2)}{2}\right) \\
&= \left[1-\cos\left(x+\tfrac{\pi}{2}\right)\right] \Big/ \sin\left(x+\tfrac{\pi}{2}\right) \\
&= \left[1-(-\sin x)\right] \Big/ \cos x \\
&= \tan x + \sec x
\end{aligned}$$

Applications

43. The viewing area of a TV screen with diagonal d and the angle between the bottom of the screen and the diagonal θ is given by $A = d^2 \sin\theta \cos\theta$.

(a) Since $\sin 2\theta = 2\sin\theta\cos\theta$, we obtain $A = \frac{1}{2}d^2 \sin 2\theta$.

(b) If the diagonal of a TV screen is $d = 27$ in, and the angle between the bottom of the screen and the diagonal is $\theta = 40^0$, then the viewing area is $A \approx 358.96$ in^2.

45. The equation that relates the angle of incidence and the angle of refraction of a ray of light passing from air to a liquid is $\frac{c_1}{c_2} = \frac{\sin\alpha}{\sin\beta}$, where α is the angle of incidence, β is the angle of refraction, c_1 and c_2 are the speed of light in air and the liquid, respectively. If $\alpha = 2\beta$, then $\frac{c_1}{c_2} = \frac{\sin 2\beta}{\sin\beta}$, so $\frac{c_1}{c_2} = \frac{2\sin\beta\cos\beta}{\sin\beta}$, and $\frac{c_1}{c_2} = 2\cos\beta$, $c_2 = \frac{1}{2}c_1 \sec\beta$.

Concepts and Critical Thinking

47. The statement: "$\sin^2 2x + \cos^2 2x = 1$ for all x" is true.

49. The statement: "$\sin 4x = 4\sin x \cos x$ for all x" is false. Take, for example $x = \pi/4$:
$\sin 4(\pi/4) = 0$, while $4\sin \pi/4 \cos \pi/4 = 2$.

51. Answers may vary. An angle x for which $\sin 2x = 2\sin x$ is $x = \pi/4$: $\sin 2(\pi/4) = 1$, while
$2\sin(\pi/4) = \sqrt{2}$.

53. A double-angle identity for secant can be obtained using a double-angle identity for cosine:
$\sec 2x = 1/\cos 2x$, therefore $\sec 2x = 1/\left(\cos^2 x - \sin^2 x\right)$.

55. If x and y are coterminal, it doesn't follow follow that $\sin x/2 = \sin y/2$: $x = \pi/3$ and $x = 7\pi/3$ are coterminal, $x/2 = \pi/6$, and $x = 7\pi/6$, $\sin x/2 = 1/2$, $\sin y/2 = -1/2$.

57. $\sin 2x = \pm\sqrt{1-\cos^2 2x}$, so $\sin 2x = \pm\sqrt{1-\left(\cos^2 x - \sin^2 x\right)^2}$, and
$\sin 2x = \pm\sqrt{1-\left(\cos^4 x - 2\sin^2 x\cos^2 x + \sin^4 x\right)}$; then, $\sin 2x = \pm\sqrt{4\sin^2 x\cos^2 x}$
since $\cos^4 x + 2\sin^2 x\cos^2 x + \sin^4 x = 1$. Therefore, $\sin 2x = \pm\sqrt{4\sin^2 x\cos^2 x}$, and
$\sin 2x = \pm 2|\sin x||\cos x|$. Now, the sign of $\sin 2x$ is the same as the sign of $\sin x\cos x$:
If $\sin 2x > 0$, then $2k\pi < 2x < (2k+1)\pi$, so $k\pi < x < k\pi + \pi/2$, that is x in the first or third quadrant, and $\sin x\cos x > 0$. If $\sin 2x < 0$, then $(2k+1)\pi < 2x < 2(k+1)\pi$, so
$k\pi + \pi/2 < x < (k+1)\pi$, that is x is in the second or fourth quadrant, and $\sin x\cos x > 0$.
We conclude that $\sin 2x = 2\sin x\cos x$.

Chapter 6. Section 6.4 Conditional Trigonometric Equations

1. The value $x = -\pi/6$ is a solution of the equation $\cos x = \sqrt{3}/2$.

3. The value $x = -5\pi/3$ is not a solution of the equation $\csc x + 1 = 4\sin x$. We have $\csc(-5\pi/3) + 1 = (2\sqrt{3}+3)/3$, while $4\sin(-5\pi/3) = 2\sqrt{3}$.

5. The solutions of the equation $\cos x = 0$ in the interval $[0, 2\pi)$ are $x = \pi/2$, and $x = 3\pi/2$.

7. The solution of the equation $\tan x - 2 = 0$ in the interval $[0, 2\pi)$ is, approximately, $x \approx 1.11$.

9. The solutions of the equation $3\sin x + \sqrt{3} = \sin x$ in the interval $[0, 2\pi)$ are $x = 4\pi/3$, and $x = 5\pi/3$.

11. The solutions of the equation $(\csc x - 2)(\sec x - 2) = 0$ in the interval $[0, 2\pi)$ are $x = \pi/6$, $x = \pi/3$, $x = 5\pi/6$, and $x = 5\pi/3$.

13. The solutions of the equation $\sin x \cos x = 0$ in the interval $[0, 2\pi)$ are $x = 0$, $x = \pi/2$, $x = \pi$, and $x = 3\pi/2$.

15. The solutions of the equation $\cos 2x = 1$ in the interval $[0, 2\pi)$ are $x = 0$, and $x = \pi$.

17. The solutions of the equation $\csc(2x + 1) = 3$ in the interval $[0, 2\pi)$ are, approximately, $x \approx 0.17$, and $x \approx 3.31$.

19. The solution of the equation $\cot x/2 = -1$ in the interval $[0, 2\pi)$ is $x = 3\pi/2$.

21. The solutions of the equation $\sin x = 0$ are $x = \pi k$, for any integer *k*.

23. The solutions of the equation $2\cos x - 1 = 0$ are $x = \pi/3 + 2\pi k$, and $x = 5\pi/3 + 2\pi k$, for any integer *k*.

25. The solutions of the equation $(\sin x + 1)(\tan x - 1) = 0$ are $x = \pi/4 + \pi k$, and $x = 3\pi/2 + 2\pi k$, for any integer *k*.

27. The solutions of the equation $\cot x \csc x - 2\cot x = 0$ are $x = \pi/6 + 2\pi k$, $x = \pi/2 + \pi k$, and $x = 5\pi/6 + 2\pi k$, for any integer *k*. ($\cot x \csc x - 2\cot x = 0$ is equivalent to $\cot x(\csc x - 2) = 0$, which implies $\cot x = 0$ or $\csc x - 2 = 0$, and equivalently: $\cos x = 0$ or $\sin x = 1/2$.)

29. The solutions of the equation $\cos^2 x = 1$ are $x = \pi k$, for any integer k. ($\cos^2 x = 1$ is equivalent to $\cos^2 x - 1 = 0$, or $(\cos x - 1)(\cos x + 1) = 0$, so $\cos x = -1$ or $\cos x = 1$.)

31. The solutions of the equation $\tan^3 x - \tan x = 0$ are $x = \pi k$, $x = \pi/4 + \pi k$, and $x = 3\pi/4 + \pi k$ for any integer k. ($\tan^3 x - \tan x = 0$ is equivalent to $(\tan^2 x - 1)\tan x = 0$, so $\tan x = -1$, $\tan x = 0$, or $\tan x = 1$.) The solutions could be further combined as $x = \pi k$, $x = \pi/4 + \pi k/2$, for any integer k.

33. The equation $2\sin^2 x = \cos x + 1$ is equivalent to $2 - 2\cos^2 x = \cos x + 1$, or $2\cos^2 x + \cos x - 1 = 0$. Equivalently, $(2\cos x - 1)(\cos x + 1) = 0$, so $\cos x = 1/2$ or $\cos x = -1$. Then, the solutions of the equation $2\sin^2 x = \cos x + 1$ are $x = \pi/3 + 2\pi k$, $x = \pi(2k+1)$, $x = 5\pi/3 + 2\pi k$, for any integer k.

35. The equation $\sec x = -\sqrt{2}\tan x$ is equivalent to $1/\cos x = -\sqrt{2}\sin x/\cos x$, or $\sin x = -\sqrt{2}/2$. Then, the solutions of the equation $\sec x = -\sqrt{2}\tan x$ in the interval $[0, 2\pi)$ are $x = 5\pi/4, x = 7\pi/4$.

37. The solutions of the equation $\sin x - \cos x = 0$ in the interval $[0, 2\pi)$ are $x = \pi/4$, and $x = 5\pi/4$.

39. The equation $3\tan^2 x - \sin x \sec^2 x = 0$ is equivalent to $3\tan^2 x - \tan x/\cos x = 0$, or $\dfrac{3\tan^2 x\cos x - \tan x}{\cos x} = 0$; then $\tan x(3\tan x\cos x - 1) = 0$, or $\tan x(3\sin x - 1) = 0$. Then $\tan x = 0$, or $\sin x = 1/3$. The solutions of $3\tan^2 x - \sin x\sec^2 x = 0$ in the interval $[0, 2\pi)$ are $x = 0$, $x \approx 0.34$, $x = \pi$, and $x \approx 2.80$.

41. The equation $\cos 2x = \cos^2 x - 1$ is equivalent to $\cos^2 x - \sin^2 x = \cos^2 x - (\sin^2 x + \cos^2 x)$, or $\cos^2 x = 0$. Then, the solutions of $\cos 2x = \cos^2 x - 1$ in the interval $[0, 2\pi)$ are $x = 0$, and $x = \pi$.

43. The equation $\cot x = \tan 2x$ is equivalent to $\cos x/\sin x = 2\sin x\cos x/(\cos^2 x - \sin^2 x)$, or $\cos x/\sin x = 2\sin x\cos x/(2\cos^2 x - 1)$, and $\cos x = 2\sin^2 x\cos x/(2\cos^2 x - 1)$. Then, $\cos x\left(1 - 2\sin^2 x/(2\cos^2 x - 1)\right) = 0$, so $\cos x = 0$, or $2\sin^2 x/(2\cos^2 x - 1) = 1$. The last equation is equivalent to $2\sin^2 x = 2\cos^2 x - 1$, or $2\left(1 - \cos^2 x\right) = 2\cos^2 x - 1$, and $4\cos^2 x = 3$. The solutions of $\cot x = \tan 2x$ are solutions of $\cos x = 0$ and $\cos x = \pm\sqrt{3}/2$, and the ones in the interval $[0, 2\pi)$ are $x = \pi/6$, $x = \pi/2$, $x = 5\pi/6$, $x = 7\pi/6$, $x = 3\pi/2$, and $x = 11\pi/6$.

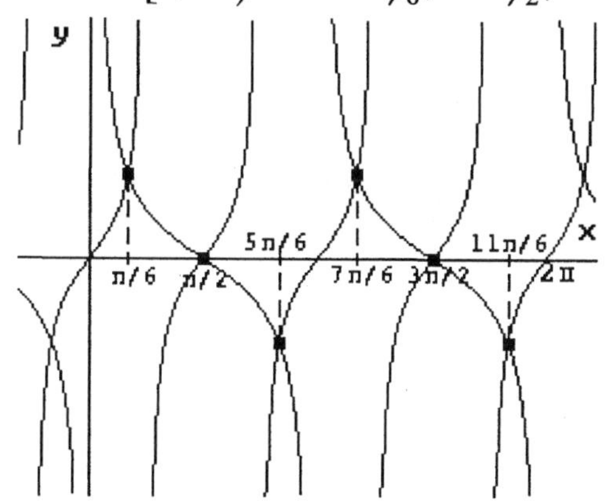

45. Using the quadratic formula for $10\sin^2 x + \sin x - 6 = 0$ we obtain $\sin x = \dfrac{-1 \pm \sqrt{241}}{20}$, that is $\sin x \approx \begin{cases} 0.72621 \\ -0.82621 \end{cases}$. The solutions of $10\sin^2 x + \sin x - 6 = 0$ in the interval $[0, 2\pi)$ are, approximately, $x \approx 0.81$, $x \approx 2.33$, $x \approx 4.11$, and $x \approx 5.31$.

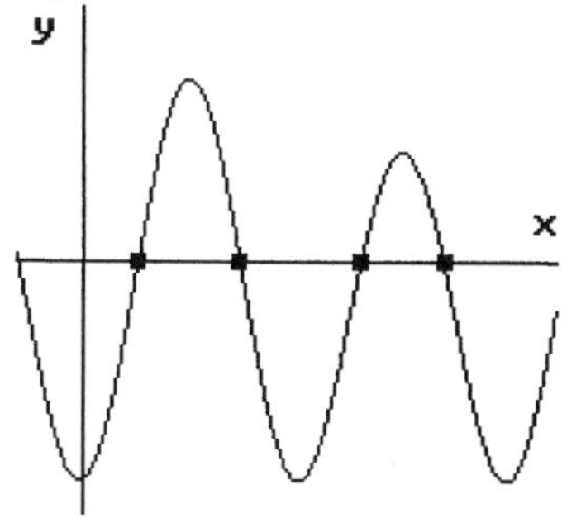

47. The equation $9\cos^2 x - 3\cos x - 2 = 0$ is equivalent to $(3\cos x + 1)(3\cos x - 2) = 0$, so $\cos x = -1/3$, or $\cos x = 2/3$. The solutions of $9\cos^2 x - 3\cos x - 2 = 0$ in the interval $[0, 2\pi)$ are, approximately, $x \approx 0.84$, $x \approx 1.91$, $x \approx 4.37$, and $x \approx 5.44$.

49. $\sec^2 x - \tan x = 3$ is equivalent to $(1+\tan^2 x) - \tan x = 3$, or $\tan^2 x - \tan x - 2 = 0$. The last equation can be factored, $(\tan x - 2)(\tan x + 1) = 0$, so $\tan x = 2$, or $\tan x = -1$. The solutions of $\sec^2 x - \tan x = 3$ in the interval $[0, 2\pi)$ are $x \approx 1.11$, $x \approx 4.25$, $x = 3\pi/4$, and $x = 7\pi/4$.

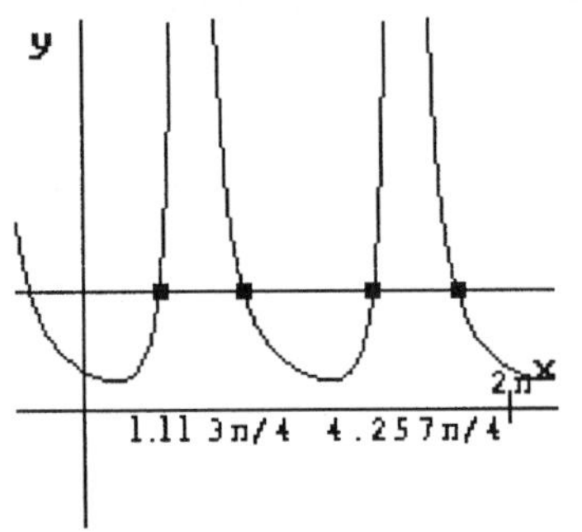

51. The equation $\cos^2 x + \sin^2 x = \dfrac{1}{2}$ has no solutions. It is a contradiction.

53. The equation $\cos x \sin x = 1/4$ is equivalent to $2\cos x \sin x = 1/2$, or $\sin 2x = 1/2$. The solutions of $2\cos x \sin x = 1/2$ are $x = \pi/12$, $x = 5\pi/12$, $x = 13\pi/12$, and $x = 17\pi/12$.

55. The solutions of the equation $\sin^3 x - 3\sin x + 1 = 0$ in the interval $[0, 2\pi)$ are, approximately, $x \approx 0.35$ and $x \approx 2.79$.

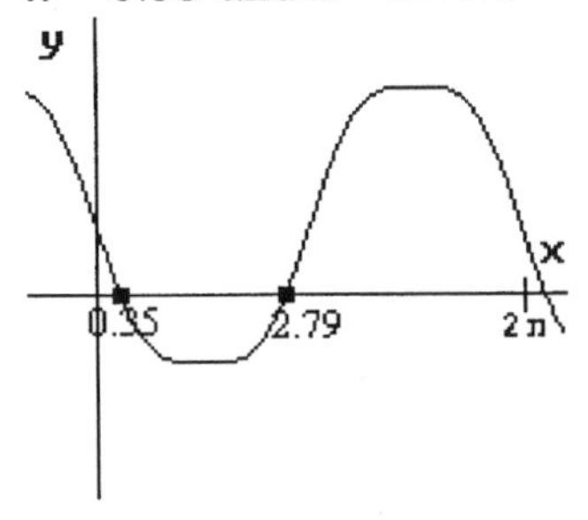

57. The equation $x = \cos x$ has only one solution in the interval $[0, 2\pi)$, approximately $x \approx 0.74$.

59. The equation $\cos^3 x = \sin x$ has two solutions in the interval $[0, 2\pi)$, approximately $x \approx 0.6$ and $x \approx 3.74$.

61. For the function $f(x) = \tan^3 \pi x - \sec^2 \pi x$:
(a) the period is $p = 1$
(b) there is only one zero of *f* in the interval [0,1), approximately $x \approx 0.31$.

63. For the function $f(x) = \ln\left(2 + \cos \pi x/6\right) - 1/2$:
(a) the period is $p = 12$
(b) the zeros of in the interval $[0, 2\pi)$ are, approximately, $x \approx 3.69$ and $x \approx 8.31$.

65. The height in feet of a passenger in a Ferris wheel is given by $y(t) = 55 + 50\sin\left(\frac{\pi t}{15} - \frac{\pi}{2}\right)$, where t is the time in seconds and $t = 0$ corresponds to the time at which the wheel was set in motion.

(a) The first two times for which the height of the passenger is 80 feet are the first two positive solutions of $55 + 50\sin\left(\frac{\pi t}{15} - \frac{\pi}{2}\right) = 80$, or $\sin\left(\frac{\pi t}{15} - \frac{\pi}{2}\right) = \frac{1}{2}$; these solutions correspond to $\frac{\pi t}{15} - \frac{\pi}{2} = \frac{\pi}{6}$ and $\frac{\pi t}{15} - \frac{\pi}{2} = \frac{5\pi}{6}$, namely $t = 10$ and $t = 20$ seconds.

(b) The time that it takes the Ferris wheel to complete one rotation is the period of y, $p = 30$ seconds.

67. The current (in amperes) in an electrical circuit t seconds after the circuit is closed is given by $E(t) = -12\cos 4t$. The first two times for which E is 8 amperes are $t \approx 0.58$ and $t \approx 1$.

In Exercises 69-71, we consider the situation of an object thrown from ground level, with an initial angle θ and an initial speed v. Ignoring air resistance, the range (maximum horizontal distance) is given by $R = \frac{v^2 \sin 2\theta}{32}$ feet, and maximum height $H = \frac{v^2 \sin^2\theta}{64}$ feet.

69. If a quarterback aims the football to reach a receiver 60 yards (180 feet) away, with an initial velocity of 80 feet per second, we have: $80^2 \sin 2\theta / 32 = 180$, so $\sin 2\theta = 0.9$, $2\theta \approx 64.16^0$, so the angle at which the football should be thrown is, approximately, 32.08^0.

71. If an object is projected into the air from ground level in such a way that its range equals its maximum height, then $v^2 \sin 2\theta / 32 = v^2 \sin^2\theta / 64$, so $2\sin 2\theta = \sin^2\theta$ and $4\sin\theta\cos\theta = \sin^2\theta$. Then $\tan\theta = 4$, so $\theta = 75.96^0$.

Concepts and Critical Thinking

73. The statement: "If $-1 \le c \le 1$, the equation $\sin x = c$ has exactly one solution in the interval $[0, \pi)$" is false. If $0 < c < 1$, then there are two solutions of $\sin x = c$ in $[0, \pi)$. If $c < 0$, there are no solutions of $\sin x = c$ in $[0, \pi)$.

75. The statement: "The equation $\tan x = c$ has a solution for any c" is true.

77. Answers may vary. An example of a number c for which the equation $\sin x = c$ has no solution is $c = 2$. Any number c that does not belong to the interval $[-1, 1]$ is such an example.

79. An example of an integer n for which $\tan nx = 1$ has exactly 4 solutions in the interval $[0, 2\pi)$ is $n = 2$. The solutions of $\tan 2x = 1$ in $[0, 2\pi)$ are $x = \frac{\pi}{8}, \frac{5\pi}{8}, \frac{9\pi}{8}$, and $\frac{13\pi}{8}$. (*Note*: $n = -2$ is another example. The solutions of $\tan(-2x) = 1$ in the interval $[0, 2\pi)$ are $x = \frac{3\pi}{8}, \frac{7\pi}{8}, \frac{11\pi}{8}$, and $\frac{15\pi}{8}$.

81. If a function f is periodic with period $\pi/3$, and $f(x) = 1/2$ has two solutions in the interval $[0, \pi/3]$, then there may be up to six solutions of this equation in the interval $[4\pi/3, 7\pi/3]$. There will be fewer than six solutions if one or both of the two solutions of $f(x) = 1/2$ in the interval $[0, \pi/3]$ happens to be 0 or $\pi/3$.

Chapter 6. Review

1. $2\sin x + \cos x = 2$ is a conditional equation; the smallest positive solution is $x \approx 0.64$.

3. The equation $\cos\left(\pi/2 - x\right) = \sin x + 1$ is a contradiction:

$$\begin{aligned}\cos\left(\pi/2 - x\right) &= \cos \pi/2 \cos x + \sin \pi/2 \sin x \\ &= \sin x\end{aligned}$$

Then, if $\cos\left(\pi/2 - x\right) = \sin x + 1$, we obtain $\sin x = \sin x + 1$, and $0 \neq 1$.

5. The equation $\dfrac{1}{1-\cos x} = \csc^2 x + \csc x \cot x$ is an identity:

$$\begin{aligned}\csc^2 x + \csc x \cot x &= 1/\sin^2 x + \left(1/\sin x\right)\left(\cos x/\sin x\right) \\ &= \frac{1+\cos x}{\sin^2 x} \\ &= \frac{1+\cos x}{1-\cos^2 x} \\ &= \frac{1+\cos x}{(1-\cos x)(1+\cos x)} \\ &= \frac{1}{1-\cos x}\end{aligned}$$

7. The equation $\cot x/\tan x = \csc^2 x - 1$ is an identity: $\cot x/\tan x = \cot^2 x$, so $\cot x/\tan x = \csc^2 x - 1$.

9. The equation $\cot^2 y + 5 = \csc^2 y + 4$ is an identity: $\csc^2 y + 4 = \left(\cot^2 y + 1\right) + 4$.

11. The equation $\cos\theta + \sin\theta\tan\theta = \sec\theta$ is an identity:

$$\begin{aligned}\cos\theta + \sin\theta\tan\theta &= \cos\theta + \frac{\sin^2\theta}{\cos\theta} \\ &= \frac{\cos^2\theta + \sin^2\theta}{\cos\theta} \\ &= \frac{1}{\cos\theta} \\ &= \sec\theta\end{aligned}$$

13. The equation $\dfrac{\sec\alpha+\tan\alpha}{\sec\alpha-\tan\alpha}=(\sec\alpha+\tan\alpha)^2$ is an identity:

$$\begin{aligned}(\sec\alpha+\tan\alpha)^2 &= \left(\frac{1}{\cos\alpha}+\frac{\sin\alpha}{\cos\alpha}\right)^2\\ &= \frac{(1+\sin\alpha)^2}{\cos^2\alpha}\\ &= \frac{(1+\sin\alpha)^2}{1-\sin^2\alpha}\\ &= \frac{(1+\sin\alpha)^2}{1-\sin^2\alpha}\\ &= \frac{1+\sin\alpha}{1-\sin\alpha}\\ &= \frac{\left(\dfrac{1+\sin\alpha}{\cos\alpha}\right)}{\left(\dfrac{1-\sin\alpha}{\cos\alpha}\right)}\\ &= \frac{\sec\alpha+\tan\alpha}{\sec\alpha-\tan\alpha}\end{aligned}$$

15. To verify that $|\sin x|=\sqrt{1-\cos^2 x}$ is an identity: we have $1-\cos^2 x=\sin^2 x$, therefore $\sqrt{1-\cos^2 x}=|\sin x|$.

17. $\cos x=\pm\sqrt{1-\sin^2 x}$; the sign of $\cos x$ is decided based on the quadrant where x terminates.

19. If $\cos u=4/5$, $0\le u<\pi/2$ and $\cos v=8/17$, $0\le v<\pi/2$, then $\sin u=3/5$ and $\sin v=15/17$; then:

(a) $\cos(u-v)=\cos u\cos v+\sin u\sin v$
$=(4/5)(8/17)+(3/5)(15/17)$
$=77/85$

(b) $\cos(u+v)=\cos u\cos v-\sin u\sin v$
$=(4/5)(8/17)-(3/5)(15/17)$
$=-13/85$

(c) $\sin(u-v)=\sin u\cos v-\sin v\cos u$
$=(3/5)(8/17)-(15/17)(4/5)$
$=-36/85$

(d) $\sin(u+v)=\sin u\cos v+\sin v\cos u$
$=(3/5)(8/17)+(15/17)(4/5)$
$=84/85$

21. $\cos 10^0\cos 50^0-\sin 10^0\sin 50^0=\cos 60^0$

23. $\sin x\cos 2x+\cos x\sin 2x=\sin 3x$

25. $\sin\left(x - \frac{3\pi}{2}\right) = \sin x \cos\frac{3\pi}{2} - \sin\frac{3\pi}{2}\cos x$
$= \sin x \cdot (0) - (-1)\cos x$
$= \cos x$

27. $\cos\left(\frac{\pi}{4} - x\right) = \cos\frac{\pi}{4}\cos x + \sin\frac{\pi}{4}\sin x$
$= \frac{\sqrt{2}}{2}\cos x + \frac{\sqrt{2}}{2}\sin x$

29. To verify that the equation $\frac{1}{(\cot x - \cot y)} = \frac{\sin x \sin y}{\sin(y-x)}$ is an identity:

$$\frac{\sin x \sin y}{\sin(y-x)} = \frac{1}{\frac{\sin(y-x)}{\sin x \sin y}}$$
$$= \frac{1}{\frac{(\sin y \cos x - \sin x \cos y)}{\sin x \sin y}}$$
$$= \frac{1}{(\cot x - \cot y)}$$

31. If $\tan\theta = \frac{3}{4}$, $\pi \le \theta < \frac{3\pi}{2}$, then $\sec\theta = -\sqrt{\tan^2\theta + 1}$, that is $\sec\theta = -\frac{5}{4}$. Therefore, $\cos\theta = -\frac{4}{5}$ and $\sin\theta = -\frac{3}{5}$. Then:

(a) $\sin 2\theta = 2\sin\theta\cos\theta$, that is $\sin 2\theta = 2\left(-\frac{3}{5}\right)\left(-\frac{4}{5}\right)$, or $\sin 2\theta = \frac{24}{25}$

(b) $\cos 2\theta = \cos^2\theta - \sin^2\theta$, that is $\cos 2\theta = \frac{16}{25} - \frac{9}{25}$, or $\cos 2\theta = \frac{7}{25}$

(c) $\tan 2\theta = \frac{24}{7}$

33. If $\sin x = -\frac{3}{5}$, $\frac{3\pi}{2} \le x < 2\pi$, then $\cos x = \sqrt{1 - (-\frac{3}{5})^2}$, that is, $\cos x = \frac{4}{5}$. We have $\frac{3\pi}{4} \le \frac{x}{2} < \pi$, so $\frac{x}{2}$ terminates in the second quadrant, therefore $\sin\frac{x}{2} > 0$ and $\cos\frac{x}{2} < 0$.

(a) $\sin\frac{x}{2} = \sqrt{\frac{1-\frac{4}{5}}{2}}$, or $\sin\frac{x}{2} = \frac{\sqrt{10}}{10}$

(b) $\cos\frac{x}{2} = -\sqrt{\frac{1+\frac{4}{5}}{2}}$, or $\cos\frac{x}{2} = -\frac{3\sqrt{10}}{10}$

(c) $\tan\frac{x}{2} = -\frac{1}{3}$

35. $\cos\frac{\pi}{12} = \sqrt{\frac{1+\cos\frac{\pi}{6}}{2}}$, or $\cos\frac{\pi}{12} = \sqrt{\frac{1+\frac{\sqrt{3}}{2}}{2}}$; that is $\cos\frac{\pi}{12} = \frac{\sqrt{2+\sqrt{3}}}{2}$

37. $\tan 75^0 = \tan(30^0 + 45^0)$; therefore $\tan 75^0 = \frac{\tan 30^0 + \tan 45^0}{1 - \tan 30^0 \tan 45^0}$, and $\tan 75^0 = \frac{\frac{\sqrt{3}}{3} + 1}{1 - \frac{\sqrt{3}}{3}(1)}$.

So, $\tan 75^0 = \frac{3+\sqrt{3}}{3-\sqrt{3}}$, or $\tan 75^0 = 2 + \sqrt{3}$.

39. To verify the identity $2\sin x = 4\sin \frac{x}{2}\cos \frac{x}{2}$:

$$\begin{aligned} 2\sin x &= 2\sin(x+x) \\ &= 2\left(2\sin \tfrac{x}{2}\cos \tfrac{x}{2}\right) \\ &= 4\sin \tfrac{x}{2}\cos \tfrac{x}{2} \end{aligned}$$

41. To verify the identity $\cos 4x = 8\cos^4 x - 8\cos^2 x + 1$:

$$\begin{aligned} \cos 4x &= \cos^2(2x) - \sin^2(2x) \\ &= \left(\cos^2 x - \sin^2 x\right)^2 - (2\sin x\cos x)^2 \\ &= \cos^4 x + \underbrace{\sin^4 x}_{\left(1-\cos^2 x\right)^2} - 6\underbrace{\sin^2 x}_{1-\cos^2 x}\cos^2 x \\ &= 8\cos^4 x - 8\cos^2 x + 1 \end{aligned}$$

43. To verify the identity $\tan x + \cot x = 2\csc 2x$:

$$\begin{aligned} 2\csc 2x &= \frac{2}{\sin 2x} \\ &= \frac{\not{2}}{\not{2}\sin x\cos x} \\ &= \frac{\sin^2 x + \cos^2 x}{\sin x\cos x} \\ &= \tan x + \cot x \end{aligned}$$

45. The value $x = \pi$ is the only one solution of the equation $\cos x = -1$ in the interval $[0, 2\pi)$.

47. The solutions of the equation $(2\sin x - 1)\left(\tan x - \sqrt{3}\right) = 0$ are solutions of $2\sin x - 1 = 0$ or $\tan x - \sqrt{3} = 0$. In the interval $[0, 2\pi)$, these solutions are $x = \frac{\pi}{6}, \frac{\pi}{3}, \frac{5\pi}{6}, \frac{4\pi}{3}$.

49. The equation $2\sin x\cos x - \sin x = 0$ is equivalent to $\sin x\left(2\cos x - 1\right) = 0$, so $\sin x = 0$ or $2\cos x - 1 = 0$. Then, the solutions of $2\sin x\cos x - \sin x = 0$ in $[0, 2\pi)$ are $x = 0, \frac{\pi}{3}, \pi, \frac{5\pi}{3}$.

51. The equation $2\cos x + \sec x - 3 = 0$ is equivalent to $\dfrac{2\cos^2 x - 3\cos x + 1}{\cos x} = 0$, so $2\cos^2 x - 3\cos x + 1 = 0$, or equivalently $(2\cos x - 1)(\cos x - 1) = 0$. Then, the solutions of $2\cos x + \sec x - 3 = 0$ in the interval $[0, 2\pi)$ are $x = 0, \frac{\pi}{3}, \frac{5\pi}{3}$.

53. The equation $\sin 2x + \sin x = 0$ is equivalent to $\sin x(2\cos x + 1) = 0$. Then, the solutions of $\sin 2x + \sin x = 0$ in the interval $[0, 2\pi)$ are $x = 0, \frac{2\pi}{3}, \pi, \frac{4\pi}{3}$.

55. The solutions of the equation $6\sin^2 x - 2\sin x - 1 = 0$ satisfy $\sin x = \dfrac{1 \pm \sqrt{7}}{6}$. In the interval $[0, 2\pi)$, these solutions are, approximately, $x \approx 0.65$, 2.49, 3.42, and 6.01.

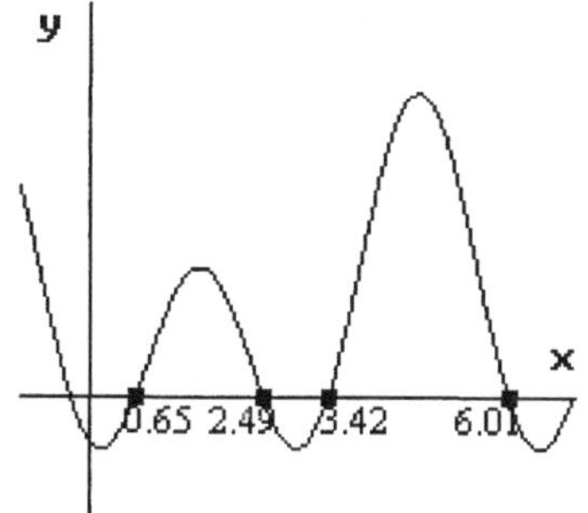

57. The equation $\cos 3x = -1$ has period $p = 2\pi/3$. The smallest positive solution satisfies $3x = \pi$, or $x = \pi/3$. The solutions of $\cos 3x = -1$ in the interval $[0, 2\pi)$ are $x = \pi/3$, π, and $5\pi/3$.

59. The solutions of the equation $\cos x = 1/2$ are $x = \pi/3 + 2\pi k$, $x = 5\pi/3 + 2\pi k$, for any integer *k*.

61. The solutions of the equation $\left(2\sin x - \sqrt{3}\right)\cos x = 0$ are solutions of $\sin x = \sqrt{3}/2$ or $\cos x = 0$. These solutions are $x = \pi/2 + \pi k$, $x = \pi/3 + 2\pi k$, $x = 2\pi/3 + 2\pi k$, for any integer *k*.

63. The equation $\cot^3 x + \cot x = 0$ is equivalent to $\left(\cot^2 x + 1\right)\cot x = 0$. The solutions of the equation $\cot^3 x + \cot x = 0$ are solutions of $\cot x = 0$, since $\cot^2 x + 1 \neq 0$ for all *x*. Then, $\cot^3 x + \cot x = 0$ has solutions $x = \pi/2 + \pi k$, for any integer *k*.

65. The equation $\sin 2x = 1$ has solutions $x = \pi/4 + \pi k$, for any integer *k*.

67. The solutions of the equation $\sin 2x + \cos 3x = 1$ in the interval $[0, 2\pi)$ are, approximately, $x \approx 0$, 0.45, 3.7, and 4.5.

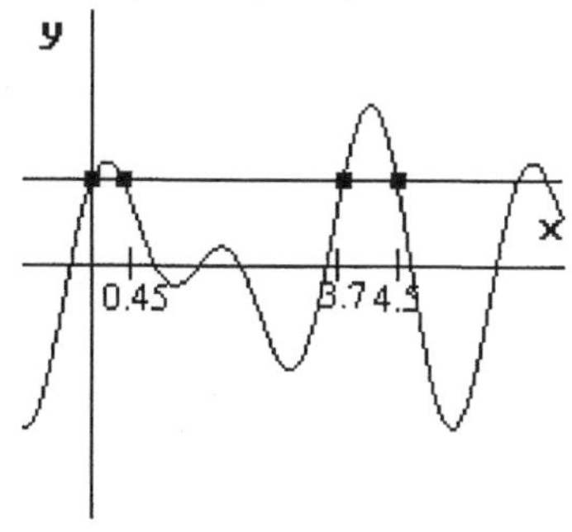

69. The function $f(x) = 2\sin(3\pi x) - 3\sin(2\pi x)$ has period $p = 2$. The zeros of f in the interval $[0, 2)$ are $x = 0$, $x = 1$ and, approximately, $x \approx 0.58$ and $x \approx 1.42$.

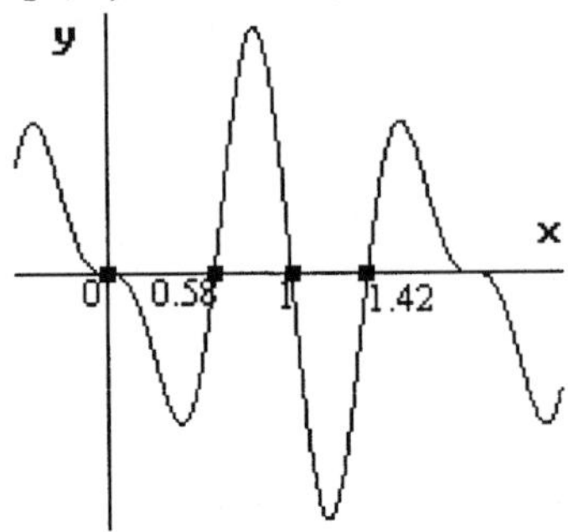

The range R of a projectile launched from ground level with an initial speed of v feet per second and at an angle θ *is* $R = {}^{v^2 \sin 2\theta}/_{32}$.

71. If a baseball is thrown with an initial speed of $v = 60$ feet per second attains a range $R = 100$ feet, then we can estimate the angle θ using the equation $\dfrac{60^2 \sin 2\theta}{32} = 100$, which implies $\sin 2\theta = {}^{3200}/_{3600}$, so $2\theta = \sin^{-1} {}^8/_9$ and $2\theta \approx 62.73^0$, or $\theta \approx 31.37^0$.

73. If the current in amperes in an electrical circuit at time t seconds is $i(t) = 20\cos\left(60t - {}^{\pi}/_{3}\right)$, the current is 10 amperes at the times $t = {}^{\pi}/_{90} + {}^{\pi k}/_{30}$, for any natural number k. That is, at the times $t \approx 0.035 + 0.105k$ seconds.

Chapter 6. Test

1. The statement: "The equation $\sin(x+y)=\sin x+\sin y$ is true for all values of x and y" is false. For example, consider $x=\pi/6$ and $y=\pi/6$. We have: $\sin(x+y)=\sqrt{3}/2$, $\sin x+\sin y=1$.

3. The statement: "The equation $\sin 2x=2\sin x$ has infinitely many solutions" is true. ($\sin 2x=2\sin x\cos x$, so $\sin 2x=2\sin x$ for all $x=\pi/2+2\pi k$ for any integer k.)

5. The statement: "The equation $1-\sin^2\left(e^x\right)=\cos^2\left(e^x\right)$ is true for all x" is true.

7. The statement: "The equation $\sin x=x$ has infinitely many solutions" is false. There is only one solution of the equation $\sin x=x$.

9. Answers may vary. An example of numbers a and b such that the equation $\cos^a x+\sin^a x=b$ is an identity is the pair $a=2,\ b=1$: $\cos^2 x+\sin^2 x=1$ for all x.

11. Answers may vary. An example of numbers a and b such that the equation $\cos^a x+\sin^a x=b$ is a contradiction is the pair $a=1,\ b=3$: $\cos x+\sin x=3$ is a contradiction.

13. To verify that $\cos x=\sec x(1-\sin^2 x)$ is an identity:

$$\sec x(1-\sin^2 x)=(1/\cos x)\cos^2 x$$
$$=\cos x$$

15.
$$\tan\left(x-\pi/4\right)=\sin\left(x-\pi/4\right)\Big/\cos\left(x-\pi/4\right)$$
$$=\frac{\sin x\cos\pi/4+\sin\pi/4\cos x}{\cos x\cos\pi/4-\sin x\sin\pi/4}$$
$$=\sqrt{2}/2(\sin x+\cos x)\Big/\sqrt{2}/2(\cos x-\sin x)$$
$$=\frac{\sin x+\cos x}{\cos x-\sin x}$$

17. The equation $2\tan x-4=2$ has two solutions in the interval $[0,2\pi)$, $x\approx 1.25$ and $x\approx 4.39$.

19. The equation $2\sin^2 x\cos x=3$ is a contradiction, since $2\sin^2 x\cos x\le 2$ for all x.

Chapter 7. APPLICATIONS OF TRIGONOMETRY

Section 7.1 The Law of Sines

1. This is an AAS case. $B = 180^0 - (75^0 + 55^0)$, that is $\underline{B = 50^0}$.

$\sin 55^0/10 = \sin 50^0/b$, therefore $b = 10\sin 50^0/\sin 55^0$, or $\underline{b \approx 9.35}$.

$\sin 55^0/10 = \sin 75^0/a$, therefore $a = 10\sin 75^0/\sin 55^0$, or $\underline{a \approx 11.79}$.

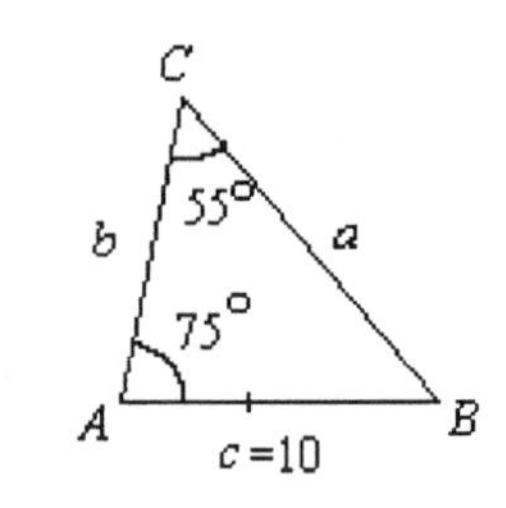

3. This is an SSA case. $\sin 120^0/10 = \sin C/8$, therefore $\sin C = 8\sin 120^0/10$, so $C = 43.85^0$ or $C = 136.15^0$, this last value being impossible. So, there is only one triangle with the given information, with $\underline{C = 43.85^0}$, and $A = 180^0 - (120^0 + 43.85^0)$, or $\underline{A = 16.15^0}$.

To obtain a, $\sin 120^0/10 = \sin 16.15^0/a$, so $a = 10\sin 16.15^0/\sin 120^0$, or $\underline{a \approx 3.21}$.

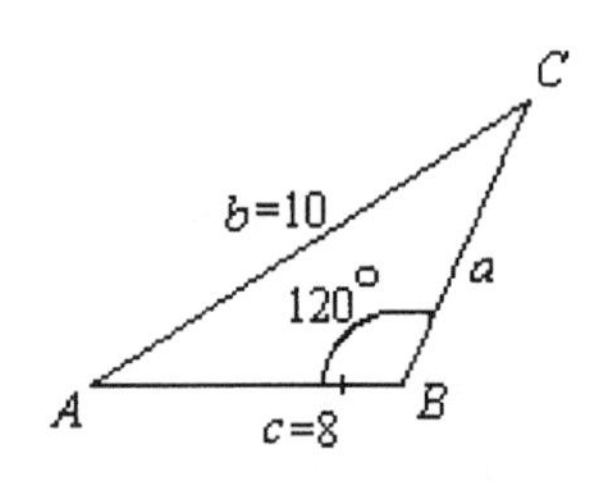

5. This is an ASA case. $A = 180^0 - (75^0 + 35^0)$, or $\underline{A = 70^0}$.

$\sin 70^0/9 = \sin 75^0/b$, so $b = 9\sin 75^0/\sin 70^0$, or $\underline{b \approx 9.25}$;

$\sin 70^0/9 = \sin 35^0/c$, so $c = 9\sin 35^0/\sin 70^0$, or $\underline{c \approx 5.49}$.

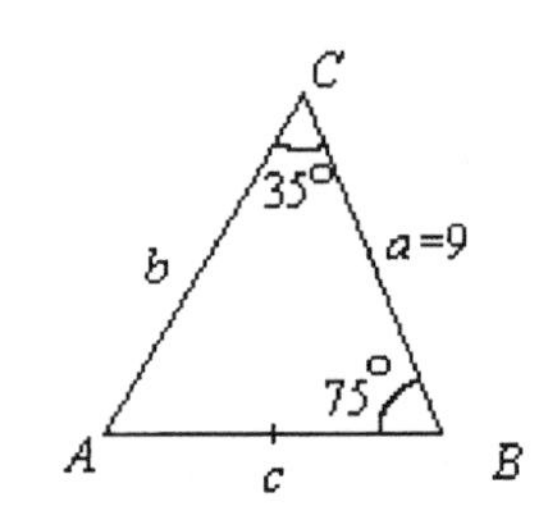

7. This is an ASA case. $C = 180^0 - (26^0 + 34^0)$, or $\underline{C = 120^0}$.

$\sin 120^0/16.2 = \sin 34^0/b$, so $b = 16.2\sin 34^0/\sin 120^0$, or $\underline{b \approx 10.33}$;

$\sin 120^0/16.2 = \sin 26^0/a$, so $a = 16.2\sin 26^0/\sin 120^0$, or $\underline{a \approx 8.2}$.

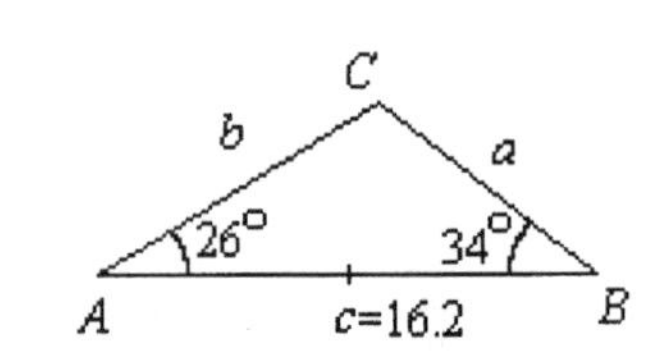

29. The area of this triangle is $A = \frac{1}{2}(20)(15)\sin 75^0$, approximately $A \approx 144.89$.

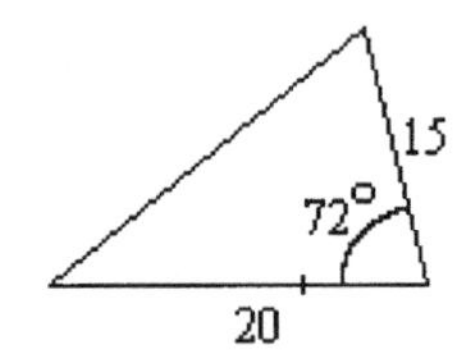

31. $A = 180^0 - (122^0 + 40^0)$, that is $A = 18^0$.
$\sin 18^0 / 2000 = \sin 122^0 / c$, therefore $c = 2000 \sin 122^0 / \sin 18^0$, or $c \approx 5488.68$. The area of this triangle is $A = \frac{1}{2}(5488.68)(2000)\sin 40^0$, approximately $A \approx 3{,}528{,}055.5$.

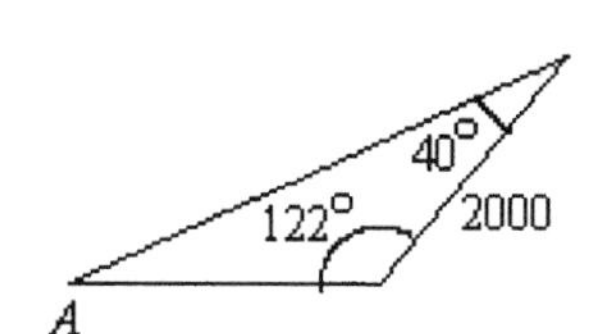

33. This is a SSA case. $\sin C / 120 = \sin 100^0 / 160$, so $\sin C = 120 \sin 100^0 / 160$, therefore $C \approx 47.61^0$ or $C \approx 132.39^0$, which is impossible. There is only one triangle with the given information, with $C \approx 47.61^0$ and $B \approx 180^0 - (100^0 + 47.61^{0)}$, or $B \approx 32.39^0$. The area of this triangle is $A \approx \frac{1}{2}(160)(120)\sin 32.39^0$, or $A \approx 5142.52$.

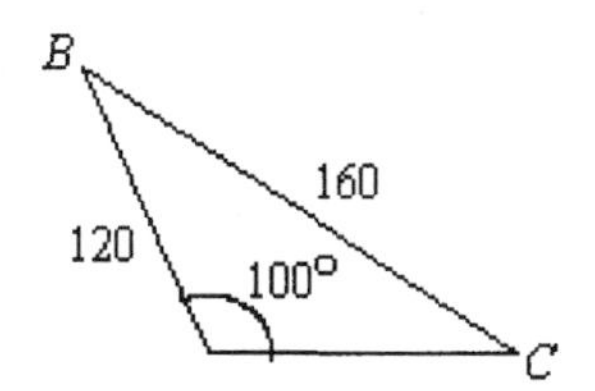

Applications

35. The angle C has measure $180^0 - \left(42^0 + 110^0\right)$, or $C = 28^0$.

The length from A to C, b, satisfies $\dfrac{b}{\sin 110^0} = \dfrac{100}{\sin 28^0}$, so

$b = \dfrac{100 \sin 110^0}{\sin 28^0}$, or $b \approx 200.16$ feet.

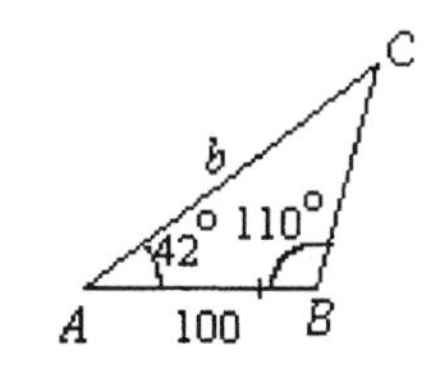

37. The distance from A to B is the distance traveled in 4 seconds, at the speed of 660 feet per second, so $\left|\overline{AB}\right| = 2640$ feet. The angle $\angle ABC$ measures $180^0 - 34^0$, that is $\angle ABC = 146^0$. The angle $\angle ACB$ measures $180^0 - \left(22^0 + 146^0\right)$, or $\angle ACB = 12^0$. The distance a from B to C satisfies $\dfrac{a}{\sin 22^0} = \dfrac{2640}{\sin 12^0}$, therefore

$a = \dfrac{2640 \sin 22^0}{\sin 12^0}$, or $a \approx 4756.64$ feet. Then, $\sin 34^0 \approx \dfrac{l}{4756.64}$, so $l \approx 4756.64 \sin 34^0$, that is $l \approx 2659.88$ feet. To obtain the height of Eiffel Tower, we subtract l from the altitude of 3650 feet at which the plane is flying, to obtain $h \approx 990.12$ feet.

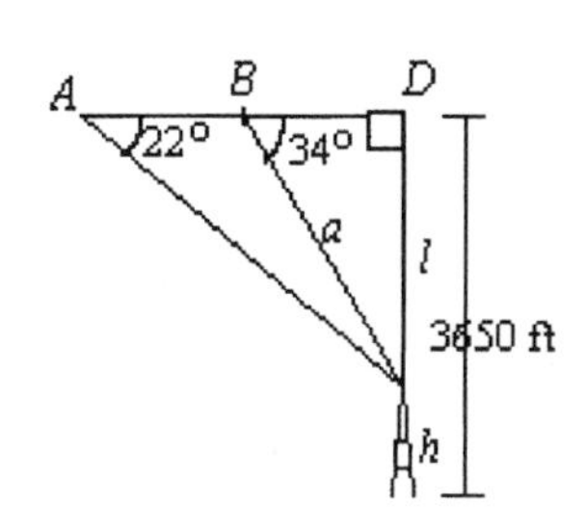

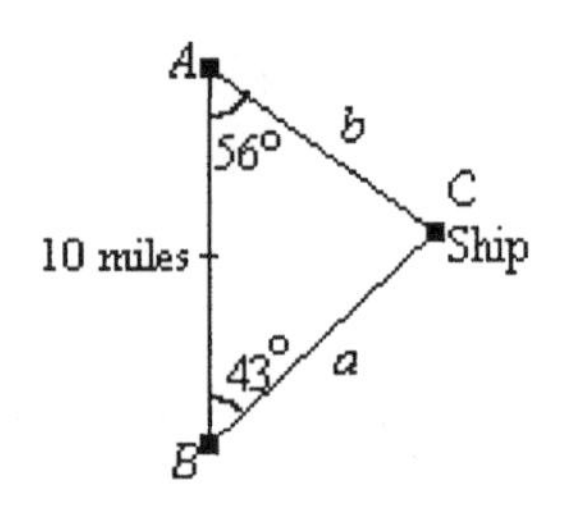

39. The measure of the angle C is $180^0-(56^0+43^0)$, that is $C=81^0$.

To obtain the distances a and b: $\dfrac{a}{\sin 56^0}=\dfrac{10}{\sin 81^0}$, therefore

$a=\dfrac{10\sin 56^0}{\sin 81^0}$, or $a\approx 8.39$ miles; now, $\dfrac{b}{\sin 43^0}=\dfrac{10}{\sin 81^0}$, so

$b=\dfrac{10\sin 43^0}{\sin 81^0}$, and $b\approx 6.9$ miles. The distance x from the ship to

the shore satisfies $\sin 43^0\approx\dfrac{x}{8.39}$, therefore $x\approx 8.39\sin 43^0$,

approximately $x\approx 5.72$ miles.

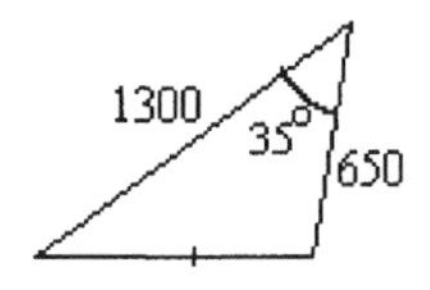

41. The area of the triangular area with two sides of 1300 feet and 650 feet each, if these sides make an angle of 35^0, is

$A=\dfrac{1}{2}(650)(1300)\left(\dfrac{35\pi}{180}\right)$, approximately $A\approx 258{,}090.56$ square

feet, or $A\approx 5.925$ acres. At \$2000 per acre, the asking price for this plot of land is \$11,850.

Concepts and Critical Thinking

43. The statement: "If two side lengths and an angle of a triangle are given, then the Law of Sines can be used to solve the triangle" is true.

45. The statement: "For a given triangle, the ratio of a side length to the sine of the angle opposite is constant" is true.

47. Answers may vary. An example of an AAS triangle is given below:

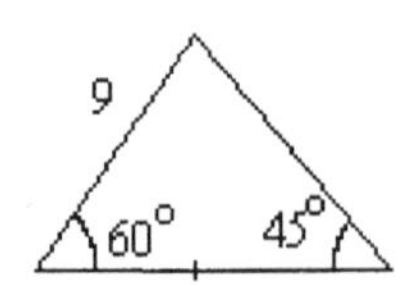

49. Answers may vary. An example of an SAS triangle is given below:

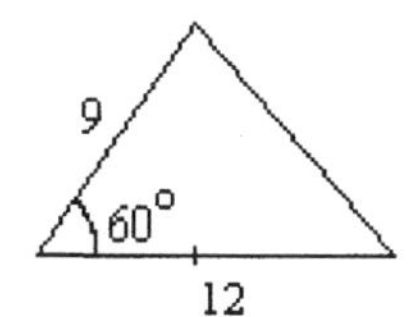

Chapter 7. Section 7.2 The Law of Cosines

In Exercises 1-27, a triangle is assumed with the standard labeling of sides and angles. That is, the angles are labeled A, B, C, and their opposite sides are labeled a, b, c, respectively. The SSA cases are solved using the Law of Sines.

1. If $a=12$, $b=9$, and $C=59^0$, then $c^2=a^2+b^2-2ab\cos C$, that is $c=\sqrt{144+81-2(12)(9)\cos 59^0}$, or $\boxed{c\approx 10.67}$. Now, $a^2=b^2+c^2-2bc\cos A$, that is $144\approx 81+113.75-2(9)(10.67)\cos A$, from where we obtain $A\approx\cos^{-1}\left(\frac{144-81-113.75}{-2(9)(10.67)}\right)$, or $\boxed{A\approx 74.68^0}$. Finally, $B\approx 180^0-\left(59^0+74.68^0\right)$, that is $\boxed{B\approx 46.32^0}$.

3. If $a=5$, $b=8$, and $c=12$, then $25=64+144-2(8)(12)\cos A$, so $\cos A\approx 0.953125$, therefore $A\approx 17.61^0$. Now, $64=25+144-2(5)(12)\cos B$, that is $B\approx\cos^{-1}0.875$, or $\boxed{B\approx 28.96^0}$. Finally, $C\approx 180^0-\left(17.61^0+28.96^0\right)$, that is $\boxed{C\approx 133.43^0}$.

5. If $b=26$, $c=4$, and $A=35^0$, then $a=\sqrt{26^2+4^2-2(26)(4)\cos 35^0}$, so $\boxed{a\approx 22.84}$. Also, $26^2\approx 22.84^2+4^2-2(22.84)(4)\cos B$, that is $B\approx\cos^{-1}\left(\frac{26^2-22.84^2-4^2}{-2(22.84)(4)}\right)$, or $\boxed{B\approx 139.21^0}$.

Finally, $C\approx 180^0-\left(139.21^0+35^0\right)$, that is $\boxed{C\approx 5.79^0}$.

7. If $b=15$, $c=20$, and $B=63^0$, then $\frac{\sin C}{20}=\frac{\sin 63^0}{15}$, so $\sin C\approx 1.18$. No such triangle exists.

9. If $b=130$, $c=180$, and $A=120^0$, then $a\approx\sqrt{130^2+180^2-2(130)(180)\cos 120^0}$, so $\boxed{a\approx 269.63}$. $\frac{\sin B}{130}\approx\frac{\sin 120^0}{130}$, from where it follows that $\boxed{B\approx 24.68^0}$ (the other possibility for B is irrelevant here.) Finally, $C\approx 180^0-24.68^0-120^0$, that is $\boxed{C\approx 35.32^0}$.

11. If $a=30$, $b=35$, and $c=50$, then $30^2\approx 35^2+50^2-2(35)(50)\cos A$, so $\boxed{A\approx 36.18^0}$. Now, $35^2\approx 30^2+50^2-2(30)(50)\cos B$, from where it follows that $\boxed{B\approx 43.53^0}$. Finally, $C\approx 180^0-43.53^0-36.18^0$, that is $\boxed{C\approx 99.99^0}$.

13. If $a=23$, $B=65^0$, and $C=44^0$, then $A=180^0-65^0-44^0$, that is $\boxed{A=71^0}$. Now, $\frac{\sin 71^0}{23}\approx\frac{\sin 44^0}{c}$, therefore $\boxed{c\approx 16.9}$. Finally, $\frac{\sin 71^0}{23}\approx\frac{\sin 65^0}{b}$, so $\boxed{b\approx 22.05}$.

15. If $a = 26$, $B = 14$, and $A = 115^o$, then $\frac{\sin B}{14} \approx \frac{\sin 115^o}{26}$, so $\boxed{B \approx 29.21^o}$ and $C \approx 180^o - (29.21^o + 115^o)$, or $\boxed{C \approx 35.79^o}$. Finally, $c \approx \sqrt{26^2 + 14^2 - 2(26)(14)\cos 35.79^o}$, so $\boxed{c \approx 16.78}$.

17. If $a = 1000$, $c = 2000$, and $C = 27^o$, then $c \approx \sqrt{1000^2 + 2000^2 - 2(1000)(2000)\cos 27^o}$, so $\boxed{c \approx 1198.32}$. Now, $\frac{\sin A}{1000} \approx \frac{\sin 27^o}{1198.32}$, so $\boxed{A \approx 22.26^o}$ (the other possibility for A is irrelevant.) Finally, $B \approx 180^o - 27^o - 22.26^o$, that is $\boxed{C \approx 130.74^o}$.

19. If $a = 240$, $b = 175$, and $c = 300$, then $240^2 = 175^2 + 300^2 - 2(175)(300)\cos A$, so $\boxed{A \approx 53.11^o}$. Now, $\frac{\sin B}{175} \approx \frac{\sin 53.11^o}{240}$, so $\boxed{B \approx 35.67^o}$ (the other possible value for B is not relevant here.) Finally, $C \approx 180^o - (53.11^o + 35.67^o)$, that is $\boxed{C \approx 91.22^o}$.

21. If $b = 14$, $c = 9$, and $B = 28^o$, then $\frac{\sin C}{9} \approx \frac{\sin 28^o}{14}$, that is $\boxed{C = 17.57^o}$ (the other possibility for C, $C \approx 162.43^o$, is irrelevant here.) Then, $A \approx 180^o - (17.57^o + 28^o)$, or $\boxed{A = 134.43^o}$. Finally, $a \approx \sqrt{14^2 + 9^2 - 2(14)(9)\cos 134.43^o}$, that is $\boxed{a \approx 21.29}$.

23. If $a = 7$, $b = 25$, and $A = 28^o$, then $\frac{\sin B}{25} \approx \frac{\sin 28^o}{7}$, and $\sin B \approx 1.68$. No such triangle exists.

25. If $a = 35$, $b = 70$, and $c = 14$, then $14^2 = 35^2 + 70^2 - 2(35)(70)\cos C$, so $\cos C = 1.21$. The triangle is not possible.

27. If $b = 1.8$, $B = 104^o$, and $C = 39^o$, then $A \approx 180^o - (104^o + 39^o)$, so $\boxed{A = 37^o}$ We have $\frac{\sin 104^o}{1.8} \approx \frac{\sin 37^o}{a}$, so $\boxed{a = 1.12}$. Also, $\frac{\sin 104^o}{1.8} \approx \frac{\sin 39^o}{c}$, so $\boxed{c \approx 1.17}$.

29. Solving for x: $\frac{139}{\sin 63^o} = \frac{140}{\sin A}$, then $\sin A = \frac{140 \sin 63^o}{139}$, and A has two possible values, $A \approx 63.82^o$ and $A \approx 116.18^o$. So, there are two possible values of x: $\boxed{x \approx 53.18^o}$, or $\boxed{x \approx 0.82^o}$.

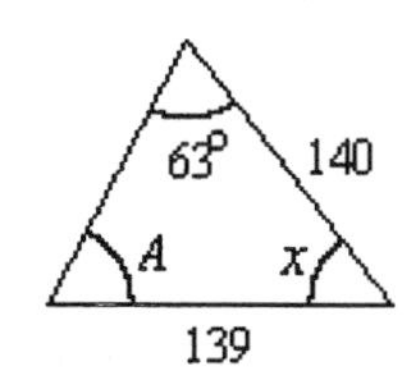

31. Solving for x:

$A = 180^o - 130^o - 23^o$, then $A = 27^o$; $\frac{x}{\sin 130^o} = \frac{700}{\sin 27^o}$, so

$x = \frac{700 \sin 130^o}{\sin 27^o}$, that is $\boxed{x \approx 1335.31}$.

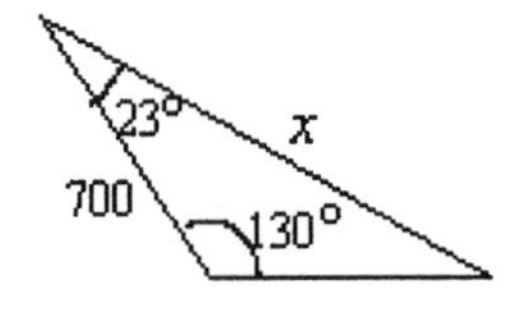

33. Solving for x:

$\frac{1.9}{\sin A} = \frac{2.3}{\sin 85^o}$, so $\sin A = \frac{1.9 \sin 85^o}{2.3}$, and $A \approx 55.38^o$. The other possibility for A, $A \approx 124.62^o$, is not relevant here. Then,

$x \approx 180^o - 85^o - 55.38^o$, that is $\boxed{x \approx 39.62^o}$.

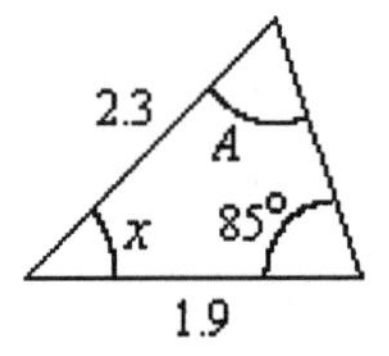

35. Solving for x:

$250^2 = 130^2 + 320^2 - 2(130)(320)\cos x$, so

$x = \cos^{-1}\left(\frac{250^2 - 130^2 - 320^2}{-2(130)(320)}\right)$, or $\boxed{x \approx 46.95^o}$.

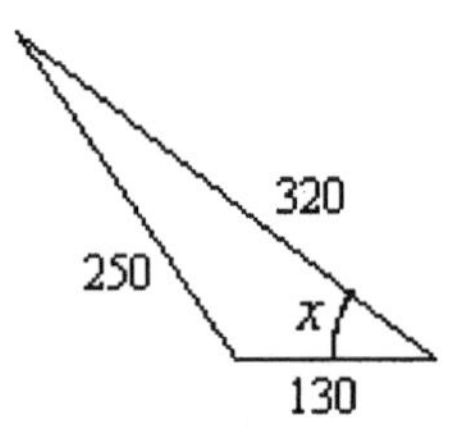

37. Solving for x:

$x = \sqrt{230^2 + 550^2 - 2(230)(550)\cos 27^o}$, that is $\boxed{x \approx 360.52}$.

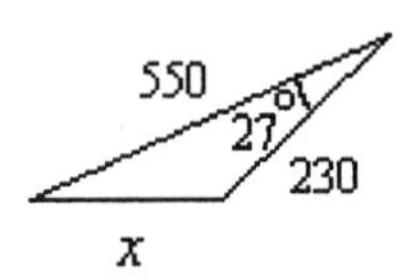

39. Solving for x:

$a = \sqrt{4000^2 + 2000^2 - 2(4000)(2000)\cos 130^o}$, that is

$a \approx 5503.14$. Now, $\frac{5503.14}{\sin 130^o} = \frac{4000}{\sin x}$, so $\boxed{x \approx 33.84^o}$.

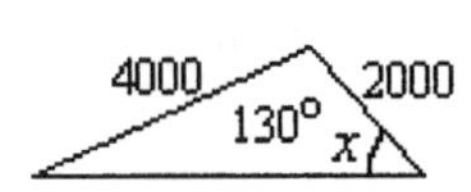

In Exercises 41-45, the area of a triangle with the given information is computed using Heron's Formula $A = \sqrt{s(s-a)(s-b)(s-c)}$, *where* $s = \frac{a+b+c}{2}$. *(The standard labeling of the sides and angles of a triangle is assumed.)*

41. If $a = 12$, $b = 14$, and $c = 20$, then $s = 23$ and the area of the triangle is

$A = \sqrt{23(23-12)(23-14)(23-20)}$, or $A \approx 82.65$.

43. If $b = 18$, $c = 15$, $B = 75^o$, then $\frac{18}{\sin 75^o} = \frac{15}{\sin C}$. Therefore, $C \approx 53.60^o$ (the other possibility for C, $C \approx 126.4^o$, is irrelevant) and $A \approx 180^o - 75^o - 53.60^o$, that is $A \approx 51.4^o$. Now, $\frac{a}{\sin 51.4^o} \approx \frac{18}{\sin 75^o}$, so $a \approx 14.56$. Then, $s \approx \frac{14.56+18+15}{2}$, and the area of the triangle is $A \approx 105.48$.

45. If $a = 1.4$, $b = 2.8$, $C = 16^o$, then $c = \sqrt{1.4^2 + 2.8^2 - 2(1.4)(2.8)\cos 16^o}$, or $c \approx 1.5$. Then, $s \approx \frac{1.4+2.8+1.5}{2}$, or approximately $s \approx 2.85$ and the are of the triangle is $A \approx 0.53$.

Applications

47. If two ships leave port at the same time, one traveling in the direction N 35° E at a rate of 25 miles per hour, and the other in the direction S 70° E at a rate of 30 miles per hour, then after 90 minutes, the distance between the first ship and the port is 37.5 miles and the distance between the second ship and the port is 45 miles. The angle that the paths of the ships make is 75°. The distance between the two ships is: $a = \sqrt{37.5^2 + 45^2 - 2(37.5)(45)\cos 75^o}$, approximately 50.57 miles.

49. If Margaret and Elizabeth are 1 mile apart, they both notice a bolt of lightning in the sky between their two homes, and Margaret hears the thunder 5 seconds after the flash, while Elizabeth hears the thunder 4 seconds after the flash, then the distances from the lightning to each of them are 5440 ft and 4352 ft. The angle of elevation A from Margaret to the lightning is $A = \cos^{-1}\left(\frac{4352^2 - 5440^2 - 5280^2}{-2(5440)(5280)}\right)$, approximately $A \approx 47.88^o$.

Lightning
5440 4352
A
Margaret 5280 Elizabeth

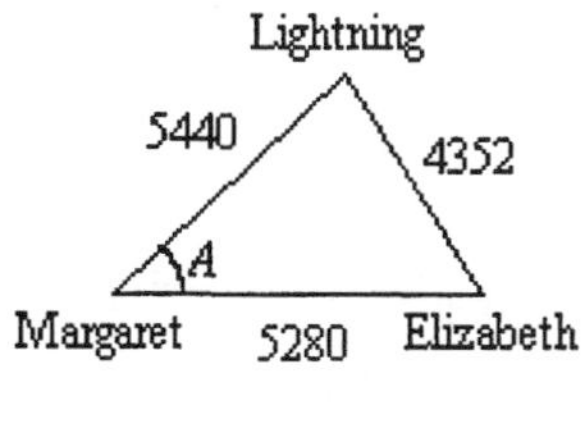

51. The area of the triangle formed by the points (3,0,0), (0,5,0), and (0,0,4) is computed using Heron's Formula, after getting the distances between each pair of points. These distances are: $a = \sqrt{34}$ (between (3,0,0) and (0,5,0)), $b = 5$ (between (3,0,0) and (0,0,4)) and $c = \sqrt{41}$ (between (0,0,4) and (0,5,0)). The semiperimeter of this triangle is $s = \frac{\sqrt{34}+5+\sqrt{41}}{2}$, that is $s \approx 8.617$. The area of the triangle is, approximately, 13.86 square inches.

53. If a train due west traveling at 80 miles per hour passes a junction at 3:00 pm, and at that moment another train approaching the junction in the direction 140° (clockwise from north) traveling at 90 miles per hour is 100 miles away, then:

(a) at 3:30 pm, the first train is 40 miles away and the second train is 55 miles away from the junction; the angle that the two paths make is 130°; therefore, the distance between the two trains at 3:30 pm is $\sqrt{40^2+55^2-2(40)(50)\cos 130^o}$, approximately 86.33 miles;

(b) the distances from the junction for each of the trains t hours after 3:00 pm are $p(t)=80t$, and $f(t)=100-90t$, where these expressions are valid for $t \le \frac{10}{9}$

(c) the distance between the two trains is $\sqrt{(80t)^2+(100-90t)^2-2(80t)(100-90t)\cos 130^o}$

(d) the minimum distance occurs approximately when $t \approx 0.74$ hours, or about 3:44 pm.

Concepts and Critical Thinking

55. The statement: "If the three angles of a triangle are given, then the Law of Cosines can be used to solve the triangle" is false.

57. The statement: "Heron's Formula is used to find the perimeter of a triangle" is false.

59. Answers may vary. An example of a triangle that can be solved more easily using the Law of Cosines than the Law of Sines is $b=3,\ c=7,\ A=25^o$. In fact, the Law of Cosines must be used here to start solving the triangle.

61. Answers may vary. An example of the side lengths of a triangle that has a semiperimeter of 12 is $a=10,\ b=8,\ c=6$.

63. Since $-1 \le \cos A \le 1$, it follows that $-2bc \le -2bc\cos A \le 2bc$, and in particular $b^2+c^2-2bc\cos A \le b^2+c^2+2bc$. Then, since $a^2=b^2+c^2-2bc\cos A$, it follows that $a^2 \le (b+c)^2$. Since *a*, *b*, and *c* are positive real numbers, it follows that $a \le b+c$ (the "Triangle Inequality".)

Chapter 7. Section 7.3 Trigonometric Form of Complex Numbers

	Complex Number z	Absolute Value $\lvert z\rvert$		Complex Number z	Absolute Value $\lvert z\rvert$
1.	$3+4i$	5	**3.**	$1+3i$	$\sqrt{10}$
5.	$1-\sqrt{3}i$	2	**7.**	-15	15
9.	$-\frac{3}{5}+\frac{4}{5}i$	1			

	Complex Number z	Graphical Representation	Trigonometric Form

11. $-4i$

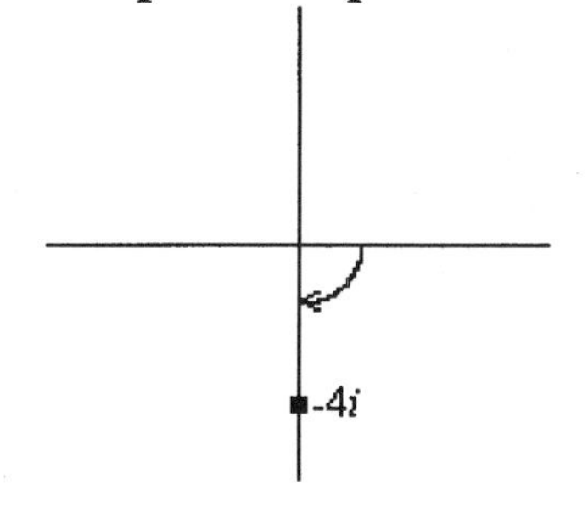

Answers may vary, one possible trigonometric form of the complex number $-4i$ is:

$$4\left(\cos\left(-\tfrac{\pi}{2}\right)+i\sin\left(-\tfrac{\pi}{2}\right)\right)$$

13. -6

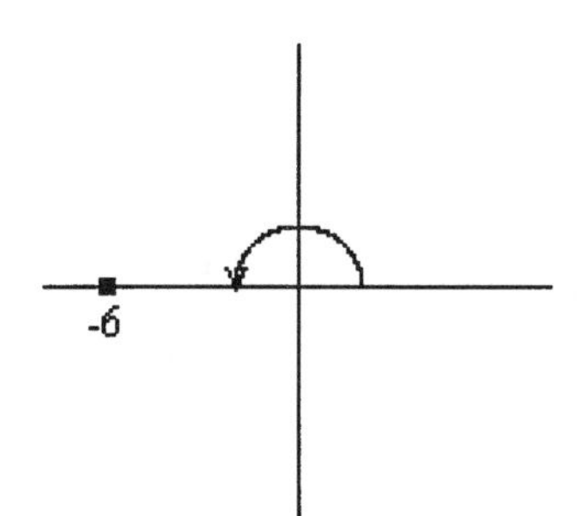

Answers may vary, one possible trigonometric form of the complex number -6 is:

$$6(\cos\pi+i\sin\pi)$$

15. $1+i$

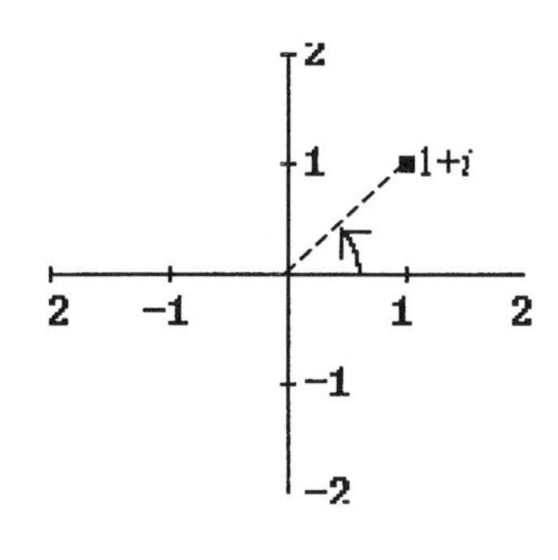

Answers may vary, one possible trigonometric form of the complex number $1+i$ is:

$$\sqrt{2}\left(\cos\tfrac{\pi}{4}+i\sin\tfrac{\pi}{4}\right)$$

17. $1-\sqrt{3}i$

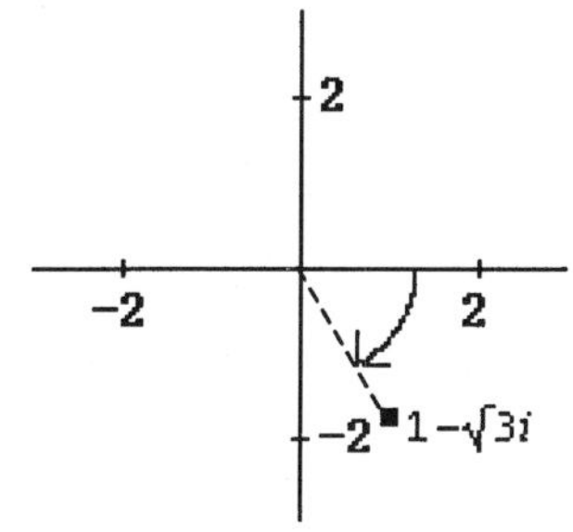

Answers may vary, one possible trigonometric form of the complex number $1-\sqrt{3}i$ is:

$$2\left(\cos\left(-\tfrac{\pi}{3}\right)+i\sin\left(-\tfrac{\pi}{3}\right)\right)$$

	Complex Number z	Graphical Representation	Trigonometric Form
19.	$-2-3i$	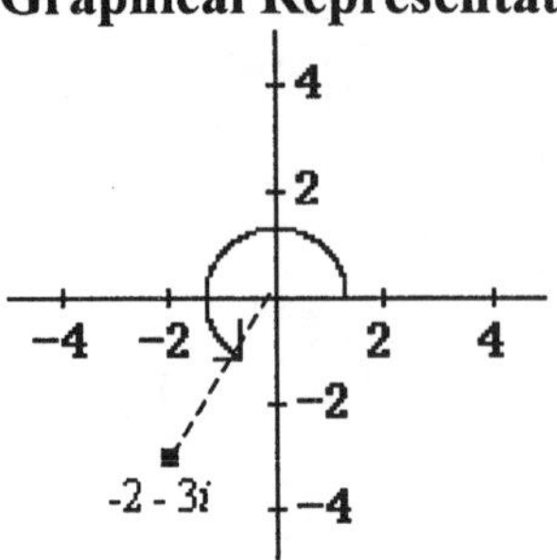	Answers may vary, one possible trigonometric form of the complex number $-2-3i$ is, approximately, $\sqrt{13}\left(\cos 4.12+i\sin 4.12\right)$

	Complex Number z in Trigonometric Form	Graphical Representation	Form $a+bi$
21.	$4\left(\cos 90^{0}+i\sin 90^{0}\right)$	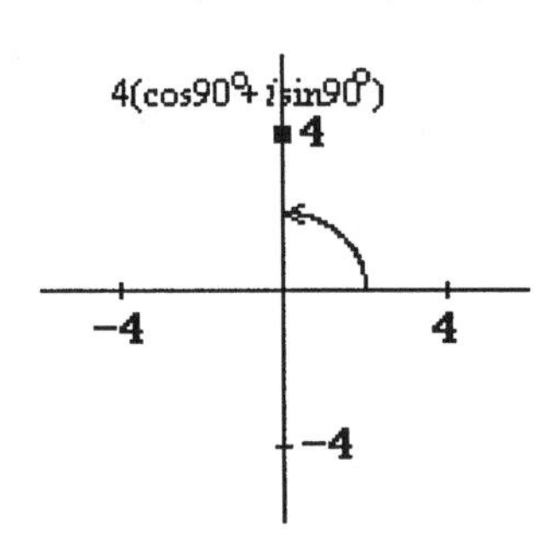	$4i$
23.	$2\left(\cos 120^{0}+i\sin 120^{0}\right)$	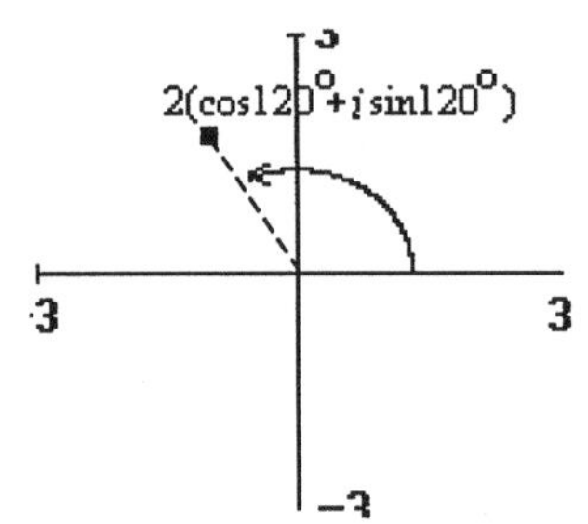	$-1+\sqrt{3}i$
25.	$9\left(\cos \frac{3\pi}{4}+i\sin \frac{3\pi}{4}\right)$	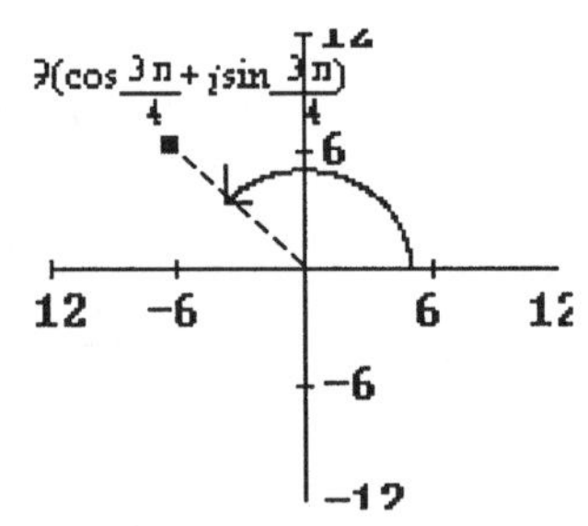	$-\frac{9\sqrt{2}}{2}+\frac{9\sqrt{2}}{2}i$
27.	$1\left(\cos\left(-\pi\right)+i\sin\left(-\pi\right)\right)$	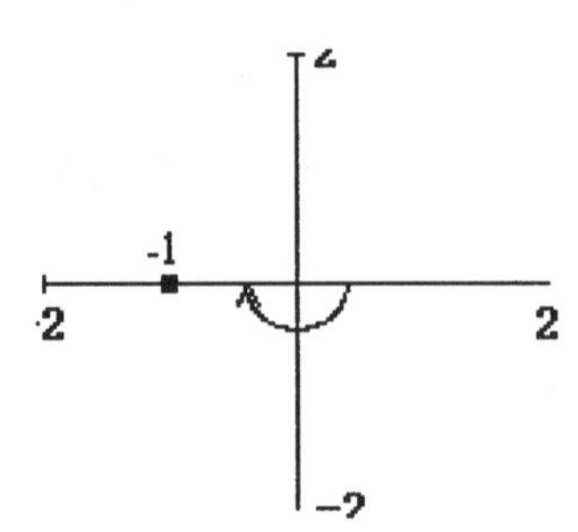	-1

29. $6\left(\cos 380^0 + i380^0\right)$ $\approx 5.64 + 2.05i$

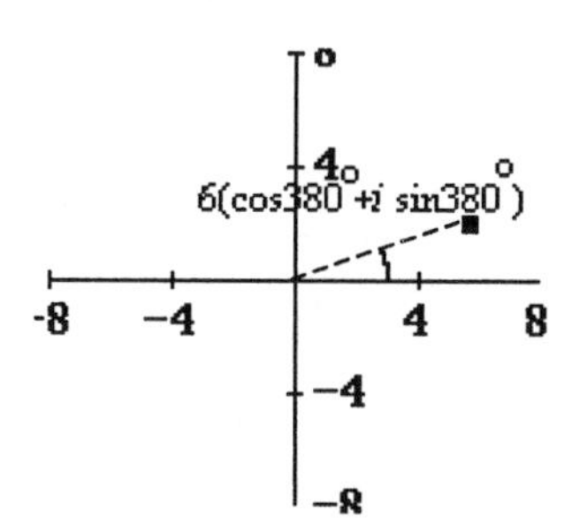

31. $\sqrt{2}\left(\cos 100^0 + i\sin 100^0\right)$ $\approx -0.25 + 1.39i$

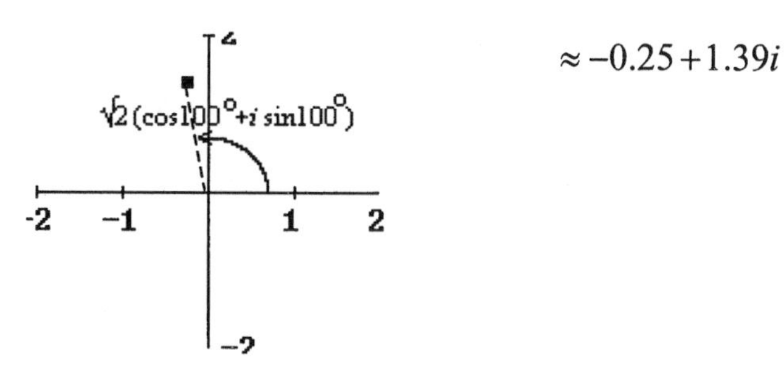

33. $\frac{3}{2}(\cos 2 + i\sin 2)$ $\approx -0.62 + 1.36i$

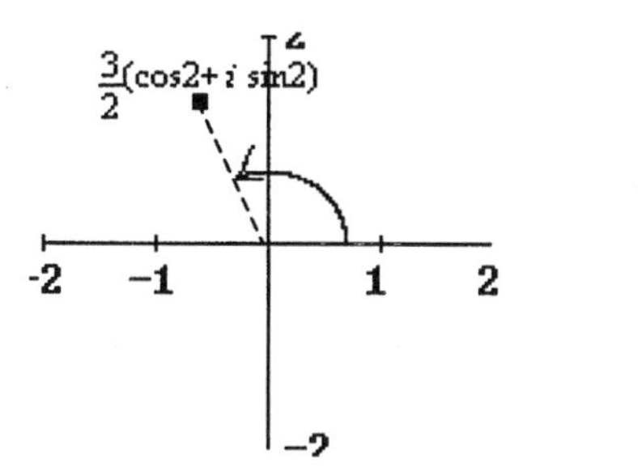

35. $2\left(\cos 30^0 + i\sin 30^0\right) \cdot 3\left(\cos 40^0 + i\sin 40^0\right)$

$= 6\left(\cos 70^0 + i\sin 70^0\right)$

37. $(3-i)(4+2i) = 14 + 2i$

39. $\dfrac{\sqrt{5}\left(\cos\frac{\pi}{4} + i\sin\frac{\pi}{4}\right)}{2\sqrt{5}\left(\cos\frac{\pi}{3} + i\sin\frac{\pi}{3}\right)} = \dfrac{1}{2}\left(\cos\left(\frac{\pi}{4} - \frac{\pi}{3}\right) + i\sin\left(\frac{\pi}{4} - \frac{\pi}{3}\right)\right)$

$= \frac{1}{2}\left(\cos\left(-\frac{\pi}{12}\right) + i\sin\left(-\frac{\pi}{12}\right)\right)$

41. $(-5+3i)(1-i) = -2 + 8i$

43. $\dfrac{3+5i}{2-4i} = \dfrac{(3+5i)(2+4i)}{(2-4i)(2+4i)}$

$= \dfrac{-14 + 22i}{20}$

$= -\frac{7}{10} + \frac{11}{10}i$

45. $5\left(\cos\frac{3\pi}{4} + i\sin\frac{3\pi}{4}\right) \cdot 3\left(\cos\frac{\pi}{2} + i\sin\frac{\pi}{2}\right) = 5\left(\cos\left(\frac{3\pi}{4} - \frac{\pi}{2}\right) + i\sin\left(\frac{3\pi}{4} - \frac{\pi}{2}\right)\right)$

$= 5\left(\cos\frac{\pi}{4} + i\sin\frac{\pi}{4}\right)$

47. $\dfrac{4.2(\cos 3\pi + i\sin 3\pi)}{0.6(\cos\frac{\pi}{4} + i\sin\frac{\pi}{4})} = 7\left(\cos\left(3\pi - \frac{\pi}{4}\right) + i\sin\left(3\pi - \frac{\pi}{4}\right)\right)$

$= 7\left(\cos\frac{11\pi}{4} + i\sin\frac{11\pi}{4}\right)$

49. $(2-2i)(-3+3i) = 12i$

51. $\dfrac{-8-8i}{2+2i} = \dfrac{-4(2+2i)}{2+2i}$, that is $\dfrac{-8-8i}{2+2i} = -4$

53. $\left(\sqrt{3}+i\right)\left(2-2\sqrt{3}i\right) = 4\sqrt{3} - 4i$

55. $\dfrac{2i}{4-3i} = \dfrac{2i(4+3i)}{(4-3i)(4+3i)}$

$= \dfrac{-6+8i}{25}$

$= -\dfrac{6}{25} + \dfrac{8}{25}i$

Concepts and Critical Thinking

57. The statement: "The absolute value of a complex number $a+bi$ is $|a|+|b|$" is false.

59. The statement: "The product of two complex numbers $z_1 = r_1(\cos\theta_1 + i\sin\theta_1)$ and $z_2 = r_2(\cos\theta_2 + i\sin\theta_2)$ is $z_1 z_2 = r_1 r_2(\cos(\theta_1\theta_2) + i\sin(\theta_1\theta_2))$" is false. A true statement is: "The product of two complex numbers $z_1 = r_1(\cos\theta_1 + i\sin\theta_1)$ and $z_2 = r_2(\cos\theta_2 + i\sin\theta_2)$ is $z_1 z_2 = r_1 r_2(\cos(\theta_1+\theta_2) + i\sin(\theta_1+\theta_2))$."

61. Answers may vary. An example of a complex number of modulus 5 that is neither real nor imaginary is $z = 4+3i$.

63. Answers may vary. An example of two nonreal complex numbers whose product is a positive real number is the pair $z_1 = 2+i,\ z_2 = 2-i$, since $z_1 z_2 = 5$.

65. If $r > 0$, then $|r(\cos\theta + i\sin\theta)| = \sqrt{r^2\cos^2\theta + r^2\sin^2\theta}$, that is $|r(\cos\theta + i\sin\theta)| = \sqrt{r^2}$, and $|r(\cos\theta + i\sin\theta)| = |r|$, therefore $|r(\cos\theta + i\sin\theta)| = r$.

67. If $z = r(\cos\theta + i\sin\theta)$, with $z \neq 0$, then $\dfrac{1}{z} = \dfrac{1(\cos 0 + i\sin 0)}{r(\cos\theta + i\sin\theta)}$. Applying exercise 66, we obtain $\dfrac{1}{z} = \dfrac{1}{r}(\cos(-\theta) + i\sin(-\theta))$, or $\dfrac{1}{z} = \dfrac{1}{r}(\cos\theta - i\sin\theta)$.

69. If $z = r(\cos\theta + i\sin\theta)$, then $\bar{z} = r\cos\theta - ir\sin\theta$, that is $\bar{z} = r(\cos\theta - i\sin\theta)$, or $\bar{z} = r(\cos(-\theta) + i\sin(-\theta))$

Chapter 7. Section 7.4 Powers and Roots of Complex Numbers

1. $\left[2(\cos 30^{\circ}+i\sin 30^{\circ})\right]^{5}=32(\cos 150^{\circ}+i\sin 150^{\circ})$

3. $\left[\sqrt{2}(\cos 10^{0}+i\sin 10^{0})\right]^{8}=16(\cos 80^{0}+i\sin 80^{0})$

5. $(1+i)^{16}=\left[\sqrt{2}\left(\cos\frac{\pi}{4}+i\sin\frac{\pi}{4}\right)\right]^{16}$
$=256(\cos 4\pi+i\sin 4\pi)$
$=256$

7. $\left(-\sqrt{3}+i\right)^{7}=\left[2\left(\cos\frac{5\pi}{6}+i\sin\frac{5\pi}{6}\right)\right]^{7}$
$=128\left(\cos\frac{35\pi}{6}+i\sin\frac{35\pi}{6}\right)$
$=128\left(\frac{\sqrt{3}}{2}-\frac{1}{2}i\right)$
$=64\sqrt{3}-64i$

9. $(1+2i)^{6}=\left[\sqrt{5}\left(\cos\left(\tan^{-1}2\right)+i\sin\left(\tan^{-1}2\right)\right)\right]^{6}$
$=125\left(\cos\left(6\tan^{-1}2\right)+i\sin\left(6\tan^{-1}2\right)\right)$
$\approx 117+44i$

11. The square roots of $\frac{1}{25}(\cos 60^{0}+i\sin 60^{0})$ are
$w_1=\frac{1}{5}(\cos 30^{0}+i\sin 30^{0})$ and $w_2=\frac{1}{5}(\cos 210^{0}+i\sin 210^{0})$,
that is $w_1=\frac{\sqrt{3}}{10}+\frac{1}{10}i$ and $w_2=-\frac{\sqrt{3}}{10}-\frac{1}{10}i$.

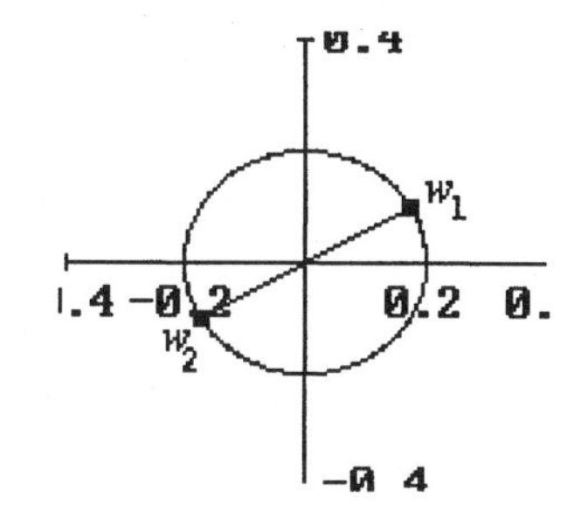

13. The cube roots of $1000\left(\cos\frac{3\pi}{2}+i\sin\frac{3\pi}{2}\right)$ are
$w_1=10\left(\cos\frac{\pi}{2}+i\sin\frac{\pi}{2}\right)$, $w_2=10\left(\cos\left(\frac{\pi}{2}+\frac{2\pi}{3}\right)+i\sin\left(\frac{\pi}{2}+\frac{2\pi}{3}\right)\right)$, and
$w_3=10\left(\cos\left(\frac{\pi}{2}+\frac{4\pi}{3}\right)+i\sin\left(\frac{\pi}{2}+\frac{4\pi}{3}\right)\right)$; that is,
$w_1=10i$, $w_2=-5\sqrt{3}-5i$, and $w_3=5\sqrt{3}-5i$.

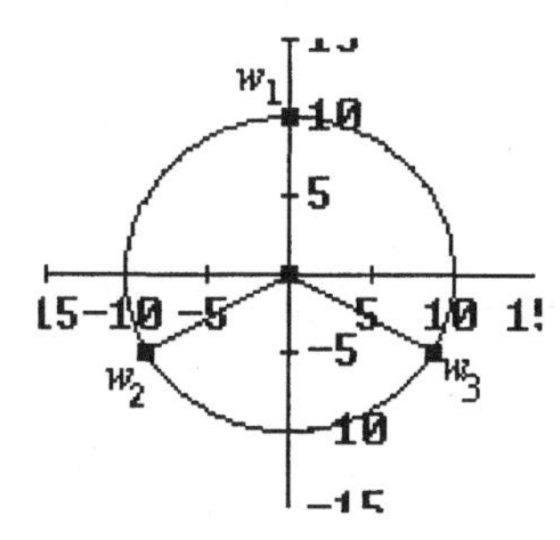

15. The square roots of $36i$ are $w_1 = 6\left(\cos\frac{\pi}{4} + i\sin\frac{\pi}{4}\right)$, and $w_2 = 6\left(\cos\frac{5\pi}{4} + i\sin\frac{5\pi}{4}\right)$, That is, $w_1 = 3\sqrt{2} + 3\sqrt{2}i$, and $w_2 = -3\sqrt{2} - 3\sqrt{2}i$,

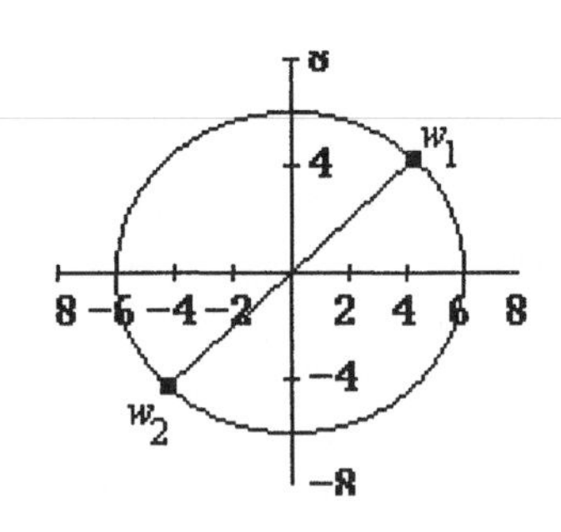

17. The cube roots of 27 are $w_1 = 3$, $w_2 = 3\left(\cos\frac{2\pi}{3} + i\sin\frac{2\pi}{3}\right)$, and $w_3 = 3\left(\cos\frac{4\pi}{3} + i\sin\frac{4\pi}{3}\right)$. That is, $w_1 = 3$, $w_2 = -\frac{3}{2} + \frac{3\sqrt{3}}{2}i$, and $w_3 = -\frac{3}{2} - \frac{3\sqrt{3}}{2}i$.

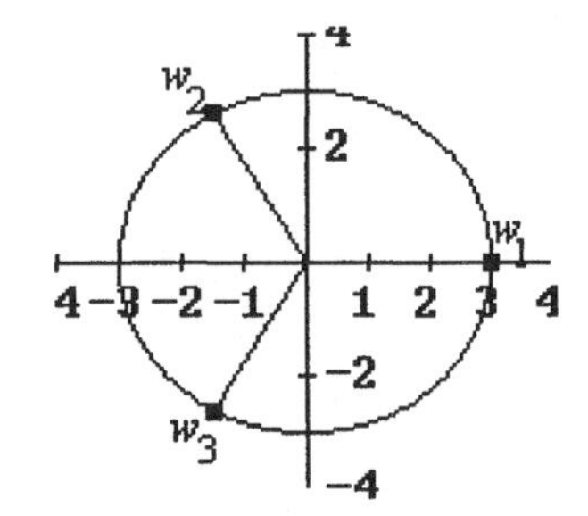

19. The sixth roots of 1 are $w_1 = 1$, $w_2 = \cos\frac{\pi}{3} + i\sin\frac{\pi}{3}$, $w_3 = \cos\frac{2\pi}{3} + i\sin\frac{2\pi}{3}$, $w_4 = -1$, $w_5 = \cos\frac{4\pi}{3} + i\sin\frac{4\pi}{3}$, and $w_6 = \cos\frac{5\pi}{3} + i\sin\frac{5\pi}{3}$. That is, $w_1 = 1$, $w_2 = \frac{1}{2} + \frac{\sqrt{3}}{2}i$, $w_3 = -\frac{1}{2} + \frac{\sqrt{3}}{2}i$, $w_4 = -1$, $w_5 = -\frac{1}{2} - \frac{\sqrt{3}}{2}i$, and $w_6 = \frac{1}{2} - \frac{\sqrt{3}}{2}i$.

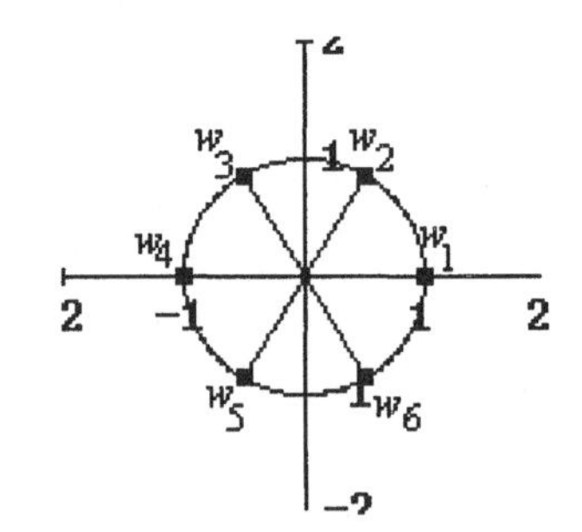

21. The cube roots of $z = -\frac{\sqrt{2}}{16} + \frac{\sqrt{2}}{16}i$, that is $z = \frac{1}{8}\left(\cos\frac{3\pi}{4} + i\sin\frac{3\pi}{4}\right)$, are $w_1 = \frac{1}{2}\left(\cos\frac{\pi}{4} + i\sin\frac{\pi}{4}\right)$, $w_2 = \frac{1}{2}\left(\cos\frac{11\pi}{12} + i\sin\frac{11\pi}{12}\right)$, and $w_3 = \frac{1}{2}\left(\cos\frac{19\pi}{12} + i\sin\frac{19\pi}{12}\right)$.

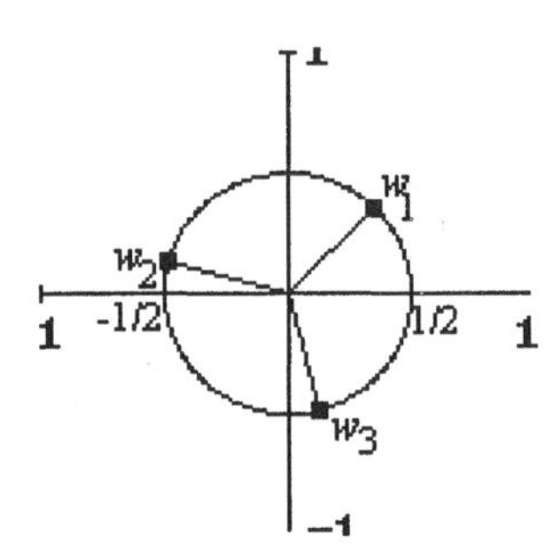

23. The square roots of $z = 3 - i$, or $z = \sqrt{10}\left(\cos\tan^{-1}\left(-\frac{1}{3}\right) + i\sin\tan^{-1}\left(-\frac{1}{3}\right)\right)$, are

$$w_1 = \sqrt[4]{10}\left(\cos\left(\frac{\tan^{-1}(-1/3)}{2}\right) + i\sin\left(\frac{\tan^{-1}(-1/3)}{2}\right)\right), \text{ and}$$

$$w_2 = \sqrt[4]{10}\left(\cos\left(\frac{\tan^{-1}(-1/3)+2\pi}{2}\right) + i\sin\left(\frac{\tan^{-1}(-1/3)+2\pi}{2}\right)\right),$$

approximately $w_1 \approx 1.755 - 0.284i$ and $w_2 \approx -1.755 + 0.284i$.

25. The fourth roots of i are $w_1 = \cos\frac{\pi}{8} + i\sin\frac{\pi}{8}$, $w_2 = \cos\frac{5\pi}{8} + i\sin\frac{5\pi}{8}$, $w_3 = \cos\frac{9\pi}{8} + i\sin\frac{9\pi}{8}$, and $w_4 = \cos\frac{13\pi}{8} + i\sin\frac{13\pi}{8}$, or approximately $w_1 \approx 0.923 + 0.382i$, $w_2 \approx -0.382 + 0.923i$, $w_3 \approx -0.923 - 0.382i$, and $w_4 \approx 0.382 - 0.923i$.

27. The cube roots of $z = 2 + 5i$, that is $z = \sqrt{29}\left(\cos\tan^{-1}\frac{5}{2} + i\sin\tan^{-1}\frac{5}{2}\right)$, are
$w_1 = \sqrt[6]{29}\left(\cos\frac{\tan^{-1}\frac{5}{2}}{3} + i\sin\frac{\tan^{-1}\frac{5}{2}}{3}\right)$, $w_2 = \sqrt[6]{29}\left(\cos\frac{\tan^{-1}\frac{5}{2}+2\pi}{3} + i\sin\frac{\tan^{-1}\frac{5}{2}+2\pi}{3}\right)$, and
$w_3 = \sqrt[6]{29}\left(\cos\frac{\tan^{-1}\frac{5}{2}+4\pi}{3} + i\sin\frac{\tan^{-1}\frac{5}{2}+4\pi}{3}\right)$, or approximately $w_1 \approx 1.616 + 0.677i$,
$w_2 \approx -1.394 + 1.061i$, and $w_2 \approx -0.221 - 1.738i$.

29. The solutions of the equation $x^4 + 1 = 0$ are solutions of $x^4 = -1$, that is, they are the fourth-roots of -1: $x_1 = \cos\frac{\pi}{4} + i\sin\frac{\pi}{4}$,
$x_2 = \cos\frac{3\pi}{4} + i\sin\frac{3\pi}{4}$, $x_3 = \cos\frac{5\pi}{4} + i\sin\frac{5\pi}{4}$, and $x_4 = \cos\frac{7\pi}{4} + i\sin\frac{7\pi}{4}$.
That is, $x_1 = \frac{\sqrt{2}}{2} + i\frac{\sqrt{2}}{2}$, $x_2 = -\frac{\sqrt{2}}{2} + i\frac{\sqrt{2}}{2}$, $x_3 = -\frac{\sqrt{2}}{2} - i\frac{\sqrt{2}}{2}$, and
$x_4 = \frac{\sqrt{2}}{2} - i\frac{\sqrt{2}}{2}$.

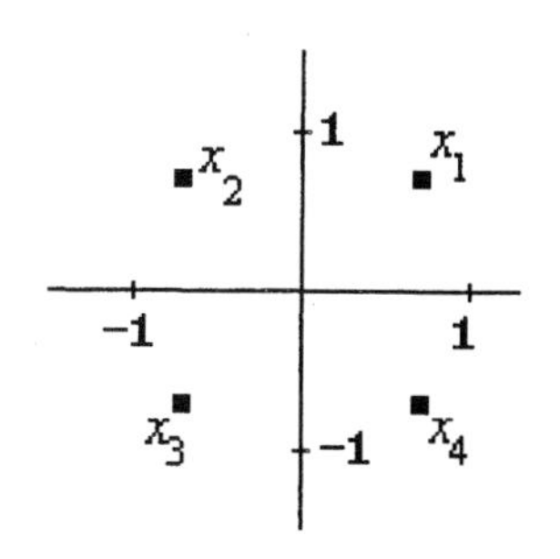

31. The solutions of the equation $x^3 + i = 0$ are solutions of $x^3 = -i$, that is, they are the cube roots of $-i$: $x_1 = \cos\frac{\pi}{2} + i\sin\frac{\pi}{2}$,
$x_2 = \cos\frac{7\pi}{6} + i\sin\frac{7\pi}{6}$, and $x_3 = \cos\frac{11\pi}{6} + i\sin\frac{11\pi}{6}$. That is, $x_1 = i$,
$x_2 = -\frac{\sqrt{3}}{2} - \frac{1}{2}i$, and $x_3 = \frac{\sqrt{3}}{2} - \frac{1}{2}i$.

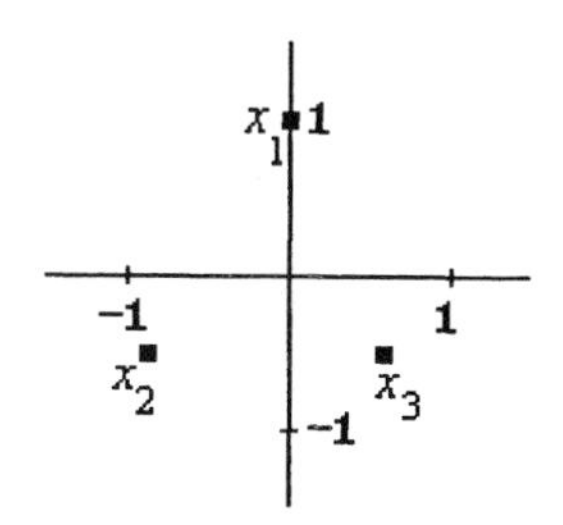

33. The solutions of the equation $x^2 - \left(2 + 2\sqrt{3}\right)i = 0$ are solutions of $x^2 = 2 + 2\sqrt{3}i$, that is, they are the square roots of $z = 2 + 2\sqrt{3}i$. The trigonometric form of z is $z = 4\left(\cos\frac{\pi}{3} + i\sin\frac{\pi}{3}\right)$. So, the square roots of z are: $x_1 = 2\left(\cos\frac{\pi}{6} + i\sin\frac{\pi}{6}\right)$, and $x_1 = 2\left(\cos\frac{7\pi}{6} + i\sin\frac{7\pi}{6}\right)$. That is,
$x_1 = \sqrt{3} + i$, and $x_2 = -\sqrt{3} - i$.

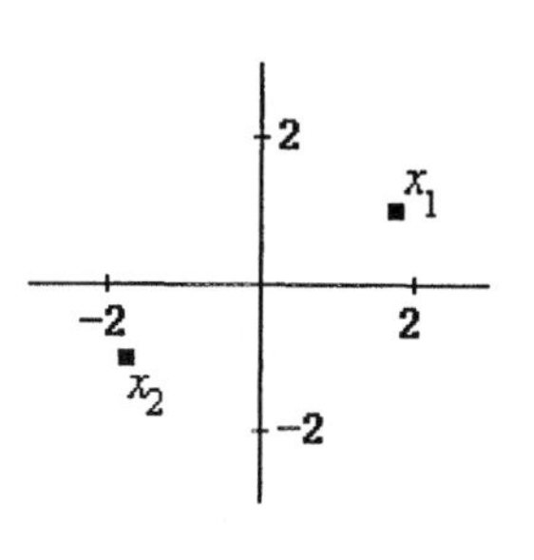

35. If $f(x) = x^6 + 64$, the zeros of f are the solutions of $x^6 = -64$, namely the sixth roots of -64:
$x_1 = 2\left(\cos\frac{\pi}{6} + i\sin\frac{\pi}{6}\right)$, $x_2 = 2\left(\cos\frac{\pi+2\pi}{6} + i\sin\frac{\pi+2\pi}{6}\right)$, $x_3 = 2\left(\cos\frac{\pi+4\pi}{6} + i\sin\frac{\pi+4\pi}{6}\right)$,
$x_4 = 2\left(\cos\frac{\pi+6\pi}{6} + i\sin\frac{\pi+6\pi}{6}\right)$, $x_5 = 2\left(\cos\frac{\pi+8\pi}{6} + i\sin\frac{\pi+8\pi}{6}\right)$, $x_6 = 2\left(\cos\frac{\pi+10\pi}{6} + i\sin\frac{\pi+10\pi}{6}\right)$.
That is, $x_1 = \sqrt{3} + i$, $x_2 = 2i$, $x_3 = -\sqrt{3} + i$, $x_4 = -\sqrt{3} - i$, $x_5 = -2i$, and $x_6 = \sqrt{3} - i$. Then,
$f(x) = (x - 2i)(x + 2i)\left(x - \left(\sqrt{3} + i\right)\right)\left(x - \left(\sqrt{3} - i\right)\right)\left(x - \left(-\sqrt{3} + i\right)\right)\left(x - \left(-\sqrt{3} - i\right)\right)$.

37. The cube roots of unity are: $x_1 = 1$, $x_2 = \cos\frac{2\pi}{3} + i\sin\frac{2\pi}{3}$, and $x_3 = \cos\frac{4\pi}{3} + i\sin\frac{4\pi}{3}$. That is, $x_1 = 1$, $x_2 = -\frac{1}{2} + \frac{\sqrt{3}}{2}i$, and $x_3 = -\frac{1}{2} - \frac{\sqrt{3}}{2}i$. We have $x_1 + x_2 + x_3 = \left(1 - \frac{1}{2} - \frac{1}{2}\right) + i\left(\frac{\sqrt{3}}{2} - \frac{\sqrt{3}}{2}\right)$, that is $x_1 + x_2 + x_3 = 0$.

Concepts and Critical Thinking

39. The statement: "According to de Moivre's Theorem, $\left[r(\cos\theta + i\sin\theta)\right]^n = r^n\left(\cos\theta^n + i\sin\theta^n\right)$ for all positive integer n" is false. A true statement is: "According to de Moivre's Theorem, $\left[r(\cos\theta + i\sin\theta)\right]^n = r^n\left(\cos n\theta + i\sin n\theta\right)$."

41. The statement: "Every nonzero complex number has three distinct third roots" is true.

43. Answers may vary. An example of a positive integer n such that $(1+i)^n$ is real and negative is $n = 4$: $(1+i)^4 = -4$.

45. Answers may vary. An example of a positive integer n such that $(1+i)^n$ is imaginary is $n = 2$: $(1+i)^2 = 2i$.

Chapter 7. Section 7.5 Polar Coordinates

In Exercises 1-10, a point in polar coordinates is given, and the corresponding point in rectangular coordinates is determined. Each point is plotted.

1. The point $\left(2,-\frac{\pi}{2}\right)$ corresponds to $(0,-2)$ in rectangular coordinates.

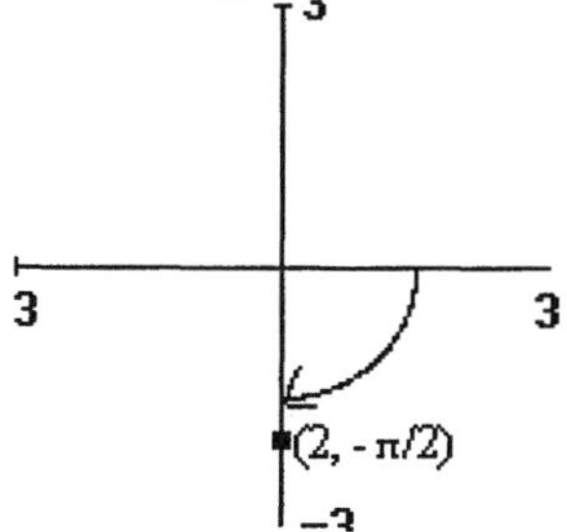

3. The point $\left(1,\frac{3\pi}{4}\right)$ corresponds to $\left(-\frac{\sqrt{2}}{2},\frac{\sqrt{2}}{2}\right)$ in rectangular coordinates.

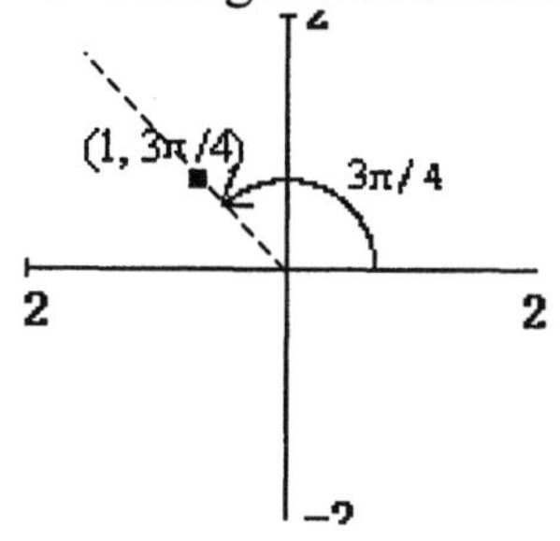

5. The point $\left(-2,\frac{\pi}{6}\right)$ corresponds to $\left(-\sqrt{3},-1\right)$ in rectangular coordinates.

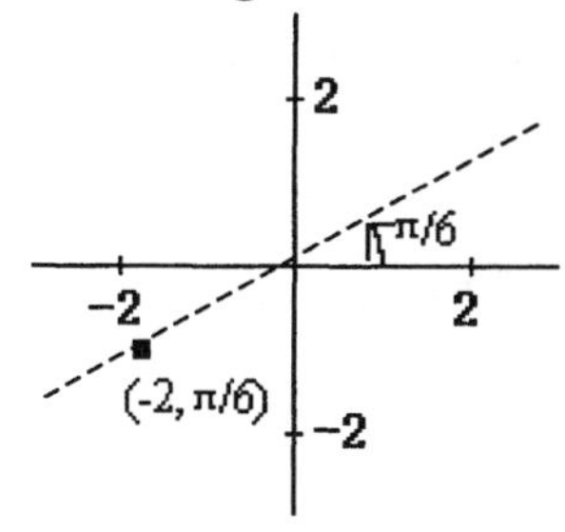

7. The point $\left(\frac{3}{2},-\frac{5\pi}{6}\right)$ corresponds to $\left(-\frac{3\sqrt{3}}{4},-\frac{3}{4}\right)$ in rectangular coordinates.

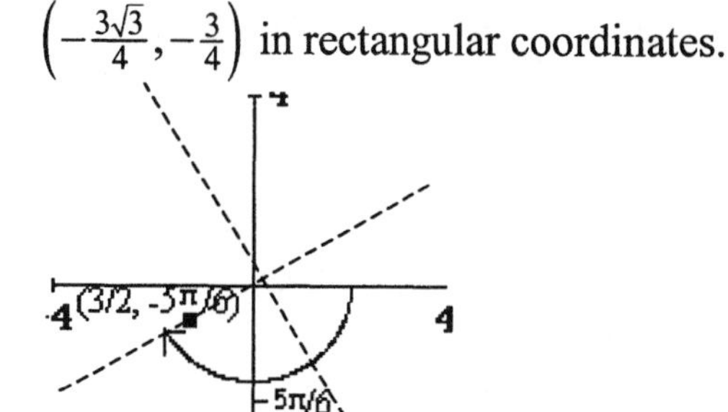

9. The point $(-3,-\pi)$ corresponds to $(3,0)$ in rectangular coordinates.

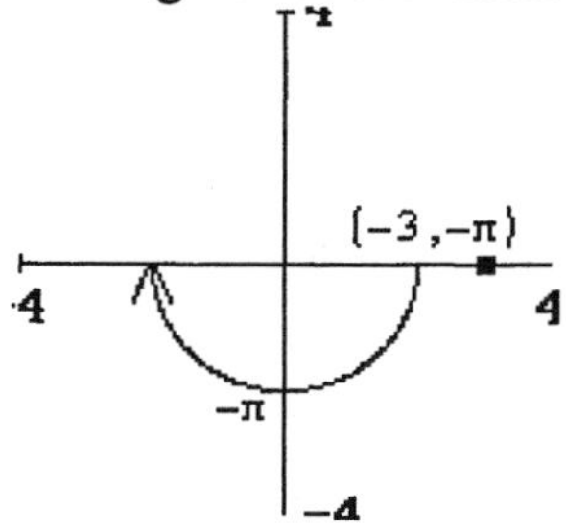

In Exercises 11-19, a point in rectangular coordinates is given, and the corresponding point in rectangular coordinates is determined. Each point is plotted.

11. The point $(4,0)$ (given in rectangular coordinates) corresponds to $(4,0)$ and $(-4,\pi)$ in polar coordinates.

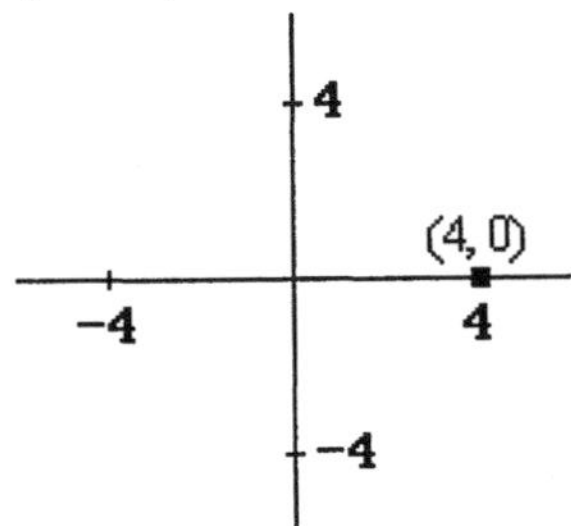

13. The point $(2,2)$ (given in rectangular coordinates) corresponds to $\left(2\sqrt{2},\frac{\pi}{4}\right)$ and $\left(-2\sqrt{2},\frac{5\pi}{4}\right)$ in polar coordinates.

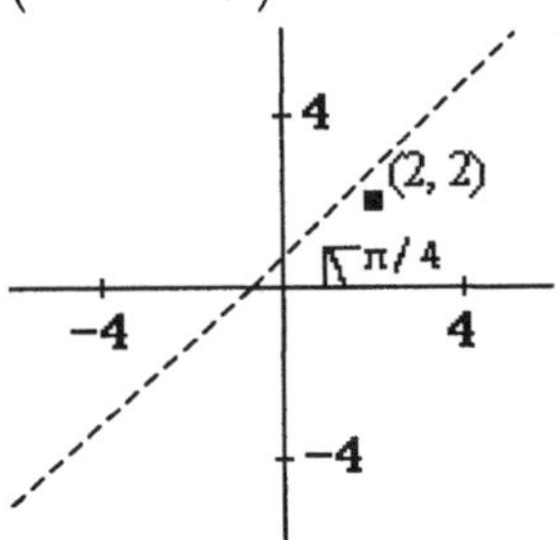

15. The point $\left(1,-\sqrt{3}\right)$ (given in rectangular coordinates) corresponds to $\left(-2,\frac{2\pi}{3}\right)$ and $\left(2,\frac{5\pi}{3}\right)$ in polar coordinates.

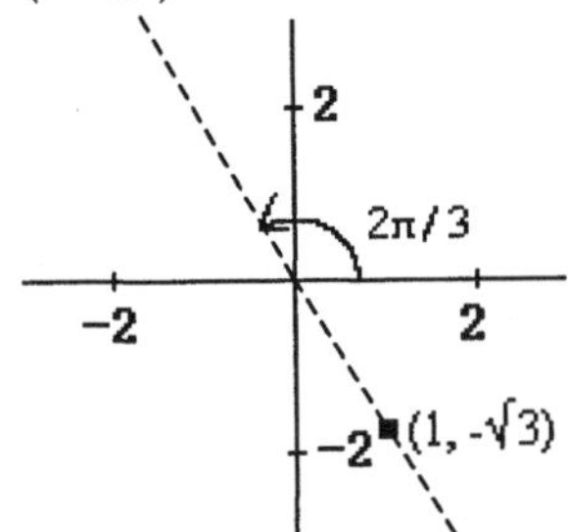

17. The point $(-4,-3)$ (given in rectangular coordinates) corresponds to $\left(-5,\tan^{-1}\frac{3}{4}\right)$ and $\left(5,\tan^{-1}\frac{3}{4}+\pi\right)$, or approximately $(-5,0.64)$ and $(5,3.78)$ in polar coordinates.

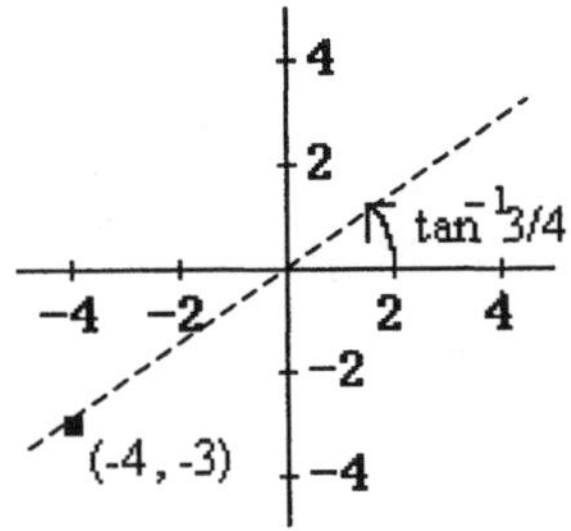

19. The point $(3,1)$ (given in rectangular coordinates) corresponds to $\left(\sqrt{10},\tan^{-1}\frac{1}{3}\right)$ and $\left(-\sqrt{10},\tan^{-1}\frac{1}{3}+\pi\right)$, approximately $\left(\sqrt{10},0.32\right)$ and $\left(-\sqrt{10},3.46\right)$ in polar coordinates.

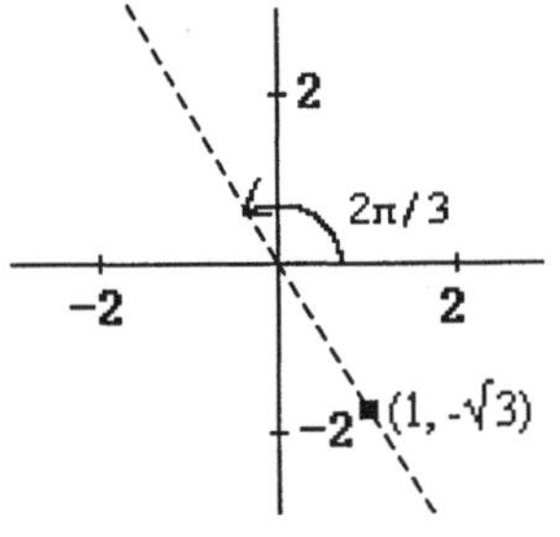

21. Answers may vary. Four points satisfying the polar equation $r=3$ are (3, 0), $\left(3,\frac{\pi}{6}\right)$, $\left(3,\frac{\pi}{4}\right)$, and $\left(3,\frac{\pi}{3}\right)$. The graph of $r=3$:

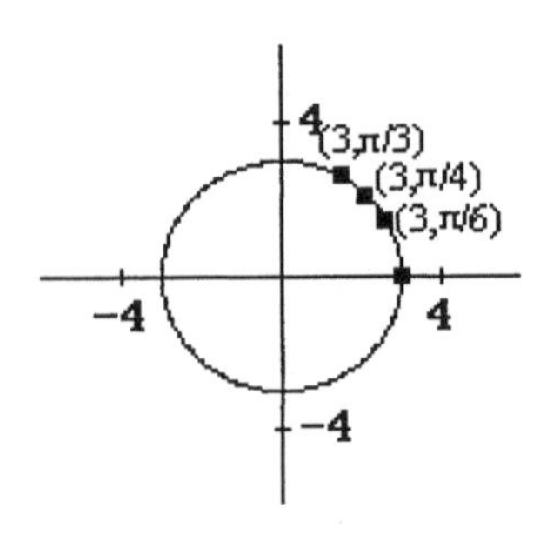

23. Answers may vary. Four points satisfying the polar equation $\theta = \frac{2\pi}{3}$ are $\left(0, \frac{2\pi}{3}\right)$, $\left(1, \frac{2\pi}{3}\right)$, $\left(2, \frac{2\pi}{3}\right)$, and $\left(3, \frac{2\pi}{3}\right)$. The graph of $\theta = \frac{2\pi}{3}$:

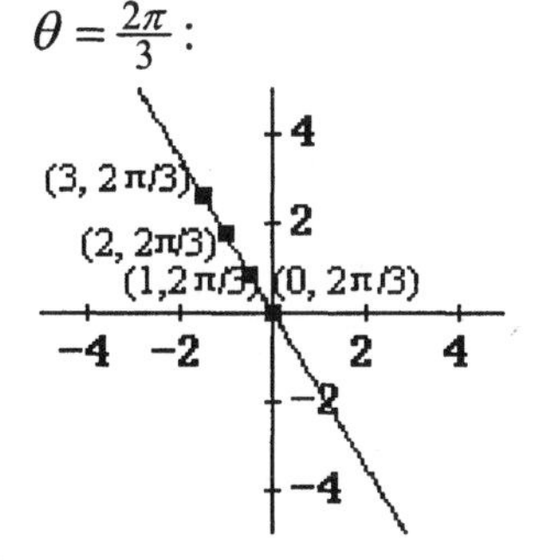

25. Answers may vary. Four points satisfying the polar equation $r = 2\theta$ are $(0,0)$, $\left(\pi, \frac{\pi}{2}\right)$, $\left(3\pi, \frac{3\pi}{2}\right)$, and $\left(5\pi, \frac{5\pi}{2}\right)$. The graph of $r = 2\theta$, $\theta \geq 0$:

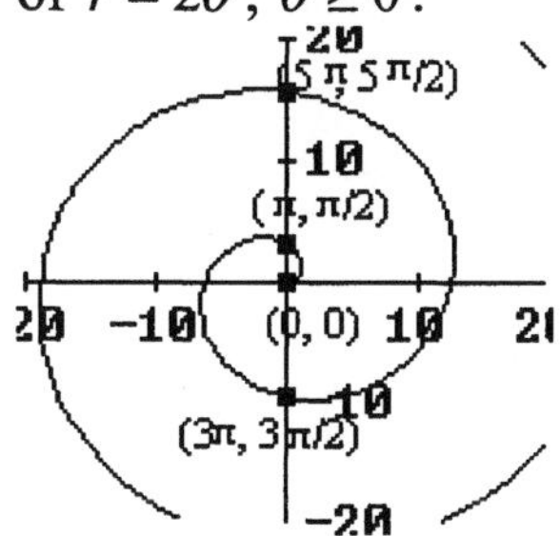

27. Answers may vary. Four points satisfying the polar equation $r = 3\sin\theta$ are $(0,0)$, $\left(3, \frac{\pi}{2}\right)$, $\left(\frac{3}{2}, \frac{5\pi}{6}\right)$, and $(0, \pi)$. The graph of $r = 3\sin\theta$:

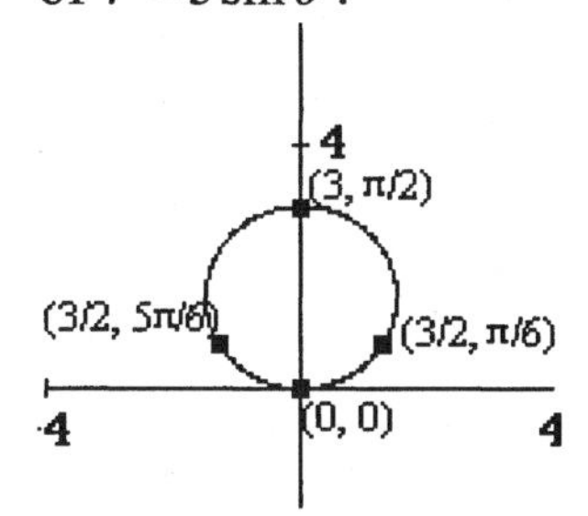

29. Answers may vary. Four points satisfying the polar equation $r = 2 - 4\cos\theta$ are $(-2,0)$, $\left(0, \frac{\pi}{3}\right)$, $\left(2, \frac{\pi}{2}\right)$, and $(6, \pi)$. The graph of $r = 2 - 4\cos\theta$:

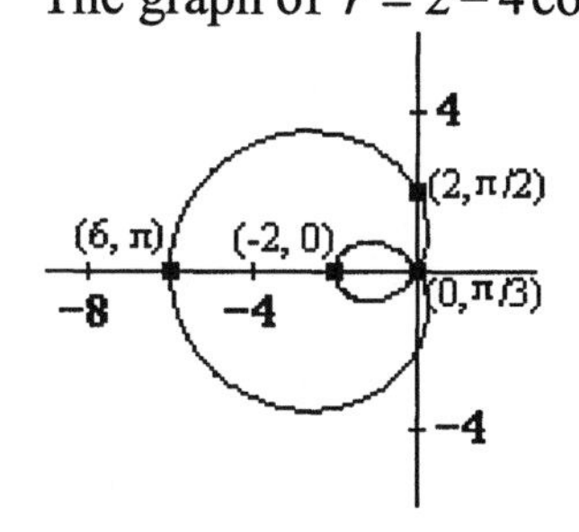

31. Answers may vary. Four points satisfying the polar equation $r = \sin 2\theta$ are $(0,0)$, $\left(1, \frac{\pi}{4}\right)$, $\left(\frac{\sqrt{3}}{2}, \frac{\pi}{3}\right)$, and $\left(-1, \frac{3\pi}{4}\right)$. The graph of $r = \sin 2\theta$:

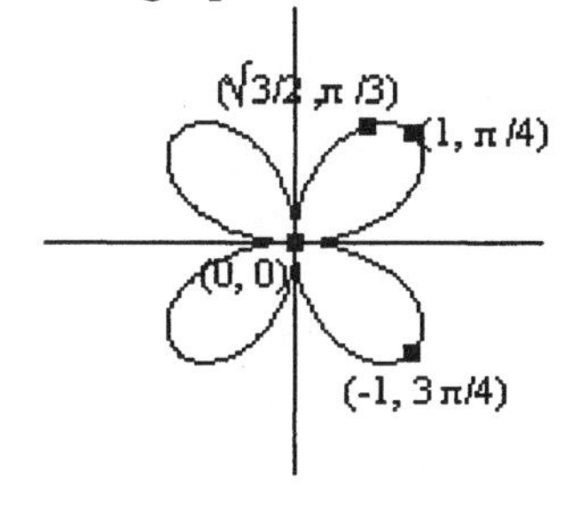

33. Answers may vary. Four points satisfying the polar equation $r = 2\cos 3\theta$ are $(2,0)$, $\left(0, \frac{\pi}{6}\right)$, $\left(-\sqrt{2}, \frac{\pi}{4}\right)$, and $\left(-2, \frac{\pi}{3}\right)$. The graph of $r = 2\cos 3\theta$:

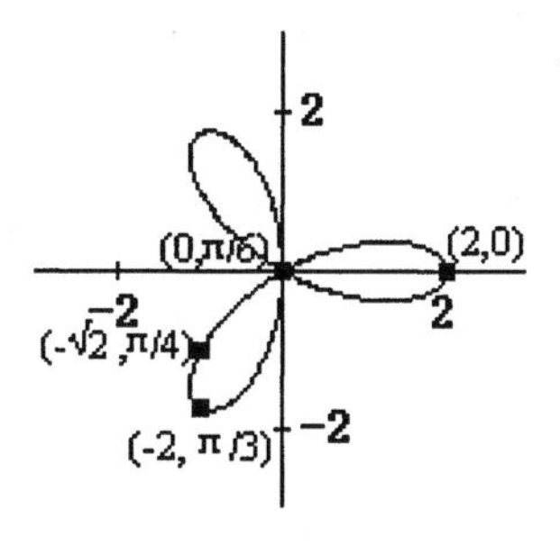

35. The rectangular equation $y = 3$ corresponds to the polar equation $r = \dfrac{3}{\sin\theta}$.

37. Answers may vary. The rectangular equation $y+x=0$ corresponds to the polar equation $r\sin\theta+r\cos\theta=0$, or equivalently $\theta=\arctan(-1)$ (Starting with $r\sin\theta+r\cos\theta=0$, we divide by $r\cos\theta$, to obtain $\tan\theta+1=0$.)

39. Answers may vary. The rectangular equation $x^2+y^2=9$ corresponds to the polar equation $r=3$.

41. Answers may vary. The rectangular equation $\ln y-\ln x=1$ corresponds to the polar equation $\ln(r\sin\theta)-\ln(r\cos\theta)=1$, that is $\ln\tan\theta=1$, or equivalently $\theta=\tan^{-1}e$.

43. The polar equation $r=7$ corresponds to the rectangular equation $x^2+y^2=49$.

45. The polar equation $\theta=\frac{2\pi}{3}$ corresponds to the rectangular equation $y=-\sqrt{3}x$.

47. The polar equation $r\cos\theta=1$ corresponds to the rectangular equation $x=1$.

49. The polar equation $r=2\sin\theta$ is equivalent $r^2=2r\sin\theta$, and corresponds to the rectangular equation $x^2+y^2=2y$, that is $x^2+(y-1)^2=1$.

51. The polar equation $r=\cos 2\theta$ is equivalent $r=\cos^2\theta-\sin^2\theta$, or $r^3=r^2\cos^2\theta-r^2\sin^2\theta$ and corresponds to the rectangular equation $\left(x^2+y^2\right)^{\frac{3}{2}}=x^2-y^2$.

53. The points of intersection of the graphs of $r=2$ and $r=4\sin\theta$ are $\left(2,\frac{\pi}{6}\right)$ and $\left(2,\frac{5\pi}{6}\right)$.

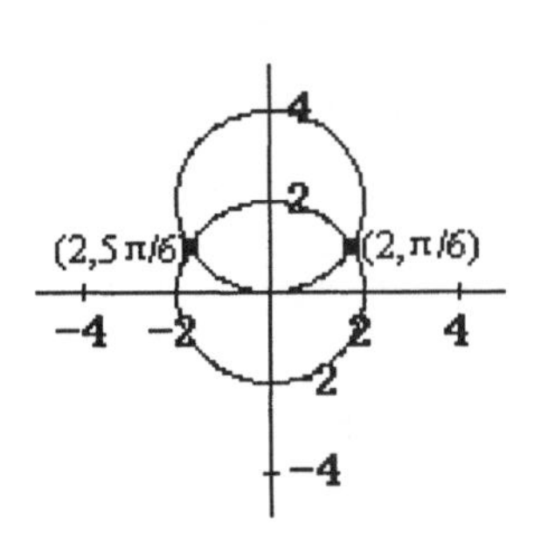

55. The points of intersection of the graphs of $r=2\cos\theta$ and $r=1-\sin\theta$ are $(0,0)$ and approximately $(1.6,-0.65)$.

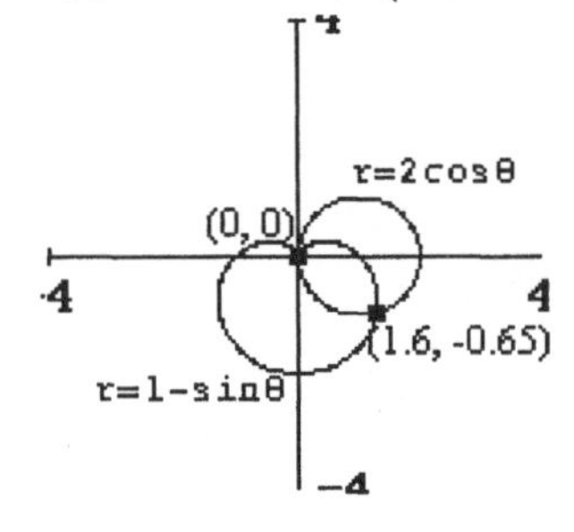

57. The points of intersection of the graphs of $r = \sin 2\theta$ and $r = \sin\theta$ are, approximately, $(0.87, 1.05)$ and $(-0.87, 5.24)$.

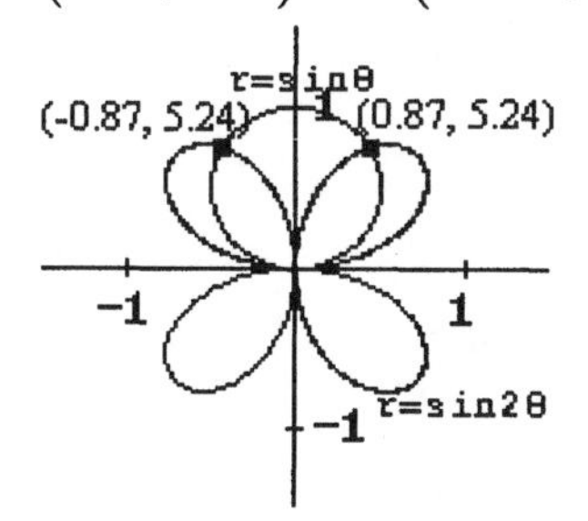

Concepts and Critical Thinking

59. The statement: "The rectangular coordinates of a point are unique" is true.

61. The statement: "If a point has rectangular coordinates (x, y) and polar coordinates (r, θ), then $r = x\cos\theta$ and $r = y\sin\theta$" is false.

63. Answers may vary. The graph of the polar equation $r = 3$ is centered at the origin.

65. Answers may vary. The graph of the polar equation $r = 2\theta$ is a spiral.

67. The point $\left(1, \frac{\pi}{2}\right)$ belongs to the graph of $r = -\sin^2\theta$, since $\left(1, \frac{\pi}{2}\right)$ corresponds to the same point as $\left(-1, \frac{3\pi}{2}\right)$, and the set of polar coordinates $\left(-1, \frac{3\pi}{2}\right)$ satisfies the equation $r = -\sin^2\theta$. Note how the set of coordinates $\left(1, \frac{\pi}{2}\right)$ does not satisfy this equation, but the point belongs to the graph. This is a disadvantage of polar coordinates: a point (r, θ) may belong to the graph of an equation $r = f(\theta)$, even if the coordinates (r, θ) do not satisfy the equation $r = f(\theta)$. Another polar representation of the point (r, θ) may satisfy this equation.

Chapter 7. Section 7.6 Vectors

1. (a) If $P=(2,5)$ and $Q=(3,7)$, the graph of $\overrightarrow{PQ}$ is shown below:

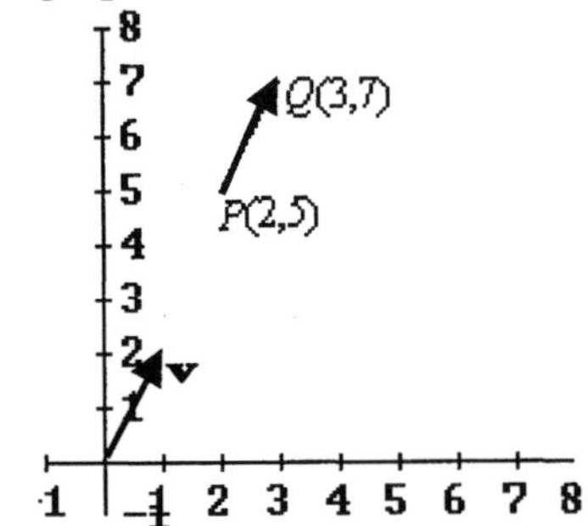

(b) The component form of **v**, the vector represented by $\overrightarrow{PQ}$ is $\mathbf{v}=\langle 1,2\rangle$.

3. (a) If $P=(-3,1)$ and $Q=(2,-2)$, the graph of $\overrightarrow{PQ}$ is shown below:

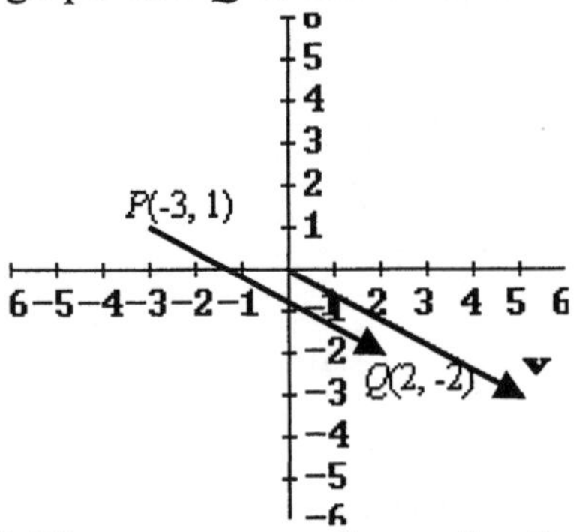

(b) The component form of **v**, the vector represented by $\overrightarrow{PQ}$ is $\mathbf{v}=\langle 5,-3\rangle$.

5. (a) If $P=(0,-\frac{1}{2})$ and $Q=(-\frac{1}{4},1)$, the graph of $\overrightarrow{PQ}$ is shown below:

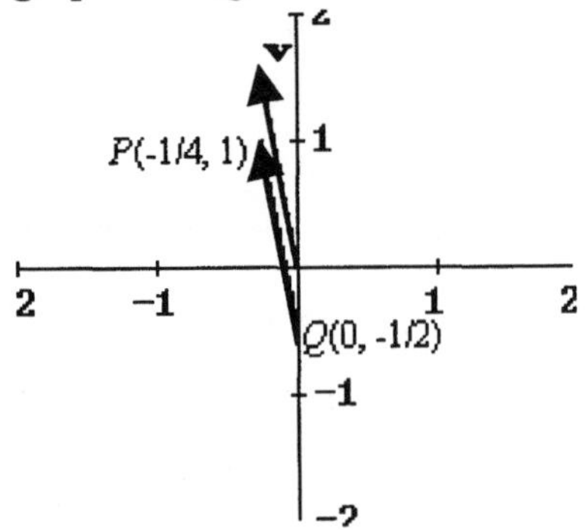

(b) The component form of **v**, the vector represented by $\overrightarrow{PQ}$ is $\mathbf{v}=\left\langle -\frac{1}{4},\frac{3}{2}\right\rangle$.

7. $\|\langle 4,0\rangle\|=4$

9. $\|-12\mathbf{i}-5\mathbf{j}\|=13$

11. $\|\langle 3,5\rangle\|=\sqrt{34}$

13. $\left\|\sqrt{5}\mathbf{i}-\sqrt{7}\mathbf{j}\right\|=2\sqrt{3}$

Exercises 15-19 refer to the vectors **u** *and* **v** *whose graph is given below*:

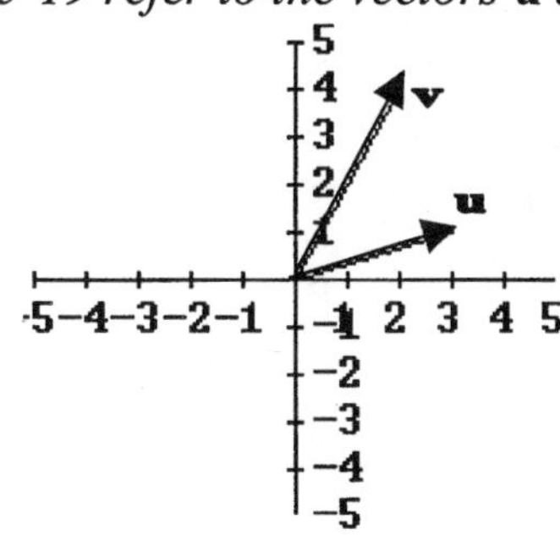

15. The graph of $-\mathbf{u}$:

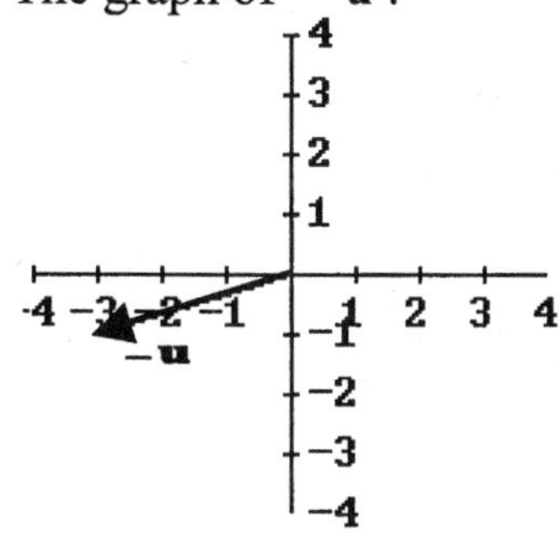

17. The graph of $\mathbf{u}+\mathbf{v}$:

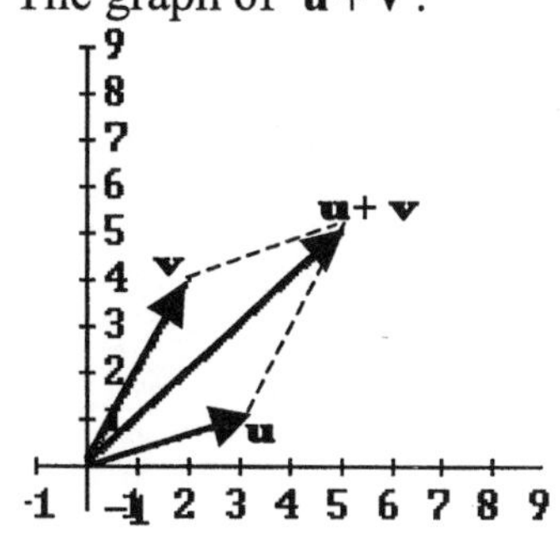

19. The graph of $3\mathbf{u}-2\mathbf{v}$:

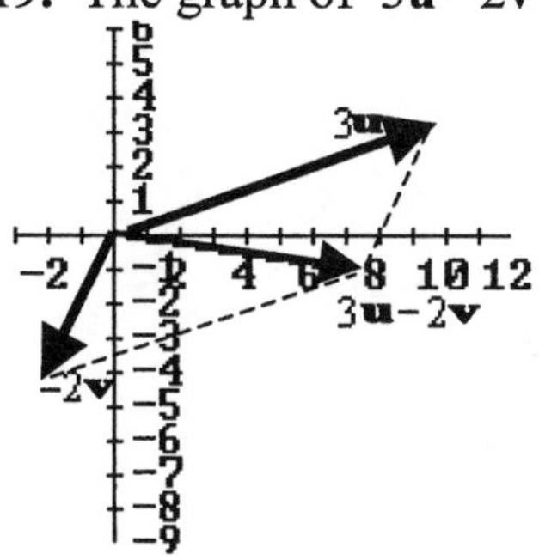

Exercises 21-25 refer to the vectors **u** *and* **v** *whose graph is given below*:

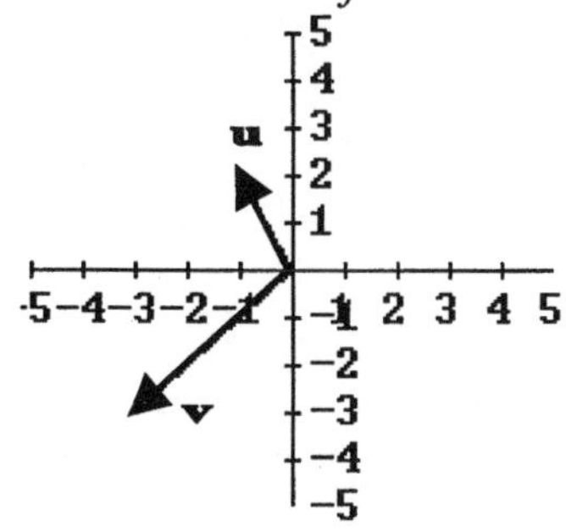

21. The graph of $2\mathbf{u}$:

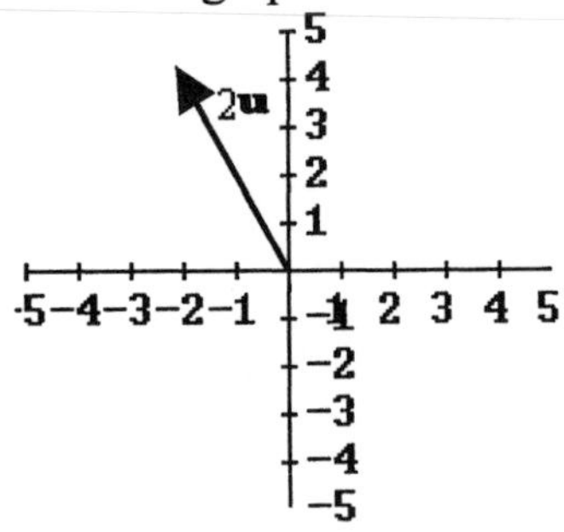

23. The graph of $\mathbf{u}+\mathbf{v}$:

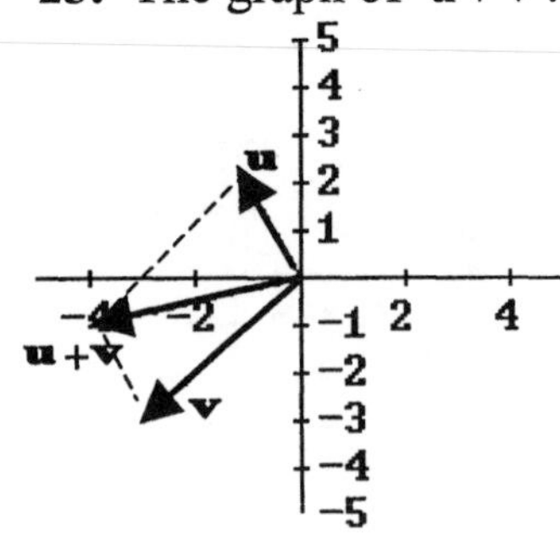

23. The graph of $\mathbf{u}+\mathbf{v}$:

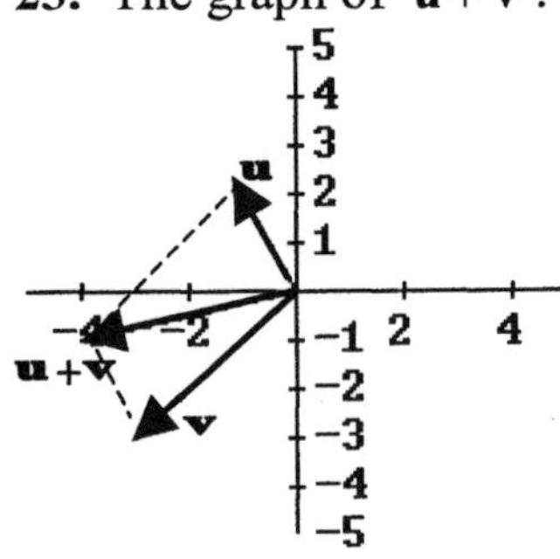

25. The graph of $3\mathbf{u}-\mathbf{v}$:

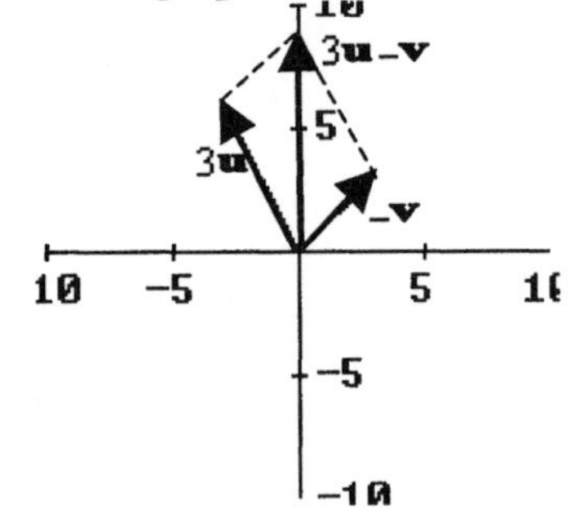

27. If $\mathbf{u}=\langle 1,3\rangle$, and $\mathbf{v}=\langle 2,5\rangle$, then:

(a) $-\mathbf{u}=\langle -1,-3\rangle$

(b) $\mathbf{u}+\mathbf{v}=\langle 3,8\rangle$

(c) $-2\mathbf{u}+\mathbf{v}=\langle 0,-1\rangle$

29. If $\mathbf{u}=2\mathbf{i}+\mathbf{j}$, and $\mathbf{v}=3\mathbf{i}-\mathbf{j}$, then:

(a) $\sqrt{2}\,\mathbf{v}=3\sqrt{2}\,\mathbf{i}-\sqrt{2}\,\mathbf{j}$

(b) $-2\mathbf{u}-\mathbf{v}=-7\mathbf{i}-\mathbf{j}$

(c) $5\mathbf{u}+7\mathbf{v}=31\mathbf{i}-2\mathbf{j}$

31. If $\mathbf{u}=\frac{1}{5}\mathbf{i}-\frac{1}{5}\mathbf{j}$, and $\mathbf{v}=-\mathbf{j}$, then:

(a) $5\mathbf{u}=\mathbf{i}-\mathbf{j}$

(b) $10\mathbf{u}-\mathbf{v}=2\mathbf{i}-\mathbf{j}$

(c) $\mathbf{u}+\frac{1}{5}\mathbf{v}=\frac{1}{5}\mathbf{i}-\frac{2}{5}\mathbf{j}$

33. The unit vector in the direction of $\sqrt{3}\,\mathbf{i}$ is $\mathbf{i}$.

35. The unit vector in the direction of $\langle -3,4\rangle$ is $\left\langle -\frac{3}{5},\frac{4}{5}\right\rangle$.

37. The unit vector in the direction of $-2\mathbf{i}-3\mathbf{j}$ is $-\frac{2\sqrt{13}}{13}\mathbf{i}-\frac{3\sqrt{13}}{13}\mathbf{j}$.

In Exercises 39-45, θ is the angle that the vector $\mathbf{v}$ makes with the positive x-axis.

39. If $\|\mathbf{v}\|=1$, and $\theta=\frac{\pi}{3}$, then $\mathbf{v}=\left\langle\frac{1}{2},\frac{\sqrt{3}}{2}\right\rangle$.

41. If $\|\mathbf{v}\|=\frac{1}{2}$, and $\theta=-120^{\circ}$, then $\mathbf{v}=\left\langle-\frac{1}{4},-\frac{\sqrt{3}}{4}\right\rangle$.

43. If $\|\mathbf{v}\|=5$, and $\theta=-\frac{\pi}{2}$, then $\mathbf{v}=\langle 0,-5\rangle$.

45. If $\|\mathbf{v}\|=2$, and $\theta=110^{\circ}$, then $\mathbf{v}\approx\langle -0.68,1.88\rangle$.

49. $7\mathbf{i}=7(\cos 0\mathbf{i}+\sin 0\mathbf{j})$

49. $\langle 2\sqrt{3},-2\rangle=4\left(\cos\frac{11\pi}{6}\mathbf{i}+\sin\frac{11\pi}{6}\mathbf{j}\right)$

51. $-\frac{1}{8}\mathbf{i}-\frac{1}{8}\mathbf{j}=\frac{\sqrt{2}}{8}\left(\cos\frac{5\pi}{4}\mathbf{i}+\sin\frac{5\pi}{4}\mathbf{j}\right)$

In Exercises 53-59, the given property of vector addition and/or scalar multiplication is verified. Assume that $\mathbf{u}=\langle u_1,u_2\rangle$, $\mathbf{v}=\langle v_1,v_2\rangle$, $\mathbf{w}=\langle w_1,w_2\rangle$, *and that r and s are scalars.*

53. $(\mathbf{u}+\mathbf{v})+\mathbf{w}=\mathbf{u}+(\mathbf{v}+\mathbf{w})$:

$$\begin{aligned}(\mathbf{u}+\mathbf{v})+\mathbf{w}&=\langle u_1+v_1,u_2+v_2\rangle+\langle w_1,w_2\rangle\\&=\langle (u_1+v_1)+w_1,(u_2+v_2)+w_2\rangle\\&=\langle u_1+(v_1+w_1)_1,u_2+(v_2+w_2)\rangle\\&=\langle u_1,u_2\rangle+\langle v_1+w_1,v_2+w_2\rangle\\&=\mathbf{u}+(\mathbf{v}+\mathbf{w})\end{aligned}$$

55. $\mathbf{u}+(-\mathbf{u})=\mathbf{0}$:

$$\begin{aligned}\mathbf{u}+(-\mathbf{u})&=\langle u_1,u_2\rangle+\langle -u_1,-u_2\rangle\\&=\langle u_1+(-u_1),u_2+(-u_2)\rangle\\&=\langle 0,0\rangle\\&=\mathbf{0}\end{aligned}$$

57. $(r+s)\mathbf{u} = r\mathbf{u} + s\mathbf{u}$:

$$\begin{aligned}(r+s)\mathbf{u} &= (r+s)\langle u_1, u_2\rangle \\ &= \langle (r+s)u_1, (r+s)u_2\rangle \\ &= \langle ru_1 + su_1, ru_2 + su_2\rangle \\ &= \langle ru_1, ru_2\rangle + \langle su_1, su_2\rangle \\ &= r\langle u_1, u_2\rangle + s\langle u_1, u_2\rangle \\ &= r\mathbf{u} + s\mathbf{u}\end{aligned}$$

59. $1\mathbf{u} = \mathbf{u}$:

$$\begin{aligned}1\mathbf{u} &= 1\langle u_1, u_2\rangle \\ &= \langle 1u_1, 1u_2\rangle \\ &= \langle u_1, u_2\rangle \\ &= \mathbf{u}\end{aligned}$$

Applications

61. If an airplane whose speed in still is 250 miles per hour is attempting to fly due north, but is being blown off course by a 50-mile-per hour wind blowing from the northeast (in the direction S 45° W), then the resulting velocity is 217.54 miles per hour in the N 9.4^o W direction.

A
50
250
x
B

Since $A = 45^o$, we have $x = \sqrt{250^2 + 50^2 - 2(250)(50)\cos 45^o}$, that is $x \approx 217.54$. The angle B satisfies $\frac{50}{\sin B} \approx \frac{217.54}{\sin 45^o}$, so $B \approx 9.4^o$.

63. If an ocean liner with a velocity of 12 miles per hour on a course of 30° (clockwise from the north) enters the Gulf Stream at a point where the velocity of the Gulf Stream is 4 miles per hour with a bearing of 45°, then the resultant velocity of the liner is 15.9 miles per hour, on a course of 33.73^o (clockwise from the north.)

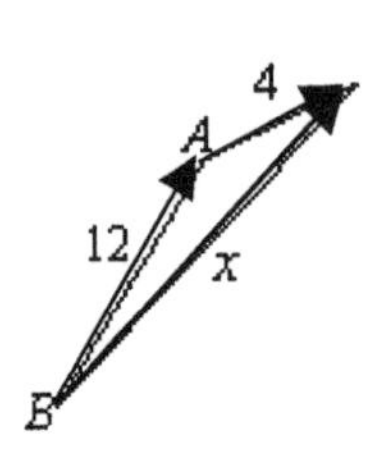

The angle A measures 165°, then $x = \sqrt{12^2 + 4^2 - 2(12)(4)\cos 165^o}$, that is $x \approx 15.9$. The angle B can be computed using the equation $\frac{15.9}{\sin 165^o} = \frac{4}{\sin B}$: $B \approx 3.73^o$.

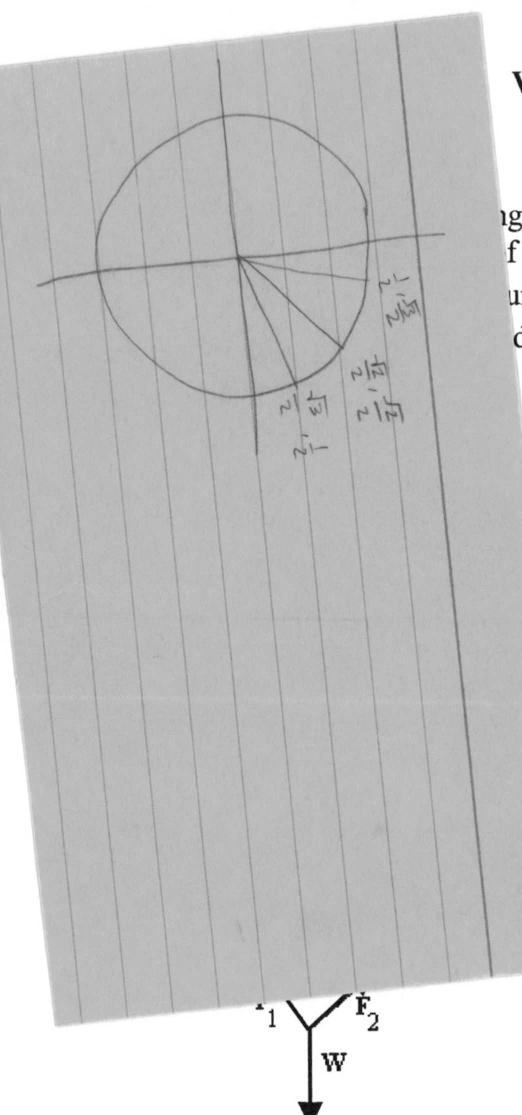

ıg on the same sled, the first with 100 pounds of force and the
f force, with the first dog pulling in the correct direction, but the
urse, then the magnitude of the net force acting on the sled is
d is 10.92^o off course.

The net force is $100\,\mathbf{i}+(120\cos 20^o\,\mathbf{i}-120\sin 20^o\,\mathbf{j})$, or $212.76\mathbf{i}-41.04\mathbf{j}$. The magnitude of this vector is 216.68, and the angle with the course of the first dog is 10.92^o.

ropes as shown below, the forces acting on the crate are the
wnward, the pull $\mathbf{F}_1$ of the rope that makes an angle of 45°
ıll $\mathbf{F}_2$ of the rope that makes an angle of 30° with the
de of 448 and $\mathbf{F}_2$ a magnitude of 366, the weight of the crate
$+|\mathbf{F}_2|\sin 30^o$, that is $|\mathbf{W}|=499.78$.

$\mathbf{F}_1$ $\mathbf{F}_2$ W

Concepts and Critical Thinking

69. The statement: "For all vectors $\mathbf{v}$, $-2\mathbf{v}$ is twice as long as $\mathbf{v}$" is true.

71. The statement: "If $\mathbf{v}\neq\mathbf{0}$ and $\mathbf{u}=\left(\frac{1}{\|\mathbf{v}\|}\right)\mathbf{v}$, then $\|\mathbf{u}\|=1$" is true.

73. Answers may vary. An example of a vector $\mathbf{v}=\langle a,b\rangle$ with $a\neq 0$ and $b\neq 0$ and such that $\|\mathbf{v}\|=2$ is $\mathbf{v}=\langle\sqrt{3},1\rangle$.

75. Answers may vary. Two vectors $\mathbf{u}$ and $\mathbf{v}$ such that $\|\mathbf{u}+\mathbf{v}\|=\|\mathbf{u}\|+\|\mathbf{v}\|$ are $\mathbf{u}=3\mathbf{i}$ and $\mathbf{v}=4\mathbf{i}$. Here, $\|\mathbf{u}+\mathbf{v}\|=7$ and $\|\mathbf{u}\|+\|\mathbf{v}\|=7$.

Chapter 7. Review

In Exercises 1-12, a triangle is labeled in the standard way.

1. If $c = 12$, $A = 47^o$, and $C = 52^o$, this is an AAS case. We have $B = 180^o - (47^o + 52^o)$, or $\boxed{B = 81^o}$. Then $\frac{a}{\sin 47^o} = \frac{12}{\sin 52^o}$, so $a = \frac{12\sin 47^o}{\sin 52^o}$, that is $a \approx 11.14$. Finally, $\frac{b}{\sin 81^o} = \frac{12}{\sin 52^o}$, and $b = \frac{12\sin 81^o}{\sin 52^o}$, or $\boxed{b \approx 15.04}$.

3. If $a = 1.2$, $b = 1$, and $B = 42^o$, we have a SSA case. It follows that $\frac{1.2}{\sin A} = \frac{1}{\sin 42^o}$, therefore $\sin A = 1.2\sin 42^o$, and A has two possible values, namely $A \approx 53.41^o$ or $A \approx 126.59^o$. Both possibilities are valid here:

(a) if $\boxed{A = 53.41^o}$, then $C \approx 180^o - (42^o + 53.41^o)$, or $\boxed{C \approx 84.59^o}$; finally $\frac{c}{\sin 84.59^o} = \frac{1}{\sin 42^o}$, so $c = \frac{\sin 84.59^o}{\sin 42^o}$, that is $\boxed{c \approx 1.49}$.

(b) if $\boxed{A = 126.59^o}$, then $C \approx 180^o - (42^o + 126.59^o)$, or $\boxed{C \approx 11.41^o}$; finally $\frac{c}{\sin 11.41^o} = \frac{1}{\sin 42^o}$, so $c = \frac{\sin 11.41^o}{\sin 42^o}$, that is $\boxed{c \approx 0.3}$.

5. If $b = 19$, $c = 15$, and $A = 10^o$, we have a SAS case. By the Law of Cosines, we have: $a^2 = 19^2 + 15^2 - 2(19)(15)\cos 10^o$, so $\boxed{a \approx 4.97}$. Now, $\frac{4.97}{\sin 10^o} \approx \frac{19}{\sin B}$, so $\sin B \approx \frac{19\sin 10^o}{4.97}$, and $\boxed{B \approx 41.59^o}$. Finally, $C \approx 180^o - (41.59^o + 10^o)$, that is $\boxed{C \approx 128.41^o}$.

7. If $a = 5$, $b = 8$, and $c = 9$, we have a SSS case. By the Law of Cosines, we have: $5^2 = 8^2 + 9^2 - 2(8)(9)\cos A$, so $\cos A = \frac{5^2 - 8^2 - 9^2}{-144}$, that is $\cos A \approx 0.8333$, and $\boxed{A \approx 33.56^o}$. Now, $\frac{5}{\sin 33.56^o} \approx \frac{8}{\sin B}$, so $\sin B \approx \frac{8\sin 33.56^o}{5}$, or $\sin B \approx 0.8845$ and $\boxed{B \approx 62.19^o}$. Finally, $C \approx 180^o - (33.56^o + 62.19^o)$, that is $\boxed{C \approx 84.25^o}$.

9. If $a = 335$, $\mathrm{b} = 260$, and $c = 540$, we have a SSS case. By the Law of Cosines, we have: $335^2 = 260^2 + 540^2 - 2(260)(540)\cos A$, so $\cos A = \frac{260^2 + 540^2 - 335^2}{2(260)(540)} = 0.8795$, and $\boxed{A \approx 28.42^o}$. Now, $\frac{335}{\sin 28.42^o} \approx \frac{260}{\sin B}$, so $\sin B = \frac{260\sin 28.42^o}{335}$, and $\boxed{B \approx 21.68^o}$. Finally, $C \approx 180^o - (21.68^o + 28.42^o)$, so $\boxed{C \approx 129.9^o}$.

11. If $a = 2$, $b = 7$, and $A = 41^o$, we have a SSA case. We have, by the Law of Sines, $\frac{2}{\sin 41^o} = \frac{7}{\sin B}$, so $\sin B = \frac{7\sin 41^o}{2}$, that is $\sin B \approx 2.2962$. There is no triangle with the given information.

21. $|4-3i|=5$

23. $|-9+i|=\sqrt{82}$

25. $-2-2i=2\sqrt{2}\left(\cos\frac{5\pi}{4}+i\sin\frac{5\pi}{4}\right)$

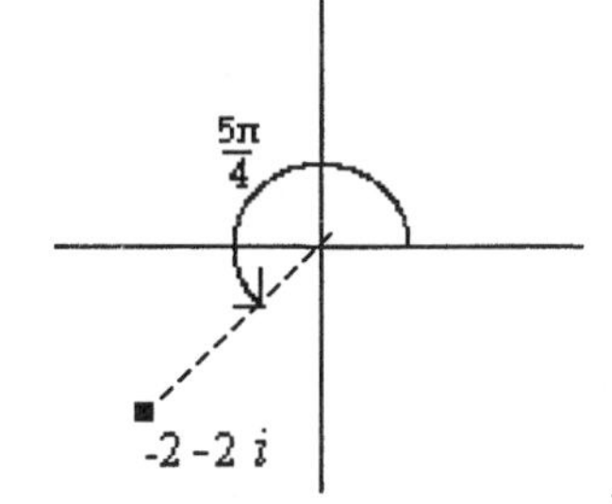

27. $\frac{3}{2}i=\frac{3}{2}\left(\cos\frac{\pi}{2}+i\sin\frac{\pi}{2}\right)$

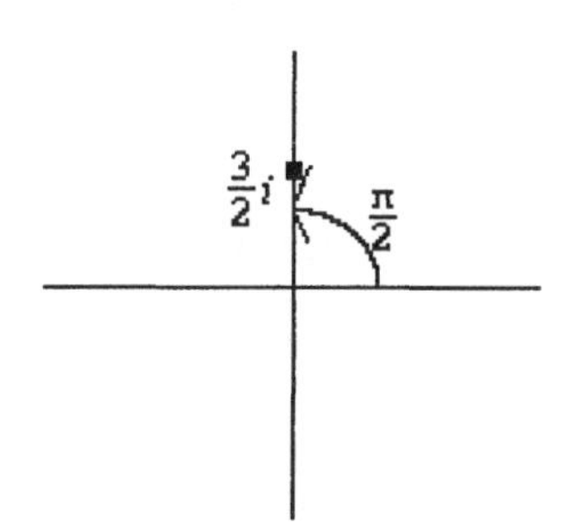

29. $2\left(\cos 330^{o}+i\sin 330^{o}\right)=\sqrt{3}-i$

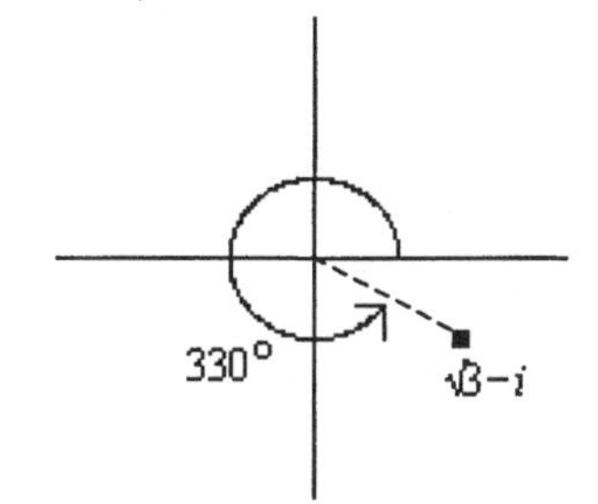

31. $\sqrt{3}\left(\cos\frac{\pi}{3}+i\sin\frac{\pi}{3}\right)=\frac{\sqrt{3}}{2}+\frac{3}{2}i$

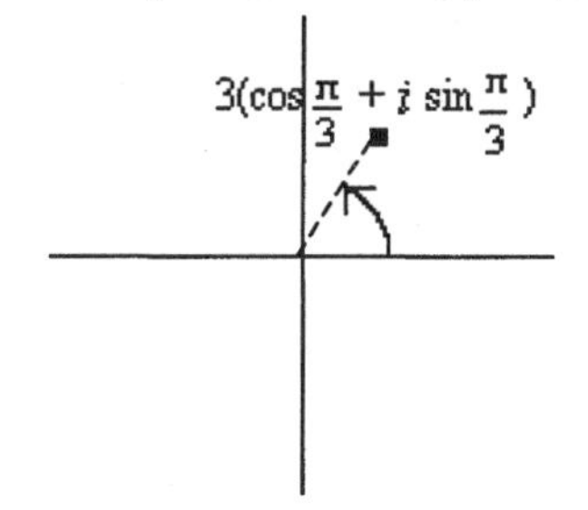

33. $5\left(\cos\frac{\pi}{3}+i\sin\frac{\pi}{3}\right)\cdot 2\left(\cos\frac{\pi}{4}+i\sin\frac{\pi}{4}\right)=10\left(\cos\frac{7\pi}{12}+i\sin\frac{7\pi}{12}\right)$

35. $\dfrac{12\left(\cos 300^{o}+i\sin 300^{o}\right)}{4\left(\cos 100^{o}+i\sin 100^{o}\right)}=3\left(\cos 200^{o}+i\sin 200^{o}\right)$

37. $7i(-3+3i)=7\left(\cos\frac{\pi}{2}+i\sin\frac{\pi}{2}\right)\cdot 3\sqrt{2}\left(\cos\frac{3\pi}{4}+i\sin\frac{3\pi}{4}\right)$

$=21\sqrt{2}\left(\cos\frac{5\pi}{4}+i\sin\frac{5\pi}{4}\right)=21\sqrt{2}\left(-\frac{\sqrt{2}}{2}-\frac{\sqrt{2}}{2}i\right)=-21-21i$

Without changing to trigonometric form, $7i(-3+3i)=-21-21i$, as expected.

Without changing to trigonometric form, $(2-2i)\left(\sqrt{3}+i\right)=\left(2\sqrt{3}+2\right)+\left(2-2\sqrt{3}\right)i$.

39. $\dfrac{-5+5\sqrt{3}i}{\sqrt{3}+i}=\dfrac{10\left(\cos\frac{2\pi}{3}+i\sin\frac{2\pi}{3}\right)}{2\left(\cos\frac{\pi}{6}+i\sin\frac{\pi}{6}\right)}=5\left(\cos\frac{\pi}{2}+i\sin\frac{\pi}{2}\right)=5i$

Without changing to trigonometric form,

$\dfrac{-5+5\sqrt{3}i}{\sqrt{3}+i}=\dfrac{\left(-5+5\sqrt{3}i\right)\left(\sqrt{3}-i\right)}{\left(\sqrt{3}+i\right)\left(\sqrt{3}-i\right)}=\dfrac{-5\sqrt{3}+5i+15i+5\sqrt{3}}{4}=\dfrac{20i}{4}=5i$.

41. $\left[5\left(\cos 10^o + i\sin 10^o\right)\right]^4 = 625\left(\cos 40^o + i\sin 40^o\right)$

43. $\left(1-\sqrt{3}i\right)^{10} = \left[2\left(\cos\frac{5\pi}{3} + i\sin\frac{5\pi}{3}\right)\right]^5 = 32\left(\cos\frac{25\pi}{3} + i\sin\frac{25\pi}{3}\right) = 32\left(\frac{1}{2} + \frac{\sqrt{3}}{2}i\right) = 16 + 16\sqrt{3}i$

45. The square roots of $4\left(\cos 120^o + i\sin 120^o\right)$ are $z_1 = 2\left(\cos 60^o + i\sin 60^o\right)$ and $z_2 = 2\left(\cos 240^o + i\sin 240^o\right)$, that is $z_1 = 1+\sqrt{3}i$ and $z_2 = -1-\sqrt{3}i$

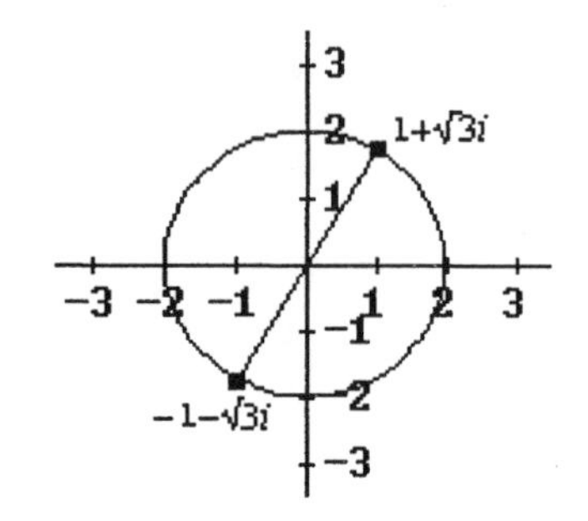

47. The fourth roots of 81 are $z_1 = 3,\ z_2 = 3i,\ z_3 = -3$, and $z_4 = -3i$.

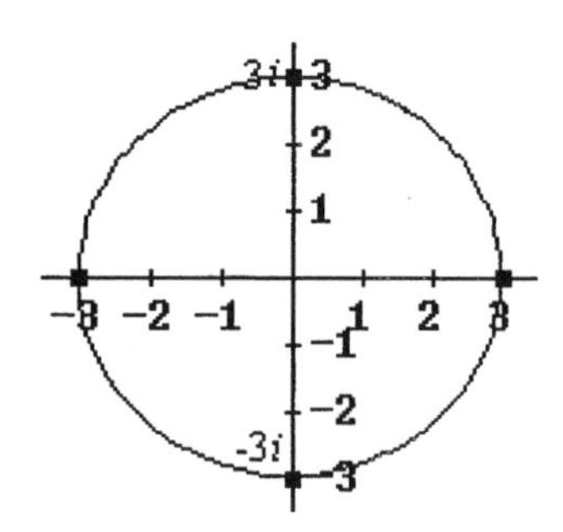

49. The square roots of $-3+4i$ are:

$$z_1 = \sqrt{5}\left[\cos\left(\frac{\tan^{-1}\left(-\frac{4}{3}\right)+\pi}{2}\right) + i\sin\left(\frac{\tan^{-1}\left(-\frac{4}{3}\right)+\pi}{2}\right)\right] = 1+2i, \text{ and}$$

$$z_2 = \sqrt{5}\left[\cos\left(\frac{\tan^{-1}\left(-\frac{4}{3}\right)+3\pi}{2}\right) + i\sin\left(\frac{\tan^{-1}\left(-\frac{4}{3}\right)+3\pi}{2}\right)\right] = -1-2i$$

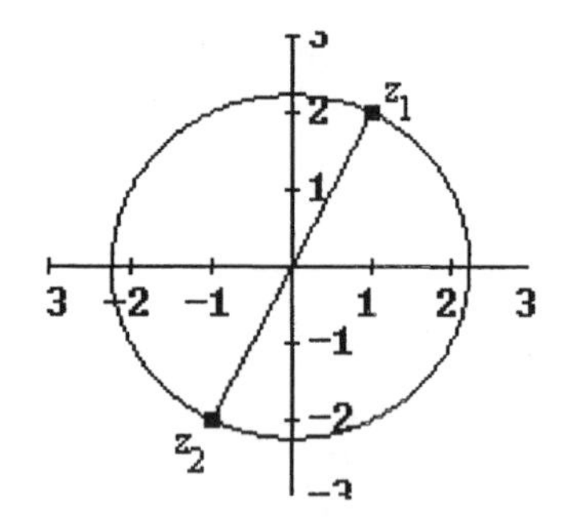

51. The solutions of $x^3+1=0$ are the cube roots of -1:

$x_1 = \cos\frac{\pi}{3} + i\sin\frac{\pi}{3} = \frac{1}{2} + \frac{\sqrt{3}}{2}i$, $x_2 = \cos\pi + i\sin\pi = -1$, and

$x_3 = \cos\frac{5\pi}{3} + i\sin\frac{5\pi}{3} = \frac{1}{2} - \frac{\sqrt{3}}{2}i$.

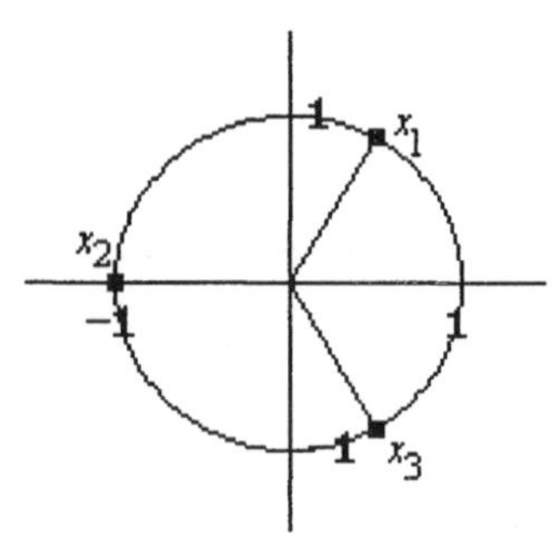

53. The point $(2,\pi)$ in polar coordinates corresponds to $(-2,0)$ in rectangular coordinates.

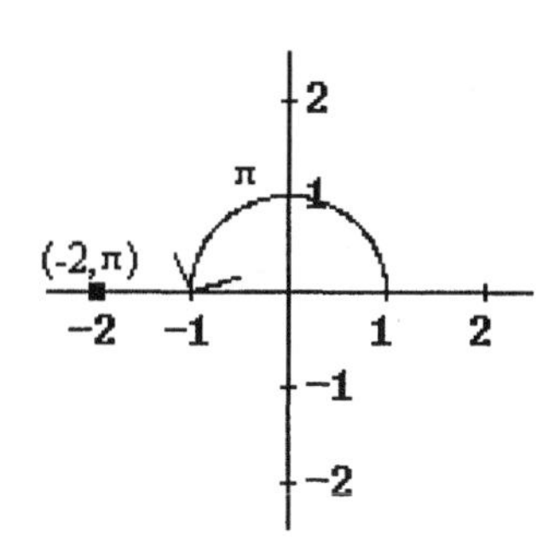

55. The point $(-3,3)$ in rectangular coordinates corresponds to $\left(3\sqrt{2},\frac{3\pi}{4}\right)$ and $\left(-3\sqrt{2},\frac{7\pi}{4}\right)$.

57. The points $A(5,0), B\left(5,\frac{\pi}{2}\right), C\left(5,\frac{3\pi}{4}\right)$, and $D(5,\pi)$ belong to the graph of the polar equation $r=5$.

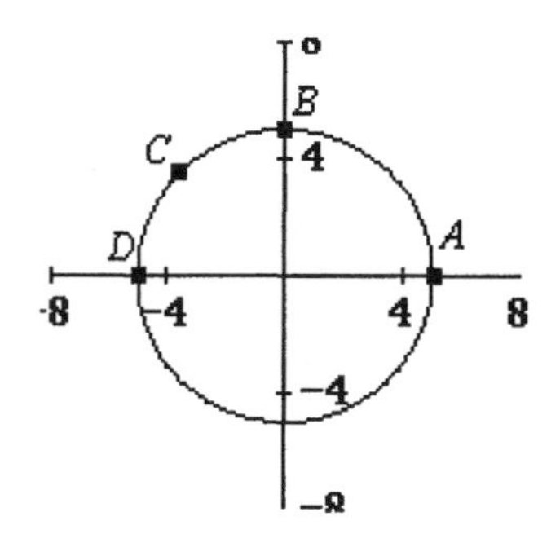

59. The points $A(0,0), B(3,1), C(9,3)$, and $D(12,4)$ belong to the graph of the polar equation $r=3\theta,\ \theta\geq 0$.

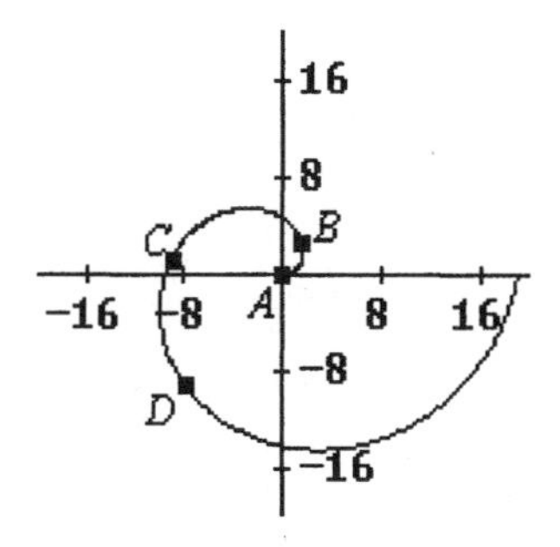

61. The points $A(0,0), B\left(\frac{3\sqrt{3}}{2},\frac{\pi}{6}\right), C\left(\frac{3\sqrt{3}}{2},\frac{\pi}{3}\right)$, and $D\left(-3,\frac{3\pi}{4}\right)$ belong to the graph of the polar equation $r=3\sin 2\theta$.

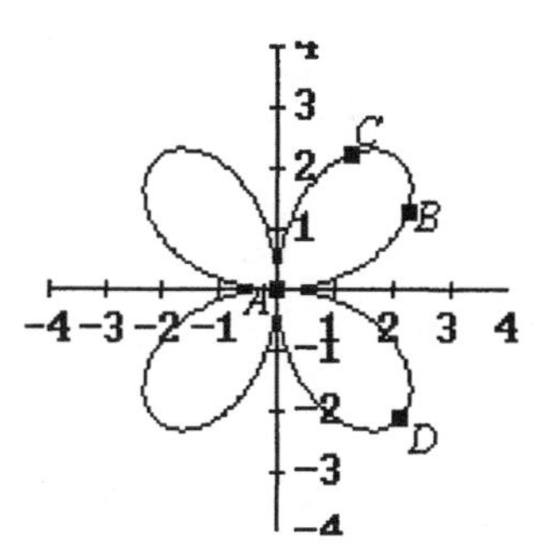

63. The rectangular equation $y=2x$ corresponds to the polar equation $\theta=\tan^{-1}2$.

65. The rectangular equation $x^2+y^2=1$ corresponds to the polar equation $r=1$.

67. The polar equation $r = -1$ corresponds to the equation $x^2 + y^2 = 1$ in rectangular coordinates.

69. The polar equation $r = 3\sin\theta$ corresponds to the rectangular equation $x^2 + y^2 - 3y = 0$.

71. The points of intersection of the graphs of the polar equations $r = 3$ and $r = 4\cos\theta$ are $P\left(\frac{9}{4}, \frac{\sqrt{63}}{4}\right)$ and $Q\left(\frac{9}{4}, -\frac{\sqrt{63}}{4}\right)$, in rectangular coordinates, or $P\left(3, \cos^{-1}\frac{3}{4}\right)$ and $Q\left(3, -\cos^{-1}\frac{3}{4}\right)$ in polar coordinates.

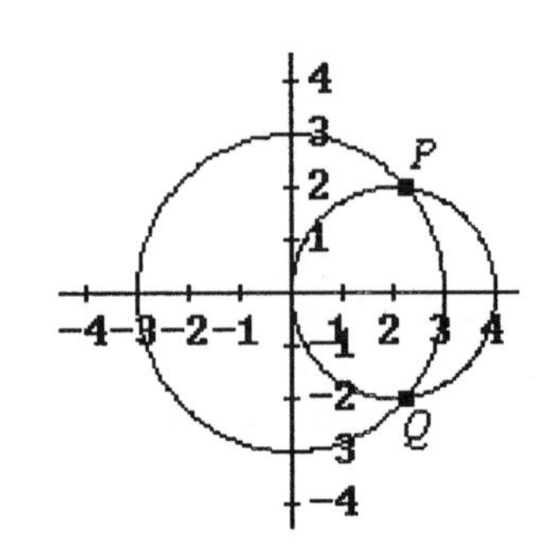

73. **(a)** If $P = (4, -2)$ and $Q = (-1, -1)$, a sketch of the directed segment $\overrightarrow{PQ}$ follows:

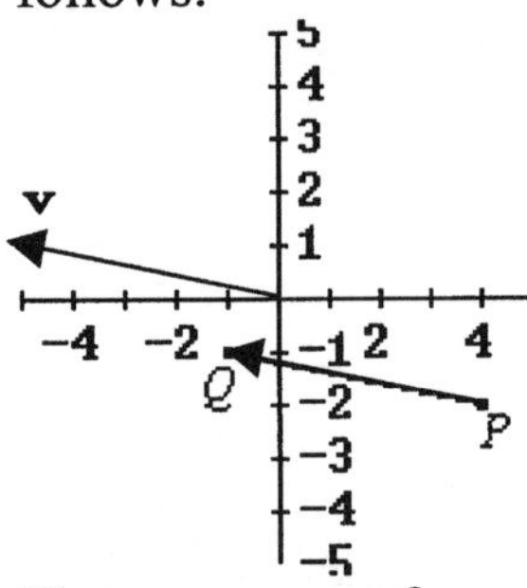

(b) The component form of the vector **v** represented by $\overrightarrow{PQ}$ is $\mathbf{v} = \langle -5, 1 \rangle$.

75. $\|\langle -6, 8 \rangle\| = \sqrt{(-6)^2 + 8^2} = 10$

77. Starting with the sketches of **u** and **v**, the graph of $\mathbf{u} - \mathbf{v}$, in standard position, is superimposed.

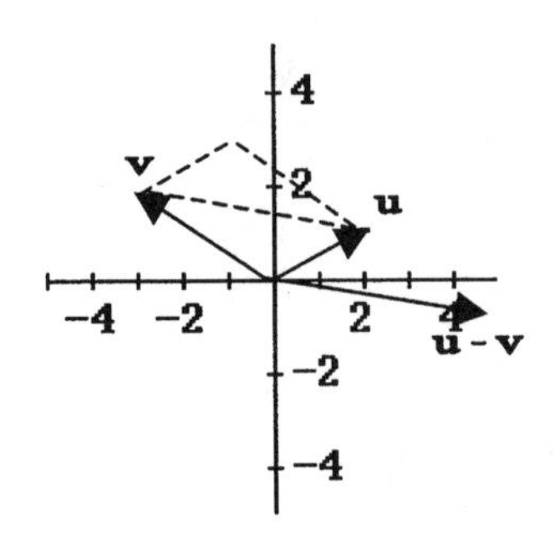

79. If $\mathbf{u} = \langle -4, 2 \rangle$, and $\mathbf{v} = \langle -2, 3 \rangle$, then:

(a) $-2\mathbf{u} = \langle 8, -4 \rangle$

(b) $\mathbf{u} - 2\mathbf{v} = \langle 0, -4 \rangle$

(c) $\frac{1}{2}\mathbf{u} + \frac{1}{2}\mathbf{v} = \left\langle -3, \frac{5}{2} \right\rangle$

81. The unit vector in the direction of $12\mathbf{i} + 5\mathbf{j}$ is $\frac{12}{13}\mathbf{i} + \frac{5}{13}\mathbf{j}$.

83. If $\|\mathbf{v}\| = 2$, and $\theta = \frac{\pi}{6}$ is the angle made by **v** with the positive x-axis, then the component form of **v** is $\mathbf{v} = \sqrt{3}\ \mathbf{i} + \mathbf{j}$.

85. The vector **4j** in trigonometric form is $4\left(\cos\frac{\pi}{2} + i\sin\frac{\pi}{2}\right)$.

87. If a boater spots a lighthouse at a bearing of $\text{N}38^{\circ}\,\text{E}$, then sails east for 10 miles, at which time the lighthouse is at a bearing of $\text{N}47^{\circ}\,\text{W}$, then $B = 42^{\circ}$, $A = 33^{\circ}$, and $C = 180^{\circ} - \left(33^{\circ} + 42^{\circ}\right)$, that is $C = 105^{\circ}$. It follows that $\frac{b}{\sin 42^{\circ}} = \frac{10}{\sin 105^{\circ}}$, so $b = \frac{10\sin 42^{\circ}}{\sin 105^{\circ}}$, approximately $b \approx 6.93$ miles. Since the boat can sail at the rate of 30 miles per hour, the time that it will take it to reach the lighthouse after sailing 10 miles due east from its starting position is 0.23 hours, or 13.8 minutes.

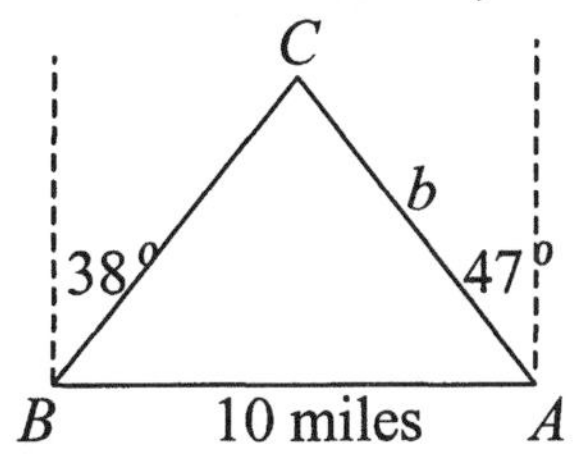

89. If a nature photographer traveling on an overhead tram that moves parallel to the ground at a constant speed spots an elk directly below the tram cable in front of her at an angle of depression of 30°, and after two minutes the elk lies in front of her at an angle of depression of 34°, at which point she estimates the distance to the elk to be 2 miles, then the speed of the tram is approximately 6.9 miles per hour.

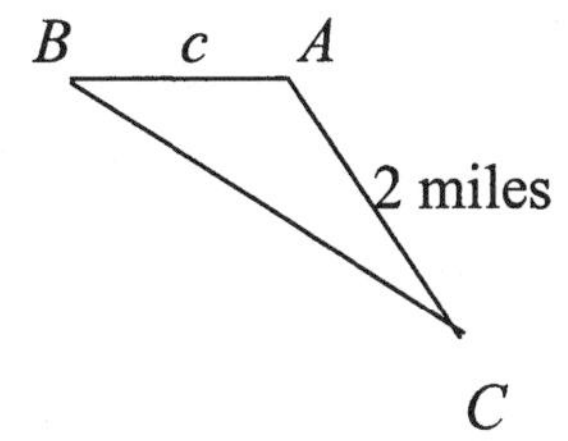

We have $B = 30^{\circ}$, $A = 146^{\circ}$, and $C = 180^{\circ} - \left(30^{\circ} + 146^{\circ}\right)$, that is $C = 4^{\circ}$.

Then $\frac{c}{\sin 4^{\circ}} = \frac{2}{\sin 30^{\circ}}$, and $c = \frac{2\sin 4^{\circ}}{\sin 30^{\circ}} \approx 0.23$ miles. The speed of the tram is, then,

$\frac{0.23 \text{ miles}}{2 \text{ min}} = \frac{0.23 \text{ miles}}{2 \text{ min}} \frac{60 \text{ min}}{1 \text{ hr}}$, approximately 6.9 miles per hour.

Chapter 7. Test

1. The statement: "Every real number has a real square root" is false.

3. The statement: "If the three angles of a triangle are known, then the side lengths can be determined" is false.

5. The statement: "The magnitude of a vector is itself a vector" is false.

7. The statement: "The length of the sum of two vectors is the sum of the lengths of the vectors" is false.

9. The statement: "If the point (x, y) in rectangular coordinates corresponds to the point (r, θ) in polar coordinates, then it follows that $\theta = \arctan \frac{y}{x}$" is false. (1,1) in corresponds to the point $\left(\sqrt{2}, \frac{9\pi}{4}\right)$ in polar coordinates, and it is not true that $\frac{9\pi}{4} = \arctan 1$.)

11. Answers may vary. An example of a triangle that has angles 60° and 70° is the triangle with angles $A = 60^o$, $B = 70^o$, $C = 50^o$ and sides $a = 3$, $b \approx 3.25$, and $c \approx 2.65$.
(*a* can be arbitrarily chosen, and the other two sides are found using the Law of Sines.)

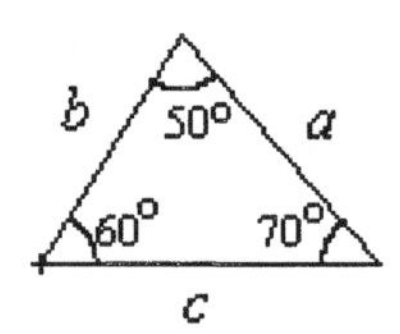

13. Answers may vary. Examples of vectors **u** and **v** such that $\|\mathbf{u}\| = \|\mathbf{v}\| = 2$ and $\mathbf{u} + \mathbf{v} = \mathbf{0}$ are $\mathbf{u} = 2\mathbf{i}$, and $\mathbf{v} = -2\mathbf{i}$.

15. If in a triangle $b = 12$, $c = 8$, and $A = 55^o$ with sides and angles labeled in the standard way, this is an SAS case. Using the Law of Cosines, $a^2 = 12^2 + 8^2 - 2(12)(8)\cos 55^o$, that is $\boxed{a \approx 9.83}$. Then, using the Law of Cosines again, $12^2 = 8^2 + 9.83^2 - 2(8)(9.83)\cos B$, so $\cos B \approx 0.1057$, and $\boxed{B \approx 83.93^o}$. Finally, $A \approx 180^o - (55^o + 83.93^o)$, that is $A \approx 41.07^o$.

17. **(a)** The complex number $z = -1 - i$, in trigonometric form, is

$$z = \sqrt{2}\left(\cos \tfrac{5\pi}{4} + i \sin \tfrac{5\pi}{4}\right)$$

(b) $z^6 = \left[\sqrt{2}\left(\cos \frac{5\pi}{4} + i \sin \frac{5\pi}{4}\right)\right]^6$, that is $z^6 = -8i$

(c) The square roots of $z = -1 - i$ are

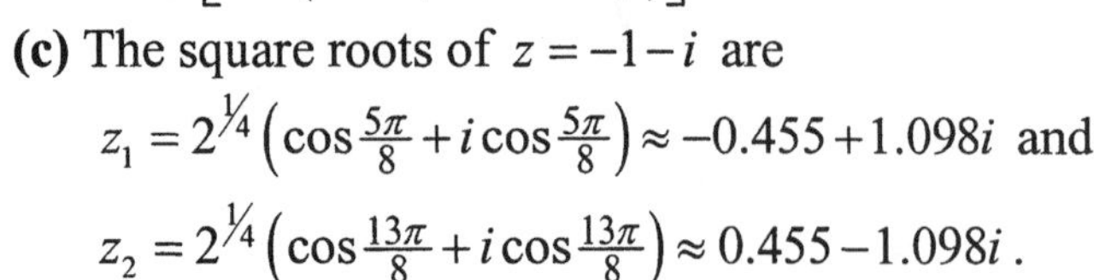

$$z_1 = 2^{1/4}\left(\cos \tfrac{5\pi}{8} + i \cos \tfrac{5\pi}{8}\right) \approx -0.455 + 1.098i \text{ and}$$

$$z_2 = 2^{1/4}\left(\cos \tfrac{13\pi}{8} + i \cos \tfrac{13\pi}{8}\right) \approx 0.455 - 1.098i.$$

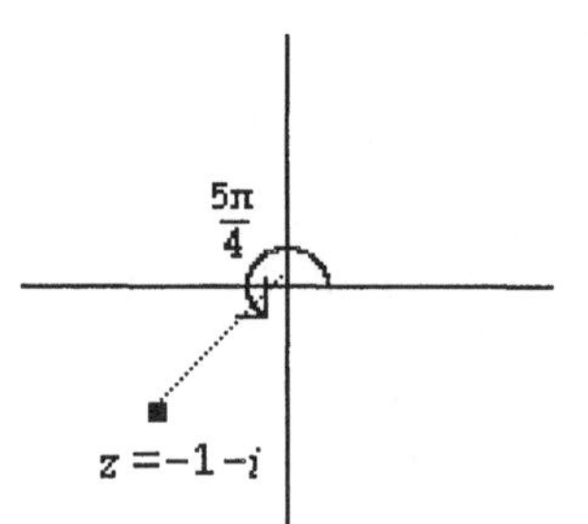

19. Given the polar equation $r = \cos 2\theta$, four points that belong to its graph are: $A(1,0), B\left(0,\frac{\pi}{4}\right), C\left(-1,\frac{\pi}{2}\right)$, and $D\left(-1,\frac{3\pi}{2}\right)$

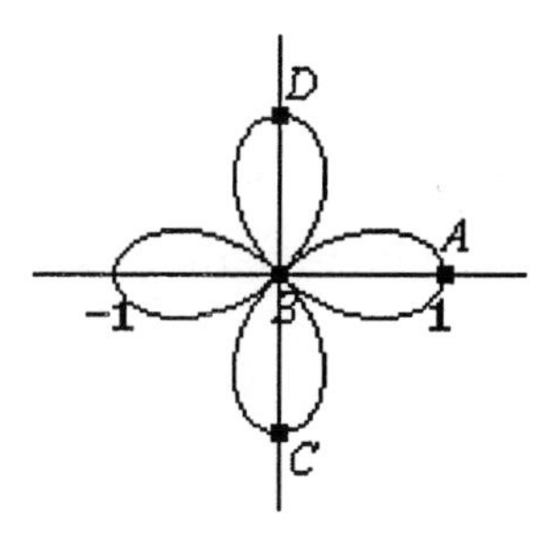

Chapter 8. RELATIONS AND CONIC SECTIONS

Section 8.1 Relations and Their Graphs

1. Based on the graph of the relation $y = x^3$:

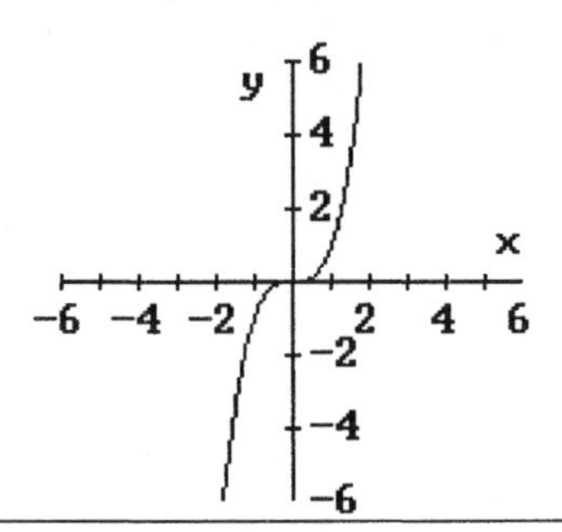

(a) The graph of $y = (x+2)^3$ is a horizontal translation, 2 units to the left.

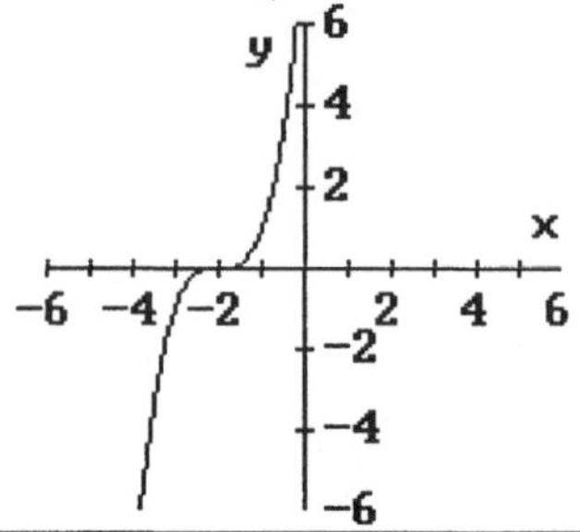

(b) The graph of $y+1 = (x-3)^3$ is a horizontal translation, 3 units to the right, combined with a vertical translation, down 1 unit.

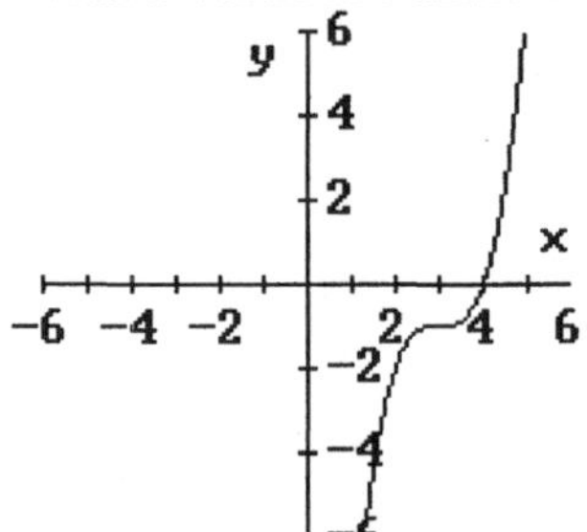

(c) The graph of $y = (x+4)^3 - 2$ is a horizontal translation, left 4 units, with a vertical translation, down 2 units.

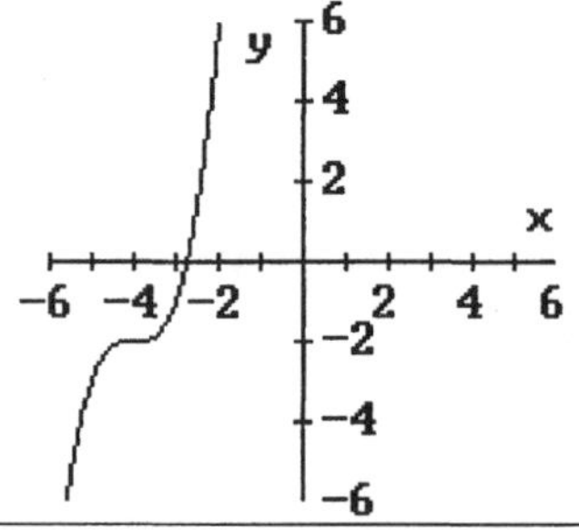

3. Based on the graph of the relation $\frac{x^2}{16}+\frac{y^2}{9}=1$:

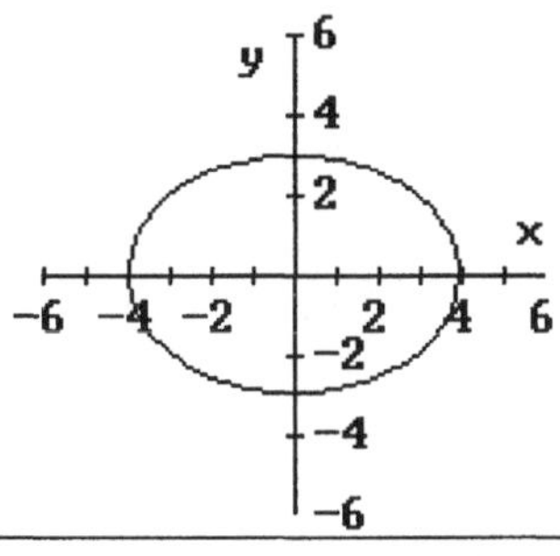

(a) The graph of $\frac{x^2}{16}+\frac{(y-1)^2}{9}=1$ is a vertical translation, up 1 unit.

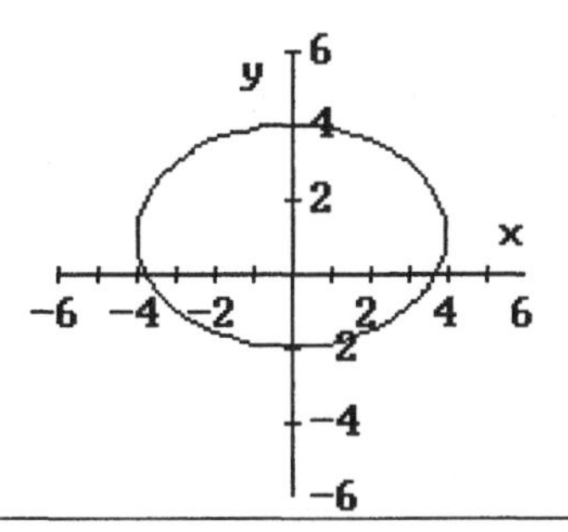

(b) The graph of $\frac{(x-4)^2}{16}+\frac{(y+2)^2}{9}=1$ is a horizontal translation, right 4 units, with a vertical translation, down 2 units.

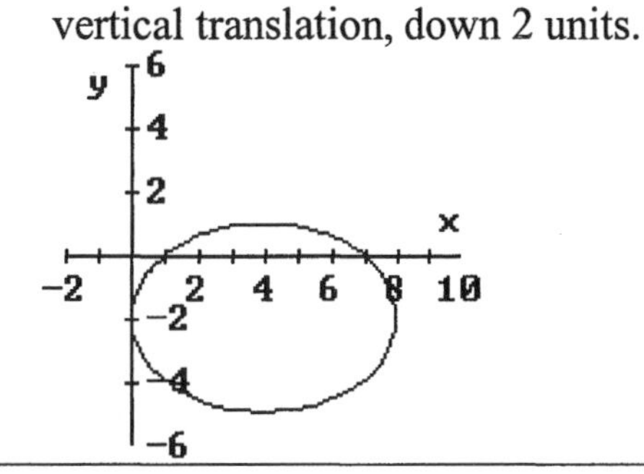

5. Given the original relation $y=\sqrt{x}$ and its graph, with the translated graph as shown:

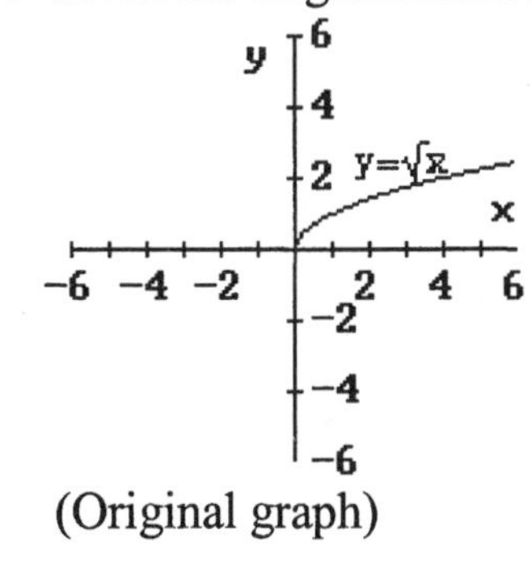

(Original graph)

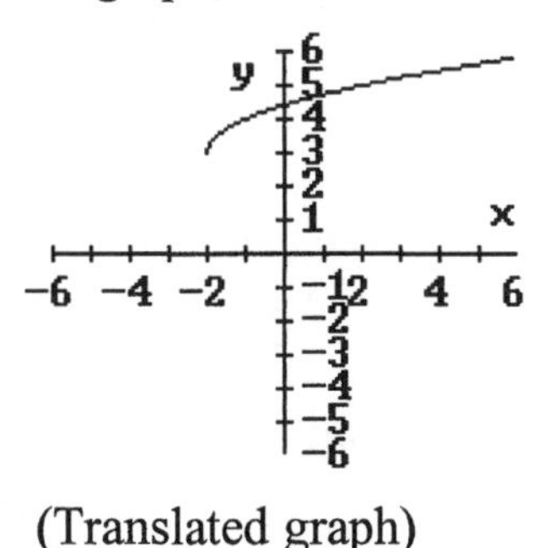

(Translated graph)

The translated graph corresponds to a horizontal translation, left 2 units, combined with a vertical translation, up 3 units. An equation for the translated graph is $y=\sqrt{x+2}+3$.

7. Given the original relation $xy=4$ and its graph, with the translated graph as shown:

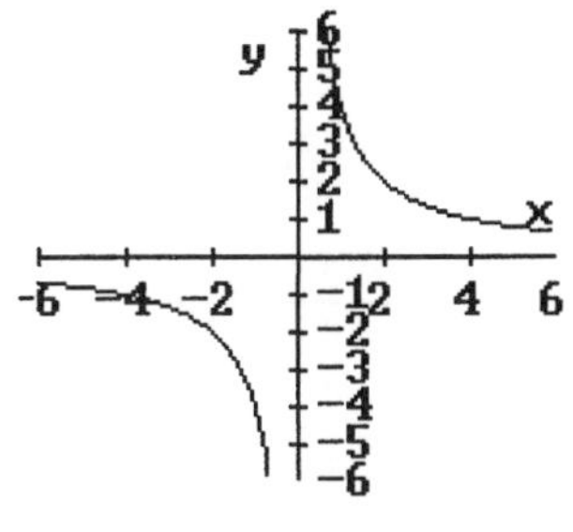

(Original graph)

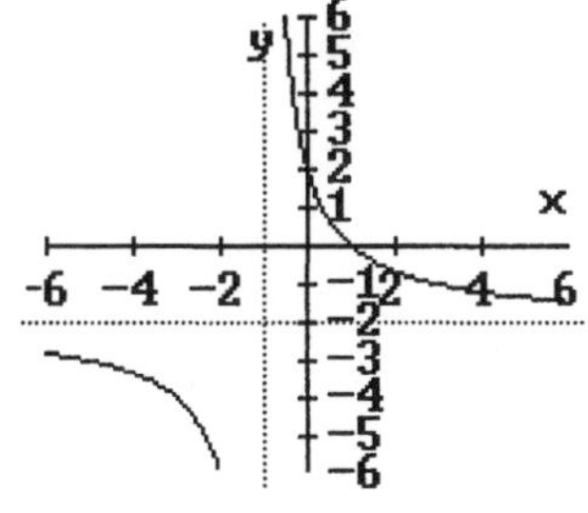

(Translated graph)

The translated graph corresponds to a horizontal translation, left 1 unit, combined with a vertical translation, down 2 units. An equation for the translated graph is $(x+1)(y+2)=4$.

9. Given the original relation $x = \sqrt{y} - 2$ and its graph,

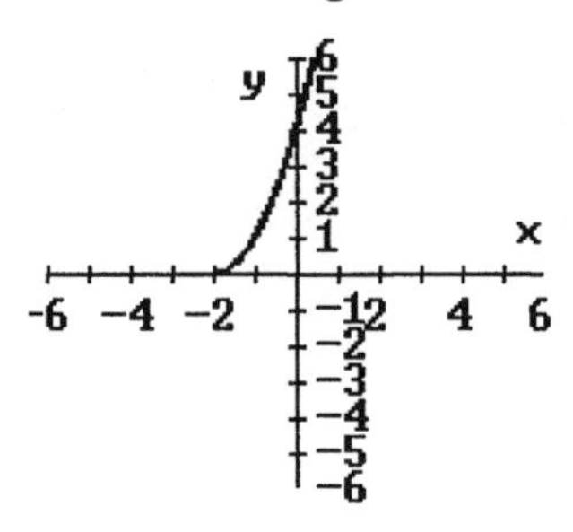

(a) The graph of $-x = \sqrt{y} - 2$ is a reflection about the y axis:

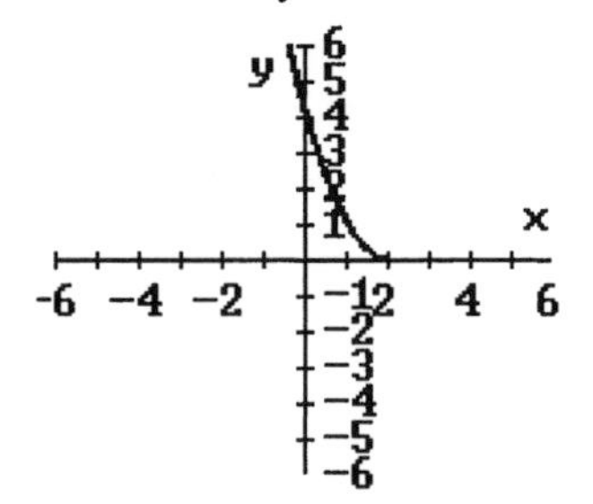

(b) The graph of $x = \sqrt{-y} - 2$ is a reflection about the x axis:

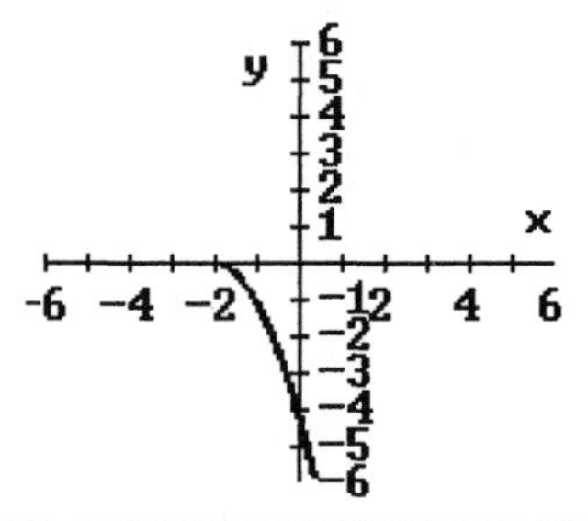

11. Given the original relation $x^3 - y^3 = 1$ and its graph,

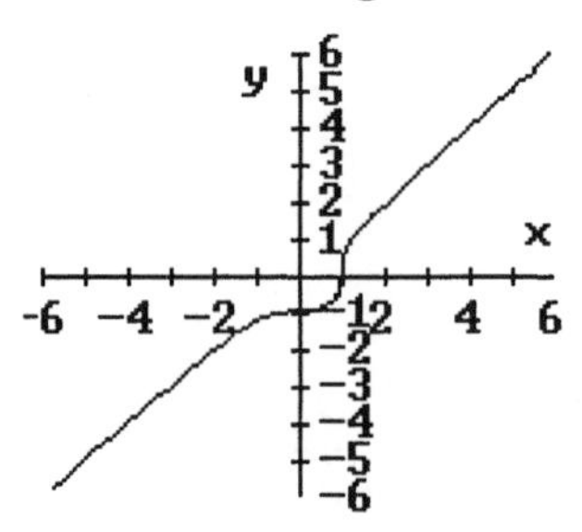

(a) The graph of $y^3 - x^3 = 1$ is a reflection about the origin.

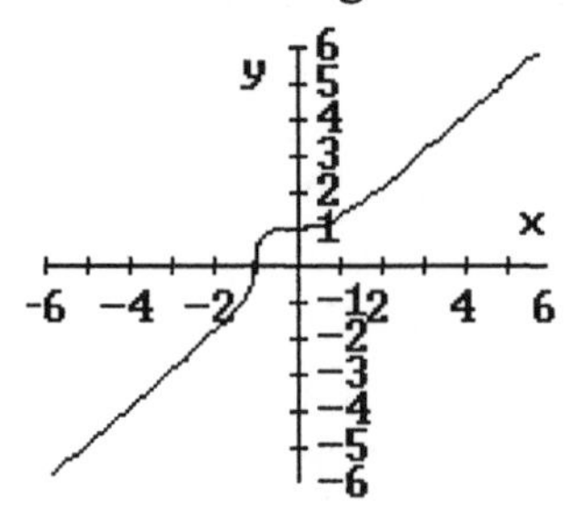

(b) The graph of $x^3 + y^3 = 1$ is a reflection about the x axis.

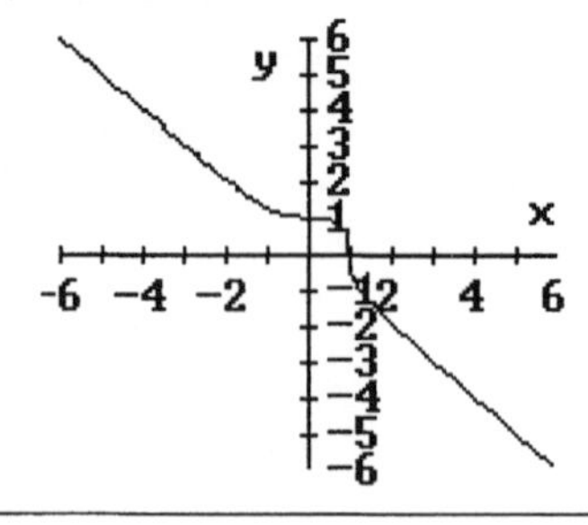

13. Given the original relation $\sqrt{-x}+\sqrt{y}=2$ and its graph, with the reflected graph as shown:

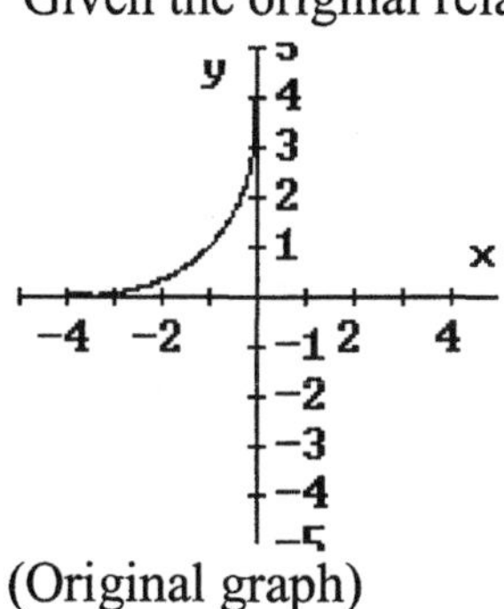

(Original graph)

(Reflected graph)

The reflection is about the y axis, so an equation for the reflected graph is: $\sqrt{x}+\sqrt{y}=2$

15. Given the original relation $x=y^3$ and its graph,

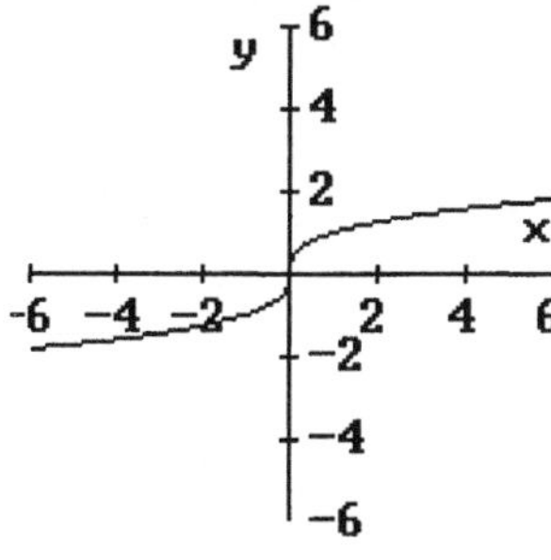

(a) The graph of $x=-(y+2)^3$ is a reflection about the x axis, combined with a vertical translation, down 2 units.

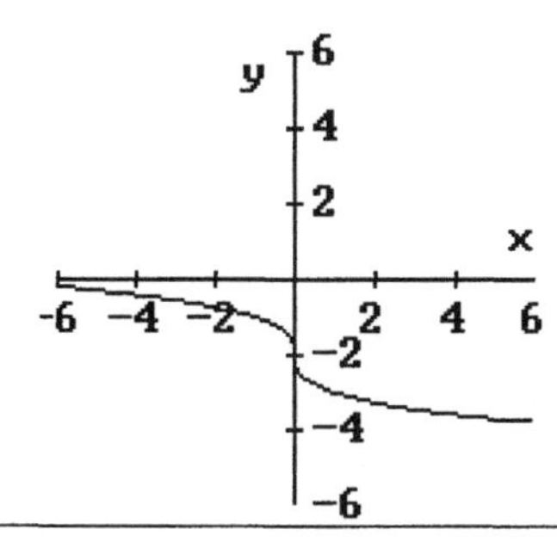

(b) The graph of $x+3=-(y-4)^3$ is a reflection about the x axis, combined with a horizontal translation, left 3 units, and a vertical translation, up 4 units.

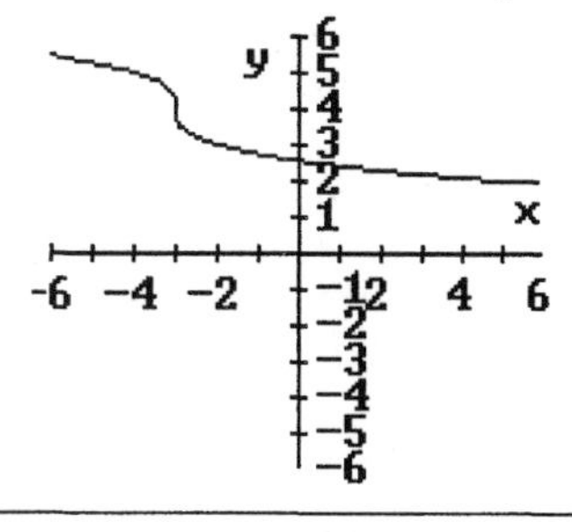

17. This graph is symmetric with respect to the x-axis.

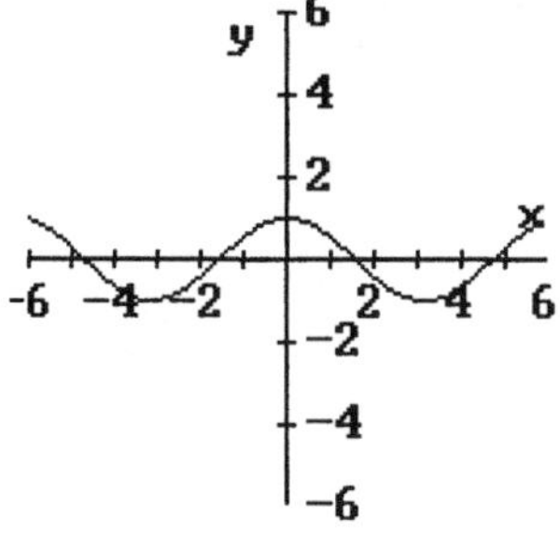

19. This graph is not symmetric with respect to the x-axis, the y-axis, or the origin.

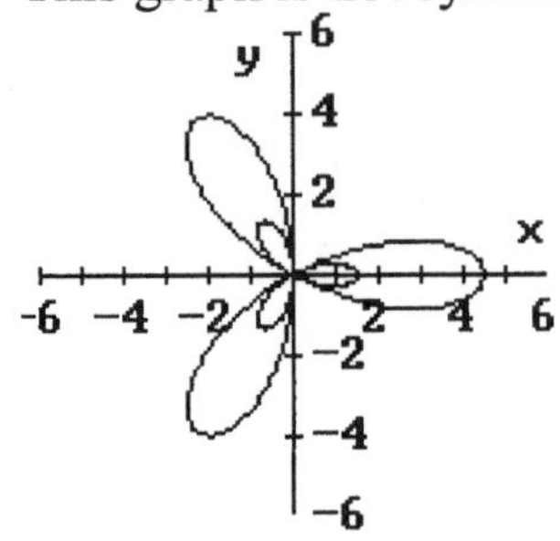

21. The graph of the relation $x = y^2$ is symmetric with respect to the x-axis: if we replace y by $-y$, we obtain $x = (-y)^2$, which is equivalent to $x = y^2$. The graph of this relation is not symmetric with respect to the y-axis, since $-x = y^2$ is not equivalent to $x = y^2$. The graph of this relation is not symmetric with respect to the origin, since $-x = (-y)^2$ is not equivalent to $x = y^2$.

23. The graph of the relation $y = 1/x$ is not symmetric with respect to the x-axis or the y-axis, since neither $y = 1/-x$ nor $-y = 1/x$ are equivalent to $y = 1/x$. The graph of this relation is symmetric with respect to the origin, since $-y = 1/-x$ is equivalent to $y = 1/x$.

25. The graph of the relation $x^3 - y^3 = 1$ is not symmetric with respect to the x-axis, the y-axis, or the origin:

$(-x)^3 - y^3 = 1$ is equivalent to $-x^3 - y^3 = 1$, which is not equivalent to $x^3 - y^3 = 1$;

$x^3 - (-y)^3 = 1$ is equivalent to $x^3 + y^3 = 1$, which is not equivalent to $x^3 - y^3 = 1$;

$(-x)^3 - (-y)^3 = 1$ is equivalent to $-x^3 + y^3 = 1$, which is not equivalent to $x^3 - y^3 = 1$.

27. The graph of the relation $y = \sin x$ is not symmetric with respect to the x-axis, since $-y = \sin x$ is not equivalent to $y = \sin x$. The graph of $y = \sin x$ is not symmetric with respect to the y-axis, since $y = \sin(-x)$ is equivalent to $y = -\sin x$, which is not equivalent to $y = \sin x$. The graph of this relation is symmetric with respect to the origin, since $-y = \sin(-x)$ is equivalent to $-y = -\sin x$, which is equivalent to $y = \sin x$.

29. Given the relation $y = x^4$ and a portion of its graph, we can complete the graph using the fact that it is symmetric with respect to the y-axis, since $y = (-x)^4$ is equivalent to $y = x^4$.

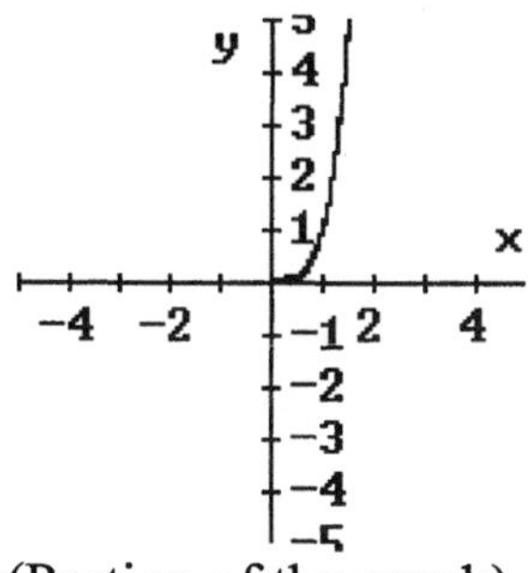

(Portion of the graph)

(Complete graph)

31. Given the relation $10y = x^5$ and a portion of its graph, we can complete the graph using the fact that it is symmetric with respect to the origin, since $10(-y) = (-x)^5$ is equivalent to $10y = x^5$.

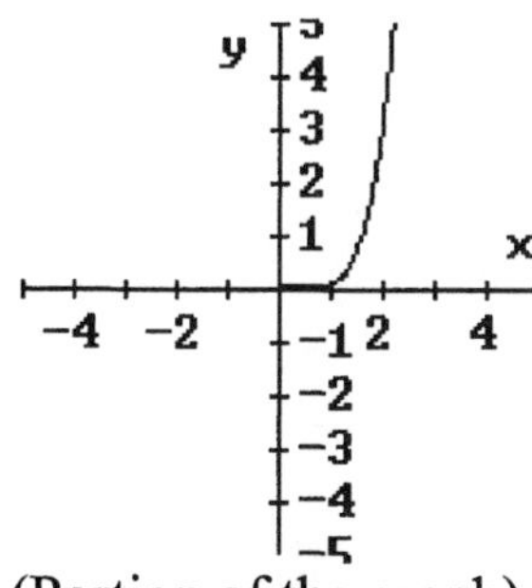

(Portion of the graph)

(Complete graph)

33. Given the relation $y^2 - x^2 = 4$ and a portion of its graph, we can complete the graph using the fact that it is symmetric with respect to the x-axis, with respect to the y-axis, and with respect to the origin: $(-y)^2 - x^2 = 4$, $y^2 - (-x)^2 = 4$, and $(-y)^2 - (-x)^2 = 4$ are equivalent to $y^2 - x^2 = 4$.

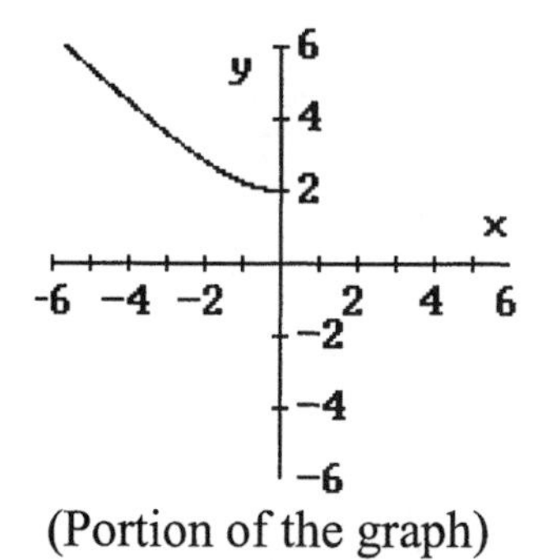

(Portion of the graph)

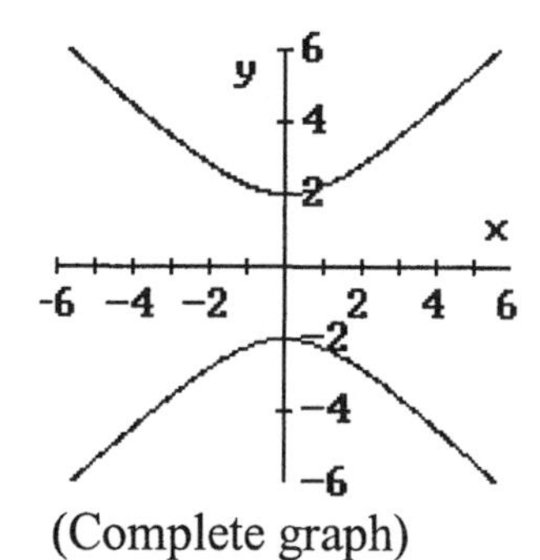

(Complete graph)

35. Given the relation $x^2y+2y=10$ and a portion of its graph, we can complete the graph using the fact that it is symmetric with respect to the y-axis, since $(-x)^2\,y+2y=10$ is equivalent to $x^2y+2y=10$.

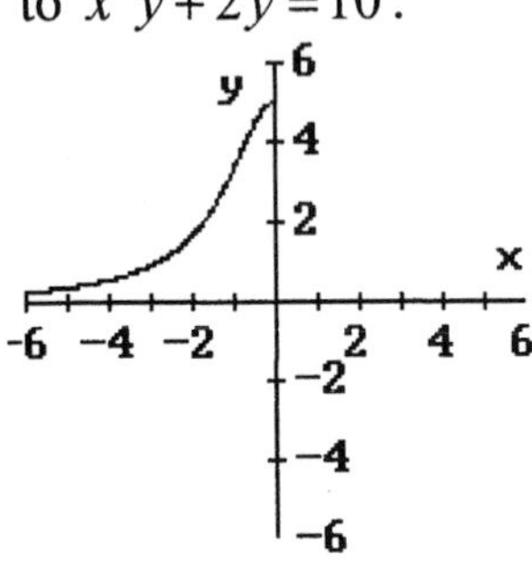

(Portion of the graph)

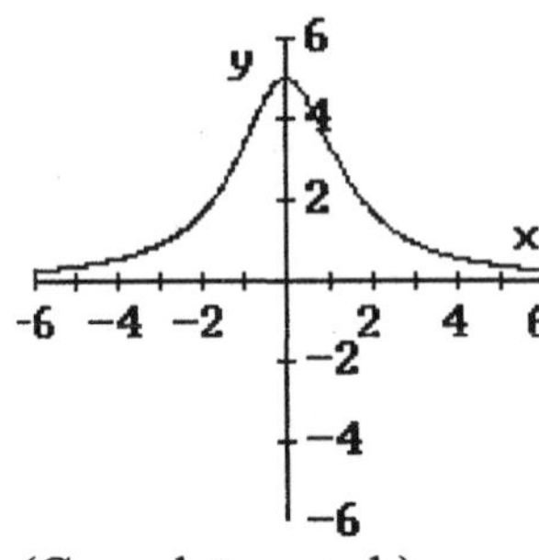

(Complete graph)

37. Based on the graph of the relation $y=x^2+x-2$:

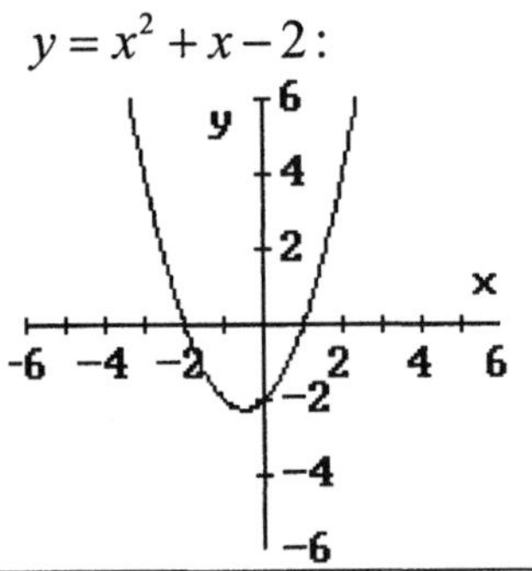

(a) The graph of $y=x^2+x-4$ is a vertical translation, down 2 units.

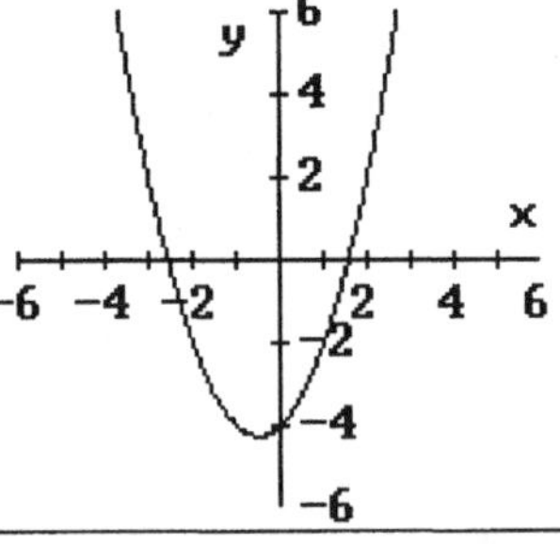

(b) The graph of $y=x^2-5x+4$ is a horizontal translation, right 3 units.

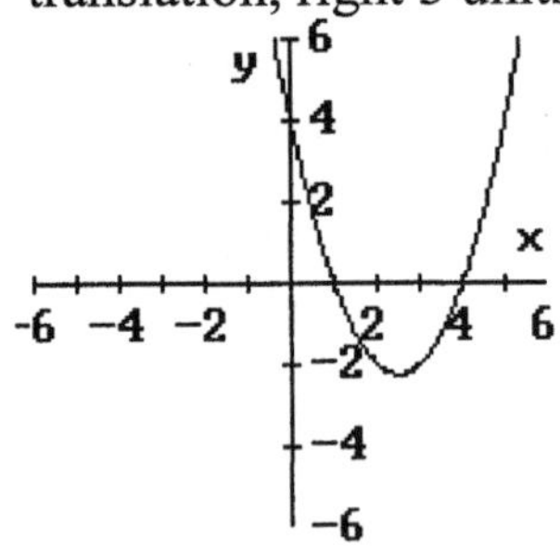

(c) The graph of $y=-x^2-x$ is a reflection about the x-axis, combined with a vertical translation, down 2 units.

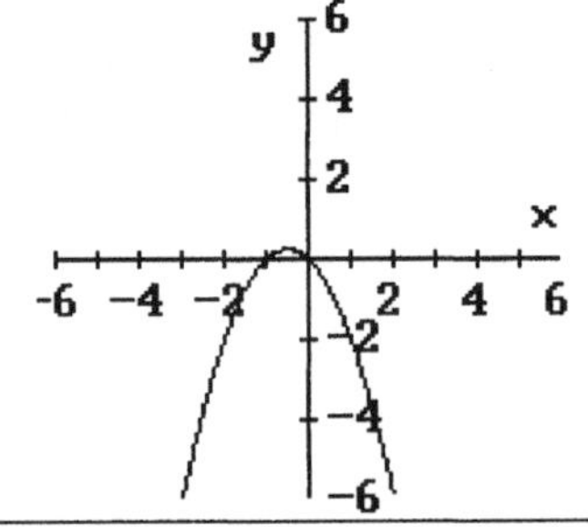

39. Based on the graph of the relation $y = \dfrac{5}{x^2+1}$:

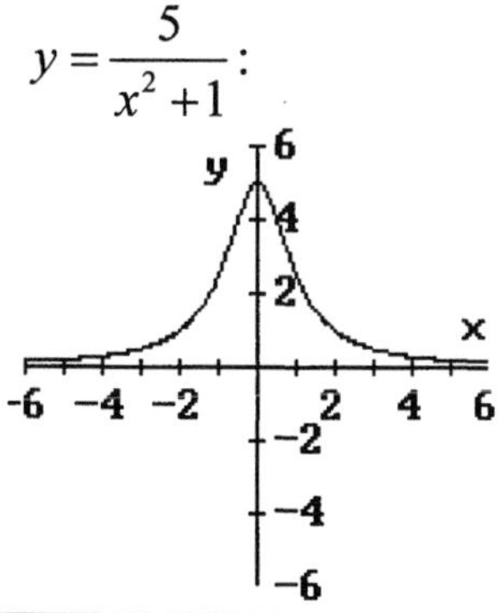

(a) The graph of $y = \dfrac{3x^2+8}{x^2+1}$ is a vertical translation, up 3 units.

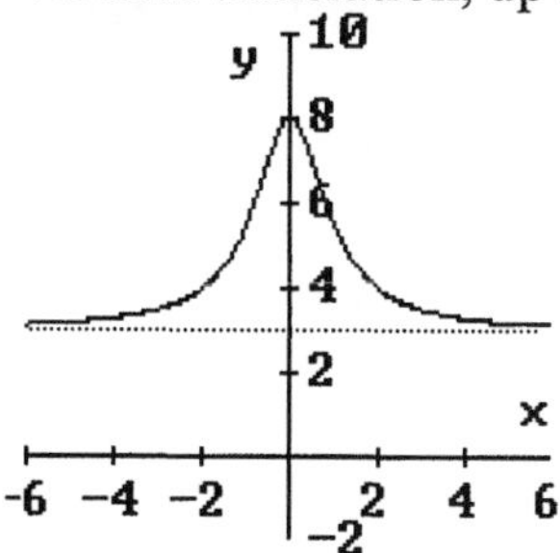

(b) The graph of $y = \dfrac{-5}{x^2+2x+2}$ is a reflection about the x-axis, combined with a horizontal translation, left 1 unit.

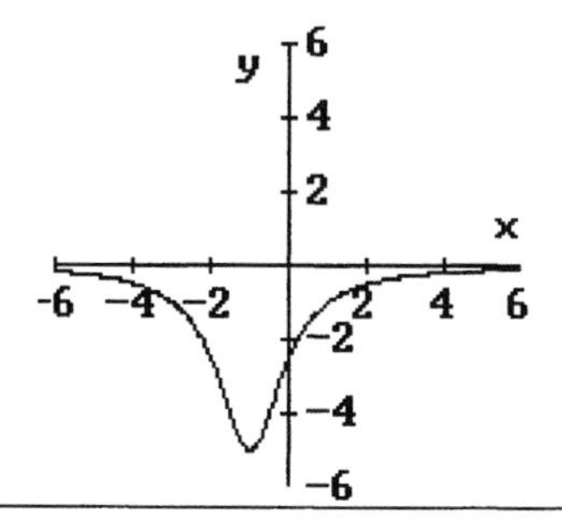

(c) The graph of $y = \dfrac{2x^2+8x+15}{x^2+4x+5}$ is a horizontal translation, left 2 units, with a vertical translation, up 2 units.

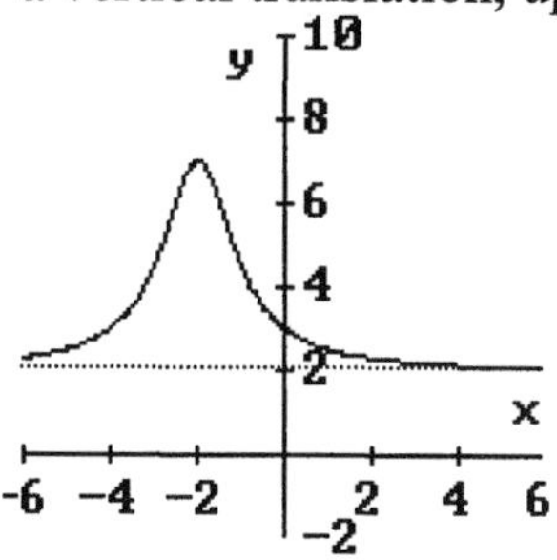

41. If $16x^2 + 25y^2 = 400$, then $y = \pm 4\sqrt{1 - x^2/25}$. The highest point on the graph is (0, 4), and the lowest point is $(0, -4)$.

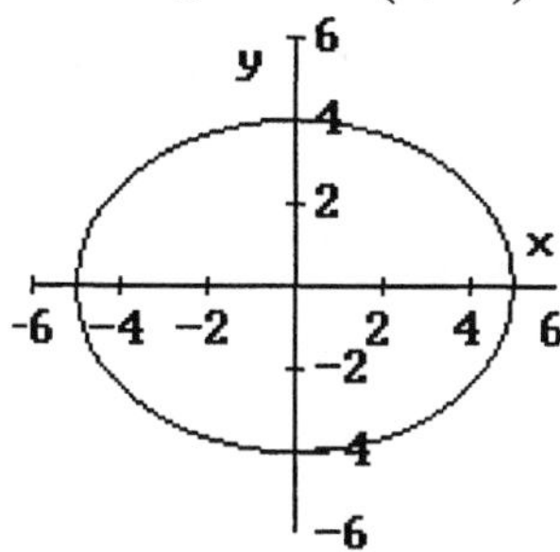

43. If $y^2+x^3+x-5=0$, then $y=\pm\sqrt{5-x-x^3}$. There is an x-intercept, $x\approx 1.52$, so 2 is the nearest integer to the x-intercept. There are two y-intercepts, $y=\pm\sqrt{5}$, or $y\approx\pm 2.24$. Then the nearest integers to the y-intercepts are ± 2.

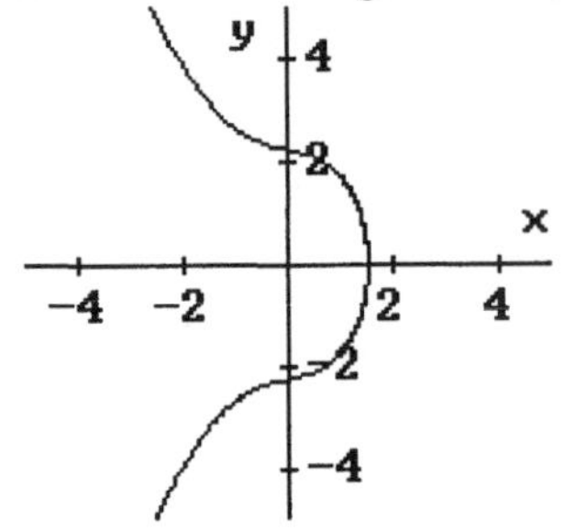

Applications

Exercises 45-47 refer to the graph of the standard normal probability function. The area under the entire curve is 1.

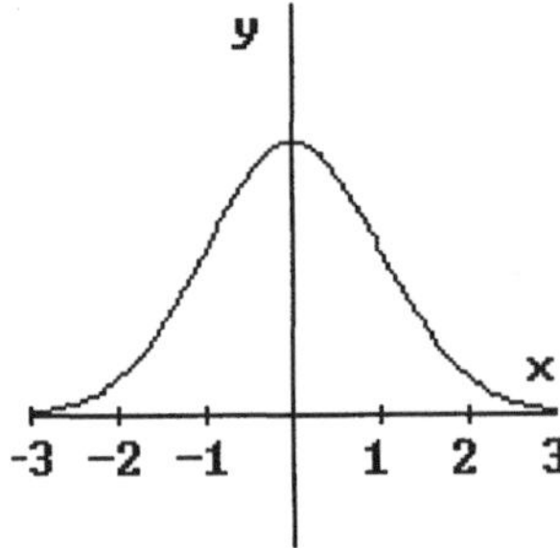

45. The area under the curve to the right of $x=0$ is $\frac{1}{2}$.

47. Given that the area under the curve between $x=0$ and $x=1$ is approximately 0.3413, then:

(a) the area under the curve over the interval $(1,\infty)$ is $\approx \frac{1}{2}-0.3413$, that is, ≈ 0.1587;

(b) the area under the curve over the interval $(-\infty,-1)$ is ≈ 0.1587;

(c) the area under the curve over the interval $(-1,1)$ is $\approx 2\times 0.3413$, that is, ≈ 0.6826;

(d) the area under the curve over the interval $(-\infty,1)$ is $\approx \frac{1}{2}+0.3413$, or ≈ 0.8413.

Concepts and Critical Thinking

49. The statement: "If the graph of the relation obtained by substituting x with $-x$ is the same as the original graph, then we say that the graph is symmetric with respect to the y-axis" is true.

51. The statement: "Substituting $x+3$ for x translates the graph 3 units to the right" is false. Such substitution produces a translation 3 units to the left.

53. The statement: "If a graph is symmetric with respect to the origin, then it will be symmetric with respect to both the x- and y-axes" is false. For example, the graph of $y=x^3$ is symmetric with respect to the origin, but not symmetric with respect to any axis.

55. Answers may vary. The relation $y = x - 5$ is a translation of $y = x$, 5 units to the right.

57. Answers may vary. The relation $y = -x$ is symmetric with respect to the line $y = x$.

59. No function of the form $y = f(x)$ can have symmetry with respect to the x-axis, except for the function $y = 0$. If the function is not the zero function, say $f(a) \neq 0$ for some a, then the point $(a, f(a))$ belongs to the graph of the function, and $(a, -f(a))$ does not belong to the graph of *f*, since *f* is a function, so there is only one value of *f* associated to a. Then the graph is not symmetric with respect to the x-axis.

Chapter 8. Section 8.2 Parabolas

1. The parabola of equation $y = 2x^2$ has its vertex at $(0,0)$, and its focus at $\left(0,\frac{1}{8}\right)$. Its directrix is $y = -\frac{1}{8}$.

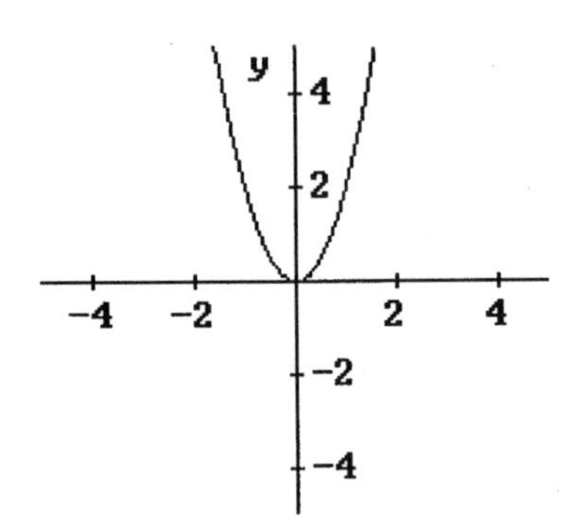

3. The parabola of equation $y^2 - 3x = 0$, or $x = \frac{1}{3}y^2$, has its vertex at $(0,0)$, and its focus at $\left(\frac{3}{4},0\right)$. Its directrix is $x = -\frac{3}{4}$.

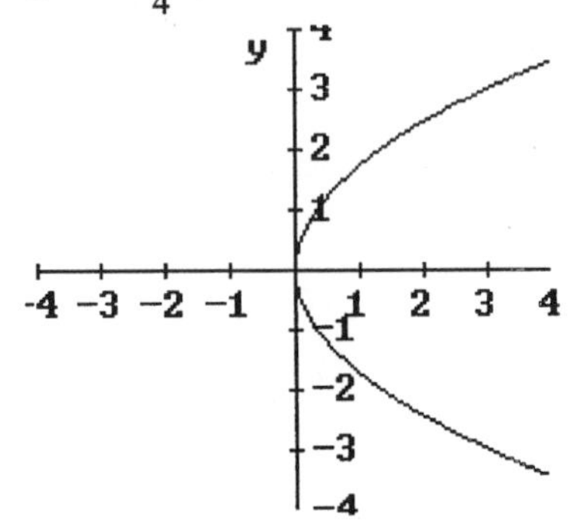

5. The parabola of equation $4y^2 + 7x = 0$, or $x = -\frac{4}{7}y^2$, has its vertex at $(0,0)$, and its focus at $\left(-\frac{7}{16},0\right)$. Its directrix is $x = \frac{7}{16}$.

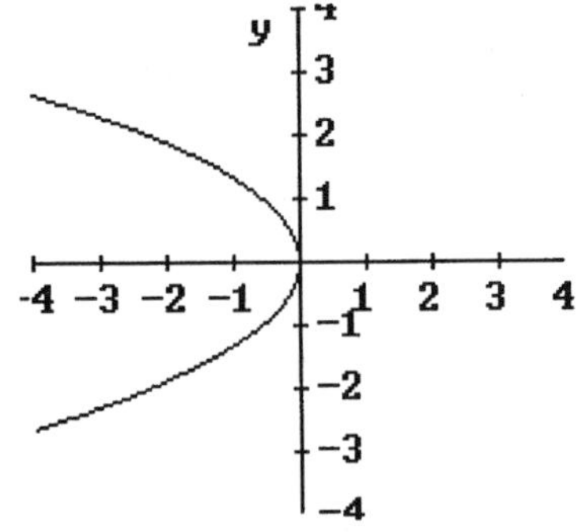

7. The parabola of equation $y - 3 = 3(x+5)^2$ has its vertex at $(-5,3)$, and its focus at $\left(-5,\frac{37}{12}\right)$. Its directrix is $y = \frac{35}{12}$.

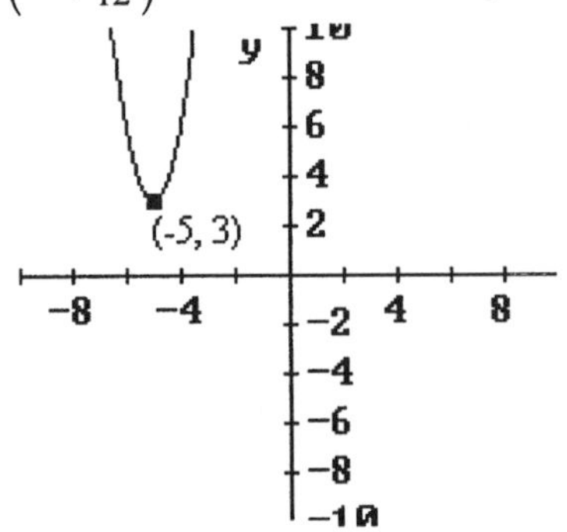

9. The parabola of equation $x^2 - 4x - 4y + 8 = 0$, or $y - 1 = \frac{1}{4}(x-2)^2$ has its vertex at $(2,1)$, and its focus at $(2,2)$. Its directrix is the x-axis ($y = 0$.)

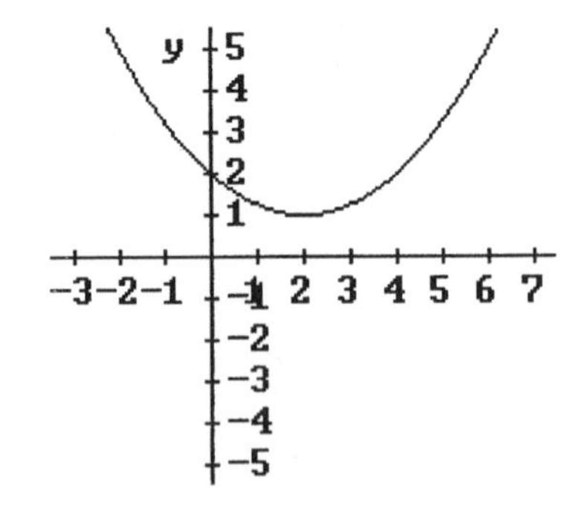

11. The parabola of equation $x = 2y^2 + 6y + 2$, or $y - 1 = \frac{1}{4}(x-2)^2$ has its vertex at $(2,1)$, and its focus at $(2,2)$. Its directrix is the x-axis ($y = 0$.)

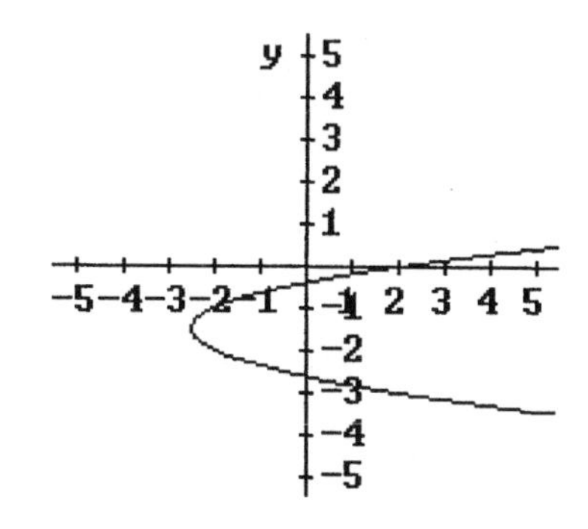

13. An equation for the parabola with vertex at $(0,0)$ and focus at $(2,0)$ is $x = \frac{1}{8}y^2$.

15. An equation for the parabola with vertex at $(0,-4)$ and directrix $y = 4$ is $y = -\frac{1}{16}x^2$.

17. An equation for the parabola with vertex at $(-2,0)$, passing through $(2,4)$, with horizontal axis of symmetry, is $x + 2 = \frac{1}{4}y^2$, or $x = \frac{1}{4}y^2 - 2$.

19. An equation for the parabola with focus at $(4,-1)$, and directrix $y = 3$, is $y + 1 = -\frac{1}{8}(x-4)^2$, or $y = -\frac{1}{8}(x-4)^2 - 1$.

21. An equation for the parabola with graph:

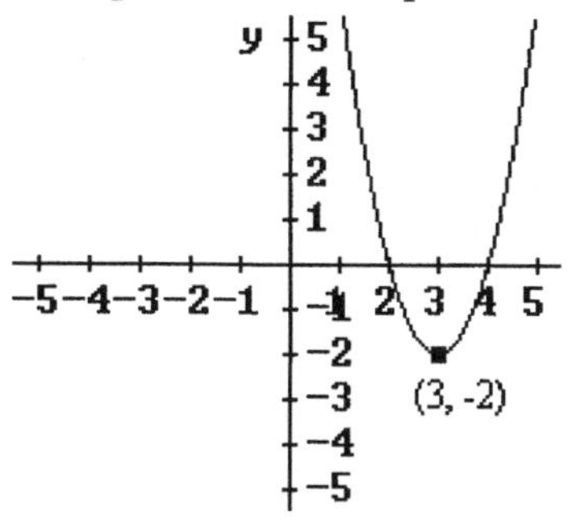

is $y + 2 = 2(x-3)^2$, or $y = 2(x-3)^2 - 2$

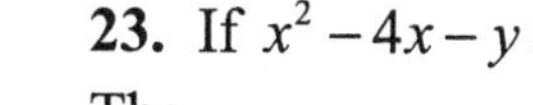

23. If $x^2 - 4x - y = 0$, then $y = x^2 - 4x$. The vertex of this parabola is $(2,-4)$.

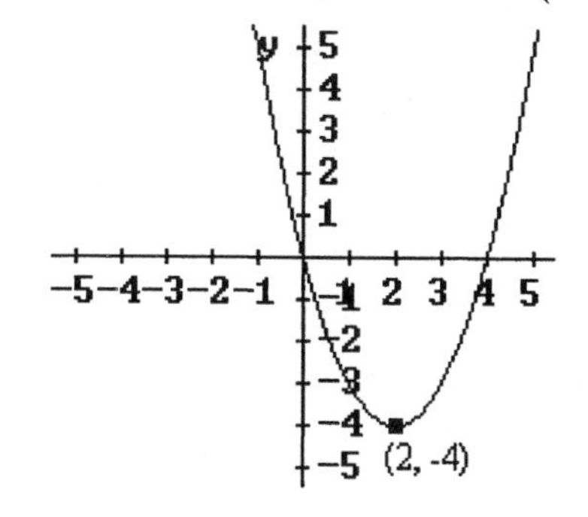

25. If $x^2+8x-2y+22=0$, then $y=\frac{1}{2}x^2+4x+11$. The vertex of this parabola is $(-4,3)$.

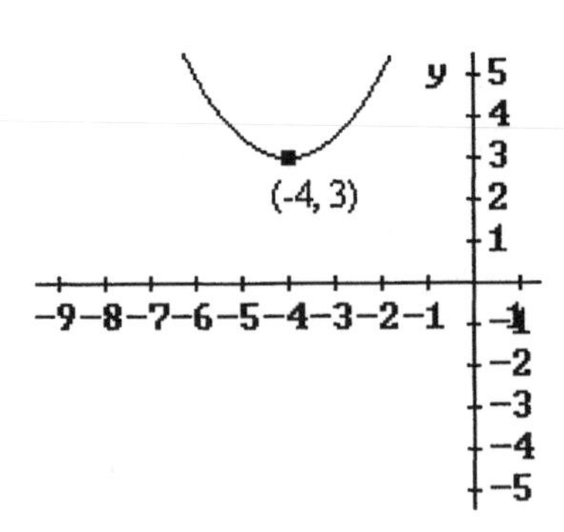

27. If $-y^2+4x+8=0$, then $y=\pm 2\sqrt{x+2}$. The vertex of this parabola is $(-2,0)$.

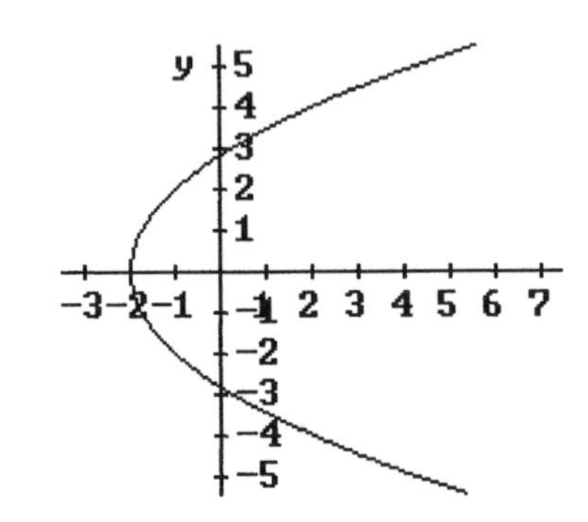

29. If $y^2-2y+x+2=0$, then $y=1\pm\sqrt{-(x+1)}$. The vertex of this parabola is $(-1,1)$.

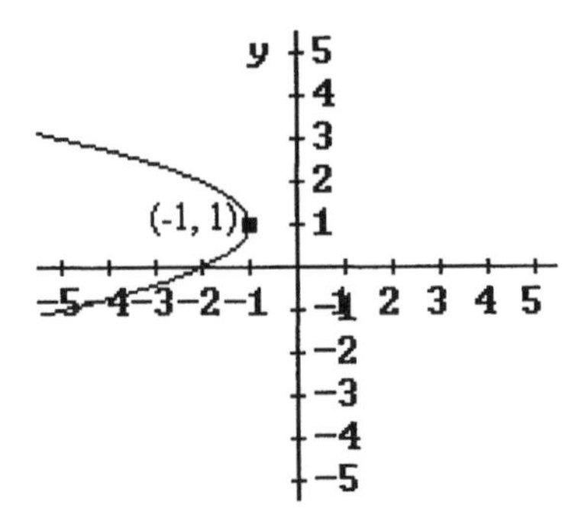

Applications

31. Given that the path of a ball thrown at an angle of 45^0 is given by $y=-\frac{x^2}{200}+x$, where y is the height in feet when the ball is a horizontal distance of x feet from where it was thrown, the maximum height of the ball is 50 feet, reached when $x=100$ feet. The total horizontal distance traveled by the ball before hitting the ball is 200 feet.

33. If a diver follows a parabolic path after taking off, where the takeoff point is the vertex of the parabola, 115 feet above the water, and if the diver must land 30 feet out from the takeoff point, then:

(a) $y=ax^2+115$, and $0=a\left(30^2\right)+115$, therefore $a=-\frac{23}{180}$ and $y=-\frac{23}{180}x^2+115$;

(b) 12 feet below the takeoff point, that is when $y=103$ feet, we have $103=-\frac{23}{180}x^2+115$, so $x^2=\frac{12(180)}{23}$, that is $x\approx 9.69$ feet. The diver passes 0.69 ft away from a rock that juts out 9 feet from the vertical line that passes through the takeoff point.

35. If the cross-sectional view of a radio telescope is parabolic, and has an equation of the form $y = ax^2$, where the vertex of the parabola is at the origin, and if the dish is 1000 feet across the top, and 167 feet deep at the center, then $500 = a(167^2)$, that is $a = \frac{500}{167^2}$, and $y = \frac{500}{27889}x^2$. The value of *p*, then is $p = \frac{1}{4a}$, or $p = \frac{167^2}{200}$, approximately $p \approx 13.94$. This is the distance from the vertex where the receiver should be located.

Concepts and Critical Thinking

37. The statement: "The graph of $x = ay^2 + by + c$ is a parabola with horizontal axis of symmetry" is true.

39. The statement: "The directrix of a parabola is a point lying on the parabola, halfway between the focus and the vertex" is false. A true statement would be: "The vertex of a parabola is a point lying on the parabola, halfway between the focus and the directrix."

41. Answers may vary. A parabola with axis of symmetry $y = 4$ is $x = (y-4)^2$.

43. Answers may vary. An equation of a parabola with its focus at the origin is $x = \frac{1}{4}(y+1)^2$.

Chapter 8. Section 8.3 Ellipses

1. The ellipse $\frac{x^2}{9}+\frac{y^2}{25}=1$ has center (0, 0), vertices $(\pm 3,0)$, and $(0,\pm 5)$. Its foci are $(0,\pm 4)$.

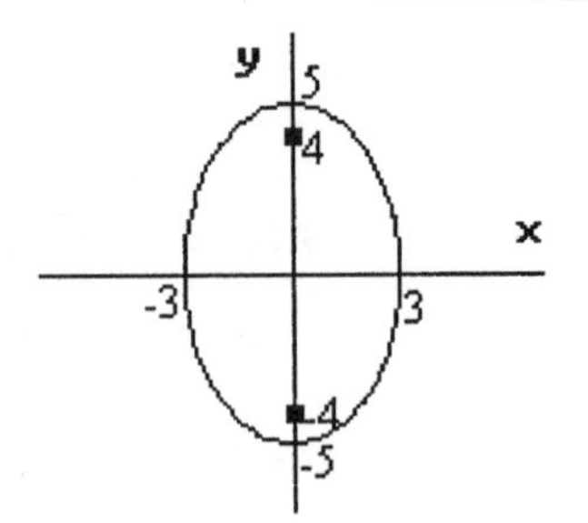

3. The ellipse $4x^2+9y^2=36$, that is $\frac{x^2}{9}+\frac{y^2}{4}=1$ has center (0, 0), vertices $(\pm 3,0)$ and $(0,\pm 2)$. Its foci are $(\pm\sqrt{5},0)$.

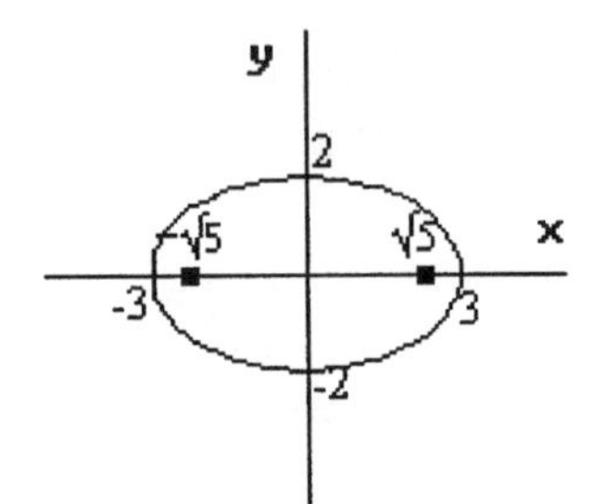

5. The ellipse $9x^2+4y^2=1$, that is $\frac{x^2}{\left(\frac{1}{3}\right)^2}+\frac{y^2}{\left(\frac{1}{2}\right)^2}=1$ has center (0, 0), vertices $\left(\pm\frac{1}{3},0\right)$ and $\left(0,\pm\frac{1}{2}\right)$. Its foci are $\left(0,\pm\frac{\sqrt{5}}{6}\right)$.

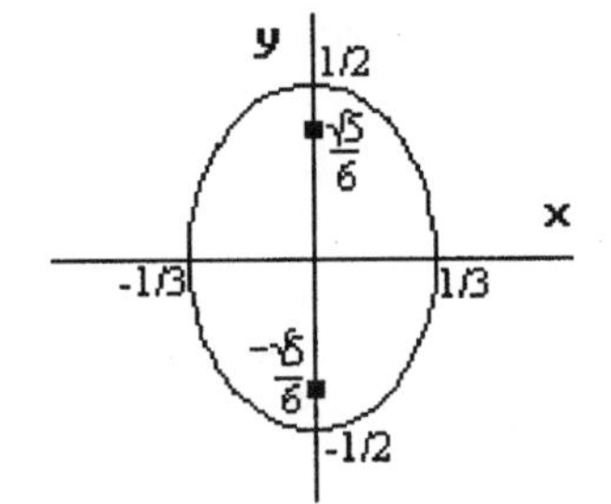

7. The ellipse $5x^2+8y^2=40$, that is $\frac{x^2}{8}+\frac{y^2}{5}=1$ has center (0, 0), vertices $(\pm\sqrt{8},0)$ and $(0,\pm\sqrt{5})$. Its foci are $(\pm\sqrt{3},0)$.

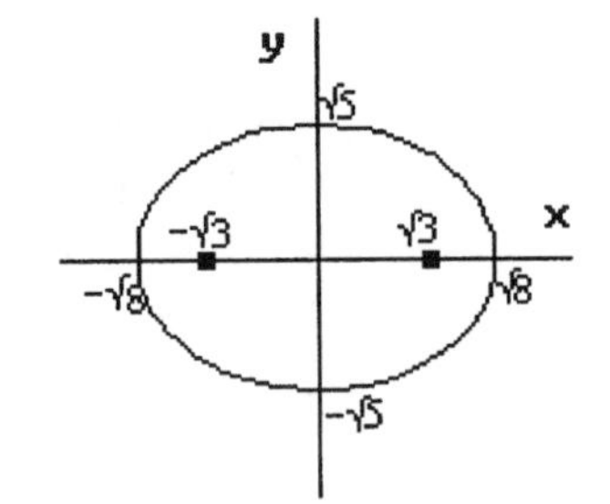

9. The ellipse $\frac{(x+3)^2}{4}+\frac{(y-1)^2}{16}=1$ has center $(-3,1)$, and vertices $(-5,1),(-1,1),(-3,-3),(-3,5)$. Its foci are $(-3,1-\sqrt{2})$ and $(-3,1+\sqrt{2})$.

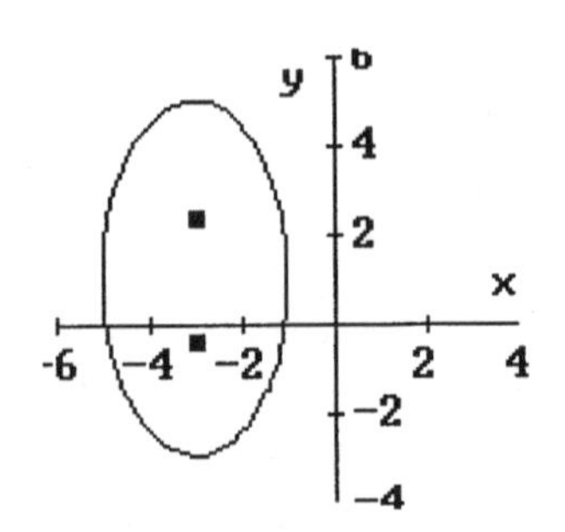

11. The equation $9x^2+4y^2-18x-24y+9=0$ is equivalent to $\frac{(x-1)^2}{4}+\frac{(y-3)^2}{9}=1$. So, this ellipse has center $(1,3)$, and vertices $(3,3),(-1,3),(1,0),(1,6)$. Its foci are $\left(1,3-\sqrt{5}\right)$ and $\left(1,3+\sqrt{5}\right)$.

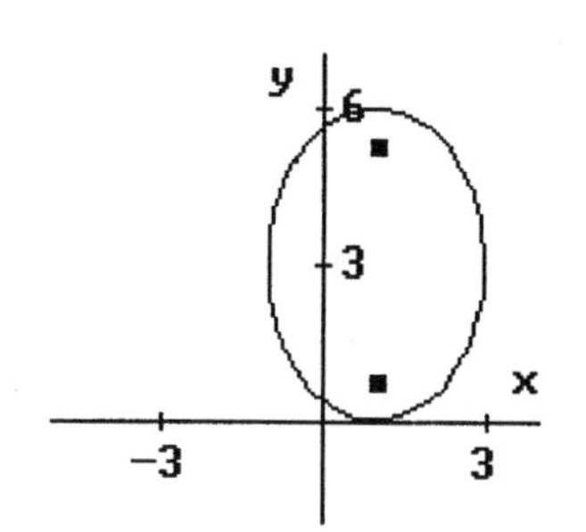

13. The equation $x^2+36y^2+4x-72y+4=0$ is equivalent to $\frac{(x+2)^2}{36}+(y-1)^2=1$. So, this ellipse has center $(-2,1)$, and vertices $(-8,1),(4,1),(-2,2),(-2,0)$. Its foci are $\left(-2-\sqrt{35},1\right)$ and $\left(-2+\sqrt{35},1\right)$.

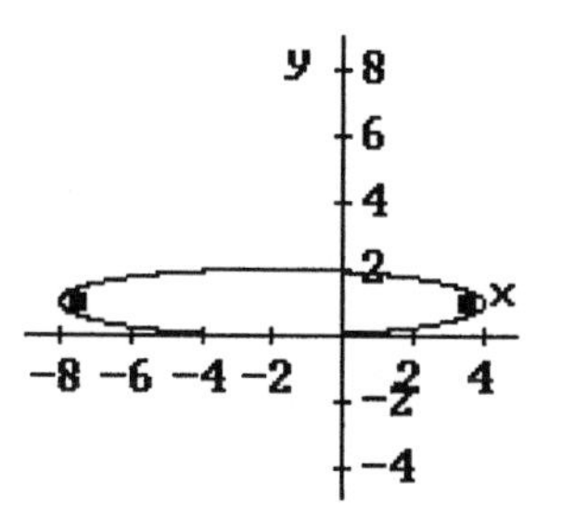

15. The ellipse of center (0,0), horizontal major axis of length 8, and vertical minor axis of length 6 has equation $\frac{x^2}{16}+\frac{y^2}{9}=1$.

17. The ellipse with vertices $\left(0,\pm\sqrt{6}\right)$ and foci $\left(0,\pm\sqrt{2}\right)$ has equation $\frac{x^2}{4}+\frac{y^2}{6}=1$.

19. The ellipse with vertices $(\pm 6,0)$, that passes through $(-4,2)$ has an equation of the form $\frac{x^2}{36}+\frac{y^2}{b^2}=1$. Now, if $x=-4$ and $y=2$, then $\frac{16}{36}+\frac{4}{b^2}=1$, so $b=\frac{6\sqrt{5}}{5}$, and the equation of this ellipse can be written as $\frac{x^2}{36}+\frac{y^2}{\left(36/5\right)}=1$, or $x^2+5y^2=36$.

21. The ellipse with center $(2,-1)$, vertical major axis of length 6 and horizontal minor axis of length 4 has equation $\frac{(x-2)^2}{4}+\frac{(y+1)^2}{9}=1$.

23. The ellipse with vertices $(-4,-2),(6,-2)$, and foci $(-2,-2),(4,-2)$ has center $(1,-2)$. We have $a=5$ and $c=3$, so $b=4$. The equation of this ellipse, in standard form, is $\frac{(x-1)^2}{25}+\frac{(y+2)^2}{16}=1$.

25. The ellipse in the graph has center $(-3,1)$, vertical major axis of length 6, and horizontal minor axis of length 2. The equation of this ellipse, in standard form, is $(x+3)^2+\frac{(y-1)^2}{9}=1$.

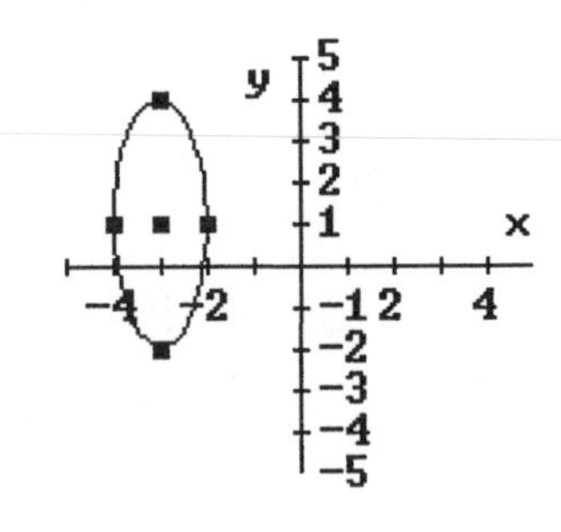

27. If $x^2+4y^2+4x=0$, then $y=\pm\frac{\sqrt{-x^2-4x}}{4}$. The vertices of this ellipse are $(-4,0),(0,0)$, $(-2,-1)$, and $(-2,1)$.

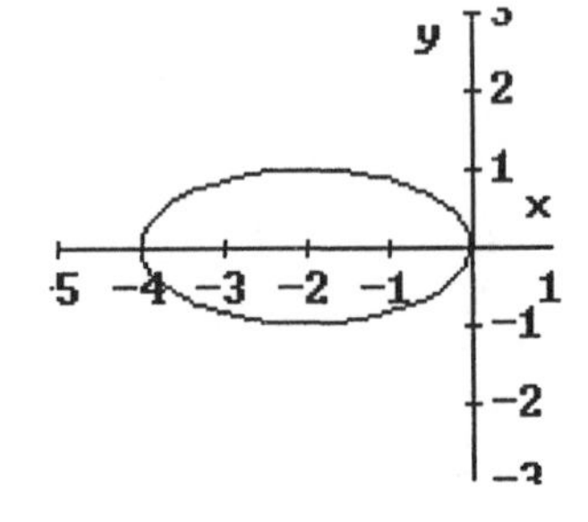

29. If $4x^2+y^2-2y-3=0$, then $y=1\pm2\sqrt{1-4x^2}$. The vertices of this ellipse are $(-1,1),(1,1)$, $(0,3)$, and $(0,-1)$.

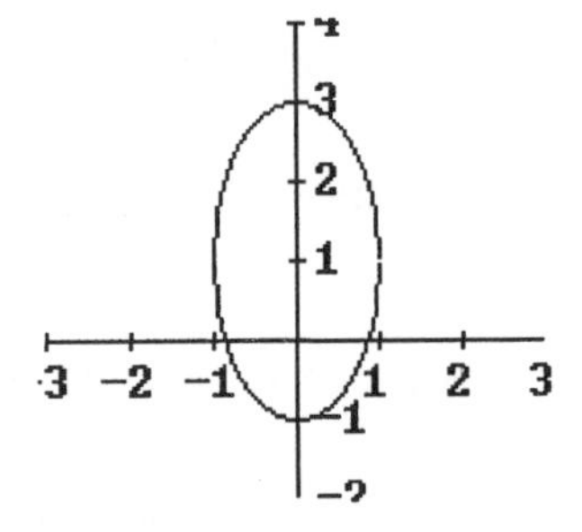

Applications

31. If the Earth's orbit is elliptical, with major axis length of 186 million miles, and minor axis length of 185.8 million miles, if the major axis is along the x-axis and the minor axis is along the y-axis, then $a=93$, $b=92.9$, and $\frac{x^2}{93^2}+\frac{y^2}{92.9^2}=1$, that is $\frac{x^2}{8649}+\frac{y^2}{8630.41}=1$ is an equation for this orbit.

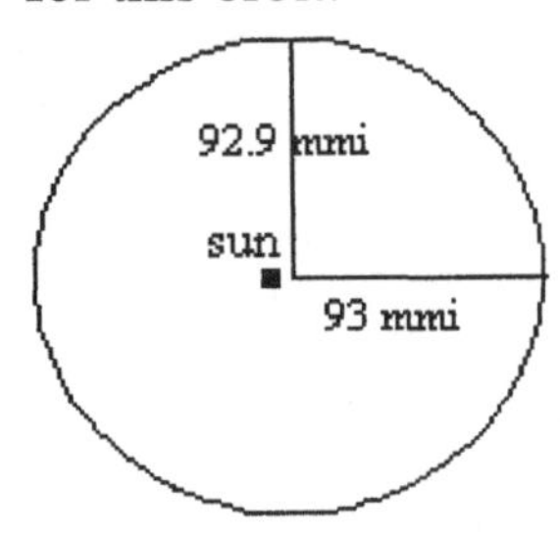

33. If the arch of a railroad bridge has the shape of a semiellipse, where the distance from the base of the arch on one side of the road to the base on the other is 50 feet, and the height of the arch at the center is 20 feet, then:

(a) the equation of the ellipse that gives the form of the arch, assuming the x-axis on the ground and perpendicular to the centerline of the highway, the origin on the centerline and the y-axis to be vertical, is $\frac{x^2}{625}+\frac{y^2}{400}=1$;

(b) the height of the arch 5 feet away from the base corresponds to $x=\pm 20$ feet away from the center, so $\frac{400}{625}+\frac{y^2}{400}=1$, that is $\frac{y^2}{400}=\frac{225}{625}$, therefore $y^2=144$, so the height of the arch 5 feet away from the base is 12 feet;

(c) if a tractor-trailer with a height of 14 feet and a width of 10 feet will no go over the centerline of the road, it will be necessary to compute the height of the arch 10 feet away from the centerline, that is the value of y when $x=\pm 10$: $\frac{100}{625}+\frac{y^2}{400}=1$, that is $\frac{y^2}{400}=\frac{525}{625}$, $y^2=\frac{(400)(525)}{625}$, so $y=\frac{20\sqrt{525}}{25}\approx 18.33\,ft$; so, as long as the tractor-trailer doesn't go over the centerline, it will be safe.

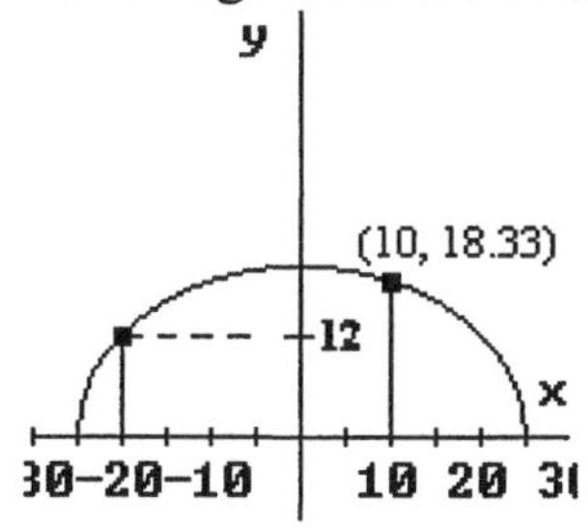

35. If the elliptical orbit of comet Encke has a major axis of length 4.4 A.U. and a minor axis of length 2.2 A.U., then an equation for this orbit is $\frac{x^2}{2.2^2}+\frac{y^2}{1.1^2}=1$. If the sun is at one of the foci, at a distance of $c=\sqrt{2.2^2-1.1^2}$, that is $c\approx 1.91$ AU from the center, then the perihelion is $2.2-1.91\approx 0.09$ A.U., and the aphelion is $2.2+1.91\approx 4.11$ A.U.

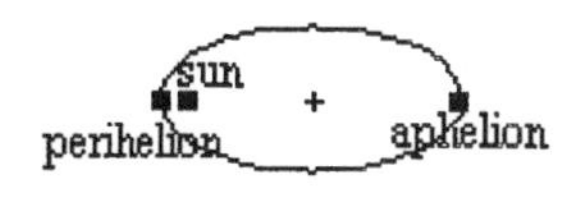

37. **(a)** For a semielliptical Anderson window, with a major axis of length 168.3 centimeters, and area of 3995 cm^2, then $\pi(84.15)b=3995$, so $b\approx 15.11\ cm$. An equation for the ellipse that forms the boundary of this window is $\frac{x^2}{84.15^2}+\frac{y^2}{15.11^2}=1$.

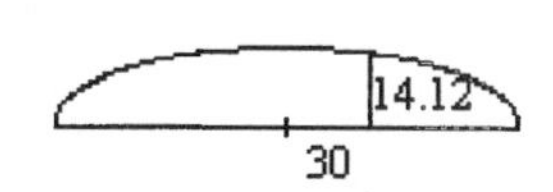

(b) The height of the window 30 cm away from the center, then $y\approx 14.12\ cm$.

Concepts and Critical Thinking

39. The statement: "It is possible for the plot of an ellipse to be entirely contained within the viewing rectangle of a graphing calculator" is true.

41. The statement: "If the foci of an ellipse are very close together, then the ellipse will appear nearly circular" is true.

43. Answers may vary. Two ellipses which are 90° rotations of one another are $\frac{x^2}{4}+\frac{y^2}{9}=1$ and $\frac{x^2}{9}+\frac{y^2}{4}=1$.

45. Answers may vary. A real-world application of the reflective property of ellipses is the design of whispering rooms, or some symphony halls.

47. If an ellipse has a horizontal major axis of lengh $2a$ and foci at $(\pm c,0)$, then $a^2-c^2>0$ because the vertices on the major axis are $(\pm a,0)$, therefore $a>c$. Now, $a^2-c^2=(a-c)(a+c)>0$, as we wanted to show.

Chapter 8. Section 8.4 Hyperbolas

1. For the hyperbola with equation $\frac{x^2}{9}-\frac{y^2}{16}=1$, the center is $(0,0)$, the vertices are $\left(\pm\sqrt{3},0\right)$, the foci are $(\pm 5,0)$, and the asymptotes are $y=\pm\frac{4}{3}x$.

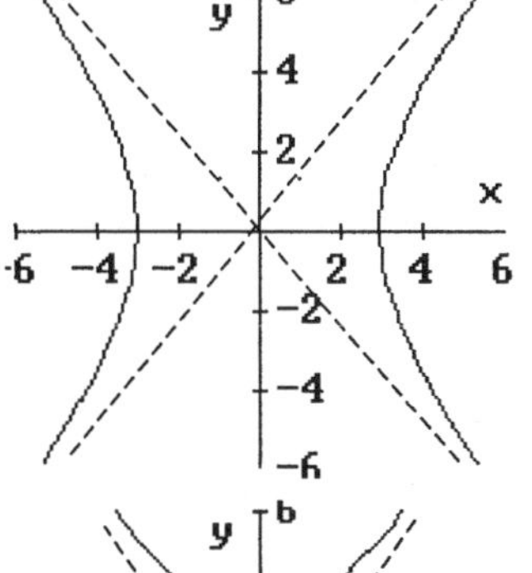

3. For the hyperbola with equation $4y^2-9x^2=36$ or, equivalently, $\frac{y^2}{9}-\frac{x^2}{4}=1$, the center is $(0,0)$, the vertices are $(0,\pm 3)$, the foci are $\left(0,\pm\sqrt{13}\right)$, and the asymptotes are $y=\pm\frac{3}{2}x$.

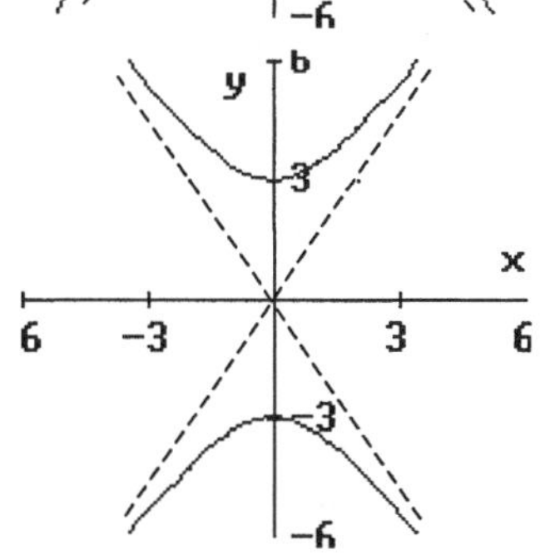

5. For the hyperbola with equation $12x^2-3y^2=-24$ or, equivalently, $\frac{y^2}{8}-\frac{x^2}{2}=1$, the center is $(0,0)$, the vertices are $\left(0,\pm 2\sqrt{2}\right)$, the foci are $\left(0,\pm\sqrt{10}\right)$, and the asymptotes are $y=\pm 2x$.

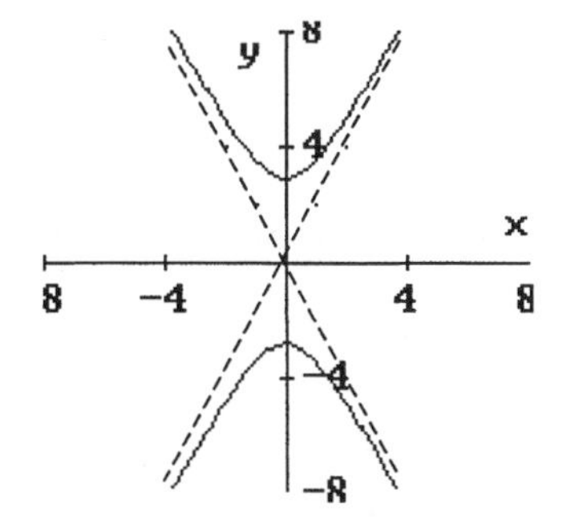

7. For the hyperbola with equation $\frac{(x-1)^2}{4}-\frac{(y-4)^2}{9}=1$, the center is $(1,4)$, the vertices are $(-1,4)$ and $(3,4)$, the foci are $\left(1\pm\sqrt{13},4\right)$, and the asymptotes are $(y-4)=\pm\frac{3}{2}(x-1)$, or $y=\frac{3}{2}x+\frac{5}{2}$, and $y=-\frac{3}{2}x+\frac{11}{2}$.

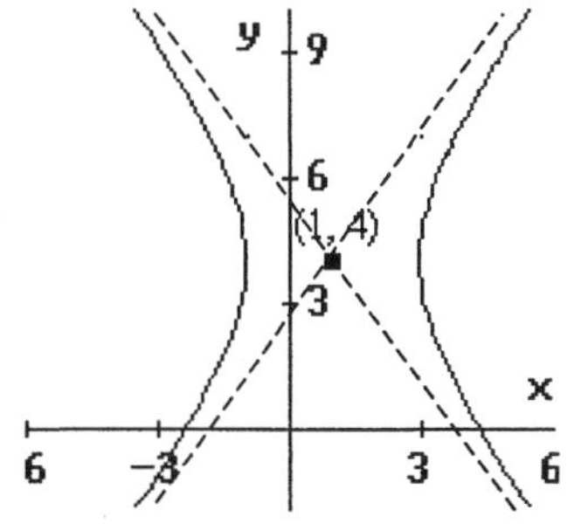

9. For the hyperbola with equation $4x^2-9y^2-16x+54x-101=0$, or equivalently, $\frac{(x-2)^2}{9}-\frac{(y-3)^2}{4}=1$, the center is $(2,3)$, the vertices are $(-1,3)$ and $(5,3)$, the foci are $\left(2\pm\sqrt{13},3\right)$, the asymptotes are $(y+3)=\pm\frac{4}{5}(x+2)$, that is $y=\frac{2}{3}x+\frac{5}{3}$, and $y=-\frac{2}{3}x+\frac{13}{3}$.

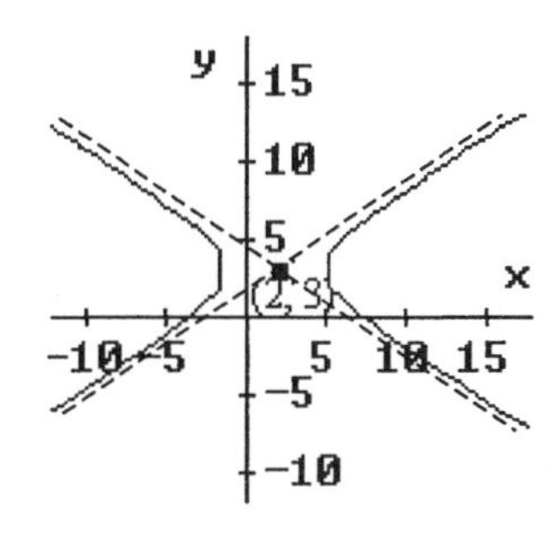

11. For the hyperbola with equation

$4x^2 - 25y^2 - 32x + 164 = 0$, or $\frac{y^2}{4} - \frac{(x-4)^2}{25} = 1$, the center is $(4,0)$, the vertices are $(4,\pm 2)$, the foci are $\left(4,\pm\sqrt{29}\right)$, and the asymptotes are $y = \pm\frac{2}{5}(x-4)$, that is $y = \frac{2}{5}x - \frac{8}{5}$, and $y = -\frac{2}{5}x + \frac{8}{5}$.

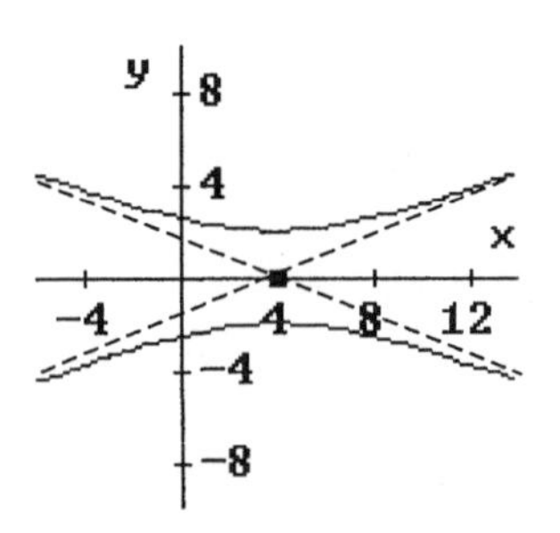

13. If a hyperbola has vertices $(0,\pm 4)$ and foci $(0,\pm 5)$, then its center is (0,0) and it has a vertical transverse axis, with $b = 4$. Now, $a^2 + 16 = 25$, so $a = 3$. Then, $\frac{y^2}{16} - \frac{x^2}{9} = 1$.

15. If a hyperbola has foci $\left(0,\pm 2\sqrt{5}\right)$ and asymptotes $y = \pm\frac{1}{2}x$ then its center is (0,0) and it has a vertical transverse axis, where $a^2 + b^2 = 20$, and $\frac{b}{a} = \frac{1}{2}$. Then $a = 2b$, $4b^2 + b^2 = 20$, so $b^2 = 4$, therefore $b = 2$ and $a = 4$. An equation for this hyperbola is $\frac{y^2}{4} - \frac{x^2}{16} = 1$.

17. If a hyperbola has vertices $(-1,3)$, $(5,3)$ and foci $(-3,3)$, $(7,3)$ then its center is (2,3) and it has a horizontal transverse axis, with $a = 3$, $a^2 + b^2 = 25$, so $b = 4$, and an equation for this hyperbola is $\frac{(x-2)^2}{9} - \frac{(y-3)^2}{16} = 1$.

Exercises 19-25 refer to nondegenerate conic sections.

19. $2x^2 - x - 3y + 4 = 0$ is the equation of a parabola ($A = 2$, $C = 0$, so $AC = 0$.)

21. $4x^2 + 4y^2 = 2x - 6y + 10$ is the equation of a circle ($A = 4$, $C = 4$.)

23. $-5x^2 - 11x = y^2$, or equivalently $5x^2 + y^2 + 11x = 0$, is the equation of an ellipse ($A = 5$, $C = 1$, so $AC > 0$.)

25. $-2x^2 + 4y^2 - 15x + 8 = 0$ is the equation of a hyperbola ($A = -2$, $C = 4$, so $AC < 0$.)

27. If $-x^2+5x+y=0$, then $y=x^2-5x$. This is the equation of a parabola of vertex $\left(\frac{5}{2},-\frac{25}{4}\right)$.

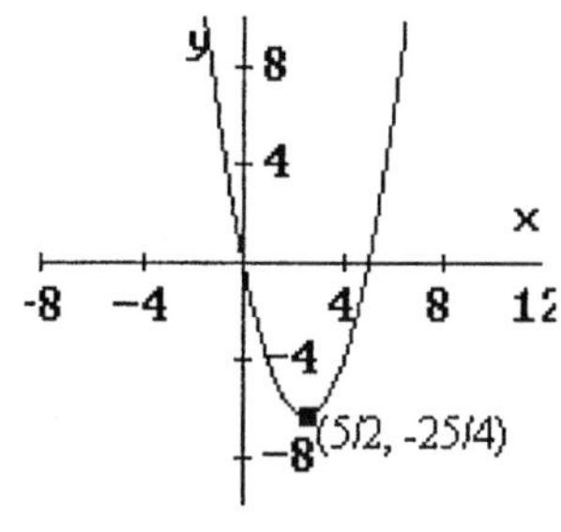

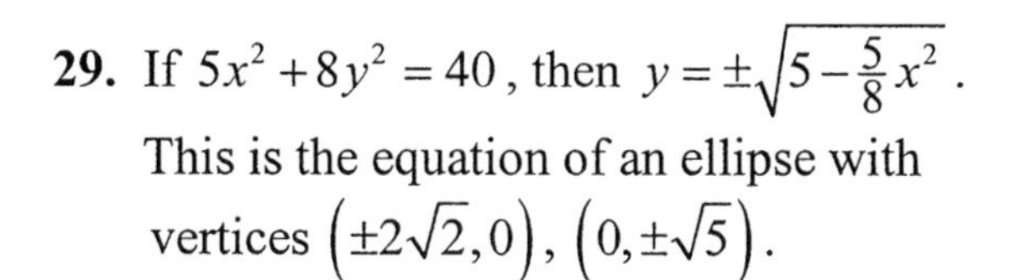

29. If $5x^2+8y^2=40$, then $y=\pm\sqrt{5-\frac{5}{8}x^2}$. This is the equation of an ellipse with vertices $\left(\pm2\sqrt{2},0\right)$, $\left(0,\pm\sqrt{5}\right)$.

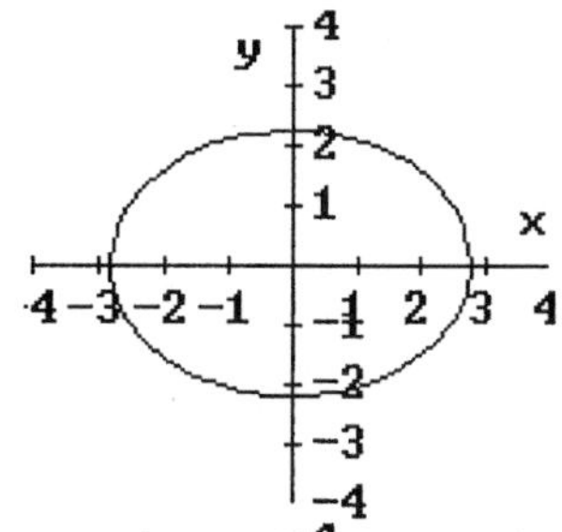

31. If $y^2-3x^2+4x+9=0$, then $y=\pm\sqrt{3x^2-4x-9}$. This is the equation of a hyperbola with vertices approximately at $(-1.19,0)$ and $(2.52,0)$.

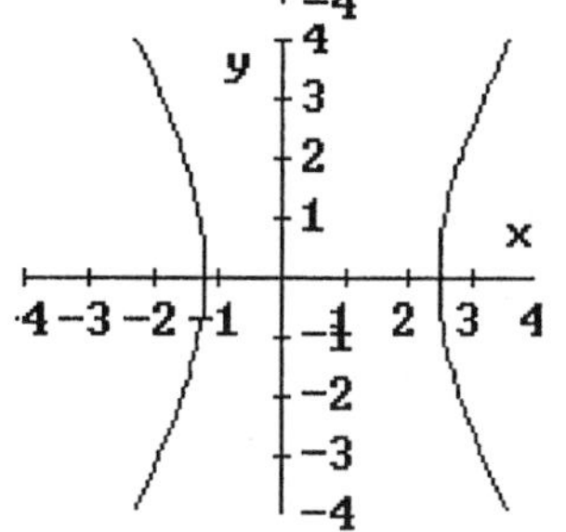

33. If $2x^2+y^2-5x-8=0$, then $y=\pm\sqrt{-2x^2+5x+8}$. This is the equation of an ellipse with vertices $\left(\frac{5\pm\sqrt{89}}{4},0\right)$, $\left(\frac{5}{4},\pm\frac{\sqrt{178}}{4}\right)$.

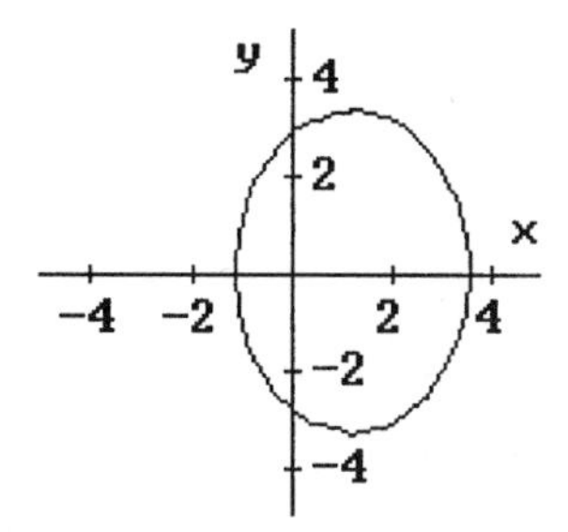

35. If $3x^2-y^2+8x-8y-7=0$, then $y=-4\pm\sqrt{3x^2+8x+9}$. This is the equation of a hyperbola with vertices $\left(-\frac{4}{3},-4\pm\frac{\sqrt{33}}{3}\right)$.

37. $y=\pm\sqrt{x+3}$ correspond to the two branches of the parabola with equation in general form $y^2-x-3=0$.

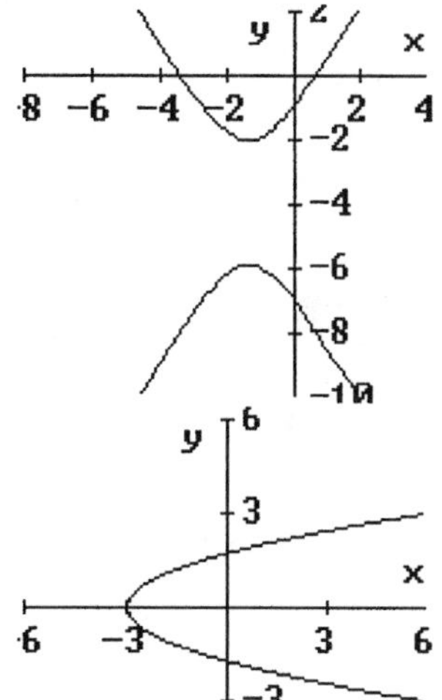

39. $y = 2 \pm \sqrt{16 - 6x - x^2}$ correspond to the two branches of the circle of equation with general form $x^2 + y^2 + 6x - 4y - 12 = 0$.

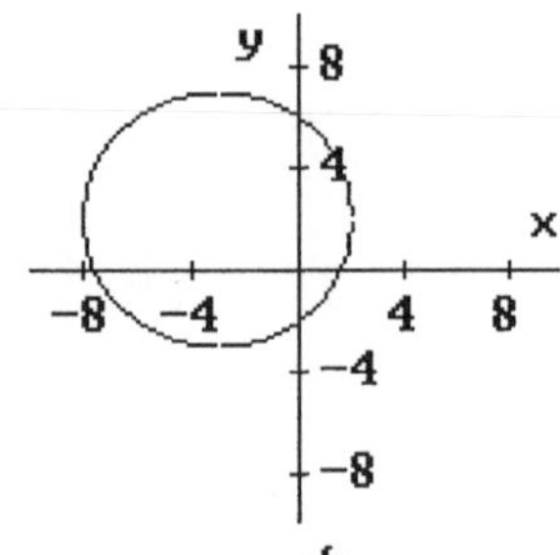

41. $y = \pm\sqrt{\frac{x^2}{3} - 2}$ correspond to the two branches of the hyperbola of equation with general form $x^2 - 3y^2 - 6 = 0$.

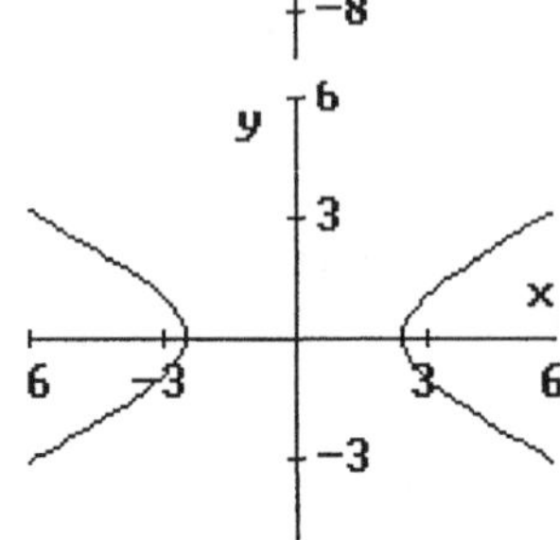

43. $y = 1 \pm \sqrt{2 - 8x - 2x^2}$ correspond to the two branches of the ellipse of equation with general form $2x^2 + y^2 + 8x - 2y - 1 = 0$.

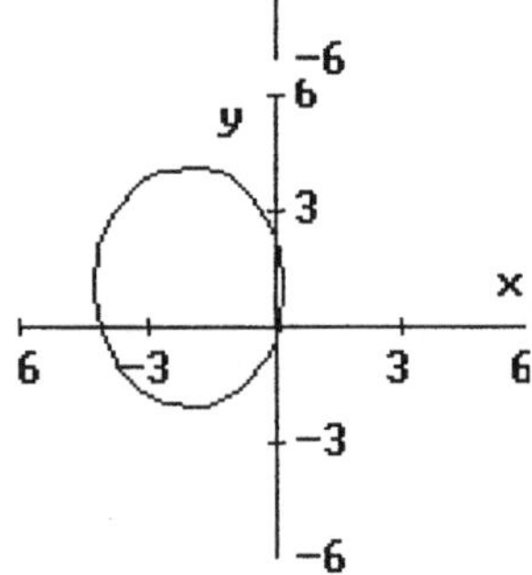

Applications

45. **(a)** If the signal from A arrives to P 800 microseconds before the signal from B, then the difference between the distance from B and the distance from A is $800\mu s \times 0.186\, ^{miles}/_{\mu s}$, or 148. 8 miles. That is, $2b = 148.8$, so $b = 74.4$. Now, $a^2 + b^2 = 100^2$, so $a \approx 66.82$. The equation of the described hyperbola is $\frac{y^2}{74.4^2} - \frac{x^2}{66.82^2} \approx 1$.

(b) If the point P where the ship is is due east from A, then $y = 100$, so $\frac{x^2}{66.82^2} \approx \frac{100^2}{74.4^2} - 1$. Then $x \approx 60.01$. The ship is approximately 60 miles east of A.

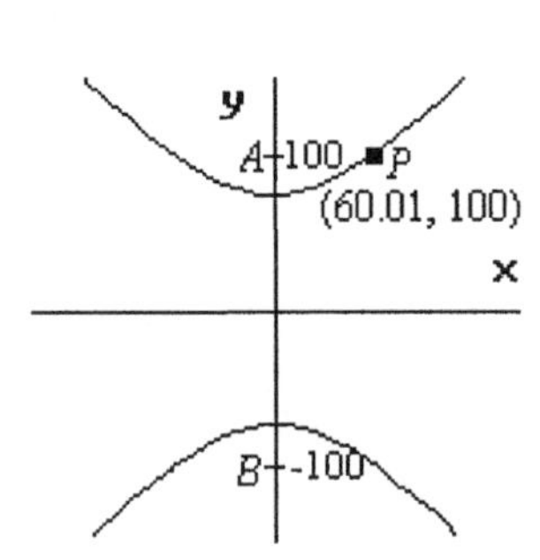

Concepts and Critical Thinking

47. The statement: "The points $(\pm a, 0)$ and $(0, \pm b)$ lie on the graph of the hyperbola with equation $\frac{x^2}{a^2} - \frac{y^2}{b^2} = 1$" is false.

49. The statement: "A hyperbola is the set of all points such that the sum of the distances from two fixed points (called the foci) is constant" is false.

51. Answers may vary. A hyperbola for which the transverse axis is horizontal is $x^2 - y^2 = 1$.

53. Answers may vary. An application of the reflective property of hyperbolas is the design of hyperbolic mirrors for telescopes.

55. If a hyperbola has a horizontal transverse axis of length $2a$, and foci at $(\pm c, 0)$, then for any point $P(x, y)$ on the hyperbola, the triangle formed by P, $F_1(-c, 0)$, and $F_2(c, 0)$ satisfies $F_1P < F_1F_2 + F_2P$, and $F_2P < F_1F_2 + F_1P$, therefore $F_1P - F_2P < F_1F_2$, and $F_2P - F_1P < F_1F_2$. So, $|F_1P - F_2P| < 2c$, that is $0 < 2a < c$. Therefore, $0 < a < c$, and $c^2 - a^2 = (c-a)(c+a) > 0$. (*Note*: if $c^2 - a^2 = 0$, then $c = a$; the hyperbola is degenerate, consisting of only the two points $(\pm c, 0)$.)

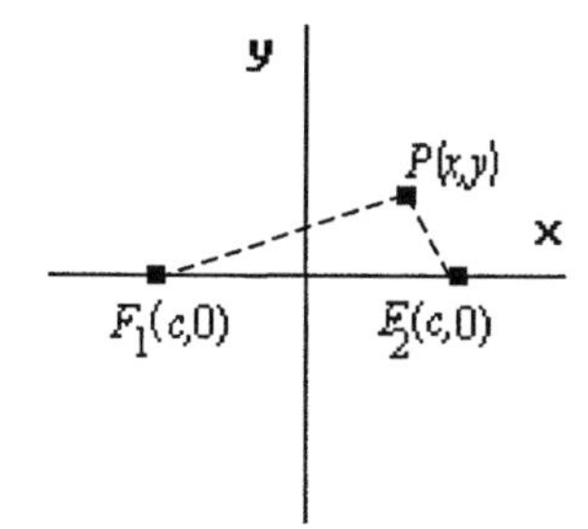

Chapter 8. Section 8.5 Rotation of Axes

In Exercises 1-5, the xy-coordinates of a point are given. The coordinates in the uv-system are obtained by rotation of the standard xy-coordinate system by the given angle θ. The transformations used are: $u = x\cos\theta + y\sin\theta,\ v = -x\sin\theta + y\cos\theta$.

1. If $(x,y)=(3,0)$ and $\theta = \frac{\pi}{4}$, then $u = \frac{\sqrt{2}}{2}x + \frac{\sqrt{2}}{2}y,\ v = -\frac{\sqrt{2}}{2}x + \frac{\sqrt{2}}{2}y$, or $u = \frac{3\sqrt{2}}{2},\ v = -\frac{3\sqrt{2}}{2}$

3. If $(x,y)=(1,-1)$ and $\theta = -\frac{\pi}{2}$, then $u = -y,\ v = x$; that is, $(u,v)=(1,1)$.

5. If $(x,y)=(3,5)$ and $\theta = \pi$, then $u = -x,\ v = -y$; that is $(u,v)=(-3,-5)$

In Exercises 7-11, a uv-coordinate system is obtained by rotating the standard xy-coordinate system by an angle θ. We use the equations: $x = u\cos\theta - v\sin\theta,\ y = u\sin\theta + v\cos\theta$.

7. If $\theta = \frac{3\pi}{4}$ and $(u,v)=(2,0)$, we have $x = -\frac{\sqrt{2}}{2}u + \frac{\sqrt{2}}{2}v,\ y = -\frac{\sqrt{2}}{2}u - \frac{\sqrt{2}}{2}v$; that is,
$(x,y)=(-\sqrt{2},-\sqrt{2})$.

9. If $\theta = \frac{3\pi}{2}$ and $(u,v)=(-4,4)$, then $x = v,\ y = -u$; that is, $(x,y)=(4,4)$.

11. If $\theta = -\frac{\pi}{6}$ and $(u,v)=(\sqrt{3},1)$, then $x = \frac{\sqrt{3}}{2}u + \frac{1}{2}v,\ y = -\frac{1}{2}u + \frac{\sqrt{3}}{2}v$; that is, $(x,y)=(2,0)$.

In Exercises 13-17, an equation in the xy-coordinate system is given. The corresponding equation in the uv-system, where this system is obtained by rotating the standard xy-coordinate system by an angle θ. The transformations used are: $x = u\cos\theta - v\sin\theta,\ y = u\sin\theta + v\cos\theta$.

13. If $\theta = \frac{\pi}{2}$, and $x = e^{-y}$, then $x = -v$ and $y = u$, therefore, $v = -e^{-u}$.

15. If $\theta = \frac{\pi}{4}$, and $x - 5y = -2$, then $x = \frac{\sqrt{2}}{2}u - \frac{\sqrt{2}}{2}v$ and $y = \frac{\sqrt{2}}{2}u + \frac{\sqrt{2}}{2}v$, therefore,
$\frac{\sqrt{2}}{2}u - \frac{\sqrt{2}}{2}v - 5\left(\frac{\sqrt{2}}{2}u + \frac{\sqrt{2}}{2}v\right) = -2$, or equivalently $2u + 3v = \sqrt{2}$.

17. If $\theta = \frac{\pi}{6}$, and $x^2 - y^2 = 1$, then $x = \frac{\sqrt{3}}{2}u - \frac{1}{2}v$ and $y = \frac{1}{2}u + \frac{\sqrt{3}}{2}v$, therefore,
$\left(\frac{\sqrt{3}}{2}u - \frac{1}{2}v\right)^2 - \left(\frac{1}{2}u + \frac{\sqrt{3}}{2}v\right)^2 = 1$, or equivalently $u^2 - v^2 - 2\sqrt{3}uv - 2 = 0$.

19. $13x^2 + 10xy + 13y^2 - 72 = 0$ corresponds to an ellipse ($B^2 - 4AC = 10^2 - 4(13)(13)$, that is $B^2 - 4AC = -576 < 0$.)

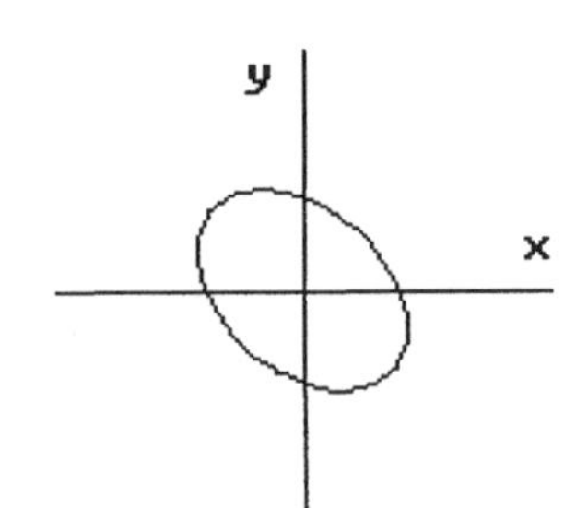

21. $11x^2 - 10\sqrt{3}xy + y^2 + 16 = 0$ corresponds to a hyperbola ($B^2 - 4AC = 300 - 4(11)(1)$, that is $B^2 - 4AC = 256 > 0$.)

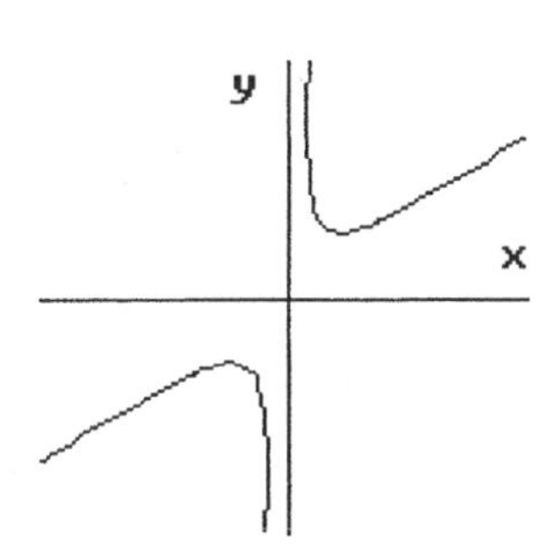

23. $x^2 - 2\sqrt{3}xy + 3y^2 - \sqrt{3}x - y = 0$ corresponds to a parabola ($B^2 - 4AC = 0$.)

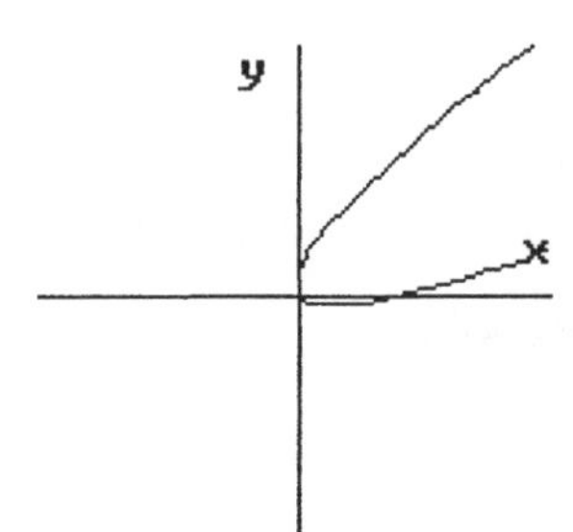

25. $7x^2 - 4\sqrt{3}xy + 3y^2 - 9 = 0$ corresponds to an ellipse ($B^2 - 4AC = -36 < 0$.)

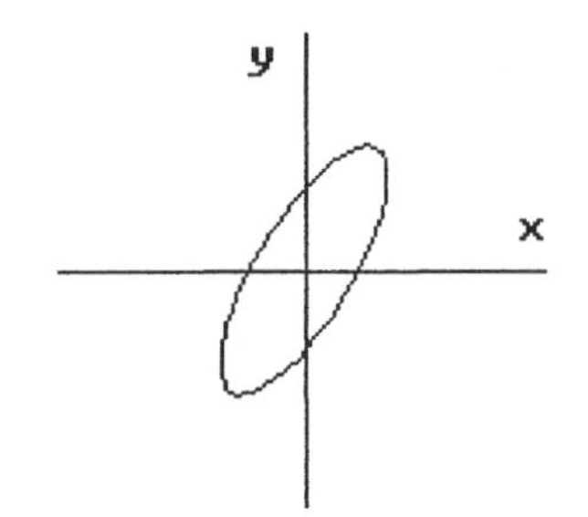

27. The equation $2xy - 1 = 0$ corresponds to a hyperbola, which has equation $u^2 - v^2 = 1$ in the *uv*-system obtained by rotating the *xy*-system by $\frac{\pi}{4}$. The vertices of this hyperbola are $(\pm 1, 0)$ in the *uv*-system, and $\left(\pm\frac{\sqrt{2}}{2}, \pm\frac{\sqrt{2}}{2}\right)$ in the *xy*-system. The foci are $\left(\pm\sqrt{2}, 0\right)$ in the *uv*-system, and $(\pm 1, \pm 1)$ in the *xy*-system. The asymptotes are $v = \pm u$, or $x = 0$ and $y = 0$.

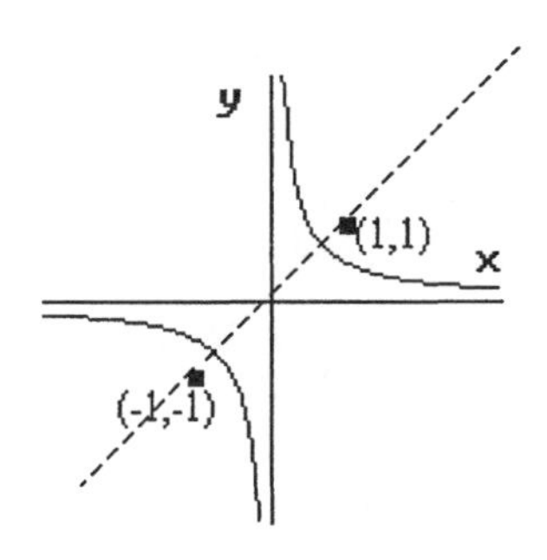

29. The equation $5x^2 + 2\sqrt{3}xy + 7y^2 - 8 = 0$ corresponds to an ellipse, which has equation $u^2 + \frac{v^2}{2} = 1$ in the *uv*-coordinate system obtained by rotating the *xy*-system by $\frac{\pi}{3}$. The vertices are $(\pm 1, 0), \left(0, \pm\sqrt{2}\right)$ in the *uv*-system, and $\left(\pm\frac{1}{2}, \pm\frac{\sqrt{3}}{2}\right), \left(\mp\frac{\sqrt{6}}{2}, \pm\frac{\sqrt{2}}{2}\right)$ in the *xy*-system. The foci are $(0, \pm 1)$ in the *uv*-system, or $\left(\mp\frac{\sqrt{3}}{2}, \pm\frac{1}{2}\right)$ in the *xy*-system.

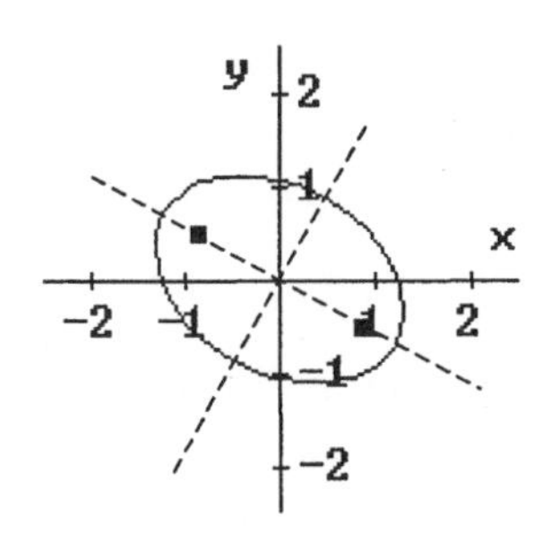

Exercises 31-36 refer to nondegenerate conic sections.

31. In the equation $xy+x-2=0$ we have $A=0,\ B=1,$ and $C=0$. Then $B^2-4AC>0$, and the equation corresponds to a hyperbola.

33. In the equation $-x^2+3xy-4y^2+x-4y-1=0$ we have $A=-1,\ B=3,$ and $C=-4$. Then $B^2-4AC=9-4(-1)(-4)=-7$, and the equation corresponds to an ellipse.

35. In the equation $\frac{(x-2)^2}{4}+(y+1)^2=3-xy$, or equivalently $\frac{1}{4}x^2+xy+y^2-x+2y-1=0$, we have $A=\frac{1}{4},\ B=1,$ and $C=1$. Then $B^2-4AC=1-4\left(\frac{1}{4}\right)(1)=0$, and the equation corresponds to a parabola.

Applications

37. If the equation of the boundary of the photograph of a soft drink is $13x^2-6\sqrt{3}xy+7y^2-16=0$, this corresponds to an ellipse with equation $\frac{u^2}{4}+v^2=1$ in the uv-system obtained by rotating the xy- system by $\frac{\pi}{3}$. The vertices of the ellipse in the xy-system are $\left(\pm1,\pm\sqrt{3}\right)$, $\left(\mp\frac{\sqrt{3}}{2},\pm\frac{1}{2}\right)$. The beginning of a sketch follows, where the vertices of parallel ellipses of the same shape are connected.

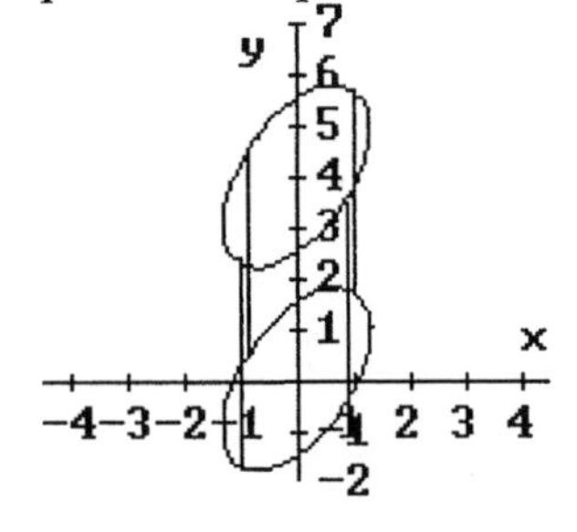

Concepts and Critical Thinking

39. The statement: "If the equation of an ellipse is given by $Ax^2+Bxy+Cy^2+Dx+Ey+F=0$, and $B\neq0$, then the major and minor axes are not parallel to the coordinate axes" is true.

41. The statement: "For some values of $A,C,D,E,$ and F, the graph of $Ax^2+xy+Cy^2+Dx+Ey+F=0$ is a circle" is false.

43. Answers may vary. The point $(1,0)$ is a 45° rotation of the point $\left(\frac{\sqrt{2}}{2},\frac{\sqrt{2}}{2}\right)$; that is, the xy-coordinate system point $\left(\frac{\sqrt{2}}{2},\frac{\sqrt{2}}{2}\right)$ corresponds to the uv-coordinate system point $(1,0)$, where the uv-system is obtained by rotating the standard xy-system by an angle of 45°.

45. Answers may vary. An example of a conic section requiring rotation by an angle θ to eliminate the *xy*-term, where $\cot 2\theta = \frac{2}{3}$ is $3x^2 + 3xy + y^2 - x - y = 0$.

47. Conic sections with vertical or horizontal axes of symmetry have equations of the form $Ax^2 + Cy^2 + Dx + Ey + F = 0$. If $AC < 0$, the conic is a hyperbola; if $AC = 0$, the conic is a parabola; if $AC > 0$, the conic is an ellipse (a circle if $A = C$.) The test for general conic sections, in this case, corresponds to looking at the sign of $B^2 - 4AC = -4AC$ ($B = 0$.) Namely, if $-4AC < 0$, the conic is an ellipse (this is equivalent to $AC > 0$); if $-4AC = 0$, the conic is a parabola; if $-4AC > 0$, the conic is a hyperbola (this is equivalent to $AC < 0$.) Therefore, the two tests are equivalent for this type of conic section.

1. $x = t - 3$
$y = 2t + 1$

t	x	y
−1	−4	−1
0	−3	1
1	−2	3
2	−1	5

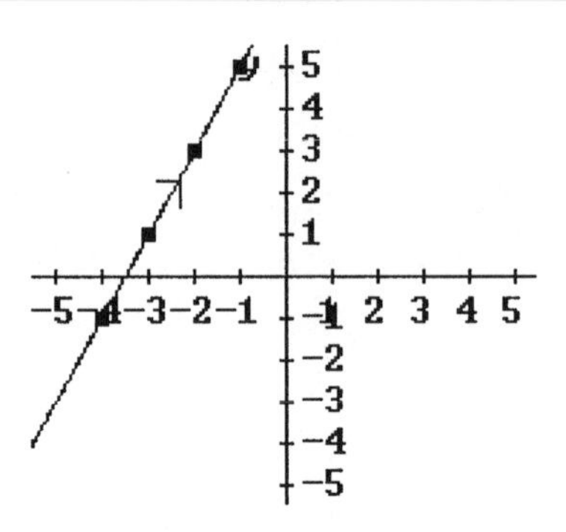

3. The graph of $x = \frac{10}{t+1}\cos t$, $y = \frac{10}{t+1}\sin t$, Tmin = 0, Tmax = 50, corresponds to (v), a spiral.

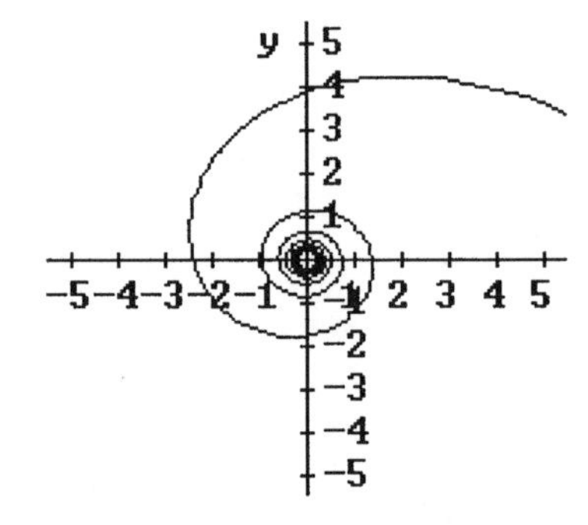

5. The graph of $x = 8 - 8\left|t - 2\left\lfloor \frac{1}{2}t \right\rfloor - 1\right|$, $y = 8 - 8\left|\frac{7}{11}t - 2\left\lfloor \frac{7}{22}t \right\rfloor - 1\right|$, Tmin = 0, Tmax = 11, corresponds to (iv), the path of a ray of light inside a mirrored room.

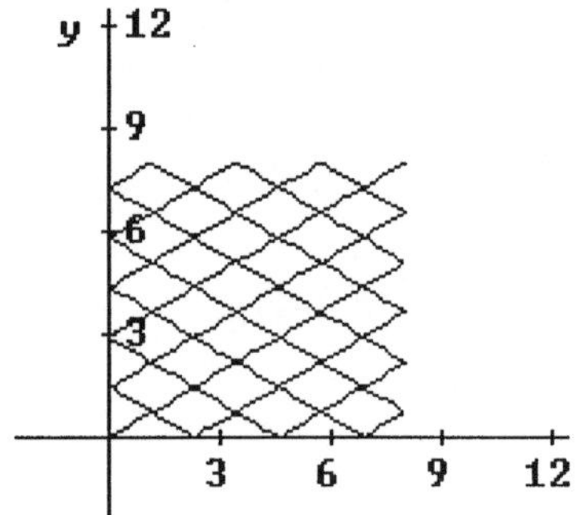

7. The graph of $x = \frac{5}{12}\cos(4\pi t) + \frac{30\sqrt{34}}{17}t$, $y = \frac{5}{12}\sin(4\pi t) - 16t^2 + \frac{76\sqrt{34}}{17}t + 6$, Tmin = 0, Tmax = 1.5, corresponds to (i), the path of a point on a basketball, shot from the free-throw line with a slight backspin.

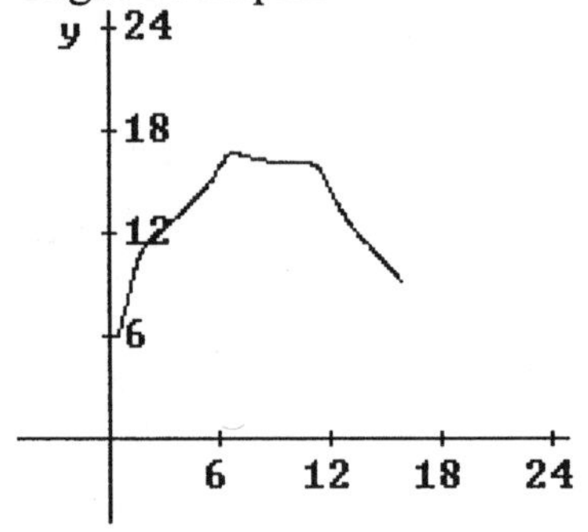

9. If $x = 3 + t$, and $y = -2t + 2$, then $y = -2x + 8$.

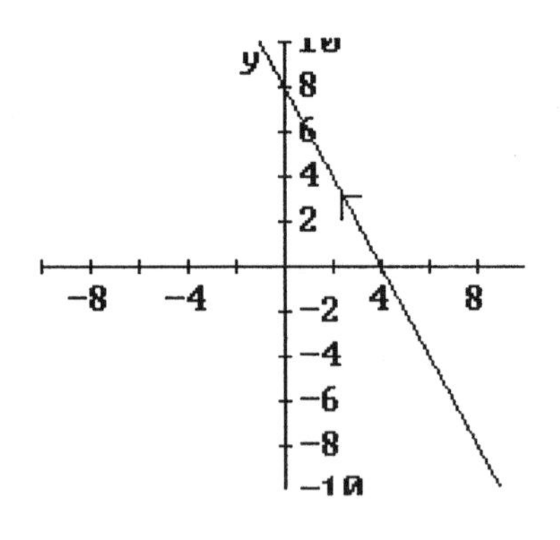

11. If $x = t + 1$, and $y = t^2 + 2t + 1$, then $y = x^2$.

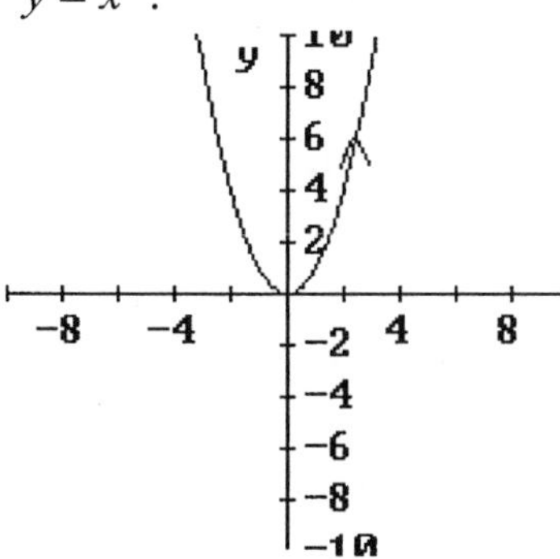

13. If $x = 4\cos t$, and $y = 4\sin t$, then $x^2 + y^2 = 16$.

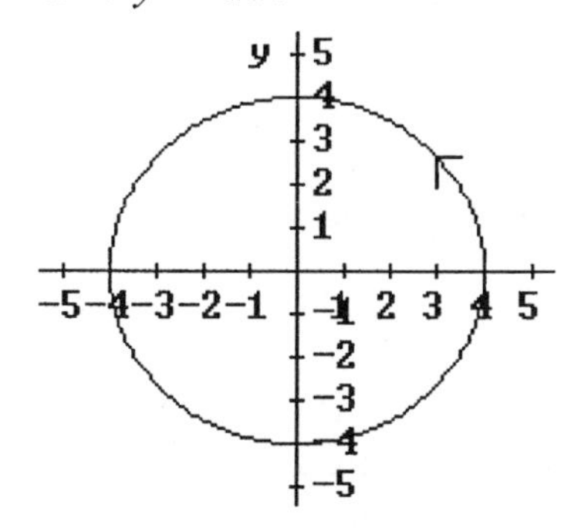

15. If $x = 1 - 2\sin t$, and $y = 2 + 3\cos t$, then $\frac{(x-1)^2}{4} + \frac{(y-2)^2}{9} = 1$.

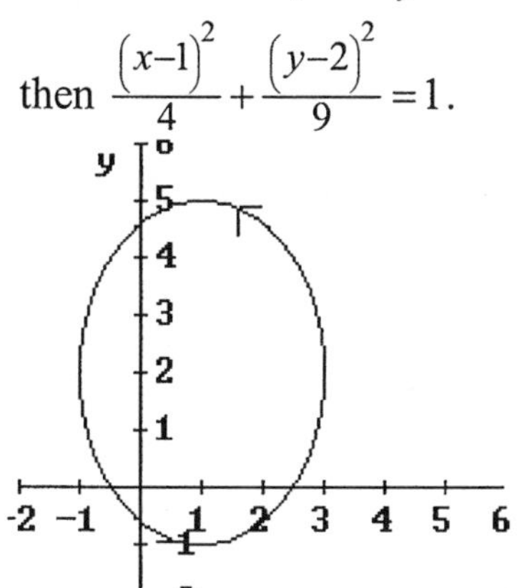

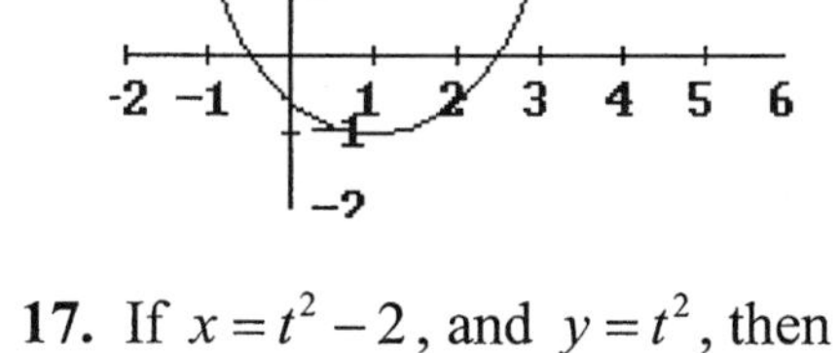

17. If $x = t^2 - 2$, and $y = t^2$, then

(a) $y = x + 2$, whose graph is:

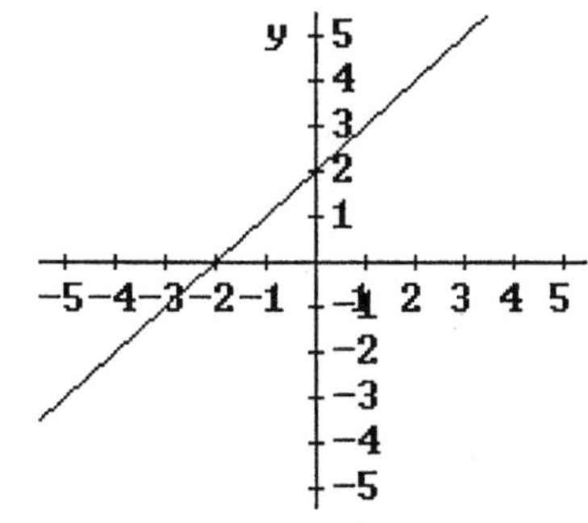

(b) The graph of $x = t^2 - 2,\ y = t^2,\ -10 \le t \le 10$:

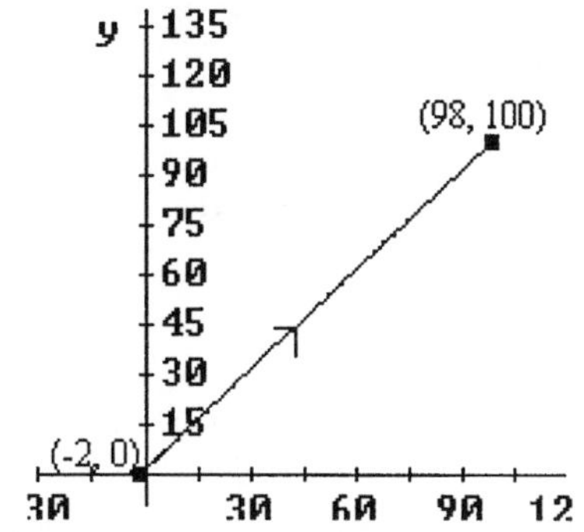

(c) The difference between the two graphs is that the second one is a segment, as opposed to a line, traversed from the initial point $(-2,0)$ to the final point $(98,100)$.

19. If $x = e^t$, and $y = e^{2t}$, then

(a) $y = x^2$, whose graph is:

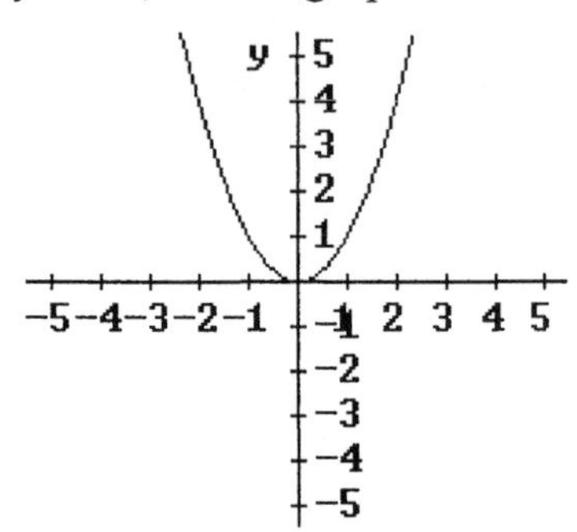

(b) The graph of $x = e^t$, $y = e^{2t}$, $-10 \le t \le 10$:

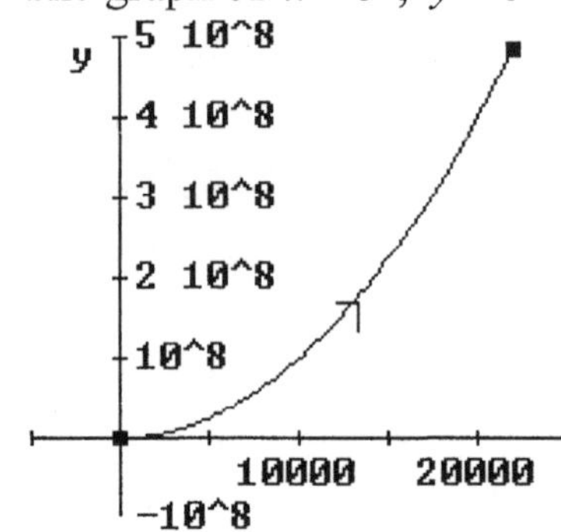

(c) The difference between the two graphs is that the second one is only a portion of the graph of $y = x^2$, the portion traversed from (e^{-10}, e^{-20}) to (e^{10}, e^{20}).

21. If $x = \cos t$, and $y = 4\cos t$, then

(a) $y = 4x$, whose graph is:

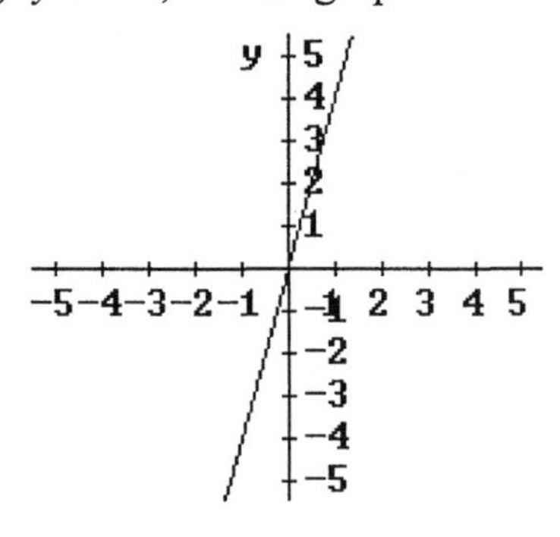

(b) The graph of $x = \cos t$, $y = 4\cos t$, $-10 \le t \le 10$:

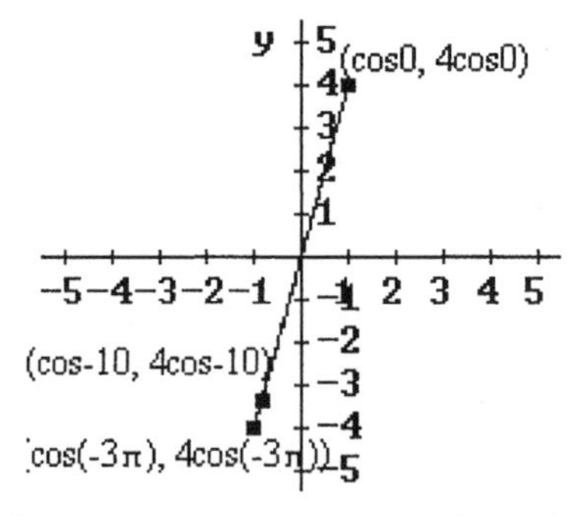

(c) The difference between the two graphs is that the second one is only a portion of the graph of $y = 4x$, the portion between $(-1, -4)$ and $(1, 4)$, traversed several times over.

23. If $x = t^3 - 3t^2 + 2t - 2$,
$y = -t^2 + 2t + 3$,
the crossing point is $(2,3)$, attained when $t = 0$ and when $t = 2$.

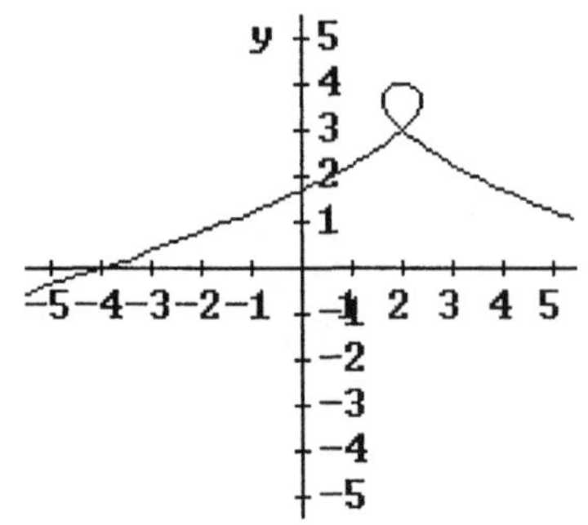

25. If $x = 2\cos^3 t$, $y = 4\sin^3 t$,
the intercepts are $(\pm 2, 0)$, $(0, \pm 4)$, attained when $t = 0, \pm\frac{\pi}{2}, \pi, \ldots$

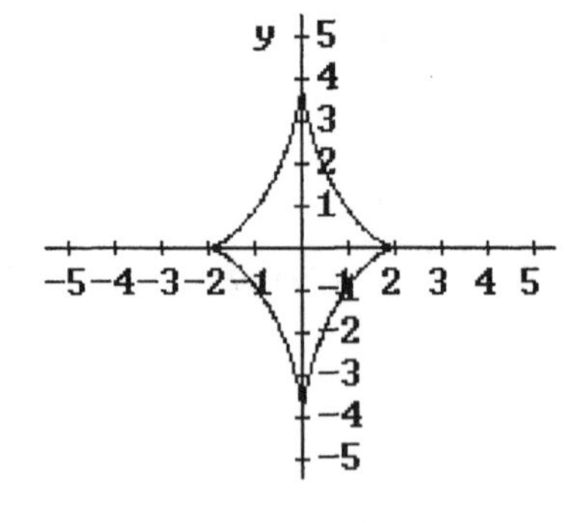

Applications

27. If the motion of the heel of the foot when using an elliptical crosstrainer is approximated by the parametric equations $x = 6.1\cos t - 0.95\sin t - 0.19$, $y = 2.1\cos t + 2.7\sin t + 7.9$, the maximum and minimum heights are, approximately 11.32 and 4.38.

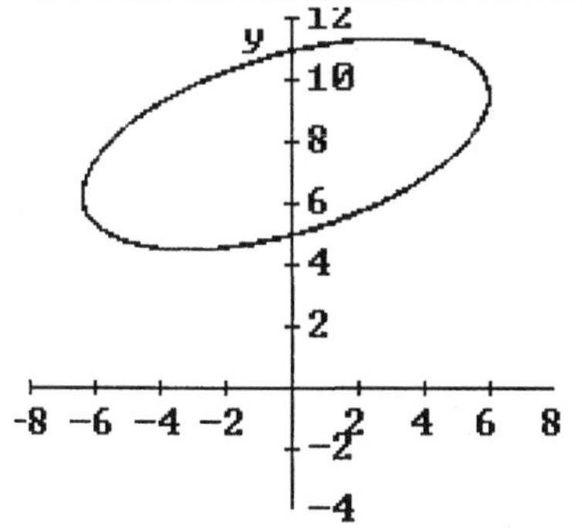

29. If the position of a rider on a double Ferris wheel is approximated by the equations

$x = 20\sin\left(\frac{\pi t}{10}\right) + 10\sin\left(\frac{2\pi t}{5}\right)$

$y = 35 - 20\cos\left(\frac{\pi t}{10}\right) - 10\cos\left(\frac{2\pi t}{5}\right)$

where x is the horizontal distance in feet from the base, y is the height in feet, and t is the elapsed time in seconds from the start of the ride, then

(a) the initial position of the rider is $x = 0$, $y = 5$;

(b) it takes 20 seconds for the rider to complete a full cycle on the ferris wheel and return to the initial position;

(c) the maximum height is approximately 59.72 feet, attained after $t \approx 7.75$ seconds and $t \approx 12.25$ seconds.

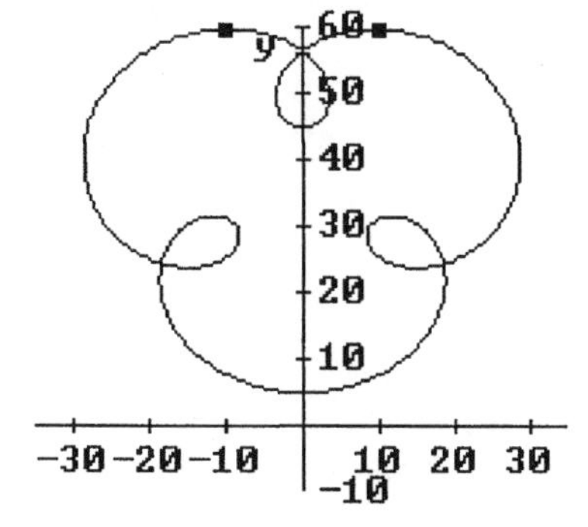

Concepts and Critical Thinking

31. The statement: "If the location of the 8-ball t seconds after the cue ball has been struck is given by $x = f_1(t)$ and $y = g_1(t)$, and the location of the 7-ball is given by $x = f_2(t)$ and $y = g_2(t)$, and the graphs of these two sets of parametric equation intersect, then the 8-ball and the 7-ball must collide" is false.

33. The statement: "No portion of the curve with parametric equations $x = t^2 + 1$ and $y = g(t)$ lies in the second quadrant" is true.

35. Answers may vary. A set of parametric equations for the portion of the graph of $y = x^2$ lying in the first quadrant is $x = t^2$, $y = t^4$.

37. Answers may vary. An example of a set of parametric equations for a curve that is not the graph of a function $y = f(x)$ is $x = \cos t$, $y = \sin t$.

Chapter 8. Review

1. Given the graph of $y = x^2$:

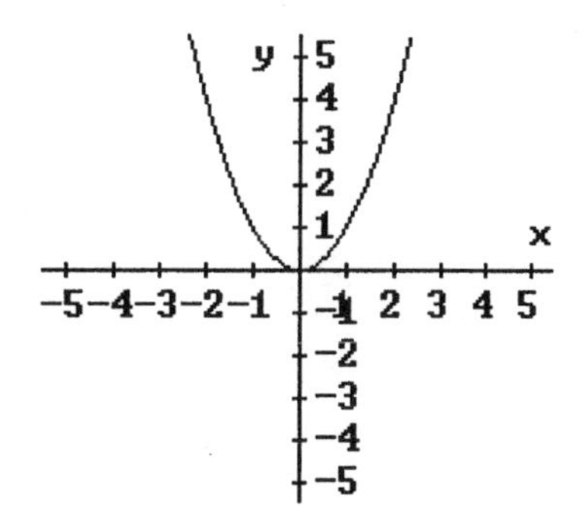

(a) The graph of $y = (x-2)^2$ is a horizontal translation to the right by 2 units.

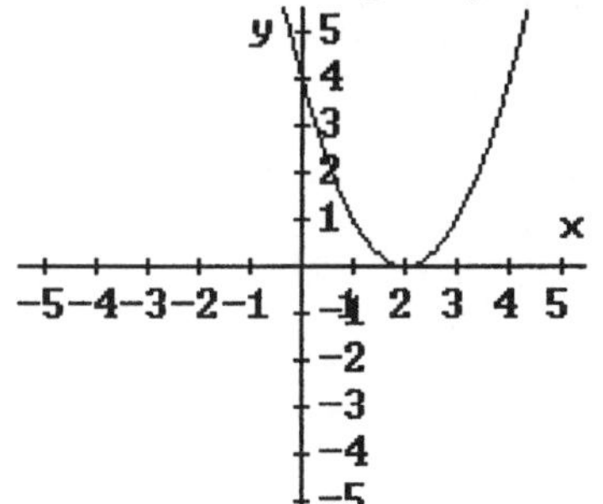

(b) The graph of $y = x^2 + 3$ is a vertical translation upwards by 3 units.

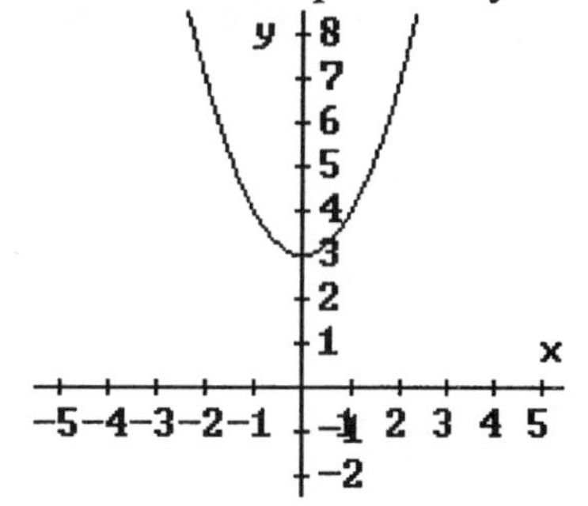

(c) The graph of $y - 4 = (x-2)^2$ is a horizontal translation to the right by 2 units, combined with a vertical translation upwards by 4 units.

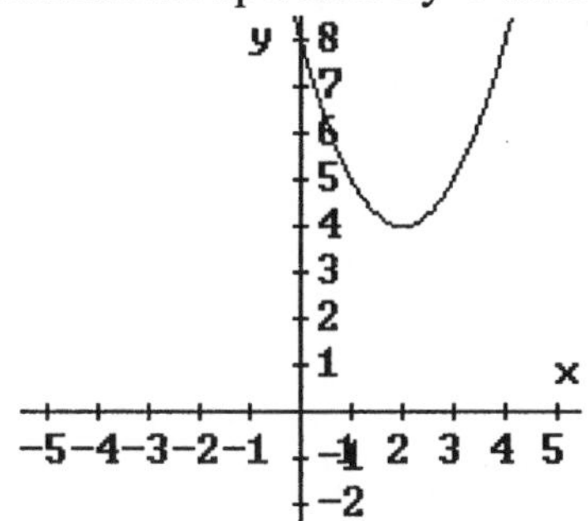

(d) The graph of $y = -x^2$ is a reflection about the *x*-axis.

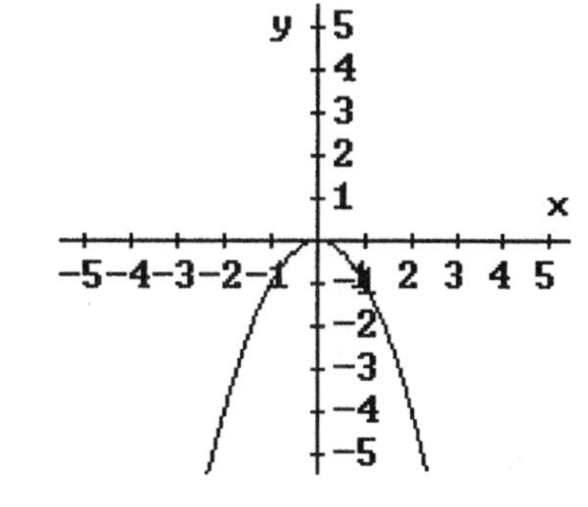

3. Given the graph of $xy^2 + x = 4$:

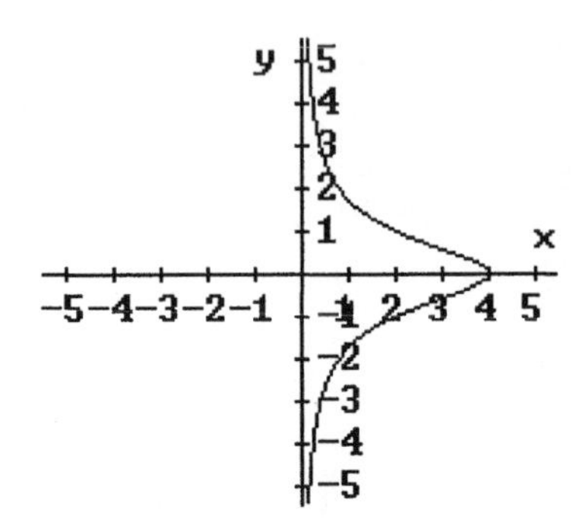

(a) The graph of $x(y+2)^2 + x = 4$ is a vertical translation down by 2 units.

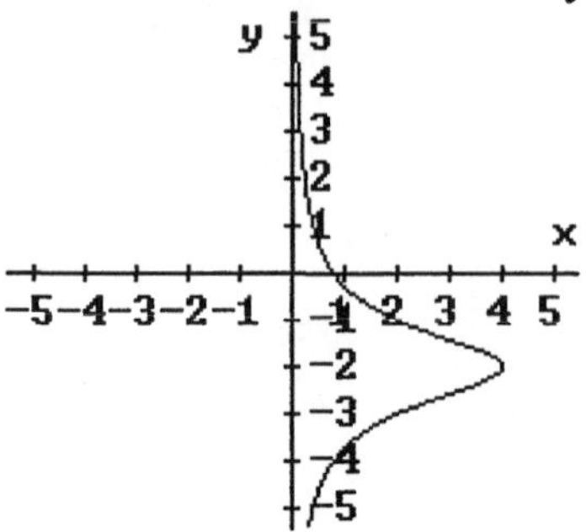

(b) The graph of $xy^2 + x = -4$ is a reflection about the $y-$axis.

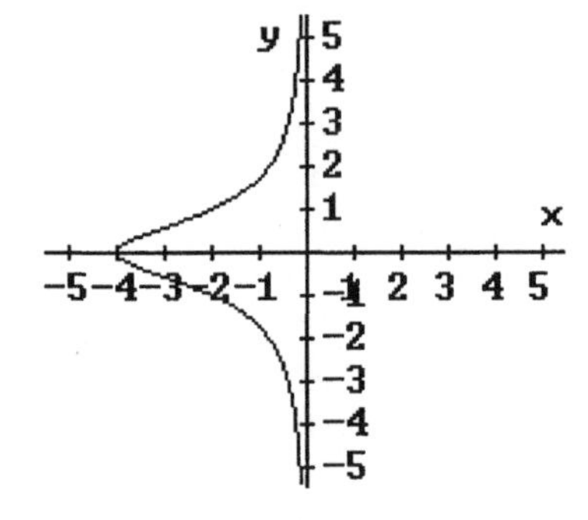

(c) The graph of $(x-1)(y-3)^2 + x - 1 = 4$ is a horizontal translation right 1 unit combined with a vertical one up 3.

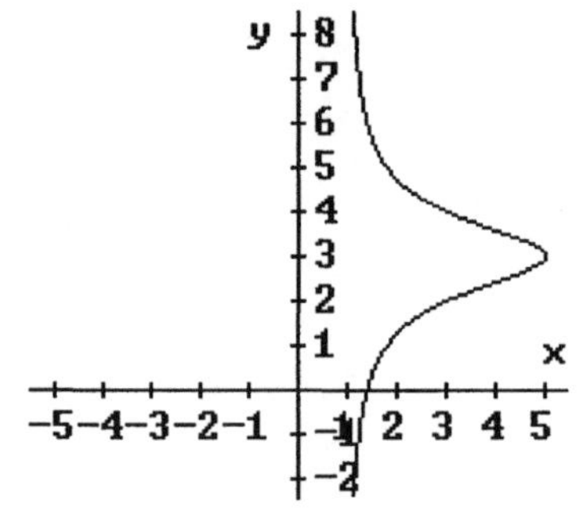

5. Given the original relation $x^2 + y^4 = 16$ and a translated graph:

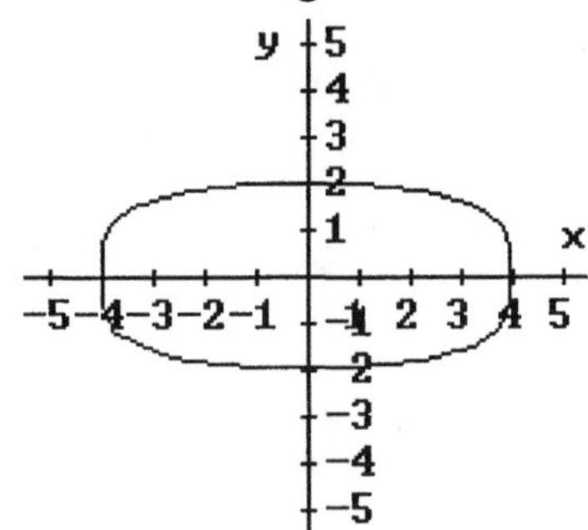

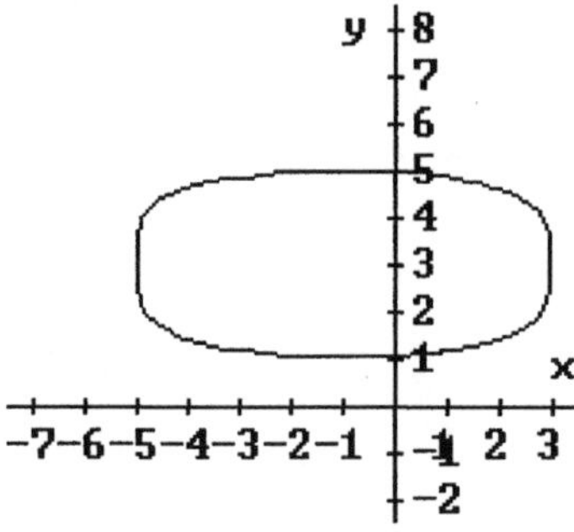

The translated graph is a horizontal shift left 1 unit, combined with a vertical shift up 3 units. An equation for the translated graph is $(x+1)^2 + (y-3)^4 = 16$.

7. Given the original relation $y = x^4 - 4x^2 + 2$ and a reflected graph:

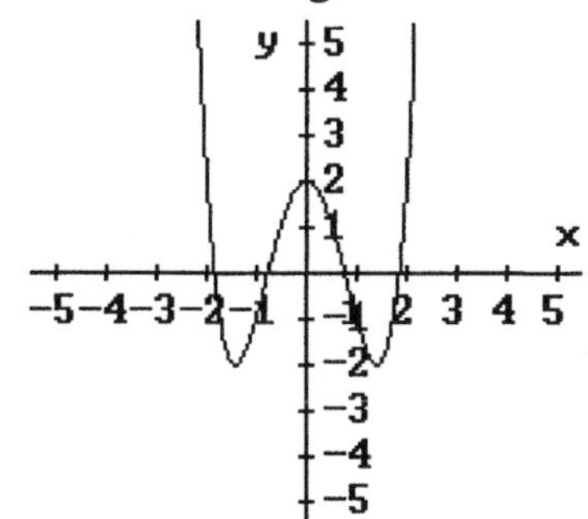

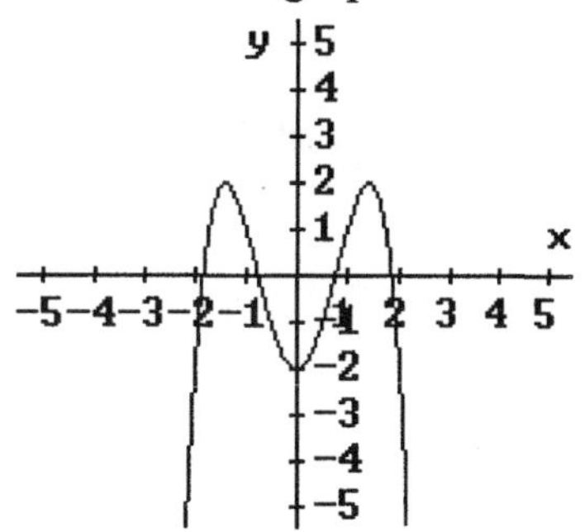

The reflected graph corresponds to a reflection about the x-axis. An equation for this reflected graph is $y = -x^4 + 4x^2 - 2$.

9. The relation $y^2 - 2x = y^4$ is symmetric with respect to the x-axis: if y is replaced by $-y$, we obtain $(-y)^2 - 2x = (-y)^4$, which is equivalent to the original relation $y^2 - 2x = y^4$. The relation $y^2 - 2x = y^4$ is not symmetric with respect to the y-axis. Replacing x by $-x$ produces the relation $y^2 + 2x = y^4$, which is not equivalent to the original. The relation $y^2 - 2x = y^4$ is not symmetric with respect to the origin. Replacing x by $-x$ and y by $-y$ produces the relation $y^2 + 2x = y^4$, which is not equivalent to the original.

11. The relation $x - y = 1$ is not symmetric with respect to the x-axis, the y-axis, or the origin. Replacing x by $-x$ produces the relation $-x - y = 1$. Replacing y by $-y$ produces the relation $x + y = 1$. Performing both replacements produces the relation $-x + y = 1$. None of these is equivalent to the original relation.

13. Given the relation $x = y^4 - 4y^2$ and a portion of its graph, we can complete the graph after we note that the relation is symmetric with respect to the x-axis, since the replacement of y by $-y$ produces the equivalent relation $x = (-y)^4 - 4(-y)^2$.

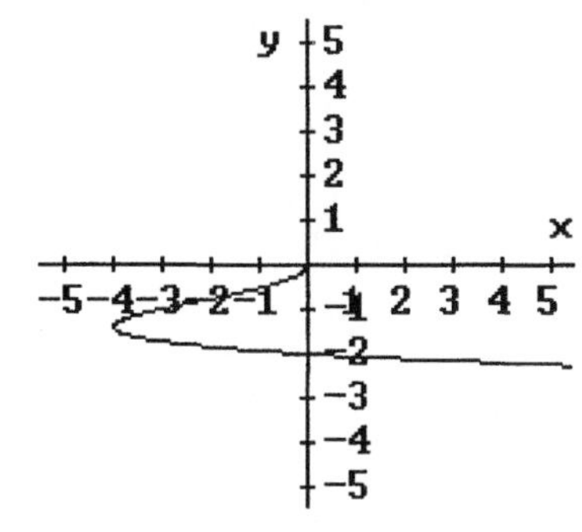

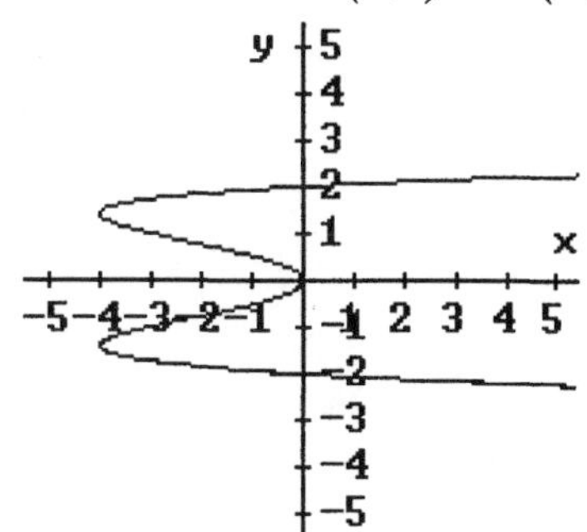

15. Given the relation $xy + x^3y = 5$ and a portion of its graph, we can complete the graph after we note that the relation is symmetric with respect to the origin: replacing x by $-x$, y by $-y$, produces the equivalent relation $(-x)(-y) + (-x)^3(-y) = 5$.

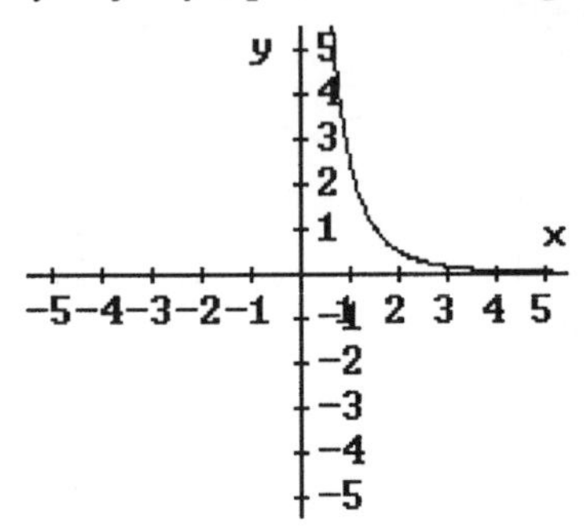

17. The parabola of equation $x = 2y^2$ has its vertex at (0, 0), and its focus at $\left(\frac{1}{8}, 0\right)$. Its directrix is $x = -\frac{1}{8}$.

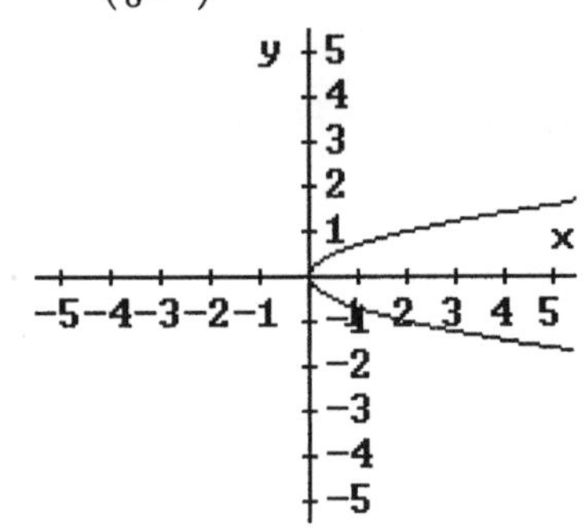

19. The parabola of equation $y + 4 = -2(x-3)^2$ has its vertex at $(3, -4)$ and its focus at $\left(3, -\frac{33}{8}\right)$. Its directrix is $y = -\frac{31}{8}$.

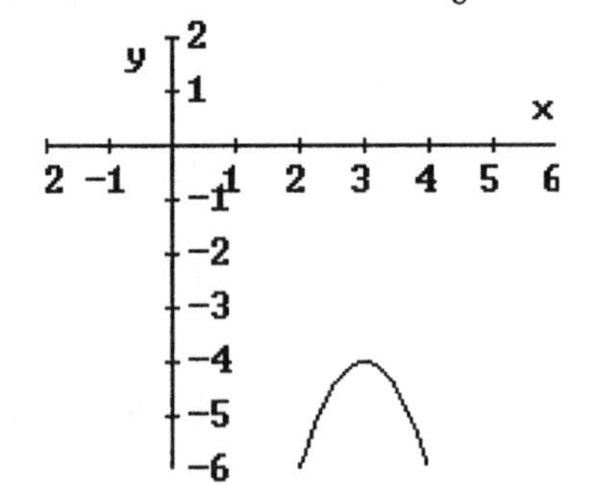

21. The parabola of equation $x = 3y^2 + 6y + 7$ has its vertex at $(4, -1)$ and its focus at $\left(\frac{49}{12}, -1\right)$. Its directrix is $x = \frac{47}{12}$.

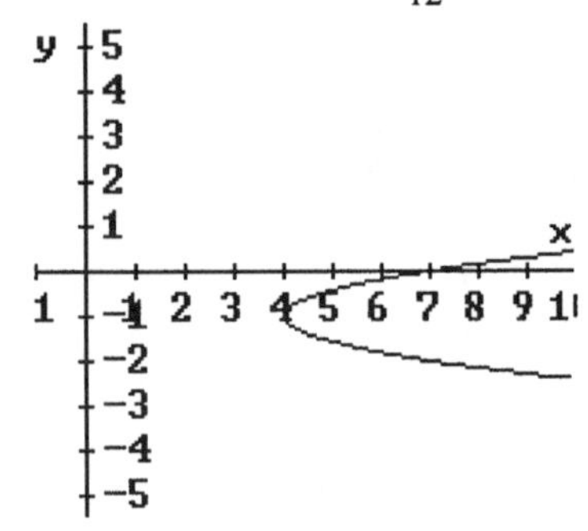

23. An equation for the parabola with vertex at the origin, and focus at $\left(0,\frac{1}{4}\right)$, is $y=x^2$. (The absolute value of p is $\frac{1}{4}$, and the parabola opens up.)

25. An equation for the parabola with vertex at $(-1,2)$, and whose directrix is $y=6$, is $y-2=-\frac{1}{16}(x+1)^2$. (The absolute value of p is 4, and the parabola opens down.)

27. The ellipse of equation $\frac{x^2}{9}+\frac{y^2}{64}=1$ has center at (0, 0). Its vertices are $(\pm 7,0)$ and $(0,\pm 8)$. Its foci are $\left(0,\pm\sqrt{15}\right)$.

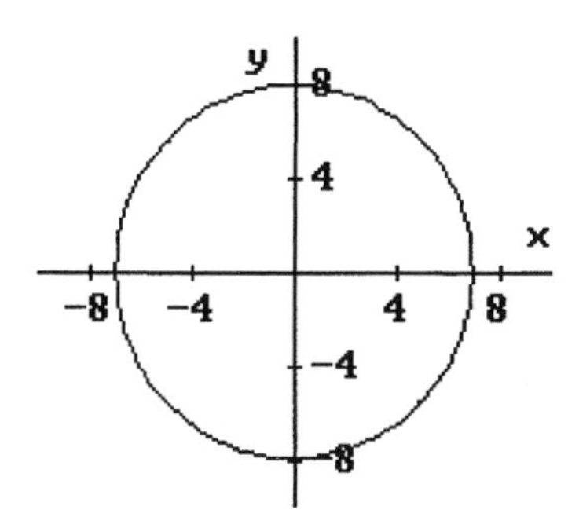

29. The ellipse $9(x-3)^2+25(y-2)^2=225$, or $\frac{(x-3)^2}{25}+\frac{(y-2)^2}{9}=1$ is centered at $(3,2)$. Its vertices are $(8,2),(-2,2),(3,5),(3,-1)$. Its foci are $(-1,2),(7,2)$.

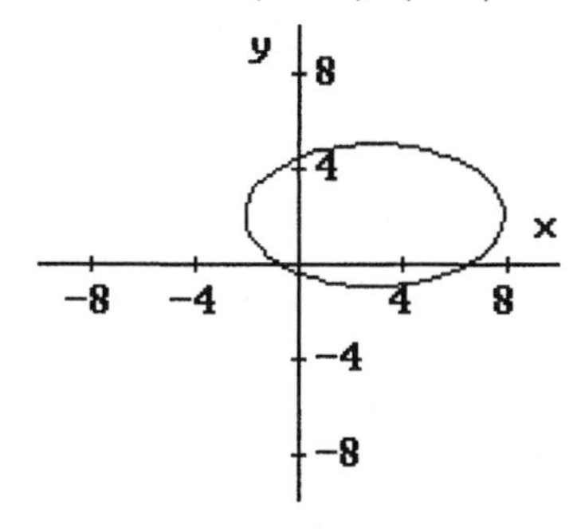

31. The ellipse $x^2+4y^2-2x+24y+21=0$, or

$\frac{(x-1)^2}{16}+\frac{(y+3)^2}{4}=1$, is centered at $(1,-3)$. Its vertices are $(5,-3),(-3,-3),(1,-1),(1,-5)$.

Its foci are $\left(1\pm 2\sqrt{3},-3\right)$.

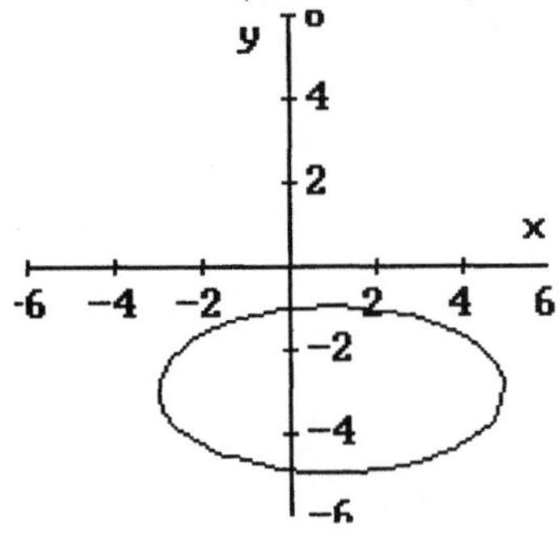

33. An equation for the ellipse with horizontal major axis of length 6, vertical minor axis of length 4, and center at the origin, is $\frac{x^2}{9}+\frac{y^2}{4}=1$.

35. Equation for the ellipse with foci at $(0,\pm 4)$ and minor axis of length 6 is $\frac{x^2}{9}+\frac{y^2}{25}=1$.

37. The hyperbola of equation $\frac{x^2}{25}-\frac{y^2}{9}=1$ has center at $(0,0)$. Its vertices are $(\pm 5,0)$, and its foci are $\left(\pm\sqrt{41},0\right)$.

Its asymptotes are $y=\pm\frac{3}{5}x$.

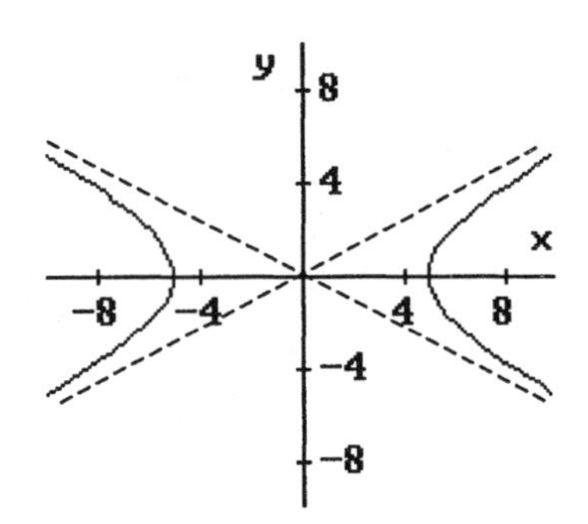

39. The hyperbola of equation $4(y-1)^2-9(x+5)^2=36$, or

$\frac{(y-1)^2}{9}-\frac{(x+5)^2}{4}=1$ has center at $(-5,1)$ and vertical transverse axis. Its vertices are $(-5,4),(-5,-2)$, and its foci are $\left(-5,1\pm\sqrt{13}\right)$. Its asymptotes are $y-1=\pm\frac{3}{2}(x+5)$, that is $y=\frac{3}{2}x+\frac{17}{2}$ and $y=-\frac{3}{2}x-\frac{13}{2}$.

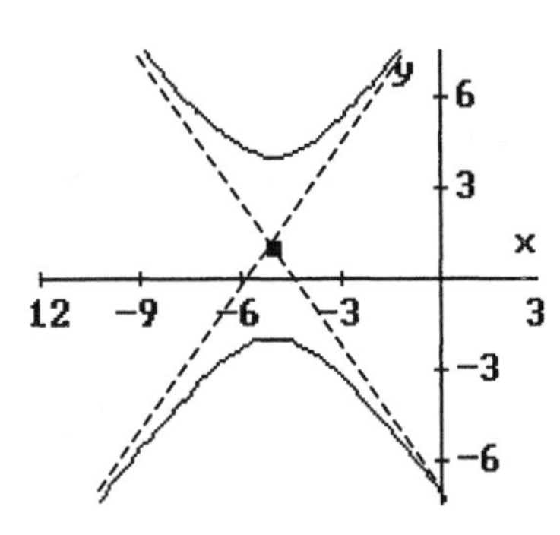

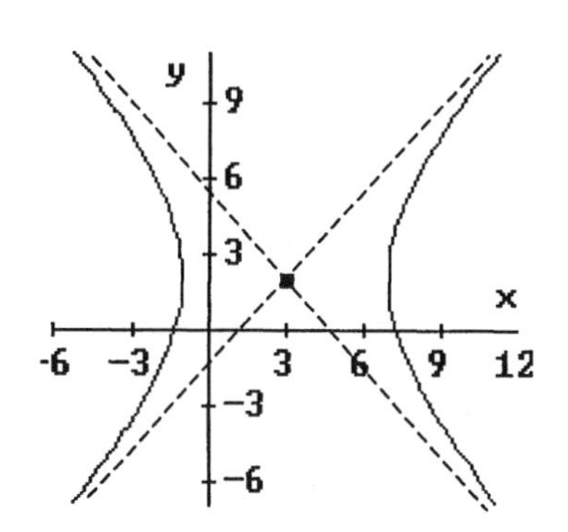

41. The hyperbola of equation

$25x^2-16y^2-150x+64y-239=0$, or $\frac{(x-3)^2}{16}-\frac{(y-2)^2}{25}=1$,

is centered at $(3,2)$. Its vertices are $(-1,2),(7,2)$, and its foci are $\left(3\pm\sqrt{41},2\right)$. Its asymptotes are

$y-2=\pm\frac{5}{4}(x-3)$, that is $y=\frac{5}{4}x-\frac{7}{4}$ and $y=-\frac{5}{4}x+\frac{23}{4}$.

43. An equation for the hyperbola with vertices at $(\pm5,0)$ and foci at $\left(\pm\sqrt{34},0\right)$ is $\frac{x^2}{25}-\frac{y^2}{9}=1$.

45. If $y^2+4x=6$, then $y=\pm\sqrt{-4x+6}$. These correspond to the two branches of a parabola. Its vertex is $\left(\frac{3}{2},0\right)$.

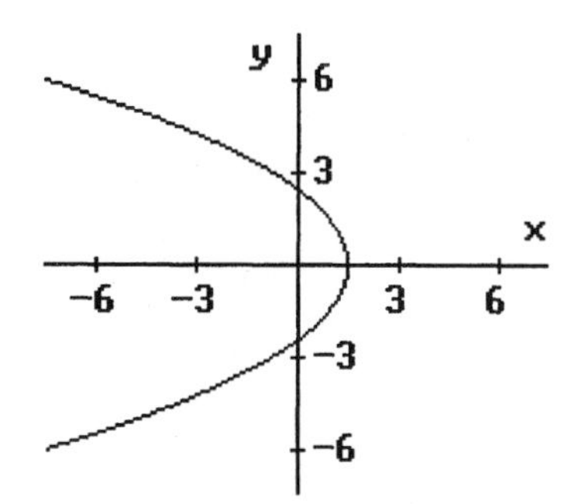

47. If $2x^2-7x+y^2=6$, then

$y=\pm\sqrt{-2\left(x-\frac{7}{4}\right)^2-\frac{97}{8}}$. These correspond to an ellipse of center $\left(\frac{7}{4},0\right)$. Its vertices are $\left(\frac{7}{4}\pm\frac{\sqrt{97}}{4},0\right)$, $\left(\frac{7}{4},\pm\frac{\sqrt{194}}{4}\right)$

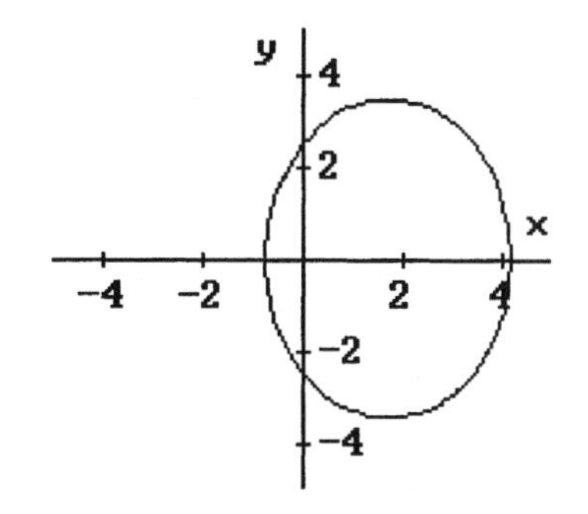

49. $y=\pm\sqrt{16-x^2}$ correspond to a circle of center (0, 0) and radius 4. Its general equation is $x^2+y^2-16=0$.

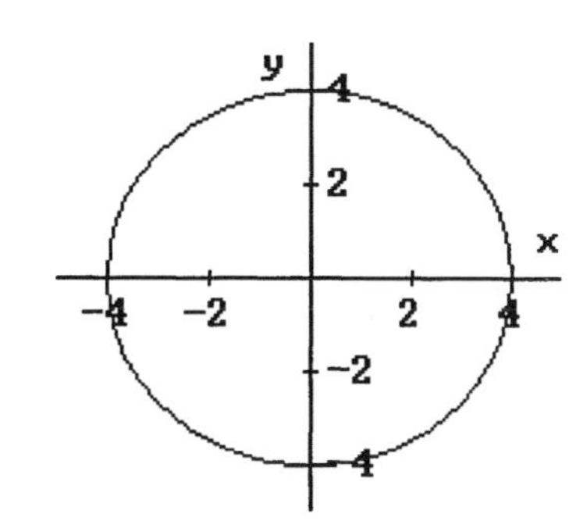

51. $y=2\pm\sqrt{x-4}$ correspond to a parabola. Its vertex is $(4,2)$. Its general equation is $y^2-x-3y+4=0$.

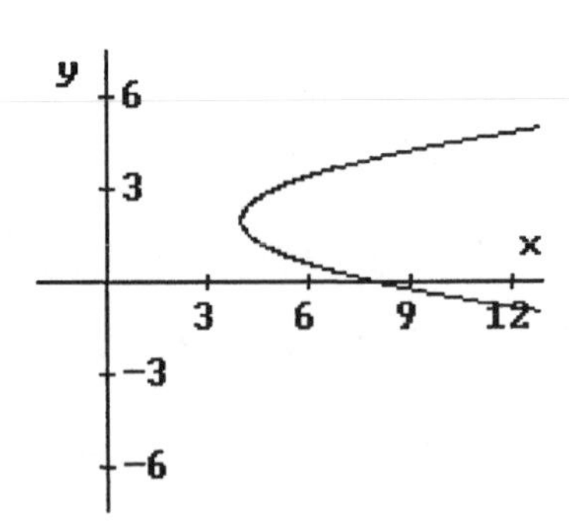

53. The point (2, 2) in the *xy*-coordinate system corresponds to the point $\left(2\sqrt{2},0\right)$ in the *uv*-system obtained by rotation of the *xy*-system by an angle $\frac{\pi}{4}$.

55. The point $(2,-1)$ in the *uv*-coordinate system obtained by rotation of the *xy*-system by an angle $\frac{\pi}{2}$.corresponds to the point $(1,2)$ in the *xy*-system.

57. The equation $y=2x^2$ corresponds to the equation $2u^2-4uv+2v^2-\sqrt{2}u-\sqrt{2}v=0$ in the *uv*-coordinate system obtained by rotation of the *xy*-system by an angle $\frac{\pi}{4}$.

59. $x^2+2xy+y^2+\frac{\sqrt{2}}{2}x-\frac{\sqrt{2}}{2}y=0$ corresponds to a parabola.

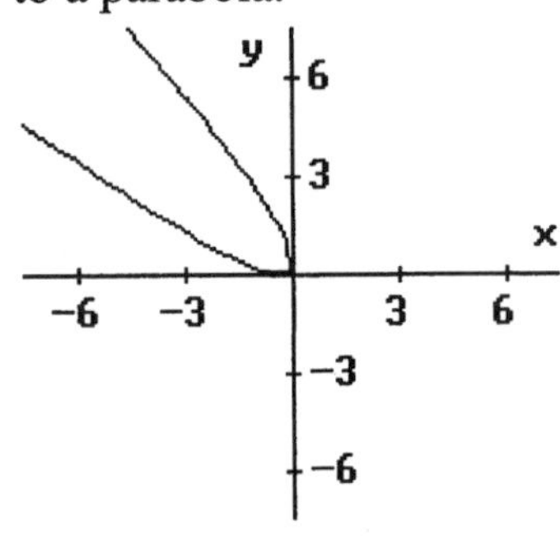

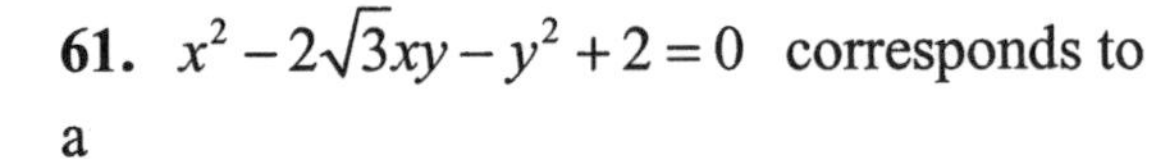

61. $x^2-2\sqrt{3}xy-y^2+2=0$ corresponds to a

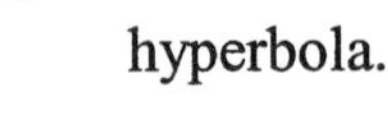

hyperbola.

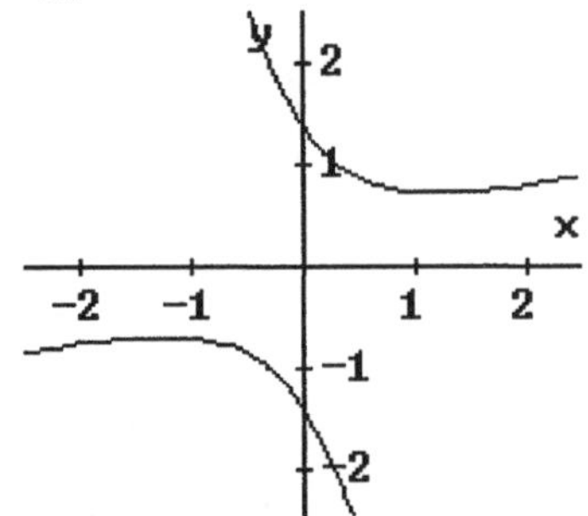

63. In the equation $4x^2+y^2-1=0$, we have $A=4,\ B=0,\ C=1$. Then $B^2-4AC=-16<0$, so $4x^2+y^2-1=0$ corresponds to an ellipse.

65. In the equation $4x^2-8x-y^2+8y-16=0$, we have $A=4,\ B=-8,\ C=-1$. Then $B^2-4AC=80>0$, so $4x^2-8x-y^2+8y-16=0$ corresponds to a hyperbola.

67. In the equation $x+4xy+2y^2-1=0$, we have $A=0,\ B=4,\ C=2$. Then $B^2-4AC=16>0$, so $-x^2-y^2+2x+4y=0$ corresponds to a hyperbola

69. In the equation $(x+y)^2 = 2xy+5$, or equivalently $x^2+y^2-5=0$, we have $A=1$, $B=0$, $C=1$. Then $B^2-4AC=-4<0$, and $A=C$, so $(x+y)^2=2xy+5$ corresponds to a circle.

71. The parametric equations $x=2t-5$, $y=-3t+2$ correspond to the line of equation $3x+2y=-11$.

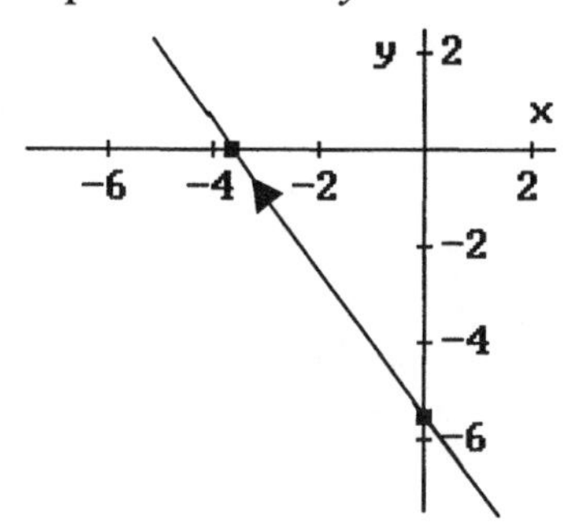

73. The parametric equations $x=t^3$, $y=2t^3+2$, correspond to the line of equation $y=2x+2$.

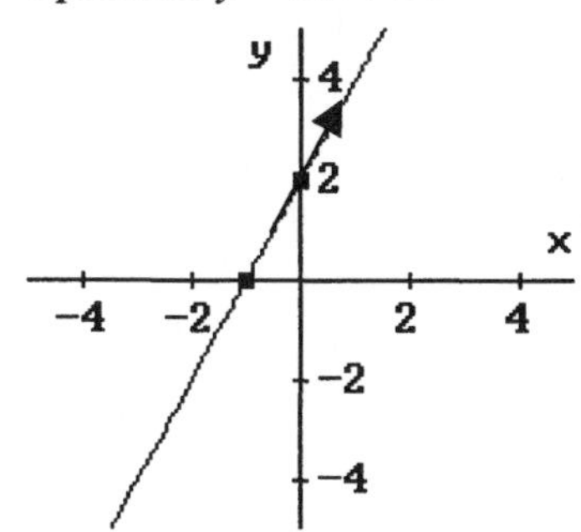

75. The parametric equations $x=2\cos t$, $y=2\sin t$ correspond to the circle of equation $x^2+y^2=4$.

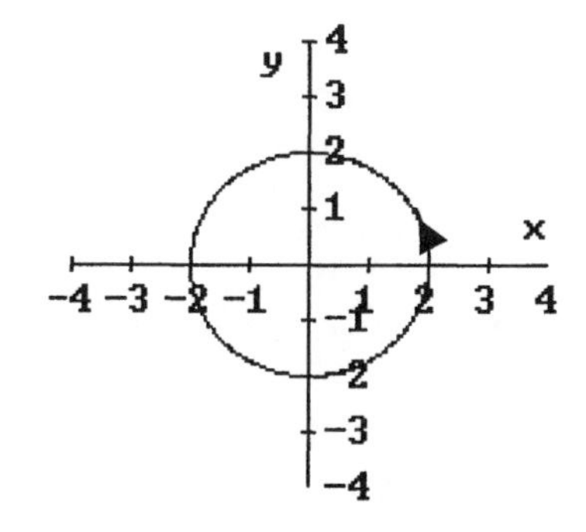

77. The equations $x = 2 - \sin t \cos(3t)$, $y = 3 + \cos t \sin(3t)$ describe a curve that intersects itself at the point (2, 3).

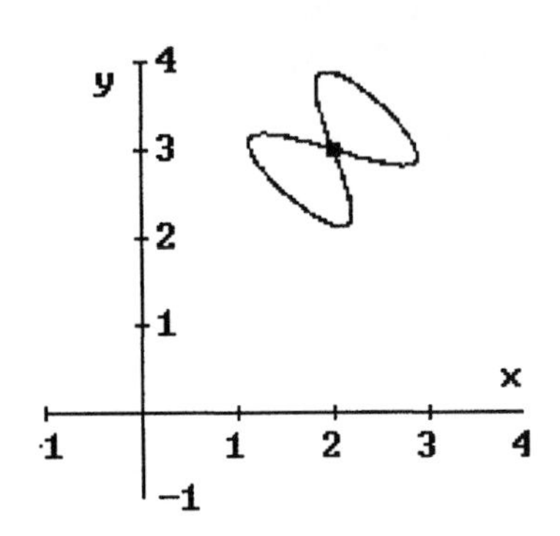

79. **(a)** $f(x) = \frac{x^2+2}{x^2+4}$ is an even function, since $f(-x) = f(x)$

(b) Given a portion of the graph of $y(x^2+4) = x^2+2$, or $y = \frac{x^2+2}{x^2+4}$:

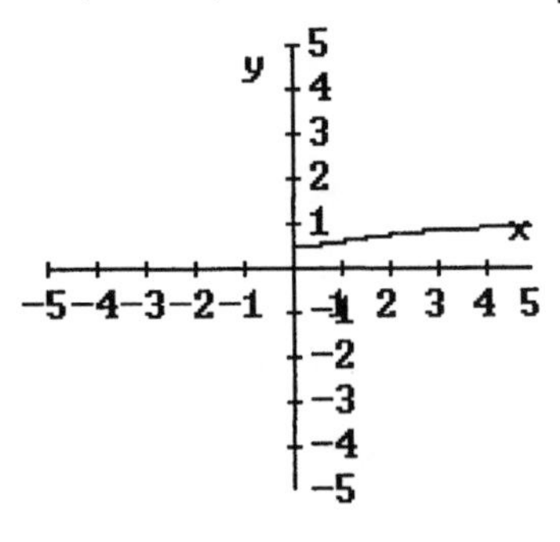

The complete graph of f is:

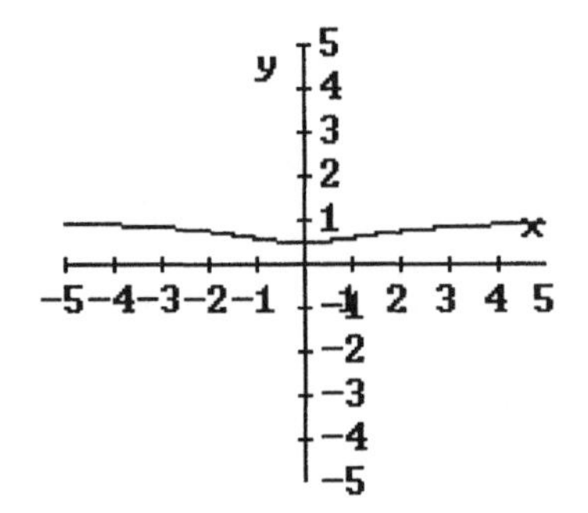

81. **(a)** The parabola of equation $y = -x^2 + 3x + 1$ has its vertex at $\left(\frac{3}{2}, \frac{13}{4}\right)$.

(b) The parabola of equation $x = -y^2 + 3y + 1$ has its vertex at $\left(\frac{13}{4}, \frac{3}{2}\right)$.

83. **(a)** $\sec^2 t - \tan^2 t = \frac{1}{\cos^2 t} - \frac{\sin^2 t}{\cos^2 t} = \frac{1-\sin^2 t}{\cos^2 t} = 1$;

(b) The equations $x = \sec t$, $y = \tan t$ describe a hyperbola ($x^2 - y^2 = 1$.)

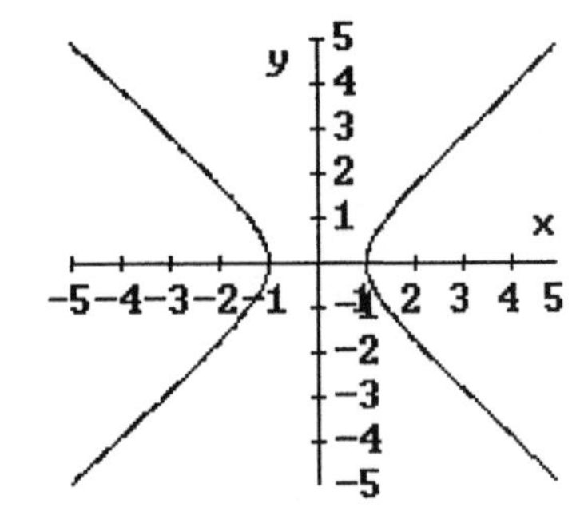

85. **(a)** The midpoint of the segment with endpoints $(-1,0)$ and (3, 4) is $(1,2)$; the slope of this segment is 1.

(b) If the vertices of an ellipse are $(-1,0)$ and (3, 4), and an endpoint of the minor axis is $(2,1)$, then the center of the ellipse is the midpoint between $(-1,0)$ and (3, 4), namely $(1,2)$. The minor axis is on the line connecting (1, 2) and (2, 1), with slope -1. The other endpoint on the minor axis is $(0,3)$

87. The graph of the equation relating pressure P and altitude h above sea level, roughly the part of a line corresponding to $h \geq 0$:

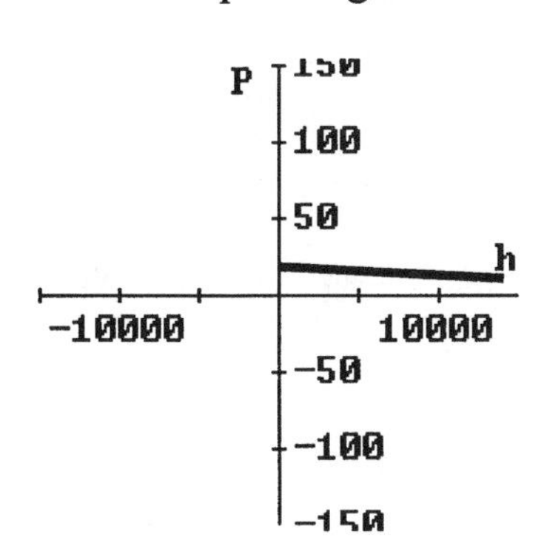

The graph of the equation relating pressure and depth below sea level d will look like (same P intercept as the previous graph.)

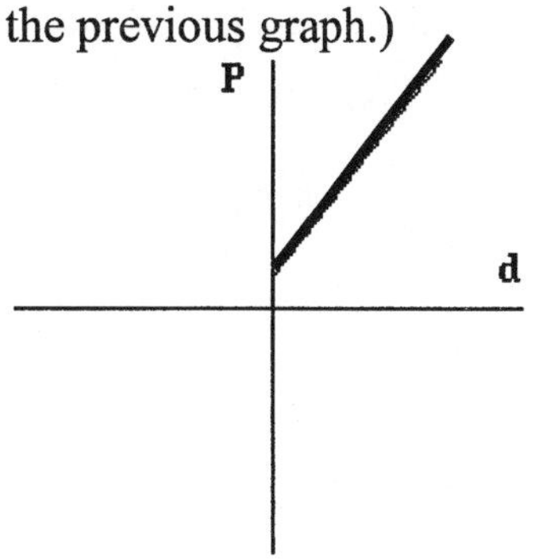

89. If the location of one pedal of an aerobic stair-stepper is given by the parametric equations $x = \frac{3}{2}(1 - \cos \pi t)$ and $y = 6(1 - \sin \pi t)$, where t is in seconds and x and y are in inches, then a full cycle takes 2 seconds.

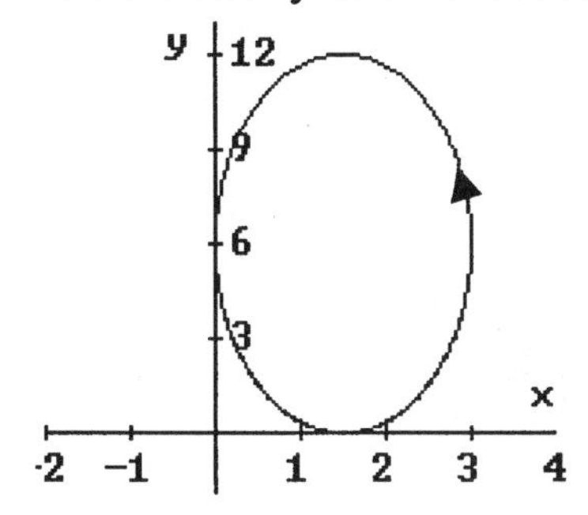

Chapter 8. Test

1. The statement: "If a graph is reflected first about the x-axis and then about the y-axis, the result is the same as if it were reflected first about the y-axis and then about the x-axis" is true.

3. The statement: "The distance from the focus to the vertex of a parabola is equal to the distance from the vertex to the directrix" is true.

5. The statement: "It is possible to find an angle θ such that in the uv-coordinate system formed by a rotation by the angle θ, the equation $y = x^2 + 3x + 4$ becomes $2u^2 + 3v^2 = 9$" is false. The first equation corresponds to a parabola, and the second to an ellipse.

7. The statement: "All conic sections with vertical or horizontal axes have equations of the form $Ax^2 + Cy^2 + Dx + Ey + F = 0$" is true..

9. Answers may vary. An example of a relation that is symmetric with respect to the x-axis is $x + y^2 = 0$.

11. Answers may vary. An example of a parabola opening to the left is $x = -y^2$.

13. Answers may vary. An example of a reflective property of a conic section is in the design of a "whispering room."

15. The relation $x^3 - y^2 = 0$ is symmetric with respect to the x-axis since $x^3 - (-y)^2 = 0$.

Based on the portion of the graph given: the complete graph is:

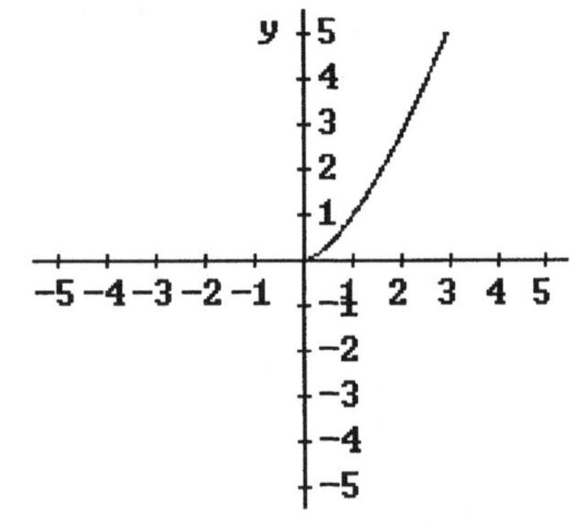

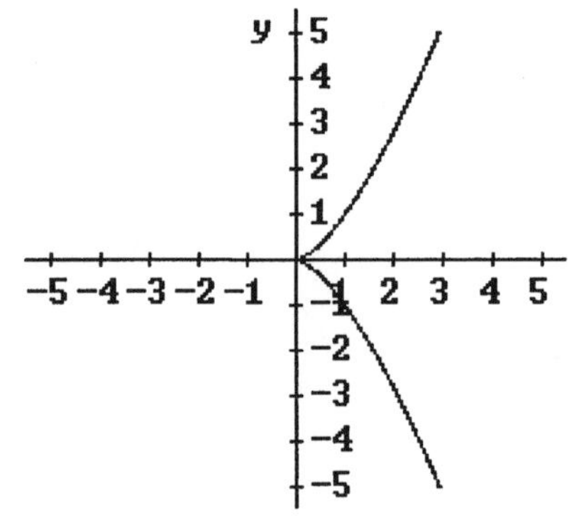

17. $4x^2 - 9y^2 = 36$, or $\frac{x^2}{9} - \frac{y^2}{4} = 1$ is hyperbola with center at $(0,0)$. Its vertices are $(\pm 3, 0)$ and its foci are $\left(-3-\sqrt{13},0\right)$, $\left(3+\sqrt{13},0\right)$. Its asymptotes are $y = \pm\frac{2}{3}x$.

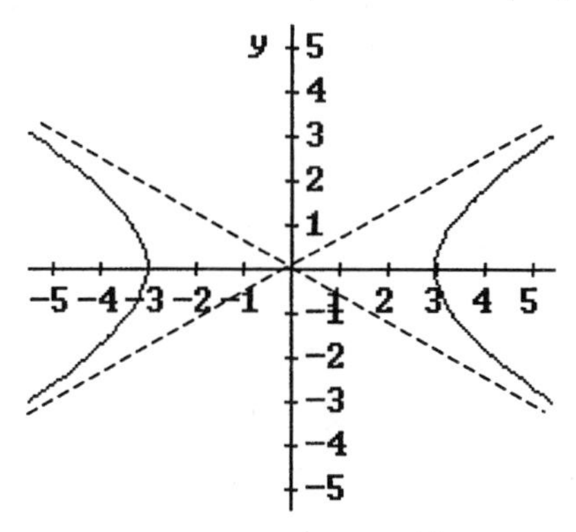

19. The graph of $3x^2 + 2\sqrt{3}xy + y^2 + 8x - 8\sqrt{3}y = 0$ is:

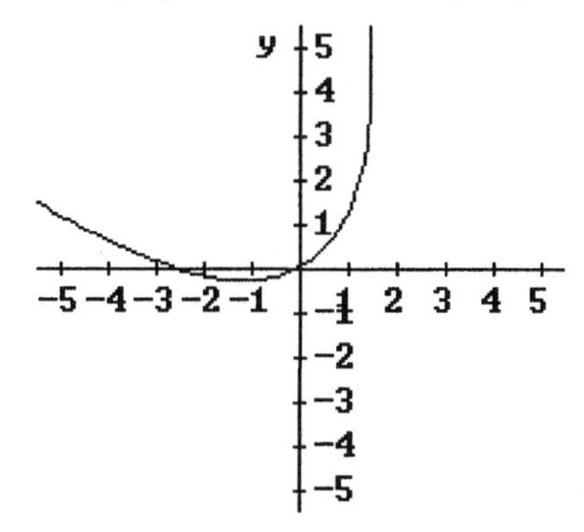

21. If a comet travels in an elliptical orbit, with the sun at one focus, and its major axis is 12 A.U. long, while the minor axis is 10 A.U. long, the equation of the orbit is $\frac{x^2}{36} + \frac{y^2}{25} = 1$. The sun is $\sqrt{11} \approx 3.32$ A.U. from the center.

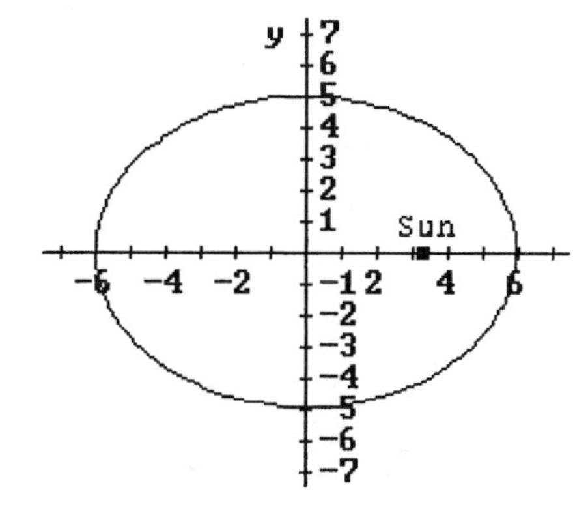

Chapter 9. SYSTEMS OF EQUATIONS AND INEQUALITIES

Section 9.1 Systems of Equations

1. The solution of the system $\begin{cases} 2x+y=10 \\ 3x-y=5 \end{cases}$ is $x=3,\ y=4$.

3. The solution of the system $\begin{cases} -3x+y=7 \\ 3x+4y=-2 \end{cases}$ is $x=-2,\ y=1$.

5. The system $\begin{cases} 2q+3r=3 \\ 4q+6r=0 \end{cases}$ has no solution.

7. The solution of the system $\begin{cases} 3m-n=2 \\ -4m+2n=2 \end{cases}$ is $m=3,\ n=7$.

9. The solution of the system $\begin{cases} x-5y=5 \\ 4x+5y=5 \end{cases}$ is $x=2,\ y=-\frac{3}{5}$.

11. The system $\begin{cases} 2z-7w=-2 \\ 10z-35w=-8 \end{cases}$ has no solution.

13. The solution of the system $\begin{cases} \frac{10}{3}a+2b=-6 \\ a+\frac{1}{5}b=1 \end{cases}$ is $a=\frac{12}{5},\ b=-7$.

15. The solution of the system $\begin{cases} 2x+3y=9.8 \\ 3x-5y=-8.1 \end{cases}$ is $x=1.3,\ y=2.4$.

17. The solution of the system $\begin{cases} 0.5x-0.2y=1.36 \\ 1.5x+0.4y=5.78 \end{cases}$ is $x=3.4,\ y=1.7$.

19. The solution of the system $\begin{cases} \left(2\times10^{-6}\right)x+10^{-6}y=33 \\ 4x-y=3\times10^{7} \end{cases}$ is $x=1.05\times10^{7},\ y=1.2\times10^{7}$.

21. The solutions of the system $\begin{cases} x+y=8 \\ xy=15 \end{cases}$ are $x=5,\ y=3$, and $x=3,\ y=5$.

23. The solutions of the system $\begin{cases} 2p^2+3q^2=30 \\ p+q=1 \end{cases}$ are $p=-\frac{9}{5},\ q=\frac{14}{5}$, and $p=3,\ q=-2$.

25. The solution of the system $\begin{cases} \frac{2}{a}+\frac{3}{b}=0 \\ \frac{4}{a}+\frac{12}{b}=-1 \end{cases}$ is $a=4,\ b=-6$.

27. The solutions of the system $\begin{cases} x^2+y^2=9 \\ 5x-y=-3 \end{cases}$ are $x=0$, $y=3$, and $x=-\frac{15}{13}$, $y=-\frac{36}{13}$

29. The solutions of the system $\begin{cases} x^4+y^4=97 \\ x^4-y^4=65 \end{cases}$ are $(x=-3,\ y=-2)$, $(x=-3,\ y=2)$, $(x=3,\ y=-2)$, and $(x=3,\ y=2)$.

31. The solution of the system $\begin{cases} 3x+\ y+2z=11 \\ \quad x+3y=\ 6 \\ \quad 3y+2z=\ 0 \end{cases}$ is $x=\frac{9}{2}$, $y=\frac{1}{2}$, $z=-\frac{3}{2}$.

33. The solution of the system $\begin{cases} x+\ y+\ z=\ 2 \\ \quad 2y-3z=-7 \\ \quad 3y+2z=\ 9 \end{cases}$ is $x=-2$, $y=1$, $z=3$.

35. The solution of the system $\begin{cases} u+\ v+w=1 \\ 2u+2v+w=2 \\ 3u+\ v-w=5 \end{cases}$ is $u=2$, $v=-1$, $w=0$.

37. The point of intersection of the pair of equations $2x-y=1$, $3x+y=9$ is $(2,3)$

39. The points of intersection of the pair of equations $y+x^2=4$, $y-x=2$ are $(-2,0)$ and $(1,3)$.

41. The points of intersection of the pair of equations $x^2+y^2=2$, $x=y^2$ are $(1,1)$ and $(1,-1)$.

43. The points of intersection of the pair of equations $y=x^3-x+2$, $y=x^2+x+1$ are, approximately, $(-1.25,1.31)$, $(0.45,1.64)$, and $(1.80,6.05)$.

45. There is no point of intersection of the pair of equations $y=15-x^2$, $y=3\sqrt{4-x^2}+4$.

47. There are infinitely many points of intersection of the pair of equations $y=3|x-2|$, $y=6-3|x-4|$, namely all the points $(x,\ 3x-6)$ where $2\le x\le 4$.

49. If a collection of dimes and nickels amounts to 80¢ and there are four more nickels than dimes, then $\begin{cases} 0.05x+0.10y=0.80 \\ x=y+4 \end{cases}$. The solution is $x=8$ nickels and $y=4$ dimes.

51. If x is the amount of 98% fat-free frozen yogurt, and y the amount of 80% fat-free dessert topping, used to prepare a 10-ounce 90% fat-free sundae, then $\begin{cases} 0.02x + 0.20y = 1 \\ x + y = 10 \end{cases}$. The solution of this system is $x = 5.56$ ounces, $y = 4.44$ ounces.

53. If Mark is now 11 times older than Ben, and in 27 years Ben will be ½ the age of Mark's, then $\begin{cases} x = 11y \\ x + 27 = 2(y + 27) \end{cases}$, where x is Mark's age and y is Ben's. The solution of this system is $x = 33$, $y = 3$. That is, Mark is 33 years old and Ben is 3 years old.

55. If 400 tickets of two kinds ($200 and $500) have been purchased for a total of $95,000, then $\begin{cases} x + y = 400 \\ 200x + 500y = 95{,}000 \end{cases}$, where x is the number of $200 tickets, and y is the number of $500 tickets. The solution of this system is $x = 350$, $y = 50$. That is, 350 tickets at $200 each and 50 tickets at $500 each were purchased.

57. If it takes Hans 10 fewer hours than it does Franz to lift 209 tons, and if it takes them 15 hours to lift 209 tons doing it together, then $\frac{1}{x} + \frac{1}{x-10} = \frac{1}{15}$. The solution is $x \approx 35.81$ hours for Franz, and $x - 10 \approx 25.81$ hours for Hans.

59. If two brothers are competing in a 50-meter race, and the first time the older brother wins by 10 meters, thereby agreeing to run another race with the older brother starting 10 meters behind, and at the end of this second race the older brother still wins by 0.5 seconds, then $\begin{cases} \frac{40}{y} = \frac{50}{x} \\ \frac{60}{x} = \frac{50}{y} - 0.5 \end{cases}$, where x is the speed of the older brother and y is the speed of the younger brother. The solution is $x = 5$ meters per second for the older brother, and $y = 4$ meters per second for the younger brother.

61. If two radar stations are located 50 miles apart along a coastline that runs north south at the points A and B, and a ship is at the point C located 20 miles from A and 40 miles from B, then the location of C is the intersection of the circles $x^2 + y^2 = 400$ and $x^2 + (y - 50)^2 = 1600$. Here we are considering the origin to be at the point A. The location of C is $x = 15.2$, $y = 13$, that is 15.2 miles east and 13 miles north of A.

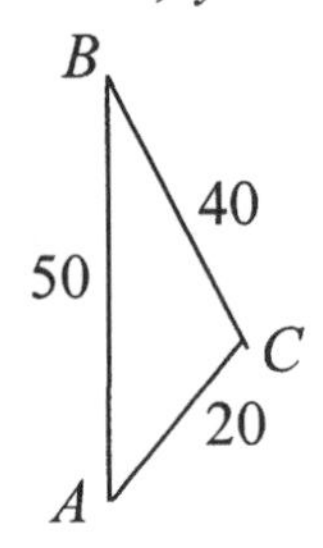

63. If x and y are related by the formula $y = a|x| + b|x-1| + c|x-2|$, and it is known that when $x = 0,\ y = 5,$ when $x = 1,\ y = 6,$ and when $x = 2,\ y = 1,$ then $\begin{cases} b + 2c = 5 \\ a + c = 6 \\ 2a + b = 1 \end{cases}$. The solution of this system is $a = 2,\ b = -3,\ c = 4.$ That is, $y = 2|x| - 3|x-1| + 4|x-2|$.

Concepts and Critical Thinking

65. The statement: "Replacing the first of two equations in a system by the sum of the two equations results in an equivalent system" is true.

67. The statement: "If even a single equation in a system is linear, then the entire system is considered to be linear" is false.

69. Answers may vary. An example of a nonlinear system of equations is $\begin{cases} xy = 4 \\ x + y = 3 \end{cases}$.

71. Answers may vary. $\begin{cases} 3x - y = 3 \\ x + y = 5 \end{cases}$ is a linear system with the only solution $x = 2,\ y = 3.$

73. If there are four more nickels than twice the number of dimes, and the value of the nickels and dimes together is \$1.00, and we let N be the number of nickels and D the number of dimes, then $\begin{cases} N = 2D + 4 \\ 5N + 10D = 100 \end{cases}$. The solution is $N = 12$ nickels and $D = 4.$ The meaning of $N =$"nickels" ought to be "number of nickels."

Chapter 9. Section 9.2 Matrices

1. The matrix $\begin{bmatrix} 1 & 2 & 3 \\ 4 & 5 & 6 \end{bmatrix}$ is of order 2×3.

3. The matrix $\begin{bmatrix} a \\ b \\ c \\ d \end{bmatrix}$ is of order 4×1.

5. $2\begin{bmatrix} 1 & 0 \\ 3 & 5 \end{bmatrix} = \begin{bmatrix} 2 & 0 \\ 6 & 10 \end{bmatrix}$

7. $\begin{bmatrix} 2 & 3 \\ 6 & 4 \end{bmatrix} + \begin{bmatrix} -4 & 8 \\ 0 & 2 \end{bmatrix} = \begin{bmatrix} -2 & 11 \\ 6 & 6 \end{bmatrix}$

9. $\begin{bmatrix} 1 & -1 \\ 0 & 3 \end{bmatrix} + \begin{bmatrix} 2 \\ 5 \end{bmatrix}$ is not defined. The matrices are not of the same order.

11. $\begin{bmatrix} -3 \\ 1 \\ 6 \end{bmatrix} + 2\begin{bmatrix} 4 \\ -1 \\ 0 \end{bmatrix} = \begin{bmatrix} 5 \\ -1 \\ 6 \end{bmatrix}$

13. $\begin{bmatrix} a & b \\ c & d \end{bmatrix} + \begin{bmatrix} 3a & 2b \\ c & 5d \end{bmatrix} = \begin{bmatrix} 4a & 3b \\ 2c & 6d \end{bmatrix}$

15. $2\begin{bmatrix} 1 & 2 & 0 \\ 2 & 6 & 9 \\ 3 & -1 & -4 \end{bmatrix} + \begin{bmatrix} -1 & 3 & 4 \\ 3 & -3 & 0 \\ 1 & 4 & 5 \end{bmatrix} = \begin{bmatrix} 1 & 7 & 4 \\ 7 & 9 & 18 \\ 7 & 2 & -3 \end{bmatrix}$

17. $\begin{bmatrix} 2 & a & 1 \\ 0 & 1 & 2 \end{bmatrix} - a\begin{bmatrix} 1 & 1 & 1 \\ 1 & 1 & 0 \end{bmatrix} = \begin{bmatrix} 2-a & 0 & 1-a \\ -a & 1-a & 2 \end{bmatrix}$

Exercises 19-21 refer to the matrices $A = \begin{bmatrix} 1 & 2 \\ 3 & 5 \end{bmatrix}$ and $B = \begin{bmatrix} 2 & 1 \\ 4 & 9 \end{bmatrix}$.

19. $A + B = \begin{bmatrix} 3 & 3 \\ 7 & 14 \end{bmatrix}$

21. $2A - 5B = \begin{bmatrix} -8 & -1 \\ -14 & -35 \end{bmatrix}$

23. If $2X = \begin{bmatrix} 4 & 6 \\ 8 & 2 \end{bmatrix}$, then $X = \begin{bmatrix} 2 & 3 \\ 4 & 1 \end{bmatrix}$

25. If $2X + \begin{bmatrix} 1 & 2 & 3 \\ 6 & 5 & 4 \end{bmatrix} = \begin{bmatrix} 0 & 0 & 0 \\ 0 & 0 & 0 \end{bmatrix}$, then $2X = \begin{bmatrix} -1 & -2 & -3 \\ -6 & -5 & -4 \end{bmatrix}$, and $X = \begin{bmatrix} -\frac{1}{2} & -1 & -\frac{3}{2} \\ -3 & -\frac{5}{2} & -2 \end{bmatrix}$

Exercises 27-31 refer to the matrices $A=\begin{bmatrix}1&2&4\\2&3&0\end{bmatrix}$, $B=\begin{bmatrix}1&2\\3&4\end{bmatrix}$, and $C=\begin{bmatrix}3&7&9\\4&5&1\\2&6&4\end{bmatrix}$.

27. The product AB is not defined.

29. The product BA is defined, of order 2×3.

31. The product CA is not defined.

33. $\begin{bmatrix}3&5\\4&2\end{bmatrix}\begin{bmatrix}3\\-3\end{bmatrix}=\begin{bmatrix}-6\\6\end{bmatrix}$

35. $\begin{bmatrix}1&0\\0&1\end{bmatrix}\begin{bmatrix}1\\2\\3\end{bmatrix}$ is not defined.

37. $\begin{bmatrix}1&2\\0&1\end{bmatrix}\begin{bmatrix}2&3\\4&6\end{bmatrix}=\begin{bmatrix}10&15\\4&6\end{bmatrix}$

38. $\begin{bmatrix}1&-2\\-5&7\end{bmatrix}\begin{bmatrix}a&b\\c&d\end{bmatrix}=\begin{bmatrix}a-2c&b-2d\\-5a+7c&-5b+7d\end{bmatrix}$

39. $\begin{bmatrix}2&1\\3&2\end{bmatrix}\begin{bmatrix}2&-1\\-3&2\end{bmatrix}=\begin{bmatrix}1&0\\0&1\end{bmatrix}$

41. $\begin{bmatrix}2&0\\1&4\\2&1\end{bmatrix}\begin{bmatrix}3&5\\1&7\end{bmatrix}=\begin{bmatrix}6&10\\7&33\\7&17\end{bmatrix}$

43. $\begin{bmatrix}1.8&3.5&4.6\\4.8&1.7&3.2\\1.7&2.5&0.7\end{bmatrix}\begin{bmatrix}2.6\\3.4\\4.9\end{bmatrix}=\begin{bmatrix}39.12\\33.94\\16.35\end{bmatrix}$

45. $\begin{bmatrix}1&0\\x&1\end{bmatrix}\begin{bmatrix}a&b\\c&d\end{bmatrix}=\begin{bmatrix}a&b\\ax+c&bx+d\end{bmatrix}$

47. $\begin{bmatrix}a\\b\end{bmatrix}\begin{bmatrix}c\\d\end{bmatrix}$ is not defined.

49. $\begin{bmatrix}0&\frac{1}{b}\\\frac{1}{a}&0\end{bmatrix}\begin{bmatrix}0&a\\b&0\end{bmatrix}=\begin{bmatrix}1&0\\0&1\end{bmatrix}$

51. $\begin{bmatrix}1&0&0\\0&1&0\\0&0&1\end{bmatrix}\begin{bmatrix}p&q\\r&s\\t&u\end{bmatrix}=\begin{bmatrix}p&q\\r&s\\t&u\end{bmatrix}$

53. $\begin{bmatrix}0&0&1\\0&1&0\\1&0&0\end{bmatrix}\begin{bmatrix}a_{11}&a_{12}&a_{13}\\a_{21}&a_{22}&a_{23}\\a_{31}&a_{32}&a_{33}\end{bmatrix}=\begin{bmatrix}a_{31}&a_{32}&a_{33}\\a_{21}&a_{22}&a_{23}\\a_{11}&a_{12}&a_{13}\end{bmatrix}$

55. $\begin{bmatrix}2&-3\\-1&1\end{bmatrix}\begin{bmatrix}x\\y\end{bmatrix}=\begin{bmatrix}5\\6\end{bmatrix}$ corresponds to the system of equations $\begin{cases}2x-3y=5\\-x+y=6\end{cases}$

57. $\begin{bmatrix}1&1&0\\-1&0&2\\1&1&1\end{bmatrix}\begin{bmatrix}r\\s\\t\end{bmatrix}=\begin{bmatrix}1\\3\\4\end{bmatrix}$ corresponds to the system of equations $\begin{cases}r+s=1\\2t-r=3\\r+s+t=4\end{cases}$

59. $\begin{bmatrix}1&1&1&1\\0&1&1&1\\0&0&1&1\end{bmatrix}\begin{bmatrix}x_1\\x_2\\x_3\\x_4\end{bmatrix}=\begin{bmatrix}4\\3\\2\end{bmatrix}$ corresponds to the system of equations $\begin{cases}x_1+x_2+x_3+x_4=4\\x_2+x_3+x_4=3\\x_3+x_4=2\end{cases}$

61. $\begin{bmatrix}1&2&3\\4&5&6\end{bmatrix}^T=\begin{bmatrix}1&4\\2&5\\3&6\end{bmatrix}$

63. $\begin{bmatrix}1&2&3\\2&5&6\\3&6&4\end{bmatrix}^T=\begin{bmatrix}1&2&3\\2&5&6\\3&6&4\end{bmatrix}$

65. $\begin{bmatrix}1&2\\0&1\end{bmatrix}^2=\begin{bmatrix}1&2\\0&1\end{bmatrix}$

67. $\begin{bmatrix}1&0\\0&0\end{bmatrix}^4=\begin{bmatrix}1&0\\0&0\end{bmatrix}$

Exercises 69-71 refer to the matrix $A=\begin{bmatrix}1&2&3\\0&1&4\\0&0&1\end{bmatrix}$.

69. $A^2=\begin{bmatrix}1&4&14\\0&1&8\\0&0&1\end{bmatrix}$

71. $A^{32}=\begin{bmatrix}1&64&4064\\0&1&128\\0&0&1\end{bmatrix}$

Exercises 73-75 refer to the 2×2 *matrices* $A=\begin{bmatrix} a_1 & a_2 \\ a_3 & a_4 \end{bmatrix}$, $B=\begin{bmatrix} b_1 & b_2 \\ b_3 & b_4 \end{bmatrix}$, and $C=\begin{bmatrix} c_1 & c_2 \\ c_3 & c_4 \end{bmatrix}$.

73. To verify that $A+B=B+A$:

$$A+B=\begin{bmatrix} a_1 & a_2 \\ a_3 & a_4 \end{bmatrix}+\begin{bmatrix} b_1 & b_2 \\ b_3 & b_4 \end{bmatrix}=\begin{bmatrix} a_1+b_1 & a_2+b_2 \\ a_3+b_3 & a_4+b_4 \end{bmatrix}=\begin{bmatrix} b_1+a_1 & b_2+a_2 \\ b_3+a_3 & b_4+a_4 \end{bmatrix}=B+A$$

75. To verify that $(A+B)C=AC+BC$: $(A+B)C=\begin{bmatrix} a_1+b_1 & a_2+b_2 \\ a_3+b_3 & a_4+b_4 \end{bmatrix}\begin{bmatrix} c_1 & c_2 \\ c_3 & c_4 \end{bmatrix}$

$$=\begin{bmatrix} (a_1+b_1)c_1+(a_2+b_2)c_3 & (a_1+b_1)c_2+(a_2+b_2)c_4 \\ (a_3+b_3)c_1+(a_4+b_4)c_3 & (a_3+b_3)c_2+(a_4+b_4)c_4 \end{bmatrix}$$

$$=\begin{bmatrix} (a_1c_1+a_2c_3)+(b_1c_1+b_2c_3) & (a_1c_2+a_2c_4)+(b_1c_2+b_2c_4) \\ (a_3c_1+a_4c_3)+(b_3c_1+b_4c_3) & (a_3c_2+a_4c_4)+(b_3c_2+b_4c_4) \end{bmatrix}$$

$$=\begin{bmatrix} a_1c_1+a_2c_3 & a_1c_2+a_2c_4 \\ a_3c_1+a_4c_3 & a_3c_2+a_4c_4 \end{bmatrix}+\begin{bmatrix} b_1c_1+b_2c_3 & b_1c_2+b_2c_4 \\ b_3c_1+b_4c_3 & b_3c_2+b_4c_4 \end{bmatrix}$$

$$=AC+BC$$

Applications

77. Consider the information on claims for 2003 for hourly and salary workers of a computer software company and their dependents, broken down by plan:

	Hourly		Salaried	
	Employees	Dependents	Employees	Dependents
Major Medical	\$230,000	\$320,000	\$125,000	\$250,000
Comprehensive	\$280,000	\$300,000	\$400,000	\$500,000

The claims filed by hourly workers and the claims filed by salaried workers as separate matrices are $D=\begin{bmatrix} 230{,}000 & 320{,}000 \\ 280{,}000 & 300{,}000 \end{bmatrix}$ and $S=\begin{bmatrix} 125{,}000 & 250{,}000 \\ 400{,}000 & 500{,}000 \end{bmatrix}$. The sum of these matrices, $D+S=\begin{bmatrix} 355{,}000 & 570{,}000 \\ 680{,}000 & 800{,}000 \end{bmatrix}$ gives the total claims (by all employees broken down by plan.) The total direct claims for employees for 2003 were \$355,000 in the Major Medical plan and \$680,000 in the Comprehensive plan. The total claims for dependents of employees were \$570,000 in the Major Medical plan and \$800,000 in the Comprehensive plan.

79. Consider the information about a game between the Dallas Mavericks and the Orlando Magic: the Mavericks made 5 three-point shots, 40 two-point field goals, and 18 free throws (one point each), while the Magic made 4 three-point shots, 37 two-point field goals, and 26 free throws. As a matrix, this information is given by $S=\begin{bmatrix}5 & 40 & 18\\ 4 & 37 & 26\end{bmatrix}$. The matrix that gives the points according to the kind of scoring play is the 3×1 matrix $T=\begin{bmatrix}3\\ 2\\ 1\end{bmatrix}$. The final score matrix is the product $ST=\begin{bmatrix}113\\ 112\end{bmatrix}$ (Mavericks 113 vs. Magic 112.)

Concepts and Critical Thinking

81. The statement: "The sum $A+B$ is defined for any pair of matrices A and B" is false. For the sum of two matrices to be defined, the matrices have to be of the same order.

83. The statement: "If A and B are of the same order, then $A+B$ is defined" is true.

85. The statement: "A 5×3 matrix has 5 rows and 3 columns" is true.

87. The statement: "Matrix multiplication is commutative; that is, for any pair of matrices A and B for which AB is defined, $AB=BA$" is false.

89. The statement: "Addition of matrices is accomplished by adding corresponding entries" is true.

91. Answers may vary. Two matrices A and B such that AB is defined but BA is not defined are $A=\begin{bmatrix}0 & 4 & 5\\ 1 & 2 & 3\end{bmatrix}$, $B=\begin{bmatrix}3\\ 0\\ 2\end{bmatrix}$. In fact, $AB=\begin{bmatrix}10\\ 7\end{bmatrix}$, BA is not defined.

93. Answers may vary. Two matrices A and B such that $A+B$ is defined, but neither AB nor BA is defined are $A=\begin{bmatrix}1\\ 7\end{bmatrix}$, $B=\begin{bmatrix}0\\ 1\end{bmatrix}$. In fact, $A+B=\begin{bmatrix}1\\ 8\end{bmatrix}$, AB and BA are not defined.

95. Answers may vary. An example of a 3×2 matrix such that all of its columns are identical but no two of its rows are the same is $A=\begin{bmatrix}1 & 1\\ 2 & 2\\ 4 & 4\end{bmatrix}$.

Chapter 9. Section 9.3 Linear Systems and Matrices

1. The system $\begin{cases} 2x-3y-2z=\frac{1}{10} \\ x+y+z=2 \end{cases}$ is linear.

3. The system $\begin{cases} \sqrt{x}-3y=7 \\ 2x+5y=9 \end{cases}$ is not linear.

5. The system $\begin{cases} x-2y=3 \\ x+3yz=1 \end{cases}$ is not linear.

7. The augmented matrix of the system $\begin{cases} 3x+4y=10 \\ 2x-7y=4 \end{cases}$ is $\left[\begin{array}{cc|c} 3 & 4 & 10 \\ 2 & -7 & 4 \end{array}\right]$.

9. The augmented matrix of the system $\begin{cases} x=3 \\ y=4 \end{cases}$ is $\left[\begin{array}{cc|c} 1 & 0 & 3 \\ 0 & 1 & 4 \end{array}\right]$.

11. The augmented matrix of the system $\begin{cases} x-y+2z=1 \\ 2y+6z=3 \end{cases}$ is $\left[\begin{array}{ccc|c} 1 & -1 & 2 & 1 \\ 0 & 2 & 6 & 3 \end{array}\right]$.

13. The system $\begin{cases} 2x-3y=4 \\ 4x-6y=7 \end{cases}$ has no solution.

The lines $2x-3y=4$ and $4x-6y=7$ are parallel.

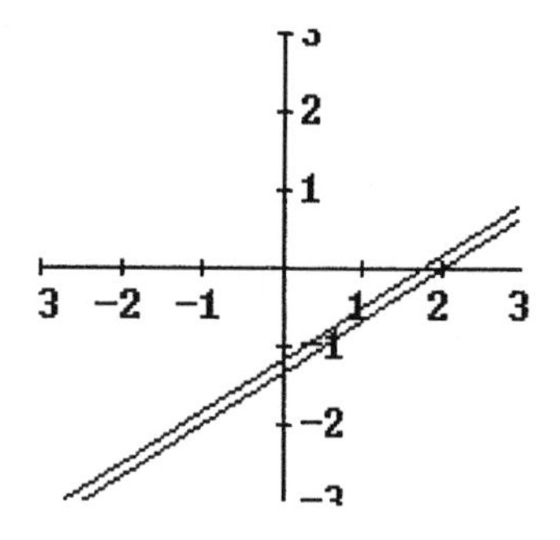

15. The system $\begin{cases} x-y=7 \\ 2x-2y=14 \end{cases}$ has infinitely many solutions. Both equations correspond to the same line. Every point on the line represents a solution of the system. For example, (7, 0), (14, 7) are solutions.

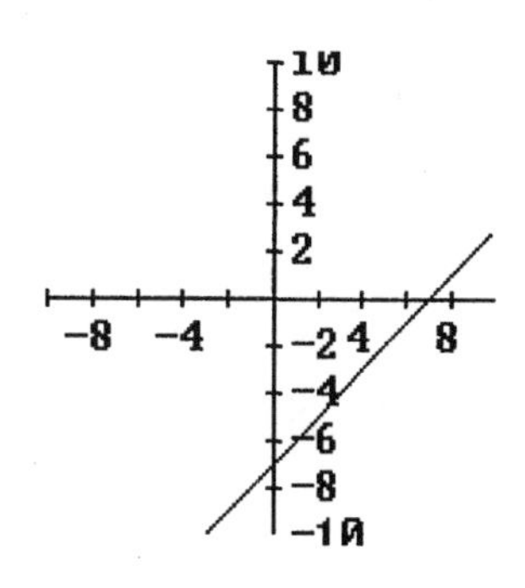

17. The system $\begin{cases} 5x-6y=1 \\ x-y=3 \end{cases}$ has only one solution.

The lines $5x-6y=1$ and $x-y=3$ intersect at only one point, namely $(17,\ 14)$.

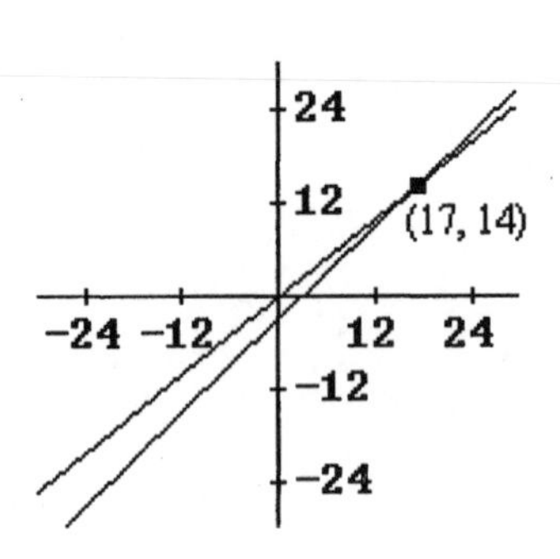

19. The solution of the system
$\begin{cases} 2x+3y=7 \\ 2y=2 \end{cases}$ is $x=2,\ y=1$.

21. The solution of the system $\begin{cases} 2x+y-z=12 \\ y+z=8 \\ 3z=3 \end{cases}$ is $x=3,\ y=7,\ z=1$.

23. The solution of the system $\begin{cases} q-2r+3s+t=3 \\ 4r+s-t=18 \\ 2s+2t=4 \\ 3t=-6 \end{cases}$ is $q=-1,\ r=3,\ s=4,\ t=-2$.

25. If the augmented matrix of a system is $\left[\begin{array}{cc|c} 2 & -3 & -5 \\ 0 & 4 & 12 \end{array}\right]$, where the variables in order are x and y, then the solution of the system is $x=2,\ y=3$.

27. If the augmented matrix of a system is $\left[\begin{array}{ccc|c} 1 & 3 & 6 & 0 \\ 0 & 1 & 2 & 2 \\ 0 & 0 & 1 & -4 \end{array}\right]$, where the variables in order are x, y, and z, then the solution of the system is $x=-6,\ y=10,\ z=-4$.

29. If the augmented matrix of a system is $\left[\begin{array}{ccc|c} 3 & -2 & 4 & 3 \\ 0 & 1 & -3 & 1 \\ 0 & 0 & 2 & -2 \end{array}\right]$, where the variables in order are x, y, and z, then the solution of the system is $x=1,\ y=-2,\ z=-1$.

Applications

31. The income of a family from three different sources, salary, municipal bonds and U.S. treasury bonds, totals $58,000. The municipal bonds and U.S. treasury bonds are free from state income tax, while the salary is subject to a 6% state income tax. The U.S. treasury bonds are free from federal tax, but both the salary and the municipal bonds income are subject to 20% federal tax. The family's state total tax is $2,400, and their federal tax totals $10,000. If *m* is the income from municipal bonds, *t* the one from the U.S. treasury bonds, and *s* is the salary, we obtain the system:

$$t + m + s = 58{,}000$$
$$0.2m + 0.2s = 10{,}000$$
$$0.6s = 2400$$

By back substitution, we obtain: the salary is $s = \$40,000$, the municipal bonds income is $10,000, and the U.S. treasury bonds income is $8000.

33. The following table gives the number of units of vitamins A, B_6, and B_{12} provided by a single serving of each of three available breakfast items:

	Milk	Bananas	Toast
Vitamin A	500	225	0
Vitamin B_6	0.1	0.6	0.04
Vitamin B_{12}	0.9	0	0

To prepare a breakfast that contains 1200 units of vitamin A, 1.5 units of vitamin B_6, and 1.35 units of vitamin B_{12}, we ought to use *x* servings of milk, *y* servings of bananas, and *z* servings of toast, where $\begin{cases} 500x + 225y = 1200 \\ 0.1x + 0.6y + 0.04z = 1.5 \\ 0.9x = 1.35 \end{cases}$. The solution of this system of equations is $x = 1.5$ servings of milk, $y = 2$ servings of bananas, and $z = 3.75$ of toast.

Concepts and Critical Thinking

35. The statement: "A system of equations is said to be linear if at least one of the equations is linear" is false.

37. The statement: "The augmented matrix $\left[\begin{array}{ccc|c} 1 & 0 & 0 & 3 \\ 2 & 1 & 0 & 2 \\ 1 & 1 & 0 & 3 \end{array}\right]$ corresponds to the system

$\begin{cases} x = 3z \\ 2x + y = 2z \\ x + y = 3z \end{cases}$" is false. The augmented matrix of this system is $\left[\begin{array}{ccc|c} 1 & 0 & 0 & 3 \\ 2 & 1 & -2 & 0 \\ 1 & 1 & -3 & 0 \end{array}\right]$.

39. Answers may vary. An example of a nonlinear system of equations is $\begin{cases} xy = 5 \\ x^2 + y = 3 \end{cases}$

41. Answers may vary. An example of an equation that when paired with $2x+3y=5$ forms a system of equations with infinitely many solutions is $\frac{2}{3}x + y = \frac{5}{3}$.

43. Answers may vary. An example of an equation that when paired with $2x+3y=5$ forms a system of equations with exactly one solution is $2x-3y=5$.

Chapter 9. Section 9.4 Gaussian and Gauss-Jordan Elimination

1. The matrix $\begin{bmatrix} 1 & 3 \\ 0 & 1 \end{bmatrix}$ is in row-echelon form.

3. The matrix $\begin{bmatrix} 1 & 2 & 3 \\ 0 & 1 & 2 \\ 1 & 0 & 3 \end{bmatrix}$ is not in row-echelon form. The entries below the leading 1 on the first row are not all zero.

5. The matrix $\left[\begin{array}{cccc|c} 1 & 2 & 0 & -1 & 0 \\ 0 & 1 & -3 & 1 & 1 \\ 0 & 1 & 2 & 0 & 0 \end{array}\right]$ is not in row-echelon form. There is a leading 1 that is not to the right of the previous leading 1. Also, there is a non-zero entry below a leading 1.

7. $\begin{bmatrix} 2 & 4 \\ 3 & 0 \end{bmatrix} \xrightarrow{\frac{1}{2}R_1} \begin{bmatrix} 1 & 2 \\ 3 & 0 \end{bmatrix}$

9. $\left[\begin{array}{cc|c} 1 & 4 & 6 \\ 2 & -1 & 3 \end{array}\right] \xrightarrow{-2R_1 + R_2} \left[\begin{array}{cc|c} 1 & 4 & 6 \\ 0 & -9 & -9 \end{array}\right]$

11. $\left[\begin{array}{cc|c} 1 & 2 & 3 \\ 0 & 1 & 4 \end{array}\right]$ is a row-echelon form of the augmented matrix $\left[\begin{array}{cc|c} 4 & 8 & 12 \\ 3 & 9 & 21 \end{array}\right]$.

13. $\left[\begin{array}{ccc|c} 1 & 0 & 3 & 4 \\ 0 & 1 & -2 & 4 \end{array}\right]$ is a row-echelon form of the augmented matrix $\left[\begin{array}{ccc|c} 0 & 2 & -4 & 8 \\ 3 & 0 & 9 & 12 \end{array}\right]$.

15. $\left[\begin{array}{ccc|c} 1 & 3 & 4 & -1 \\ 0 & 1 & -2 & -\frac{7}{2} \\ 0 & 0 & 1 & 3 \end{array}\right]$ is a row-echelon form of the augmented matrix $\left[\begin{array}{ccc|c} 2 & 6 & 8 & -2 \\ 2 & 4 & 12 & 5 \\ 1 & 7 & -6 & -21 \end{array}\right]$.

17. $\left[\begin{array}{ccc|c} 1 & -3 & -2 & -\frac{3}{2} \\ 0 & 1 & \frac{6}{5} & \frac{8}{5} \\ 0 & 0 & 1 & \frac{7}{2} \end{array}\right]$ is a row-echelon form of the augmented matrix $\left[\begin{array}{ccc|c} 1 & -3 & -1 & 2 \\ 2 & 4 & 12 & 5 \\ 1 & 7 & -6 & -21 \end{array}\right]$.

19. $\left[\begin{array}{ccc|c} 1 & 3 & 0 & 4 \\ 0 & 1 & 2 & -2 \end{array}\right]$ is a row-echelon form of the augmented matrix $\left[\begin{array}{ccc|c} 1 & 3 & 0 & 4 \\ \frac{1}{3} & -1 & -4 & \frac{16}{3} \end{array}\right]$.

21. $\left[\begin{array}{cccc|c} 1 & 0 & 1 & 2 & 1 \\ 0 & 1 & \frac{1}{2} & \frac{3}{2} & 0 \\ 0 & 0 & 1 & 2 & 2 \\ 0 & 0 & 0 & 1 & 2 \end{array}\right]$ is a row-echelon form of the augmented matrix $\left[\begin{array}{cccc|c} 1 & 0 & 1 & 2 & 1 \\ -1 & 1 & -1 & -1 & -1 \\ 0 & 2 & 1 & 3 & 0 \\ 1 & 0 & 2 & 4 & 3 \end{array}\right]$.

23. The augmented matrix of the system $\begin{cases} 2x-6y=-2 \\ 3x+5y=11 \end{cases}$ is $\left[\begin{array}{cc|c} 2 & -6 & -2 \\ 3 & 5 & 11 \end{array}\right]$. A row-echelon form for this matrix is $\left[\begin{array}{cc|c} 1 & 11 & 13 \\ 0 & 1 & 1 \end{array}\right]$, which corresponds to the system $\begin{cases} x+11y=13 \\ y=1 \end{cases}$.
The solution of the system is: $x=2,\ y=1$.

25. The augmented matrix of the system $\begin{cases} 2p+q = \frac{5}{3} \\ 3p-6q = -\frac{5}{2} \end{cases}$ is $\left[\begin{array}{cc|c} 2 & 1 & \frac{5}{3} \\ 3 & -6 & -\frac{5}{2} \end{array}\right]$. A row-echelon form for this matrix is $\left[\begin{array}{cc|c} 1 & \frac{1}{2} & \frac{5}{6} \\ 0 & 1 & \frac{2}{3} \end{array}\right]$, which corresponds to the system $\begin{cases} p+\frac{1}{2}q=\frac{5}{6} \\ q=\frac{2}{3} \end{cases}$.
The solution of the system is: $p=\frac{1}{2},\ q=\frac{2}{3}$.

27. The augmented matrix of the system $\begin{cases} x+y+2z=7 \\ x+2y+3z=10 \\ x-4y+z=4 \end{cases}$ is $\left[\begin{array}{ccc|c} 1 & 1 & 2 & 7 \\ 1 & 2 & 3 & 10 \\ 1 & -4 & 1 & 4 \end{array}\right]$. A row-echelon form for this matrix is $\left[\begin{array}{ccc|c} 1 & 1 & 2 & 7 \\ 0 & 1 & \frac{1}{5} & \frac{3}{5} \\ 0 & 0 & 1 & 3 \end{array}\right]$, which corresponds to the system $\begin{cases} x+y+2z=7 \\ y+\frac{1}{5}z=\frac{3}{5} \\ z=3 \end{cases}$.
The solution of the system is: $x=1,\ y=0,\ z=3$.

29. The augmented matrix of the system $\begin{cases} 2p+r=2 \\ 2p+q+4r=9 \\ 3p+q+8r=17 \end{cases}$ is $\left[\begin{array}{ccc|c} 2 & 0 & 1 & 2 \\ 2 & 1 & 4 & 9 \\ 3 & 1 & 8 & 17 \end{array}\right]$. A row-echelon form for this matrix is $\left[\begin{array}{ccc|c} 1 & \frac{1}{3} & \frac{8}{3} & \frac{17}{3} \\ 0 & 1 & \frac{13}{2} & 14 \\ 0 & 0 & 1 & 2 \end{array}\right]$, which corresponds to the system $\begin{cases} p+\frac{1}{3}q+\frac{8}{3}r=\frac{17}{3} \\ q+\frac{13}{2}r=14 \\ r=2 \end{cases}$.
The solution of the system is: $p=0,\ q=1,\ r=2$.

31. The augmented matrix of the system $\begin{cases} w+x+y+2z=1 \\ -w+x-y-z=-2 \\ 2x+y+3z=-1 \\ w+x+2y+5z=1 \end{cases}$ is $\left[\begin{array}{cccc|c} 1 & 1 & 1 & 2 & 1 \\ -1 & 1 & -1 & -1 & -2 \\ 0 & 2 & 1 & 3 & -1 \\ 1 & 1 & 2 & 5 & 1 \end{array}\right]$.

A row-echelon form for this matrix is $\left[\begin{array}{cccc|c} 1 & 1 & 1 & 2 & 1 \\ 0 & 1 & 0 & \frac{1}{2} & -\frac{1}{2} \\ 0 & 0 & 1 & 2 & 0 \\ 0 & 0 & 0 & 1 & 0 \end{array}\right]$, which corresponds to the system

$\begin{cases} w+x+y+2z=1 \\ x+\frac{1}{2}z=-\frac{1}{2} \\ y+2z=0 \\ z=0 \end{cases}$. The solution of the system is: $w=\frac{3}{2}$, $x=-\frac{1}{2}$, $y=0$, $z=0$.

33. The reduced row-echelon form of the matrix $\left[\begin{array}{cc|c} 1 & 2 & 3 \\ 0 & 1 & 2 \end{array}\right]$ is $\left[\begin{array}{cc|c} 1 & 0 & -1 \\ 0 & 1 & 2 \end{array}\right]$.

35. Reduced row-echelon form of the matrix $\left[\begin{array}{cc|c} 2 & 4 & -8 \\ 3 & 3 & -3 \end{array}\right]$ is $\left[\begin{array}{cc|c} 1 & 0 & 2 \\ 0 & 1 & -3 \end{array}\right]$.

37. The reduced row-echelon form of the matrix $\left[\begin{array}{ccc|c} 1 & 2 & 2 & 2 \\ 0 & 1 & 3 & -2 \\ 0 & 0 & 1 & -1 \end{array}\right]$ is $\left[\begin{array}{ccc|c} 1 & 0 & 0 & 2 \\ 0 & 1 & 0 & 1 \\ 0 & 0 & 1 & -1 \end{array}\right]$.

39. The solution of the system $\begin{cases} 2x-3y=12 \\ x+2y=-1 \end{cases}$ is $x=3$, $y=-2$.

41. The solution of the system $\begin{cases} 4x+6y=0 \\ x-y=\frac{5}{6} \end{cases}$ is $x=\frac{1}{2}$, $y=-\frac{1}{3}$.

43. The solution of the system $\begin{cases} x-2y+z=9 \\ y-3z=-10 \\ z=3 \end{cases}$ is $x=4$, $y=-1$, $z=3$.

45. The solution of the system $\begin{cases} x+2y+4z=-10 \\ 2x+3y+6z=-15 \\ x-y+z=-7 \end{cases}$ is $x=0$, $y=3$, $z=-4$.

47. The augmented matrix of the system $\begin{cases} 2x+\ y=5 \\ 4x+2y=10 \end{cases}$ is $\left[\begin{array}{cc|c} 2 & 1 & 5 \\ 4 & 2 & 10 \end{array}\right]$. The reduced row-echelon form of this matrix is $\left[\begin{array}{cc|c} 1 & \frac{1}{2} & \frac{5}{2} \\ 0 & 0 & 0 \end{array}\right]$. The system has infinitely many solutions, namely any pair (x, y) where $y=-2x+5$.

49. The augmented matrix of the system $\begin{cases} -x+3y=-7 \\ 3x-2y=\ 7 \end{cases}$ is $\left[\begin{array}{cc|c} -1 & 3 & -7 \\ 3 & -2 & 7 \end{array}\right]$. The reduced row-echelon form of this matrix is $\left[\begin{array}{cc|c} 1 & 0 & 1 \\ 0 & 1 & -2 \end{array}\right]$. The system has a unique solution, $x=1,\ y=-2$.

51. The augmented matrix of the system $\begin{cases} -3x-\ y+z=-1 \\ x+4y-z=\ 3 \\ -5x+2y+z=\ 2 \end{cases}$ is $\left[\begin{array}{ccc|c} -3 & -1 & 1 & -1 \\ 1 & 4 & -1 & 3 \\ -5 & 2 & 1 & 2 \end{array}\right]$.

The reduced row-echelon form of this matrix is $\left[\begin{array}{ccc|c} 1 & 0 & -\frac{3}{11} & 0 \\ 0 & 1 & -\frac{2}{11} & 0 \\ 0 & 0 & 0 & 1 \end{array}\right]$. No solutions exist.

53. The augmented matrix of the system $\begin{cases} 2x+4y+2z=0 \\ 3x-\ y+\ z=1 \\ x-5y-\ z=1 \end{cases}$ is $\left[\begin{array}{ccc|c} 2 & 4 & 2 & 0 \\ 3 & -1 & 1 & 1 \\ 1 & -5 & -1 & 1 \end{array}\right]$.

The reduced row-echelon form of this matrix is $\left[\begin{array}{ccc|c} 1 & 0 & \frac{3}{7} & \frac{2}{7} \\ 0 & 1 & \frac{2}{7} & -\frac{1}{7} \\ 0 & 0 & 0 & 0 \end{array}\right]$. There are infinitely many solutions (x, y, z), of the form $\left(-\frac{1}{7}+\frac{2}{7}a,\ \frac{2}{7}+\frac{3}{7}a,\ a\right)$, where a is any real number.

Applications

55. If a parabola of equation $y=ax^2+bx+c$ passes through the points $(1,5), (2,7)$, and $(3,13)$, then $\begin{cases} a+b+c\ =5 \\ 4a+2b+c=7 \\ 9a+3b+c=13 \end{cases}$. The solution of this system is $a=2,\ b=-4,\ c=7$. The parabola that passes through the given points has equation $y=2x^2-4x+7$.

57. If the total value of 15 coins (nickels, dimes and quarters) is \$2.10, and there are two more quarters than nickels, then (if a is the number of quarters, b the number of dimes, and c the number of nickels):

$$\begin{cases} a+b+c = 15 \\ 0.25a+0.10b+0.05c=2.10 \\ a-c = 2 \end{cases}.$$

The solution of this system is $a=5$, $b=7$, and $c=3$ (that is, 5 quarters, 7 dimes and 3 nickels.)

59. A barbell with 2 large disks and 1 small disk on each side weighs 70 pounds; a barbell with 1 small disk on each side weighs 30 pounds; a barbell with 3 large disks on each side weighs 75 pounds. Then, if b is the weight of the barbell, l is the weight of a large disk, and s is the weight of a small disk, we have: $\begin{cases} b+4l+2s=70 \\ b+\quad 2s=30 \\ b+6l \quad =75 \end{cases}$. The solution is $b=15$, $l=10$, $s=7.5$.

A barbell with 4 large disks and 2 small disks on each side weighs 125 pounds.

61. In 1517, Buddhism and Hinduism were a combined 5059 years old. In 1827, Islam and Buddhism were a combined 3557 years old. Hinduism is 975 years older than Buddhism.

Then, $\begin{cases} b+h=5059 \\ b+i+620=3557 \\ h=b+975 \end{cases}$. Equivalently, $\begin{cases} b+h=5059 \\ b+i=2937 \\ -b+h=975 \end{cases}$, with solution $b=2042$, $h=3017$, and $i=895$. Then, the years when each of these religions were founded are:
525 B.C.E for Buddhism, 1500 B.C.E. for Hinduism, and 622 C.E. for Islam.

63. Using the information on the ethnic composition of the middle schools in a community:

(1) Schools of about 500 students, with roughly half white and white black students,
(2) Schools with about 450 students, with approximately half black and half Hispanic students,
(3) Schools with about 400 students that are 80% white and 20% black,

to form a high school with 3750 students that are 46.8% black, 35.2% white, and 18% Hispanic, the number of middle schools of each type that should feed into the new school can be obtained solving the following system:

$$500x+450y+400z=3750$$

$$0.5(500x)+0.8(400z)=0.352(3750)$$

$$0.5(450y)=0.18(3750)$$

The solution of this system is: $x=4$, $y=3$, $z=1$. That is, 4 schools of type (1), 3 schools of type (2) and 1 school of type (3) should feed into the new high school.

Concepts and Critical Thinking

65. The statement: "Any matrix in which each row begins with a 1 is in row-echelon form" is false.

67. The statement: "Multiplying each entry in a column by 5 is an example of an elementary row operation" is false.

69. Answers may vary. An example of a matrix in row-echelon but not reduced row-echelon form is $\begin{bmatrix} 1 & 2 \\ 0 & 1 \end{bmatrix}$.

71. Answers may vary. An example of a matrix consisting only of 1s and 0s that is not in row-echelon form is $\begin{bmatrix} 1 & 0 & 0 \\ 1 & 1 & 0 \end{bmatrix}$.

73. **(a)** If $\left[\begin{array}{cccc|c} 0 & 0 & \cdots & 0\,1 & 0 \end{array}\right]$ is the last row of the reduced row-echelon form of an augmented matrix, then the system is consistent, with a unique solution.

(b) If $\left[\begin{array}{cccc|c} 0 & 0 & \cdots & 0\,0 & 0 \end{array}\right]$ is the last row of the reduced row-echelon form of an augmented matrix, then the system is consistent, with infinitely many solutions.

(c) If $\left[\begin{array}{cccc|c} 0 & 0 & \cdots & 0\,0 & 1 \end{array}\right]$ is the last row of the reduced row-echelon form of an augmented matrix, then the system is inconsistent. That is, there are no solutions for the system.

Chapter 9. Section 9.5 Inverses of Square Matrices

1. If $A = \begin{bmatrix} 1 & -1 \\ 2 & -3 \end{bmatrix}$, and $B = \begin{bmatrix} 3 & -1 \\ 2 & -1 \end{bmatrix}$, then $AB = \begin{bmatrix} 1 & 0 \\ 0 & -1 \end{bmatrix}$; A and B are inverses of each other.

3. If $A = \begin{bmatrix} 1 & 0 & -1 \\ 0 & 2 & 1 \\ 1 & 1 & 0 \end{bmatrix}$, and $B = \begin{bmatrix} -1 & -1 & 2 \\ 1 & 1 & -1 \\ -2 & -1 & 3 \end{bmatrix}$, then $AB = \begin{bmatrix} 1 & 0 & -1 \\ 0 & 1 & 1 \\ 0 & 0 & 1 \end{bmatrix}$; A and B are not inverses of each other.

5. If $A = \begin{bmatrix} 1 & 0 \\ 2 & 1 \end{bmatrix}$, then $A^{-1} = \begin{bmatrix} 1 & 0 \\ -2 & 1 \end{bmatrix}$.

7. If $A = \begin{bmatrix} 0.2 & 0.4 \\ 0.5 & 0.5 \end{bmatrix}$, then $A^{-1} = \begin{bmatrix} -5 & 4 \\ 5 & -2 \end{bmatrix}$.

9. $A = \begin{bmatrix} 1 & 2 \\ 2 & 4 \end{bmatrix}$ is noninvertible.

11. If $A = \begin{bmatrix} 0 & a \\ b & 0 \end{bmatrix}$, then $A^{-1} = \begin{bmatrix} 0 & \frac{1}{b} \\ \frac{1}{a} & 0 \end{bmatrix}$.

13. If $A = \begin{bmatrix} x & x+1 \\ x & x-1 \end{bmatrix}$, then $A^{-1} = \begin{bmatrix} \frac{1-x}{2x} & \frac{1+x}{2x} \\ \frac{1}{2} & -\frac{1}{2} \end{bmatrix}$.

15. $A = \begin{bmatrix} -\frac{1}{5} & 0 & 1 \\ \frac{2}{5} & -1 & -1 \\ -\frac{1}{5} & 1 & 0 \end{bmatrix}$ is noninvertible.

17. If $A = \begin{bmatrix} 1 & 4 & -2 \\ 0 & 2 & -1 \\ -3 & 0 & 2 \end{bmatrix}$, then $A^{-1} = \begin{bmatrix} 1 & -2 & 0 \\ \frac{3}{4} & -1 & \frac{1}{4} \\ \frac{3}{2} & -3 & \frac{1}{2} \end{bmatrix}$.

19. If $A = \begin{bmatrix} -\frac{1}{12} & \frac{7}{12} & -\frac{1}{3} \\ -\frac{1}{12} & -\frac{5}{12} & \frac{2}{3} \\ \frac{5}{12} & \frac{1}{12} & -\frac{1}{3} \end{bmatrix}$, then $A^{-1} = \begin{bmatrix} 1 & 2 & 3 \\ 3 & 2 & 1 \\ 2 & 3 & 1 \end{bmatrix}$.

21. If $A = \begin{bmatrix} 1 & x & x^2 \\ 0 & 1 & x \\ 0 & 0 & 1 \end{bmatrix}$, then $A^{-1} = \begin{bmatrix} 1 & -x & x^2 \\ 0 & 1 & -x \\ 0 & 0 & 1 \end{bmatrix}$.

23. $A = \begin{bmatrix} 1 & 0 & 0 & 1 \\ 0 & 1 & 0 & 1 \\ 0 & 0 & 1 & 1 \\ 1 & 1 & 1 & 3 \end{bmatrix}$ is noninvertible.

25. If $A = \begin{bmatrix} a & 0 & 0 & 0 & 0 \\ -a & a & 0 & 0 & 0 \\ 0 & -a & a & 0 & 0 \\ 0 & 0 & -a & a & 0 \\ 0 & 0 & 0 & -a & a \end{bmatrix}$, then $A^{-1} = \begin{bmatrix} \frac{1}{a} & 0 & 0 & 0 & 0 \\ \frac{1}{a} & \frac{1}{a} & 0 & 0 & 0 \\ \frac{1}{a} & \frac{1}{a} & \frac{1}{a} & 0 & 0 \\ \frac{1}{a} & \frac{1}{a} & \frac{1}{a} & \frac{1}{a} & 0 \\ \frac{1}{a} & \frac{1}{a} & \frac{1}{a} & \frac{1}{a} & \frac{1}{a} \end{bmatrix}$.

27. If $A = \begin{bmatrix} 1 & -1 \\ 1 & 2 \end{bmatrix}$, then $A^{-1} = \begin{bmatrix} \frac{2}{3} & \frac{1}{3} \\ -\frac{1}{3} & \frac{1}{3} \end{bmatrix}$.

(a) The system $\begin{cases} x - y = -1 \\ x + 2y = 12 \end{cases}$ in matrix form is $\begin{bmatrix} 1 & -1 \\ 1 & 2 \end{bmatrix}\begin{bmatrix} x \\ y \end{bmatrix} = \begin{bmatrix} -1 \\ 12 \end{bmatrix}$. The solution is

$\begin{bmatrix} x \\ y \end{bmatrix} = \begin{bmatrix} \frac{2}{3} & \frac{1}{3} \\ -\frac{1}{3} & \frac{1}{3} \end{bmatrix}\begin{bmatrix} -1 \\ 12 \end{bmatrix}$, or $\begin{bmatrix} x \\ y \end{bmatrix} = \begin{bmatrix} \frac{10}{3} \\ \frac{13}{3} \end{bmatrix}$, that is $x = \frac{10}{3}$, $y = \frac{13}{3}$.

(b) The system $\begin{cases} x - y = -4 \\ x + 2y = -1 \end{cases}$ in matrix form is $\begin{bmatrix} 1 & -1 \\ 1 & 2 \end{bmatrix}\begin{bmatrix} x \\ y \end{bmatrix} = \begin{bmatrix} -4 \\ -1 \end{bmatrix}$. The solution is

$\begin{bmatrix} x \\ y \end{bmatrix} = \begin{bmatrix} \frac{2}{3} & \frac{1}{3} \\ -\frac{1}{3} & \frac{1}{3} \end{bmatrix}\begin{bmatrix} -4 \\ -1 \end{bmatrix}$, or $\begin{bmatrix} x \\ y \end{bmatrix} = \begin{bmatrix} -3 \\ 1 \end{bmatrix}$, that is $x = -3$, $y = 1$.

(c) The system $\begin{cases} x - y = 1 \\ x + 2y = -\frac{1}{2} \end{cases}$ in matrix form is $\begin{bmatrix} 1 & -1 \\ 1 & 2 \end{bmatrix}\begin{bmatrix} x \\ y \end{bmatrix} = \begin{bmatrix} 1 \\ -\frac{1}{2} \end{bmatrix}$. The solution is

$\begin{bmatrix} x \\ y \end{bmatrix} = \begin{bmatrix} \frac{2}{3} & \frac{1}{3} \\ -\frac{1}{3} & \frac{1}{3} \end{bmatrix}\begin{bmatrix} 1 \\ -\frac{1}{2} \end{bmatrix}$, or $\begin{bmatrix} x \\ y \end{bmatrix} = \begin{bmatrix} \frac{1}{2} \\ -\frac{1}{2} \end{bmatrix}$, that is $x = \frac{1}{2}$, $y = -\frac{1}{2}$.

29. If $A = \begin{bmatrix} 1.5 & -4 \\ 2.3 & 6 \end{bmatrix}$, then $A^{-1} = \begin{bmatrix} \frac{30}{91} & \frac{20}{91} \\ -\frac{23}{182} & \frac{15}{182} \end{bmatrix}$.

(a) The system $\begin{cases} 1.5x - 4y = -12.5 \\ 2.3x + 6y = 41.5 \end{cases}$ in matrix form is $\begin{bmatrix} 1.5 & -4 \\ 2.3 & 6 \end{bmatrix}\begin{bmatrix} x \\ y \end{bmatrix} = \begin{bmatrix} -12.5 \\ 41.5 \end{bmatrix}$. The solution is $\begin{bmatrix} x \\ y \end{bmatrix} = \begin{bmatrix} \frac{30}{91} & \frac{20}{91} \\ -\frac{23}{182} & \frac{15}{182} \end{bmatrix}\begin{bmatrix} -12.5 \\ 41.5 \end{bmatrix}$, or $\begin{bmatrix} x \\ y \end{bmatrix} = \begin{bmatrix} 5 \\ 5 \end{bmatrix}$, that is $x = 5$, $y = 5$.

(b) The system $\begin{cases} 1.5x - 4y = 2 \\ 2.3x + 6y = 15.2 \end{cases}$ in matrix form is $\begin{bmatrix} 1.5 & -4 \\ 2.3 & 6 \end{bmatrix}\begin{bmatrix} x \\ y \end{bmatrix} = \begin{bmatrix} 2 \\ 15.2 \end{bmatrix}$. The solution is $\begin{bmatrix} x \\ y \end{bmatrix} = \begin{bmatrix} \frac{30}{91} & \frac{20}{91} \\ -\frac{23}{182} & \frac{15}{182} \end{bmatrix}\begin{bmatrix} 2 \\ 15.2 \end{bmatrix}$, or $\begin{bmatrix} x \\ y \end{bmatrix} = \begin{bmatrix} 4 \\ 1 \end{bmatrix}$, that is $x = 4$, $y = 1$.

(c) The system $\begin{cases} 1.5x - 4y = -15 \\ 2.3x + 6y = 13.4 \end{cases}$ in matrix form is $\begin{bmatrix} 1.5 & -4 \\ 2.3 & 6 \end{bmatrix}\begin{bmatrix} x \\ y \end{bmatrix} = \begin{bmatrix} -15 \\ 13.4 \end{bmatrix}$. The solution is $\begin{bmatrix} x \\ y \end{bmatrix} = \begin{bmatrix} \frac{30}{91} & \frac{20}{91} \\ -\frac{23}{182} & \frac{15}{182} \end{bmatrix}\begin{bmatrix} -15 \\ 13.4 \end{bmatrix}$, or $\begin{bmatrix} x \\ y \end{bmatrix} = \begin{bmatrix} -2 \\ 3 \end{bmatrix}$, that is $x = -2$, $y = 3$.

31. **(a)** The solution of the system $\begin{cases} x + 2y - 4z = -8 \\ 3x - y + z = 7 \\ 2x + 2y - z = 6 \end{cases}$ is $x = 2$, $y = 3$, $z = 4$.

(b) The solution of the system $\begin{cases} x + 2y - 4z = -6 \\ 3x - y + z = -5 \\ 2x + 2y - z = -5 \end{cases}$ is $x = -2$, $y = 0$, $z = 1$.

(c) The solution of the system $\begin{cases} x + 2y - 4z = -5 \\ 3x - y + z = 15 \\ 2x + 2y - z = 15 \end{cases}$ is $x = 5$, $y = 5$, $z = 5$.

33. A man walks at 20 minutes per mile, jogs at 11 minutes per mile, and runs at 8 minutes per mile. Walking burn 5.8 calories per minute, jogging burns 9.1 calories per minute, and running burns 14.1 calories per minute.

(a) If he wants to burn 1200 calories, work out for 2 hours and travel 11 miles, then he should walk 35.06 minute, jog 40.2 minutes, and run 44.74 minutes.

(b) If he wants to burn 750 calories, work out for 1.5 hours, and travel 7 miles, he should walk 38.64 minutes jog 39.65 minutes, and run 11.7 minutes.

35. The cubic function $f(x)=2x^3-3x^2+5x+1$ passes through the points $(1,5)$, $(3,43)$, $(5,201)$, and $(7,575)$.

Concepts and Critical Thinking

37. The statement: "Only square matrices can have inverses" is true.

39. The statement: "Every square matrix has an inverse" is false.

41. There exists only one noninvertible 1×1 matrix, namely the matrix $A=[0]$.

43. Answers may vary. A system of equations that cannot be solved using the technique of this section is $\begin{cases} x+2y=3 \\ 2x+4y=6 \end{cases}$.

45. If A and B are invertible matrices of the same size, and $C=AB$, $D=B^{-1}A^{-1}$, then $CD=(AB)(B^{-1}A^{-1})$, that is $CD=A(B(B^{-1}A^{-1}))=A((BB^{-1})A^{-1})=A(IA^{-1})=AA^{-1}=I$. That is, C and D are inverses of one another.

47. If A is an $n\times n$ matrix, the (i, j) entry of I_nA is $0\cdot a_{1j}+0\cdot a_{2j}+\ldots+1\cdot a_{ij}+\ldots 0\cdot a_{nj}=a_{ij}$. Then, $I_nA=A$.

49. If $A=\begin{bmatrix} a & 0 & 0 & 0 \\ 0 & b & 0 & 0 \\ 0 & 0 & c & 0 \\ 0 & 0 & 0 & d \end{bmatrix}$, then A is invertible if and only if none of the diagonal elements is zero. In such case, the inverse is $A^{-1}=\begin{bmatrix} \frac{1}{a} & 0 & 0 & 0 \\ 0 & \frac{1}{b} & 0 & 0 \\ 0 & 0 & \frac{1}{c} & 0 \\ 0 & 0 & 0 & \frac{1}{d} \end{bmatrix}$.

51. **(a)** The trace of $\begin{bmatrix} 2 & -1 \\ 3 & -4 \end{bmatrix}$ is -2.

(b) The trace of $\begin{bmatrix} 1 & 2 & 4 \\ 6 & 9 & 8 \\ 2 & 4 & 8 \end{bmatrix}$ is 18.

(c) The trace of aI_n is na.

53. $I^{-1}=I$

Chapter 9. Section 9.6 Determinants and Cramer's Rule

1. $\begin{vmatrix} 2 & 3 \\ 4 & 5 \end{vmatrix} = -2$

3. $\begin{vmatrix} 1 & 6 \\ 2 & 12 \end{vmatrix} = 0$

5. $\begin{vmatrix} 3 & 7 \\ 4 & 1 \end{vmatrix} = -25$

7. $\begin{vmatrix} x & x-1 \\ x & x \end{vmatrix} = x$

9. $\begin{vmatrix} 2 & 1 & 1 \\ 0 & 1 & 4 \\ 0 & 2 & 3 \end{vmatrix} = -10$

11. $\begin{vmatrix} -4 & 2 & 1 \\ -2 & 1 & 0 \\ 3 & -1 & 5 \end{vmatrix} = -19$

13. $\begin{vmatrix} 1.2 & 3.9 & 2.5 \\ 3.7 & 4.1 & 3.6 \\ 1.0 & 2.5 & 7.1 \end{vmatrix} = -51.406$

15. $\begin{vmatrix} a & b & c \\ 0 & 1 & d \\ 0 & 0 & 1 \end{vmatrix} = a$

17. $\begin{vmatrix} i & j & k \\ a_1 & a_2 & a_3 \\ b_1 & b_2 & b_3 \end{vmatrix} = (a_2b_3 - a_3b_2)i + (a_3b_1 - a_1b_3)j + (a_1b_2 - a_2b_1)k$

19. $\begin{vmatrix} 1 & 2 & 3 & 4 \\ 2 & 2 & 5 & 7 \\ 8 & 7 & 0 & 2 \\ 8 & 3 & 9 & 0 \end{vmatrix} = -561$

21. $\begin{vmatrix} a & 1 & 1 & 1 & 1 \\ 0 & a & 1 & 1 & 1 \\ 0 & 0 & a & 1 & 1 \\ 0 & 0 & 0 & a & 1 \\ 0 & 0 & 0 & 0 & a \end{vmatrix} = a^5$

23. $\begin{vmatrix} -3 & 5 \\ 1 & 0 \end{vmatrix} = -5$

25. $\begin{vmatrix} 2 & 5 \\ 4 & 10 \end{vmatrix} = 0$

27. $\begin{vmatrix} 2 & 0 & 5 \\ -1 & 3 & 0 \\ -1 & 9 & 5 \end{vmatrix} = 0$

29. $\begin{vmatrix} -12 & 1 & 14 \\ 1 & -16 & 2 \\ 15 & 0 & 0 \end{vmatrix} = 3390$

31. The matrix $\begin{bmatrix} x & -1 \\ 1 & 2 \end{bmatrix}$ is not invertible if $x = -\frac{1}{2}$.

33. The matrix $\begin{bmatrix} -3 & x \\ x & -3 \end{bmatrix}$ is not invertible if $x = \pm 3$.

35. The matrix $\begin{bmatrix} x & 0 & 1 \\ 1 & x & 0 \\ x & 1 & 1 \end{bmatrix}$ is invertible for all real values of x.

37. The solution of the system

$$\begin{aligned} x+2y &= 19 \\ 3x-7y &= -8 \end{aligned}$$

is $x = \dfrac{\begin{vmatrix} 19 & 2 \\ -8 & -7 \end{vmatrix}}{\begin{vmatrix} 1 & 2 \\ 3 & -7 \end{vmatrix}} = 9, \quad y = \dfrac{\begin{vmatrix} 1 & 19 \\ 3 & -8 \end{vmatrix}}{\begin{vmatrix} 1 & 2 \\ 3 & -7 \end{vmatrix}} = 5.$

39. The solution of the system

$$\begin{aligned} 2x+z &= 6 \\ x-y &= 4 \\ -y+z &= 7 \end{aligned}$$

is $x = \dfrac{\begin{vmatrix} 6 & 0 & 1 \\ 4 & -1 & 0 \\ 7 & -1 & 1 \end{vmatrix}}{\begin{vmatrix} 2 & 0 & 1 \\ 1 & -1 & 0 \\ 0 & -1 & 1 \end{vmatrix}} = 1, \quad y = \dfrac{\begin{vmatrix} 2 & 6 & 1 \\ 1 & 4 & 0 \\ 0 & 7 & 1 \end{vmatrix}}{\begin{vmatrix} 2 & 0 & 1 \\ 1 & -1 & 0 \\ 0 & -1 & 1 \end{vmatrix}} = -3, \quad z = \dfrac{\begin{vmatrix} 2 & 0 & 6 \\ 1 & -1 & 4 \\ 0 & -1 & 7 \end{vmatrix}}{\begin{vmatrix} 2 & 0 & 1 \\ 1 & -1 & 0 \\ 0 & -1 & 1 \end{vmatrix}} = 4.$

41. The solution of the system

$$\begin{aligned} x+2y-\ z &= -3 \\ 3x-8y-5z &= -3 \\ 2x+2y+\ z &= 10 \end{aligned}$$

is, approximately, $x \approx 3.783$, $y \approx -1.087$, $z \approx 4.609$.

43. The solution of the system

$$\begin{aligned} x+2y+3z+4w &= 1 \\ y-\ z+\ w &= 0 \\ 2z+3w &= 0 \\ z-\ w &= 0 \end{aligned}$$

is $x = 1$, $y = 0$, $z = 0$, $w = 0$.

45. If $ae - bd \neq 0$, the solution of the system

$$ax + by = c$$

$$dx + ey = f$$

is $x = \frac{ce - bf}{ae - bd}$, $y = \frac{af - cd}{ae - bd}$.

Applications

47. $C = ax^2 + bx + c$ is the cost of producing x units, and we know C for three values of x, as shown below:

x (units)	10	50	100
C (cost)	\$8000	\$5000	\$10,000

Then, we obtain the system in a, b, c:

$$100a + 10b + c = 8000$$

$$2500a + 50b + c = 5000$$

$$10{,}000a + 100b + c = 10{,}000$$

The solution of this system is, approximately, $a \approx 1.944$, $b \approx -191.66$, $c \approx 9722.22$. That is, the fixed costs are \$9722.22 (when $x = 0$.)

49. A breakfast consists of milk, oatmeal and English muffins. The following table shows the caloric and fat content per gram of each item.

	Oatmeal	Milk	Muffins
Calories per gram	0.54	0.42	2.37
Grams of fat per gram	0.0097	0.011	0.01

If the mass, caloric value, and fat content from each of the components of this breakfast are 427 grams, 311 calories, and 4.4 grams, respectively, then we have the system:

$$x + y + z = 427$$

$$0.54x + 0.42y + 2.37z = 311$$

$$0.0096x + 0.011y + 0.01z = 4.4$$

where x, y, z are the amounts of oatmeal, milk, and English muffins. The amount of milk is $y = 185.58$ grams.

Concepts and Critical Thinking

51. The statement: "The determinant of a matrix is itself a matrix" is false.

53. The statement: "If a is a real number, then the matrix $\begin{bmatrix} a & -1 \\ 1 & a \end{bmatrix}$ is invertible no matter what the value of a" is true.

55. Answers may vary. An example of a 3×3 matrix such that the easiest method of computing the determinant is probably expansion about the second column is $\begin{bmatrix} 1 & 1 & 3 \\ 4 & 0 & 0 \\ -8 & 0 & 3 \end{bmatrix}$.

57. Answers may vary. An example of a 3×3 matrix with determinant 125 is $\begin{bmatrix} 25 & 1 & 1 \\ 0 & 5 & 1 \\ 0 & 0 & 1 \end{bmatrix}$

59. If A is a 3×3 matrix such that the matrix equation $AX=0$ has a nonzero solution X, and $\det A \neq 0$, then A is invertible. Now, $A^{-1}(AX)=A^{-1}(0)=0$, that is $\left(A^{-1}A\right)(X)=0$, so $I_3X=0$. This means $X=0$, which is a contradiction. Therefore, $\det A = 0$.

61. Consider the matrix $A=\begin{bmatrix} a & b \\ c & d \end{bmatrix}$.

(a) If $c=ka$, and $d=kb$, then $ad-bc=a(kb)-b(ka)=0$ (that is, $\det(A)=0$);

(b) If $ad-bc=0$, then $\dfrac{ad-bc}{cd}=0$, that is $\dfrac{a}{c}-\dfrac{b}{d}=0$, and $\dfrac{a}{c}=\dfrac{b}{d}=k$, for some real number k. Then $a=kc$, and $b=kd$. (If $c=0$, then $ad=0$, so $d=0$ or $a=0$, and in either case the first row is a multiple of the second.)

(a) and **(b)** show that one row is a multiple of the other if and only if $ad-bc=0$.

Chapter 9. Section 9.7 Systems of Inequalities

1. $(0,0)$ is not a solution of the inequality $x-2y\geq 5$; $(4,1)$ is not a solution of $x-2y\geq 5$ either; $(-1,-3)$ is a solution of $x-2y\geq 5$ (in fact, $(-1)-2(-3)=5$.)

3. $(1,4)$ is a solution of $4x+3y>12$; $(3,0)$ and $(-2,-1)$ are not solutions of $4x+3y>12$.

5. $(0,2)$ is a solution of $y+x^2\leq 4$ (in fact, $2+0^2\leq 4$); $(-1,3)$ is a solution of $y+x^2\leq 4$ (because $3+(-1)^2\leq 4$); $(5,-1)$ is not a solution of $y+x^2\leq 4$.

7. The graph of $3x-2y\geq 6$ corresponds to (iv):

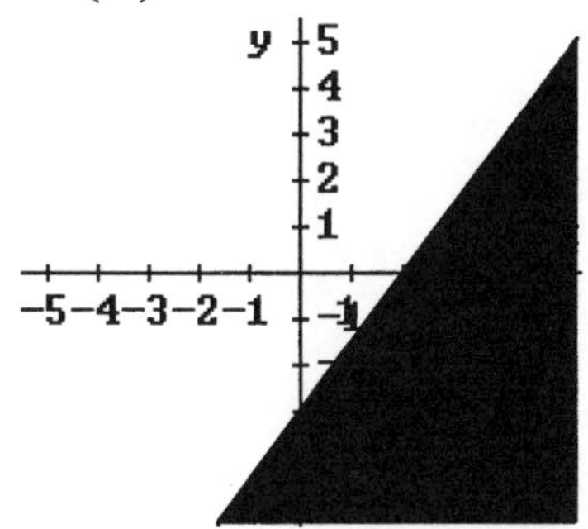

9. The graph of $x<3$ corresponds to (v):

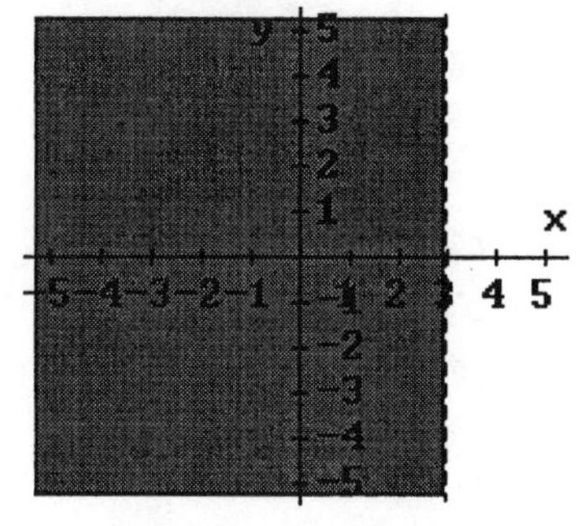

11. The graph of $y-x^2\geq 0$ corresponds to (vi):

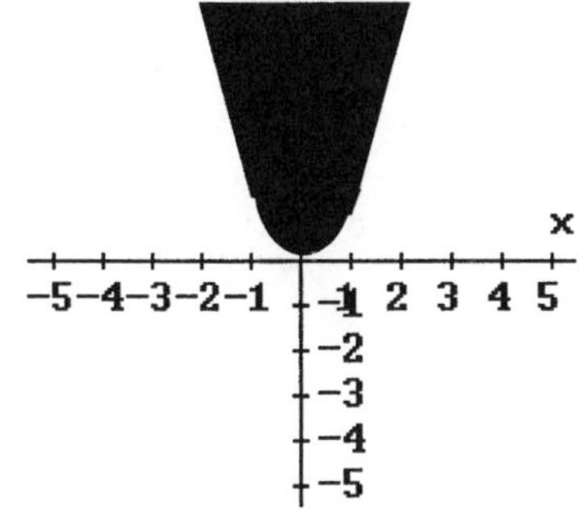

13. The graph of $x+y<8$:

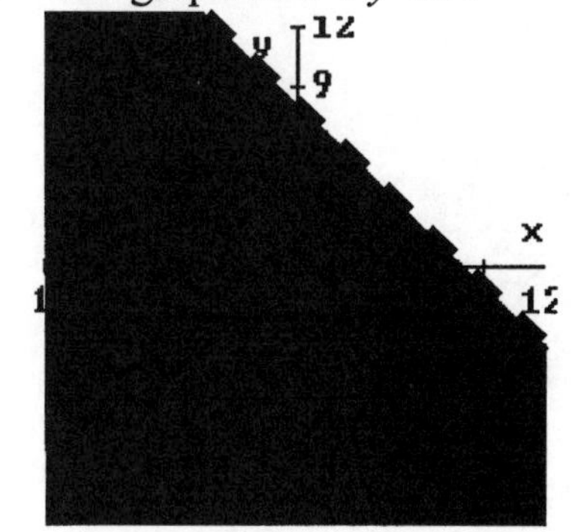

15. The graph of $6x-2y\le 12$:

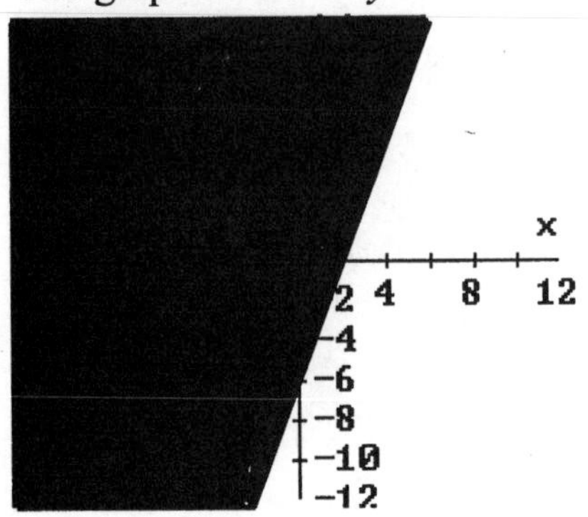

17. The graph of $x\ge -5$:

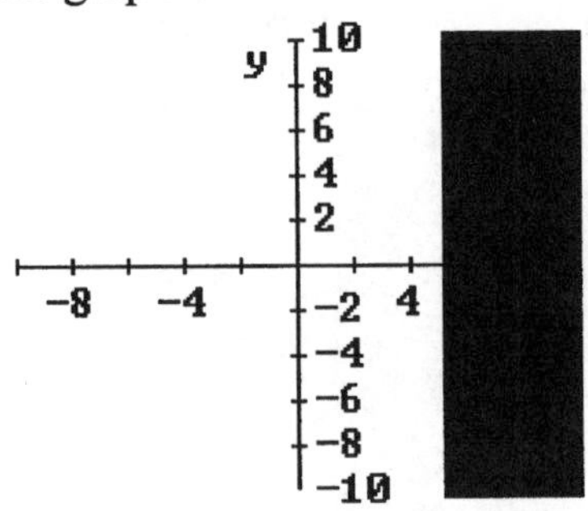

19. The graph of $x>-3y$:

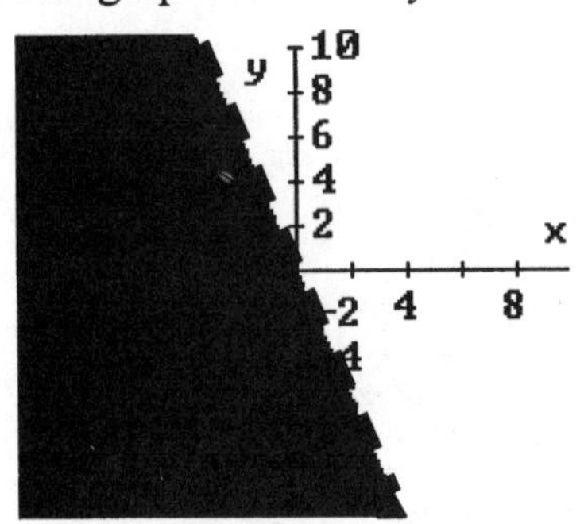

21. The graph of $y<x^2+1$:

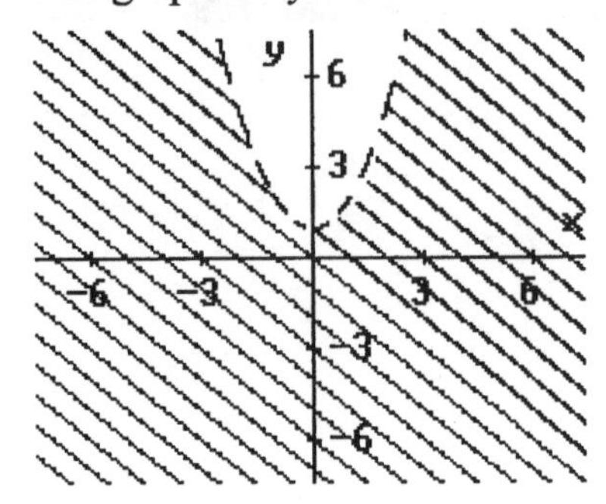

23. The graph of $y^2 \le 1 - x^2$:

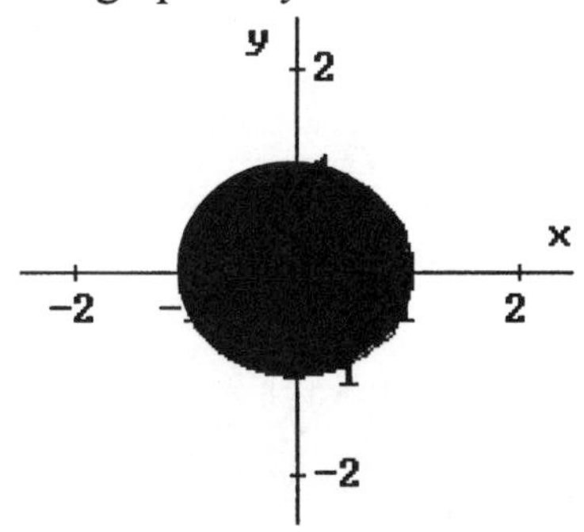

25. The graph of $y \ge \frac{1}{x}$:

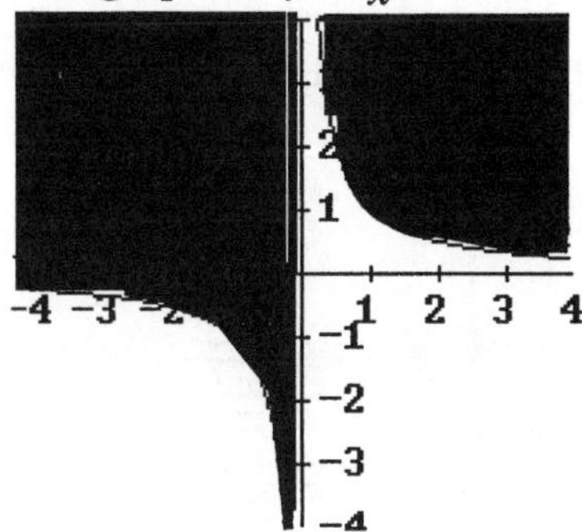

27. The graph of $y > e^x$:

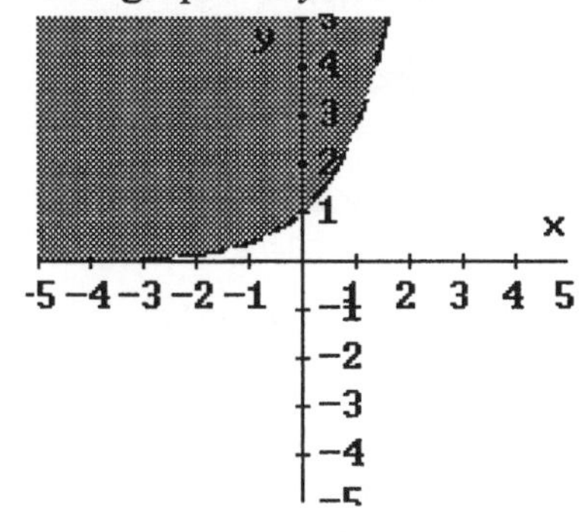

29. The graph of $y \le -x^2 - 6x - 7$ corresponds to (ii).

31. The graph of $y \le x^3 - 9x$ corresponds to (i).

33. The graph of the system

$x + y \le 5$

$x - y \le 1$

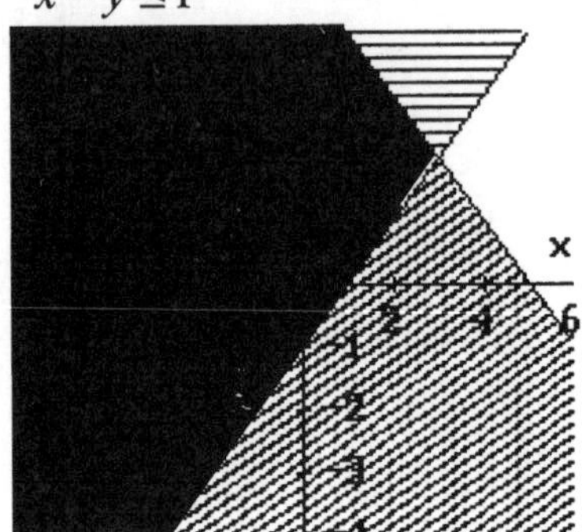

35. The graph of the system

$y < 2x$

$x + 2y < 5$

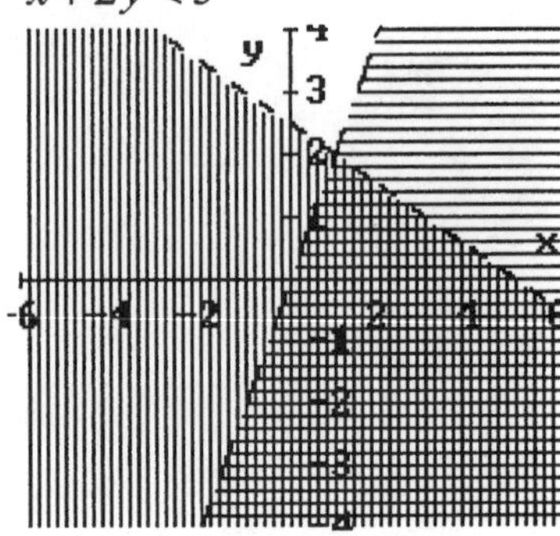

37. The graph of the system

$4x + y \le 6$

$-4x - y \le 4$

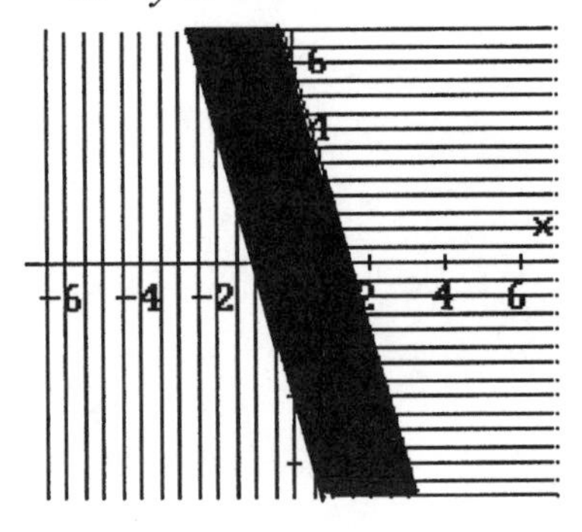

39. The graph of the system

$x + y \le 2$

$-x + y \ge 2$

$x \ge -2$

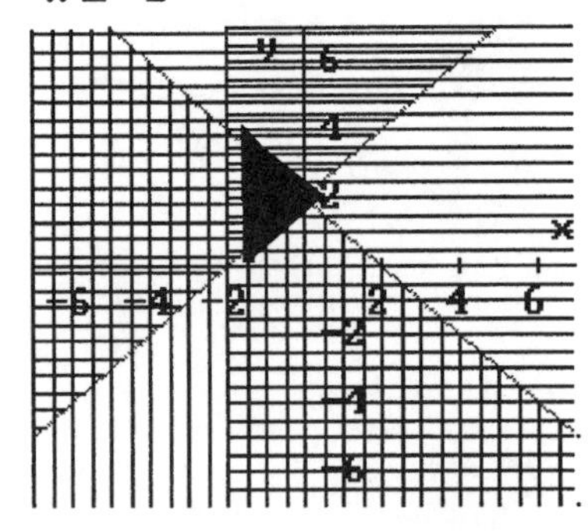

41. The graph of the system

$-3x + 2y < 6$

$-x + 3y > 2$

$2x + y < 3$

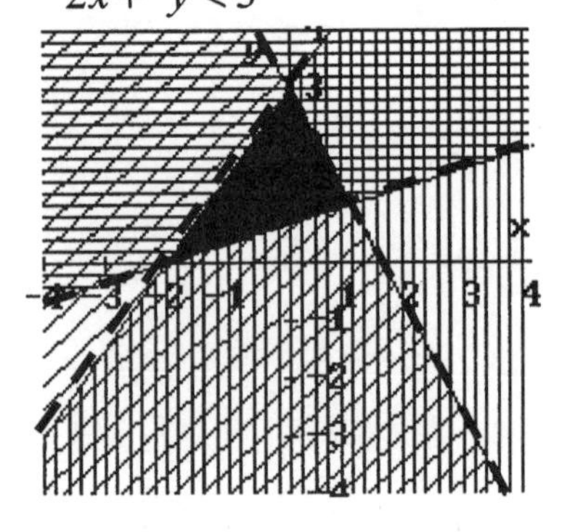

43. The graph of the system

$2x + 5y \le 10$

$x + y \le 3$

$x \ge 0$

$y \ge 0$

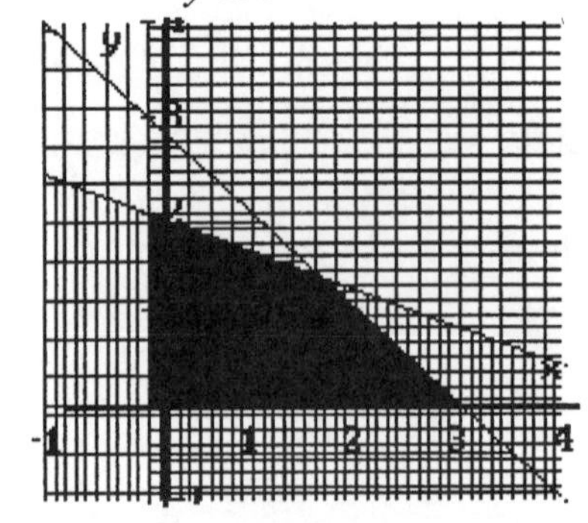

45. The graph of the system

$$2x + y \geq 8$$
$$2x - y \leq 8$$
$$x \geq 3$$
$$y \leq 4$$

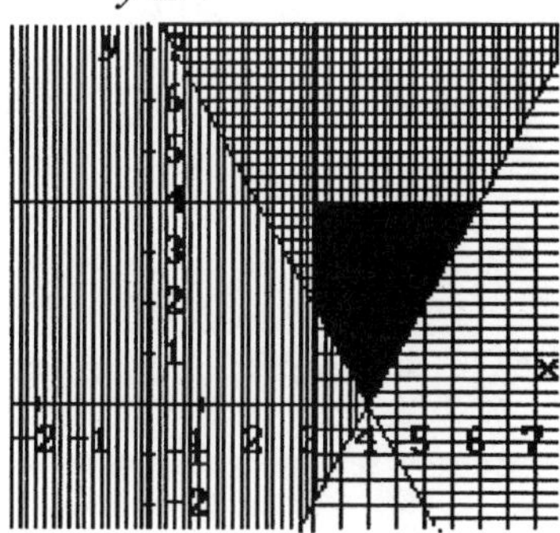

47. The graph of the system

$$-2x + y \geq 1$$
$$y \leq 4 - x^2$$

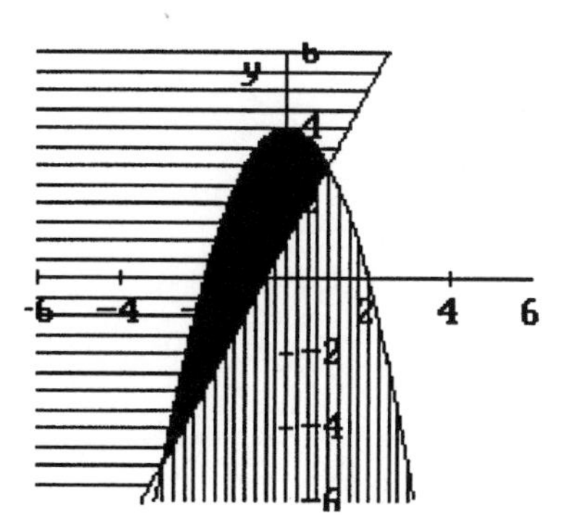

49. The graph of the system

$$xy < 1$$
$$y < x$$

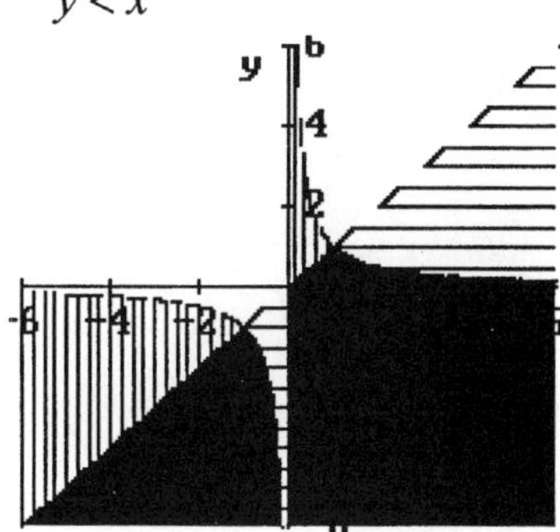

51. The graph of the system

$$x^2 + y^2 \leq 16$$
$$x^2 + y^2 \geq 4$$

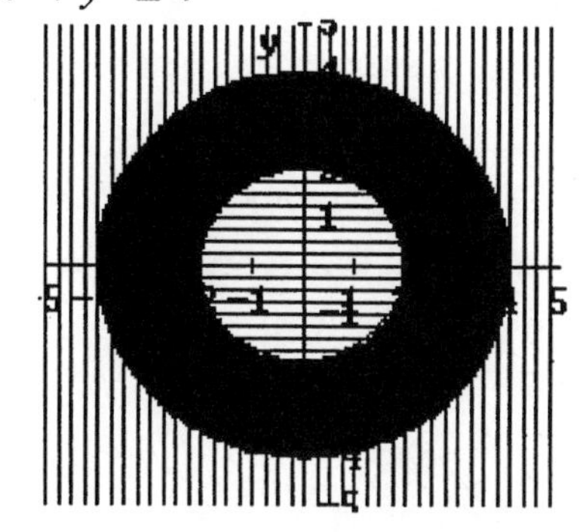

53. The graph of the system

$$y > x^3 - x$$
$$y < 6x$$
$$x \geq 0$$

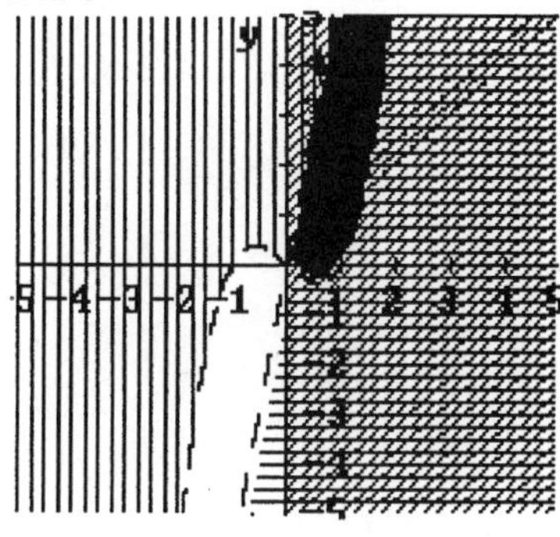

55. The inequality $y \leq \frac{3}{4}x - 3$ has the solution set:

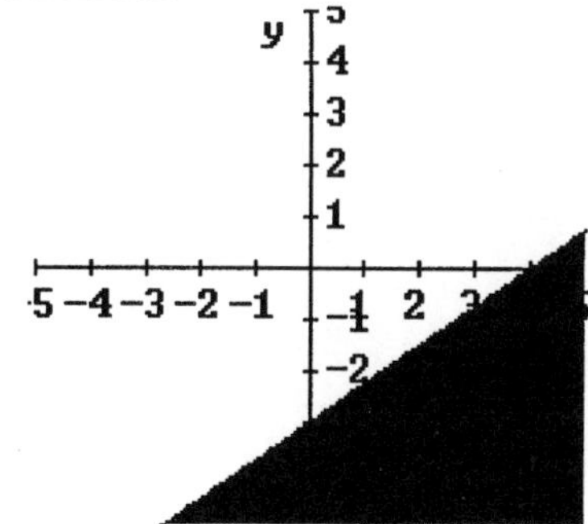

57. The system of inequalities

$y \geq x+2$

$y \leq -\frac{2}{5}x+2$

has the solution set:

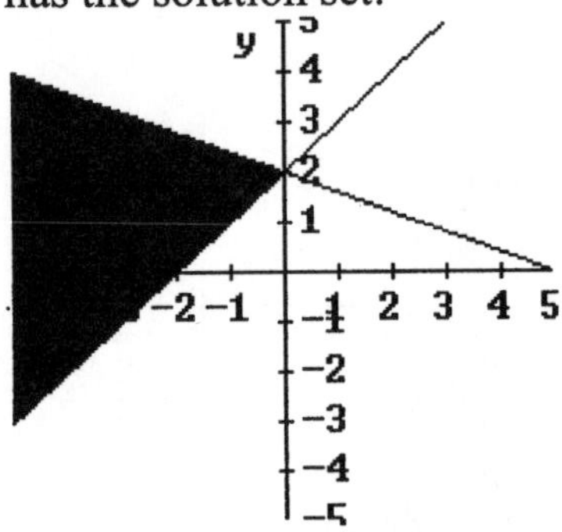

59. The system of inequalities

$y \leq -\frac{2}{3}x+4$

$y \geq \frac{2}{3}x$

$x \geq 0$

has the solution set:

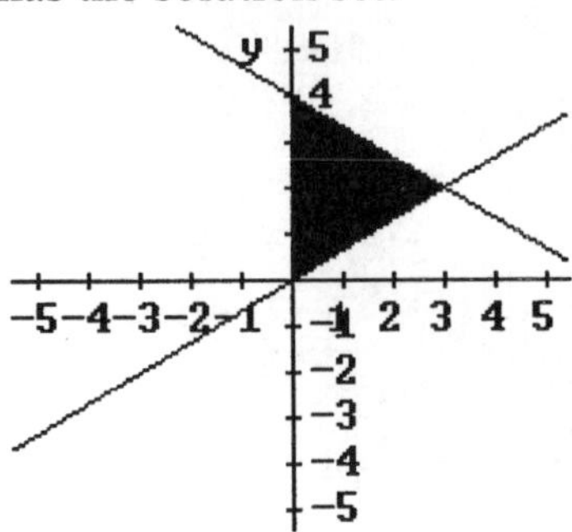

61. The system of inequalities

$y \leq -\frac{3}{2}x+\frac{19}{2}$

$y \geq -\frac{3}{2}x+3$

$y \geq \frac{2}{3}x-\frac{4}{3}$

$y \leq \frac{2}{3}x+3$

has the solution set:

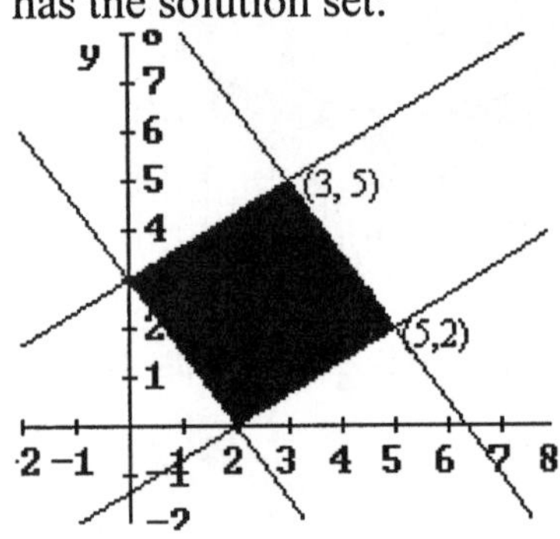

63. The system of inequalities

$x^2+y^2 \leq 9$

$x \geq 0$

has the solution set:

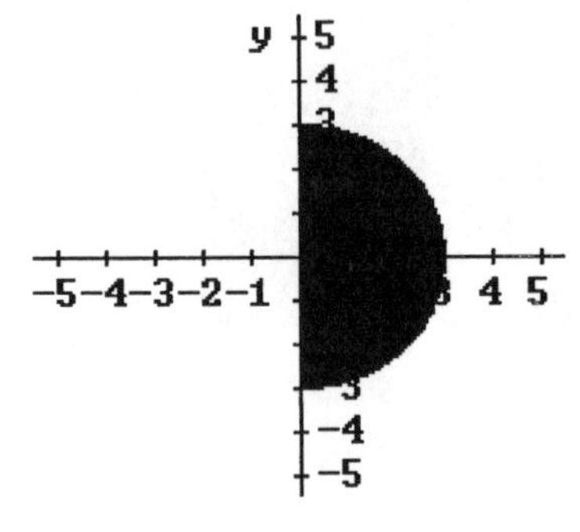

Applications

65. If Juan and Rosa are to invest up to \$24,000 of their savings in two retirement funds, Redbook Growth and Redbook Small Cap, in such a manner that no more than half the amount invested in Redbook Growth is invested in Redbook Small Cap, and if at least \$6000 is invested in each fund, then we have the following system of inequalities relating x, the amount invested in Redbook Growth, and y, the amount invested in Redbook Small Cap.

$$x + y \le 24{,}000$$
$$y \le \tfrac{1}{2}x$$
$$x \ge 6000$$
$$y \ge 6000$$

The vertices of the solution set are (12000, 6000), (16,000, 8000), and (18,000, 6000).

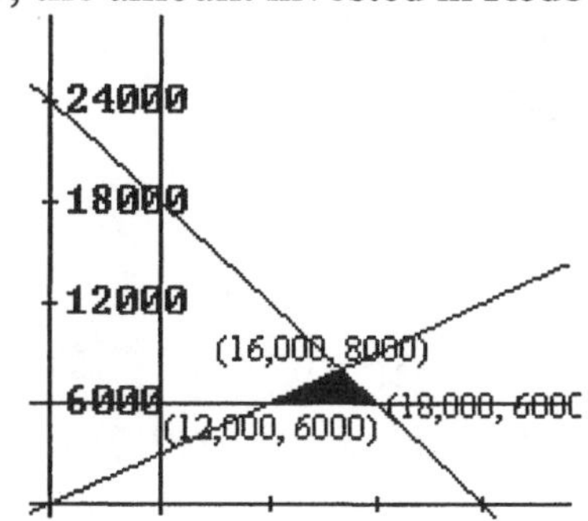

67. If 1000 acres are available to plant corn and soybeans, each acre of corn costs \$100 and requires 2 hours of labor, each acre of soybeans costs \$80 and requires 1 hour of labor, no more than \$88,000 should be spent, and 1600 hours of labor are available, then we have the system of inequalities:

$$x + y \le 1000$$
$$100x + 80y \le 88{,}000$$
$$2x + y \le 1600$$
$$x \ge 0$$
$$y \ge 0$$

The vertices of the solution set are

$$\left(666\tfrac{2}{3}, 266\tfrac{2}{3}\right),\ (600, 400),\ (400, 600)$$

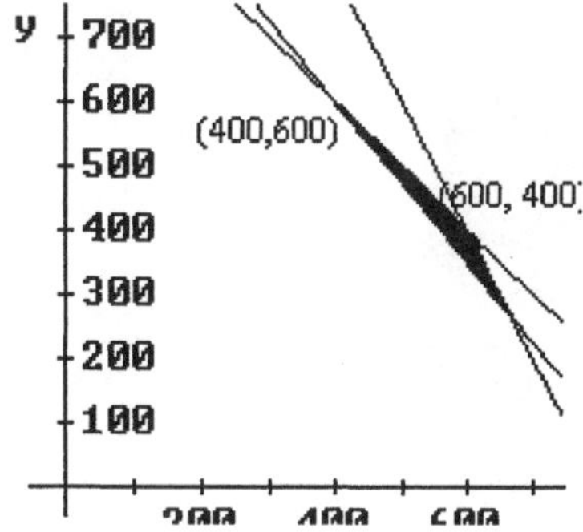

Concepts and Critical Thinking

69. The statement: "It is possible that the inside of a circle could be the solution set to a system of linear inequalities" is false.

71. The statement: "The solution set to a linear inequality is a half-plane" is true, if we are considering linear inequalities in two variables.

72. The statement: "The solution set of $mx + b < y$ is the half-plane lying entirely above the line $y = mx + b$" is true.

73. Answers may vary. An example of a system of inequalities, the graph of which consists of all points in the first quadrant is the system :

$$x > 0$$
$$y > 0$$

75. Answers may vary. An example of a system of inequalities, the graph of which is a right triangle is:

$$y \le -x + 1$$
$$x \ge 0,\ y \ge 0$$

Chapter 9. Section 9.8 Linear Programming

1. Over the feasible set:

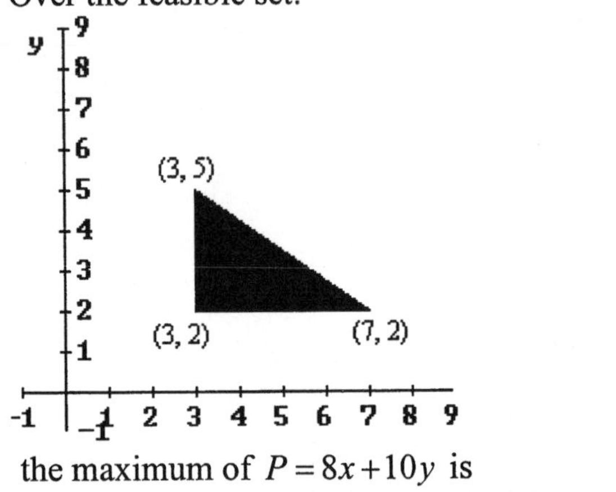

the maximum of $P = 8x + 10y$ is $P(7,2) = 76$, the minimum is $P(3,2) = 54$.

3. Over the feasible set:

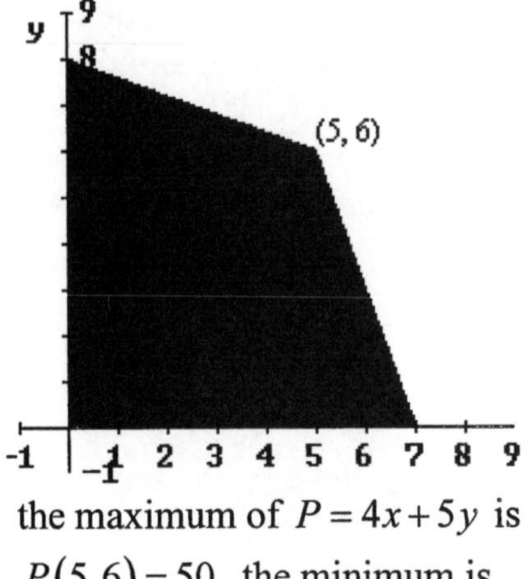

the maximum of $P = 4x + 5y$ is $P(5,6) = 50$, the minimum is $P(0,0) = 0$.

5. Over the feasible set:

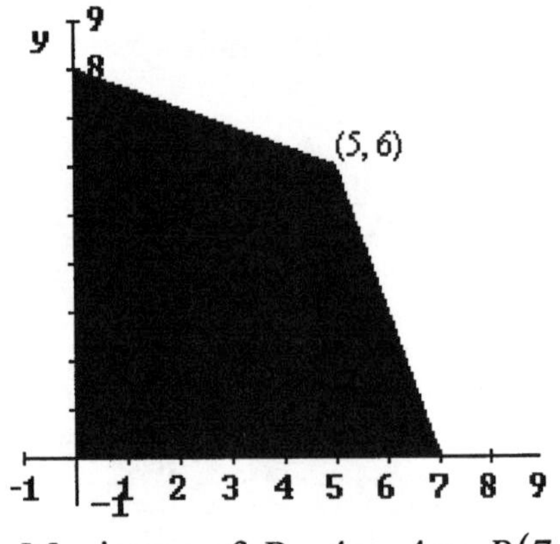

Maximum of $P = 4x - 4y$: $P(7,0) = 28$; minimum: $P(0,8) = -32$.

7. The feasible set of $P = 10x + 15y$ subject to:

$$3x + 4y \le 24$$
$$x + 2y \le 10$$
$$x \ge 0$$
$$y \ge 0:$$

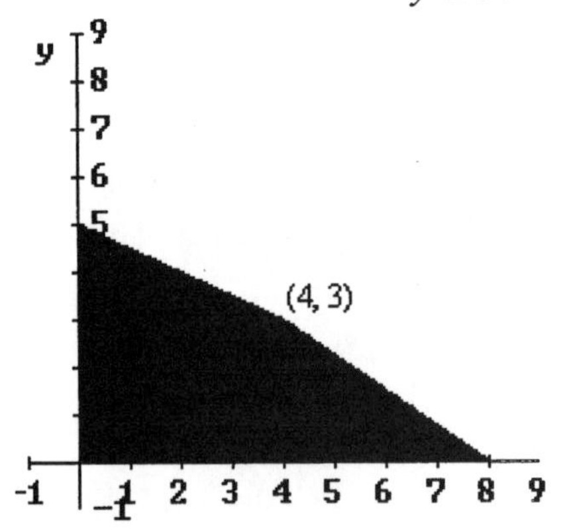

9. The feasible set of $P = 15x + 40y$ subject to:

$$\begin{aligned} 2x + 5y &\geq 20 \\ 3x + 2y &\geq 19 \\ x &\geq 0 \\ y &\geq 0: \end{aligned}$$

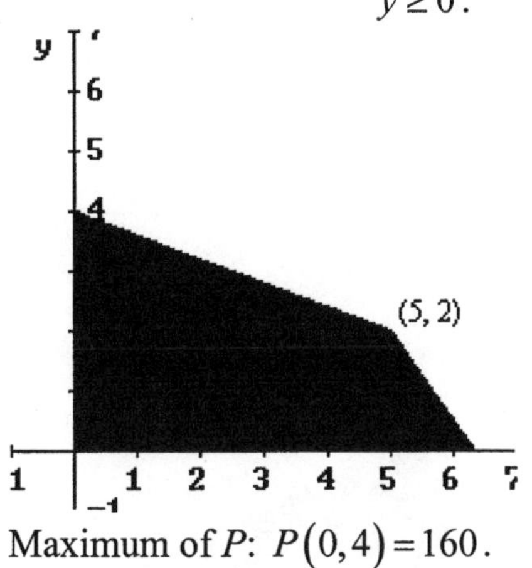

Maximum of P: $P(0,4) = 160$.

11. The feasible set of $P = 10x + 12y$ subject to:

$$\begin{aligned} x + 2y &\leq 17 \\ 2x + 3y &\leq 30 \\ x &\geq 3 \\ y &\geq 2: \end{aligned}$$

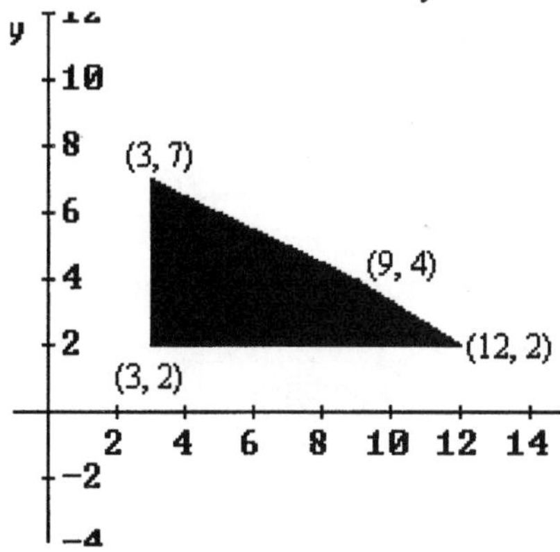

Maximum of P: $P(12,2) = 144$.

13. The feasible set of $P = 20x + 24y$ subject to:

$$\begin{aligned} 2x + 3y &\leq 12 \\ x + y &\geq 5 \\ x + 4y &\geq 8 \\ x &\geq 0 \\ y &\geq 0 \end{aligned}$$

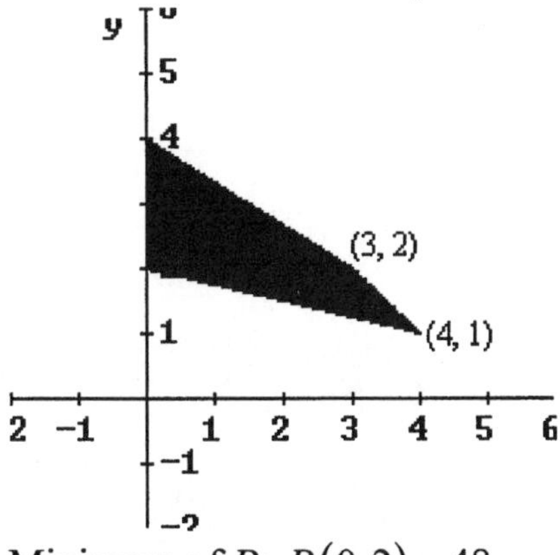

Minimum of P: $P(0,2) = 48$.

15. The feasible set of $P = 2x + 10y$ subject to:

$$\begin{aligned} x + y &\geq 9 \\ 3x - 2y &\leq 24 \\ x + 4y &\leq 36 \\ x &\geq 4 \\ y &\geq 3 \end{aligned}$$

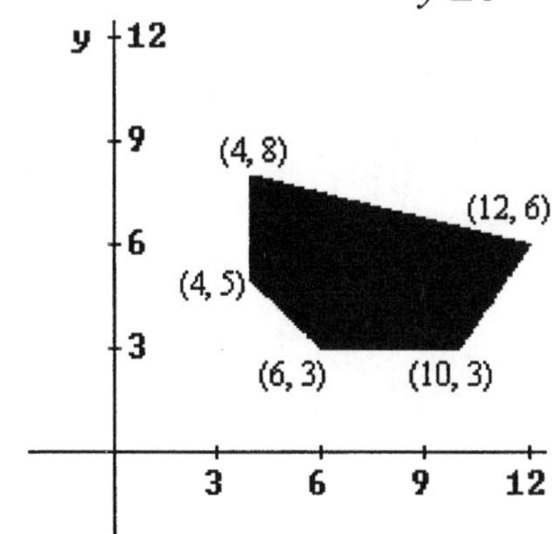

Maximum of P: $P(4,8) = 88$.

Applications

17. A clothing company makes two styles of tailored suits, with the requirements in labor and profit margins per suit summarized in the following table.

	Style A	Style B
Cutting (hours)	2	4
Sewing (hours	3	2
Profit (dollars)	35	40

If 56 cutting hours and 72 sewing hours are available, and x, y are the number of Style A suits and the number of Style B suits, respectively, then we are to maximize the total profit $P = 35x + 40y$ subject to: $2x + 4y \le 56$

$$3x + 2y \le 72$$
$$x \ge 0$$
$$y \ge 0$$

The maximum profit occurs when $x = 22,\ y = 3$, that is, 22 Style A suits and 3 Style B suits are tailored.

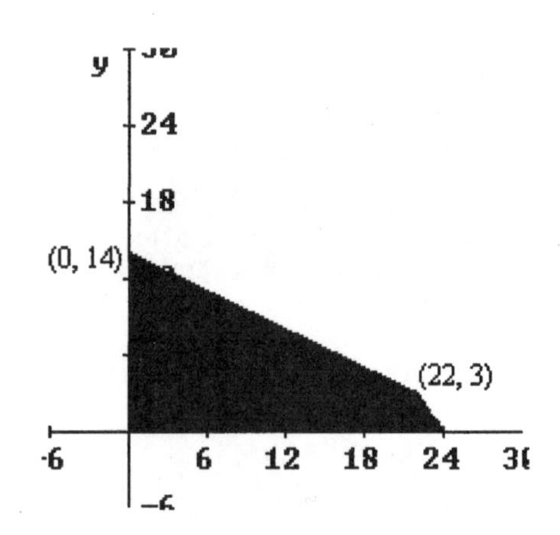

19. The following table summarizes the production limits and operating costs per day of two refineries owned by an oil company.

	Refinery 1	Refinery 2
High-grade oil (barrels)	100	200
Medium-grade oil (barrels)	200	100
Low-grade oil (barrels)	300	200
Operating Cost (dollars)	10,000	9000

If an order is received for 1000 barrels of high-grade oil, 1000 barrels of medium-grade oil, and 1800 barrels of low-grade oil, we want to minimize the operating costs required to satisfy the order. Minimize $C = 10{,}000x + 9000y$ subject to:

$$100x + 200y \ge 1000$$
$$200x + 100y \ge 1000$$
$$300x + 200y \ge 1800$$
$$x \ge 0$$
$$y \ge 0$$

where x, y are the number of days that Refinery 1, Refinery 2 should be operated. The minimum cost of \$67,000 under these conditions is attained if Refinery 1 is operated $x = 4$ days, and Refinery 2 is operated $y = 3$ days.

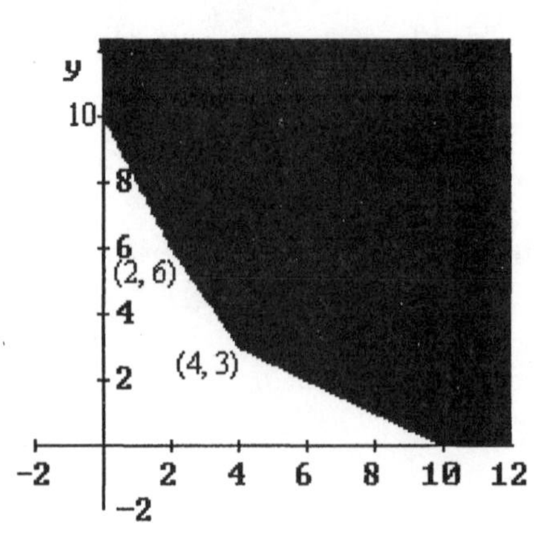

21. Steve can borrow his mother's car, which will cost him 4¢ per mile, or his father's truck, which will cost him 6¢ per mile. He travels at least 300 miles each month, but no more than 350. He must drive his father's truck at least twice as far as his mother's car. He is not to drive his father's truck more than 250 miles per month. We are to minimize his cost $C = 4x + 6y$ subject to:

$$x + y \geq 300$$
$$x + y \leq 350$$
$$y \geq 2x$$
$$y \leq 250$$
$$x \geq 0$$
$$y \geq 0$$

where x and y are the number of miles using the car and the truck, respectively. The vertices of the feasible set are $(50, 250)$, $(100, 200)$, $(100, 250)$, and $(116\frac{2}{3},\ 233\frac{1}{3})$. The minimum cost of \$16 occurs if he drives his mother's car $x = 100$ miles and his father's car $y = 200$ miles.

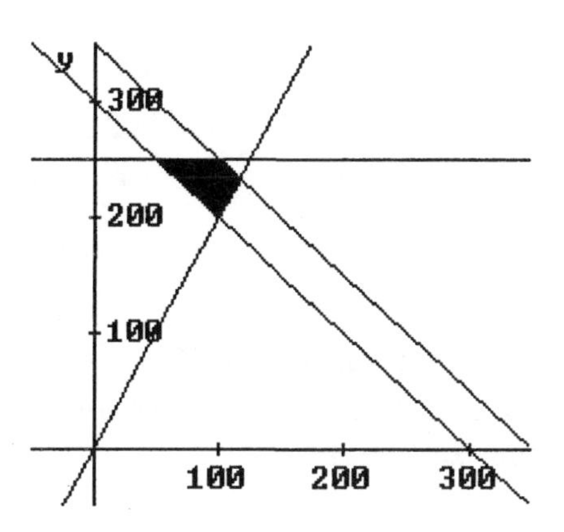

23. When planning a ½ hour television show as part of the political campaign for a senator and a governor, a political party estimates that 40,000 viewers will watch the show for every minute that the senator is on, and that 60,000 viewers will watch for every minute that the governor is on. The senator has to be at least 1 ½ times as long as the governor. The governor will not participate if her time is less than 10 minutes. We are to maximize the number of viewers $N = 40{,}000x + 60{,}000y$ subject to:

$$x + y \leq 30$$
$$x \geq \frac{3}{2}y$$
$$y \geq 10$$
$$x \geq 0$$
$$y \geq 0$$

The vertices of the feasible set are $(15, 10)$, $(18, 12)$, and $(20, 10)$. A maximum number of viewers of 1,440,000 is attained if 18 minutes are dedicated to the senator and 12 minutes to the governor.

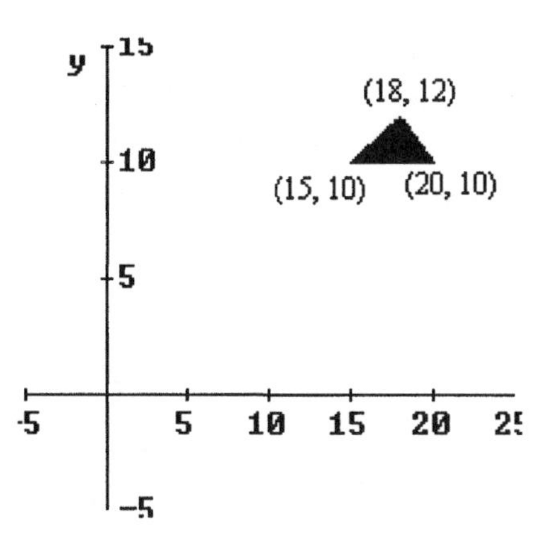

25. The following table summarizes the vitamin content of each serving of swiss steak and peas:

	Swiss Steak	**Peas**
Vitamin A (units)	150	120
Vitamin B_6 (units)	0.06	0.02
Vitamin C	3	9
Cost (cents)	9	4

A meal is to be planned by a hospital, where the swiss steak and peas served should provide at least 1260 units of vitamin A, 0.35 units of vitamin B_6, and 45 units of vitamin C. We are to minimize the cost $C = 9x + 4y$ subject to:

$$150x + 120y \geq 1260$$
$$0.06x + 0.02y \geq 0.35$$
$$3x + 9y \geq 45$$
$$x \geq 0$$
$$y \geq 0$$

The minimum cost of 62.5¢ is attained when the meal includes 4.5 ounces of Swiss steak and 5 ounces of peas.

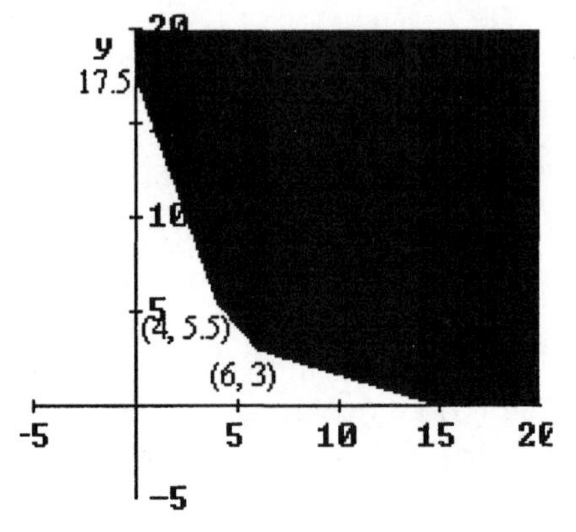

27. The following table gives the round-trip air fare from New Haven and Bozeman to Orlando and Columbus:

	New Haven	Bozeman
Orlando	500	750
Columbus	300	450

A minimum of 32,000 frequent flyer miles should be accumulated, and only the Orlando destination provides frequent flyer miles, 2000 in the case of New Haven, and 4000 in the case of Bozeman. At least 15 employees should travel to Columbus, and at least 10 to Orlando. There are 15 employees in Bozeman and 20 in New Haven, and all should travel. of the trips. We are to minimize the air fare $P = 500x + 700y + 300(15 - x) + 450(20 - y)$, that is $P = 200x + 300y + 13{,}500$ subject to:

$$2000x + 4000y \geq 32{,}000$$
$$x + y \geq 10$$
$$(15 - x) + (20 - y) \geq 15 \quad (\text{or } x + y \leq 20)$$
$$x \geq 0, \quad x \leq 20$$
$$y \geq 0, \quad y \leq 15$$

where x is the number of New Haven employees and y the number of Bozeman's going to Orlando. The minimum air fare \$16,100 is attained if $x = 4$ New Haven employees go to Orlando (therefore, 11 employees from this office go to Columbus), and $y = 6$ Bozeman employees go to Orlando (therefore, 14 from this office go to Columbus.)

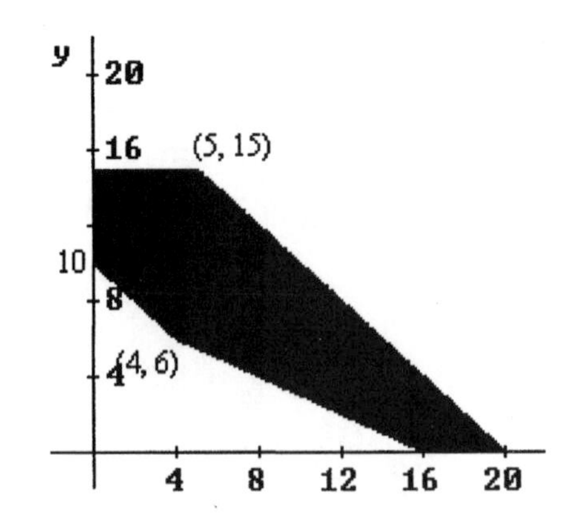

Concepts and Critical Thinking

29. The statement: "If the objective function of a linear programming problem has a maximum, then the maximum occurs at one of the vertices of the feasible set" is true.

31. The statement: "According to the Fundamental Theorem of Linear Programming, the minimum value of the function $R = x^2 + y^2$ subject to the constraints $-1 \le x \le 1$ and $-1 \le y \le 1$ will be obtained at one of the points $(1,1), (-1,1), (-1,-1)$, or $(1,-1)$" is false. This is not a linear programming problem. In fact, such a minimum is attained at $(0,0)$.

33. Answers may vary. A linear programming problem for which the optimal solution is the origin is to minimize $C = 2x + 3y$ subject to $x \ge 0$, $y \ge 0$.

35. Answers may vary. A linear programming problem for which the feasible set has just one vertex is to minimize $C = 2x + 3y$ subject to $x \ge 0$, $y \ge 0$.

Chapter 9. Review

1. The system

$$x+5y=13$$
$$2x+3y=12$$

has a unique solution, $x=3,\ y=2.$

3. The system

$$2x-4y=5$$
$$-x+2y=6$$

has no solution.

5. The system

$$x+2y+z=11$$
$$y-z=-1$$
$$x+z=\ 5$$

has a unique solution, $x=1,\ y=3,\ z=4.$

7. The equations

$$2x+3y=6$$
$$4x-\ y=4$$

have one point of intersection, $\left(\frac{9}{7},\frac{8}{7}\right).$

9. The equations

$$y=2x^3-3x^2-11x$$
$$y=x+7$$

have two points of intersection, $(3.5, 10.5)$ and $(-1,6)$.

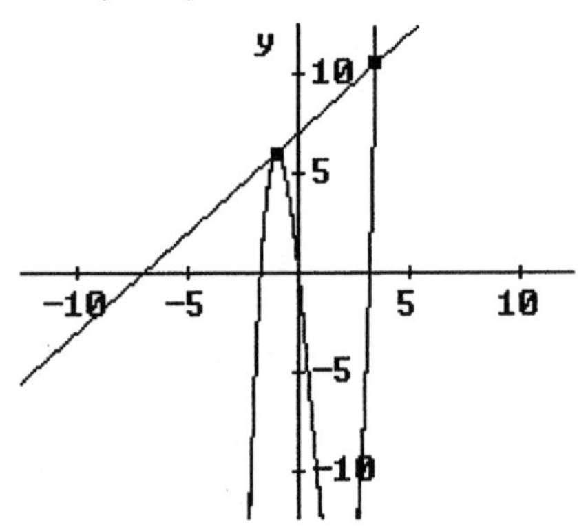

11. $3\begin{bmatrix} 2 & -1 \\ -5 & 3 \end{bmatrix}=\begin{bmatrix} 6 & -3 \\ -15 & 9 \end{bmatrix}$

13. $\begin{bmatrix} 3 & 0 & -2 \\ 5 & -1 & 3 \end{bmatrix}+\begin{bmatrix} 0 & -1 & 4 \\ -2 & 6 & 2 \end{bmatrix}=\begin{bmatrix} 3 & -1 & 2 \\ 3 & 5 & 5 \end{bmatrix}$

In Exercises 15-21, $A=\begin{bmatrix} 1 & 0 \\ -2 & 3 \end{bmatrix}$, $B=\begin{bmatrix} 2 & -1 \\ 4 & 3 \end{bmatrix}$, $C=\begin{bmatrix} 4 & 0 \\ -3 & 1 \\ 0 & 2 \end{bmatrix}$, and $D=\begin{bmatrix} 3 & 1 & 0 \\ -2 & 0 & 4 \\ 1 & -2 & 1 \end{bmatrix}$.

15. $A+2B=\begin{bmatrix} 5 & -2 \\ 6 & 9 \end{bmatrix}$

17. CD is undefined.

19. $DC = \begin{bmatrix} 9 & 1 \\ -8 & 8 \\ 10 & 0 \end{bmatrix}$

21. C^2 is undefined.

23. $\begin{bmatrix} -2 & 3 \\ 1 & 4 \end{bmatrix}\begin{bmatrix} 3 \\ -1 \end{bmatrix} = \begin{bmatrix} -9 \\ -1 \end{bmatrix}$

25. $\begin{bmatrix} 2 & 5 \\ 3 & 7 \end{bmatrix}\begin{bmatrix} 4 & 0 \end{bmatrix}$ is undefined.

27. $\begin{bmatrix} 3 & 4 \end{bmatrix}\begin{bmatrix} 2 \\ -1 \end{bmatrix} = \begin{bmatrix} 2 \end{bmatrix}$

29. $\begin{bmatrix} 0 & 4 \\ -1 & 2 \\ 3 & 0 \end{bmatrix}\begin{bmatrix} 2 & 3 \\ -3 & 1 \end{bmatrix} = \begin{bmatrix} -12 & 4 \\ -8 & -1 \\ 6 & 9 \end{bmatrix}$

31. $\begin{bmatrix} 1 & -1 & 2 \\ 3 & 0 & -1 \\ 4 & 2 & 0 \end{bmatrix}\begin{bmatrix} b \\ c \\ a \end{bmatrix} = \begin{bmatrix} b-c+2a \\ 3b-a \\ 4b+2c \end{bmatrix}$

33. The augmented matrix of the system:

$$\begin{aligned} 5x+\ y&=3 \\ x-3y&=4 \end{aligned}$$

is $\left[\begin{array}{cc|c} 5 & 1 & 3 \\ 1 & -3 & 4 \end{array}\right]$.

35. The matrix

$$\left[\begin{array}{ccc|c} 1 & -4 & 1 & 3 \\ 0 & 1 & -2 & 5 \\ 0 & 0 & 1 & -2 \end{array}\right].$$

is the augmented matrix of the system

$$\begin{aligned} x-4y+\ z&=\ 3 \\ y-2z&=\ 5 \\ z&=-2 \end{aligned}$$

37. A row-echelon form of the matrix $\left[\begin{array}{cc|c} -1 & 3 & -2 \\ 4 & -9 & 1 \end{array}\right]$ is $\left[\begin{array}{cc|c} 1 & -\frac{9}{4} & \frac{1}{4} \\ 0 & 1 & 3 \end{array}\right]$.

39. A row-echelon form of the matrix $\left[\begin{array}{ccc|c} 2 & 4 & -3 & 1 \\ 3 & 0 & \frac{9}{2} & -\frac{3}{2} \end{array}\right]$ is the matrix

$$\left[\begin{array}{ccc|c} 1 & 0 & \frac{3}{2} & -\frac{1}{2} \\ 0 & 1 & -\frac{3}{2} & \frac{1}{2} \end{array}\right].$$

41. The augmented matrix of the system

$$\begin{aligned} -x+2y&=\ 7 \\ 2x-3y&=-9 \end{aligned}$$

is $\left[\begin{array}{cc|c} -1 & 2 & 7 \\ 2 & -3 & -9 \end{array}\right]$. The system has one solution, $x=3,\ y=5$.

43. The augmented matrix of the system

$$\begin{aligned} -2x+3y&=\ 6 \\ 4x-6y&=-12 \end{aligned}$$

is $\left[\begin{array}{cc|c} -2 & 3 & 6 \\ 4 & -6 & -12 \end{array}\right]$. The system has infinitely many solutions.

45. The augmented matrix of the system
$$\begin{aligned}-x-\ y+z&=-4\\ 3x-2y-z&=-4\\ x-2z&=\ 1\end{aligned}$$
is $\left[\begin{array}{ccc|c}-1 & -1 & 1 & -4\\ 3 & -2 & -1 & -4\\ 1 & 0 & -2 & 1\end{array}\right]$. The system has one solution, $x=\frac{5}{7}$, $y=\frac{22}{7}$, $z=-\frac{1}{7}$.

47. The augmented matrix of the system
$$\begin{aligned}x-y&=4\\ 2x-z&=5\\ y-z&=1\end{aligned}$$
is $\left[\begin{array}{ccc|c}1 & -1 & 0 & 4\\ 2 & 0 & -1 & 5\\ 0 & 1 & -1 & 1\end{array}\right]$. The system has one solution, $x=0$, $y=-4$, $z=-5$.

49. The row-echelon form of the matrix $\left[\begin{array}{cc|c}1 & 2 & 4\\ 2 & 3 & 5\end{array}\right]$ is $\left[\begin{array}{cc|c}1 & 0 & -2\\ 0 & 1 & 3\end{array}\right]$.

50. The row-echelon form of the matrix $\left[\begin{array}{ccc|c}1 & -1 & 2 & 8\\ 0 & 1 & -4 & -9\\ 0 & 0 & 1 & 3\end{array}\right]$ is $\left[\begin{array}{ccc|c}1 & 0 & 0 & 5\\ 0 & 1 & 0 & 3\\ 0 & 0 & 1 & 3\end{array}\right]$.

51. The system
$$\begin{aligned}3x+5y&=-5\\ -x-2y&=\ 3\end{aligned}$$
in matrix form is $\begin{bmatrix}3 & 5\\ -1 & -2\end{bmatrix}\begin{bmatrix}x\\ y\end{bmatrix}=\begin{bmatrix}-5\\ 3\end{bmatrix}$. The system has one solution, $x=5$, $y=-4$.

53. The system
$$\begin{aligned}x-\ y+3z&=\ 9\\ y-2z&=-3\\ z&=\ 4\end{aligned}$$
in matrix form is $\begin{bmatrix}1 & -1 & 3\\ 0 & 1 & -2\\ 0 & 0 & 1\end{bmatrix}\begin{bmatrix}x\\ y\\ z\end{bmatrix}=\begin{bmatrix}9\\ -3\\ 4\end{bmatrix}$. The system has one solution, $x=2$, $y=5$, $z=4$.

55. $\begin{bmatrix}2 & 1\\ -3 & -1\end{bmatrix}^{-1}=\begin{bmatrix}-1 & -1\\ 3 & 2\end{bmatrix}$

57. $\begin{bmatrix}a & b\\ b & a\end{bmatrix}^{-1}=\begin{bmatrix}\frac{a}{a^2-b^2} & -\frac{b}{a^2-b^2}\\ -\frac{b}{a^2-b^2} & \frac{a}{a^2-b^2}\end{bmatrix}$

59. $\begin{bmatrix}1 & -2 & 3\\ 0 & 1 & -2\\ 0 & 0 & 1\end{bmatrix}^{-1}=\begin{bmatrix}1 & 2 & 1\\ 0 & 1 & 2\\ 0 & 0 & 1\end{bmatrix}$

61. $\begin{bmatrix}0 & 0 & a\\ 0 & 0 & 0\\ a & 0 & 0\end{bmatrix}$ is not invertible.

63. The matrix of coefficients of the systems

(a) $\begin{aligned} x+2y &= 3 \\ 2x+y &= -4 \end{aligned}$ **(b)** $\begin{aligned} x+2y &= 2 \\ 2x+y &= 0 \end{aligned}$

is $A=\begin{bmatrix} 1 & 1 \\ 2 & 1 \end{bmatrix}$. The systems in matrix form are $\begin{bmatrix} 1 & 1 \\ 2 & 1 \end{bmatrix}\begin{bmatrix} x \\ y \end{bmatrix}=\begin{bmatrix} 3 \\ -4 \end{bmatrix}$, $\begin{bmatrix} 1 & 1 \\ 2 & 1 \end{bmatrix}\begin{bmatrix} x \\ y \end{bmatrix}=\begin{bmatrix} 2 \\ 0 \end{bmatrix}$

The solution of system **(a)** is:

$\begin{bmatrix} x \\ y \end{bmatrix}=\begin{bmatrix} 1 & 1 \\ 2 & 1 \end{bmatrix}^{-1}\begin{bmatrix} 3 \\ -4 \end{bmatrix}=\begin{bmatrix} -1 & 1 \\ 2 & -1 \end{bmatrix}\begin{bmatrix} 3 \\ -4 \end{bmatrix}=\begin{bmatrix} -7 \\ 10 \end{bmatrix}$; that is $x=-7,\ y=10$.

The solution of system **(b)** is:

$\begin{bmatrix} x \\ y \end{bmatrix}=\begin{bmatrix} 1 & 1 \\ 2 & 1 \end{bmatrix}^{-1}\begin{bmatrix} 2 \\ 0 \end{bmatrix}=\begin{bmatrix} -1 & 1 \\ 2 & -1 \end{bmatrix}\begin{bmatrix} 2 \\ 0 \end{bmatrix}=\begin{bmatrix} -2 \\ 4 \end{bmatrix}$; that is $x=-2,\ y=4$.

65. The matrix of coefficients of the systems

(a) $\begin{aligned} \tfrac{1}{2}x+\tfrac{3}{2}y &= 3 \\ -\tfrac{1}{4}x+\tfrac{5}{4}y &= -4 \end{aligned}$ **(b)** $\begin{aligned} \tfrac{1}{2}x+\tfrac{3}{2}y &= 0 \\ -\tfrac{1}{4}x+\tfrac{5}{4}y &= -3 \end{aligned}$

is $A=\begin{bmatrix} \frac{1}{2} & \frac{3}{2} \\ -\frac{1}{4} & \frac{5}{4} \end{bmatrix}$. In matrix form: $\begin{bmatrix} \frac{1}{2} & \frac{3}{2} \\ -\frac{1}{4} & \frac{5}{4} \end{bmatrix}\begin{bmatrix} x \\ y \end{bmatrix}=\begin{bmatrix} 3 \\ -4 \end{bmatrix}$, $\begin{bmatrix} \frac{1}{2} & \frac{3}{2} \\ -\frac{1}{4} & \frac{5}{4} \end{bmatrix}\begin{bmatrix} x \\ y \end{bmatrix}=\begin{bmatrix} 0 \\ -3 \end{bmatrix}$

The solution of system **(a)** is:

$\begin{bmatrix} x \\ y \end{bmatrix}=\begin{bmatrix} \frac{1}{2} & \frac{3}{2} \\ -\frac{1}{4} & \frac{5}{4} \end{bmatrix}^{-1}\begin{bmatrix} 3 \\ -4 \end{bmatrix}=\begin{bmatrix} \frac{5}{4} & -\frac{3}{2} \\ \frac{1}{4} & \frac{1}{2} \end{bmatrix}\begin{bmatrix} 3 \\ -4 \end{bmatrix}=\begin{bmatrix} \frac{39}{4} \\ -\frac{5}{4} \end{bmatrix}$; that is $x=\frac{39}{4},\ y=-\frac{5}{4}$.

The solution of system **(b)** is:

$\begin{bmatrix} x \\ y \end{bmatrix}=\begin{bmatrix} \frac{1}{2} & \frac{3}{2} \\ -\frac{1}{4} & \frac{5}{4} \end{bmatrix}^{-1}\begin{bmatrix} 0 \\ -3 \end{bmatrix}=\begin{bmatrix} \frac{5}{4} & -\frac{3}{2} \\ \frac{1}{4} & \frac{1}{2} \end{bmatrix}\begin{bmatrix} 0 \\ -3 \end{bmatrix}=\begin{bmatrix} \frac{9}{2} \\ -\frac{3}{2} \end{bmatrix}$; that is $x=\frac{9}{2},\ y=-\frac{3}{2}$.

67. $\begin{vmatrix} 4 & 6 \\ 3 & 2 \end{vmatrix}=-10$

69. $\begin{vmatrix} \sqrt{5} & 2 \\ 2 & \sqrt{5} \end{vmatrix}=1$

71. $\begin{vmatrix} x & 0 & 1 \\ 0 & x & 0 \\ 1 & 0 & x \end{vmatrix}=x^3-x$

73. $\begin{vmatrix} 1 & -1 & 0 & 0 \\ 3 & 4 & 0 & 0 \\ 0 & 0 & 2 & -3 \\ 0 & 0 & 1 & 1 \end{vmatrix} = 35$

75. The matrix $\begin{bmatrix} 2 & 2 \\ 5 & x \end{bmatrix}$ is not invertible if $x = 5$. (The determinant of such a matrix is $2x - 10$.)

77. The matrix $\begin{bmatrix} 1 & 2x & -4 \\ 0 & x & 2 \\ 0 & 3 & x-5 \end{bmatrix}$ is not invertible if $x = -2$ or $x = 3$. (Determinant $= x^2 - 5x - 6$.)

79. The solution of the system

$$3x + y = 0$$
$$5x + 3y = -1$$

is $x = \dfrac{\begin{vmatrix} 0 & 1 \\ -1 & 3 \end{vmatrix}}{\begin{vmatrix} 3 & 1 \\ 5 & 3 \end{vmatrix}} = \frac{1}{4}$, $y = \dfrac{\begin{vmatrix} 3 & 0 \\ 5 & -1 \end{vmatrix}}{\begin{vmatrix} 3 & 1 \\ 5 & 3 \end{vmatrix}} = -\frac{3}{4}$.

81. The solution of the system

$$x + y = 0$$
$$x + 2y - z = -6$$
$$-4x - 3y = -1$$

is $x = \dfrac{\begin{vmatrix} 0 & 1 & 0 \\ -6 & 2 & -1 \\ -1 & -3 & 0 \end{vmatrix}}{\begin{vmatrix} 1 & 1 & 0 \\ 1 & 2 & -1 \\ -4 & -3 & 0 \end{vmatrix}} = \frac{1}{1} = 1$, $y = \dfrac{\begin{vmatrix} 1 & 0 & 0 \\ 1 & -6 & -1 \\ -4 & -1 & 0 \end{vmatrix}}{\begin{vmatrix} 1 & 1 & 0 \\ 1 & 2 & -1 \\ -4 & -3 & 0 \end{vmatrix}} = -1$, $z = \dfrac{\begin{vmatrix} 1 & 1 & 0 \\ 1 & 2 & -6 \\ -4 & -3 & -1 \end{vmatrix}}{\begin{vmatrix} 1 & 1 & 0 \\ 1 & 2 & -1 \\ -4 & -3 & 0 \end{vmatrix}} = 5$.

83. The graph of $x + y \geq 4$:

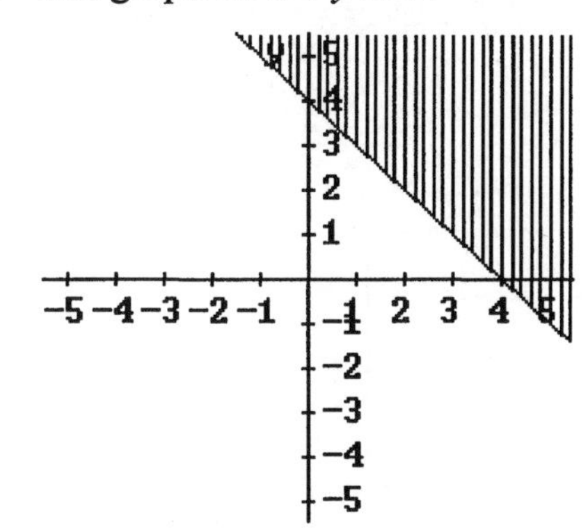

85. The graph of $x < 2y$:

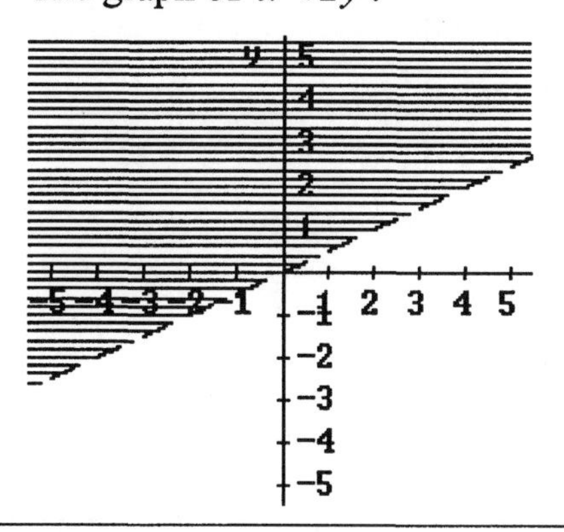

87. The graph of $x \geq -2$:

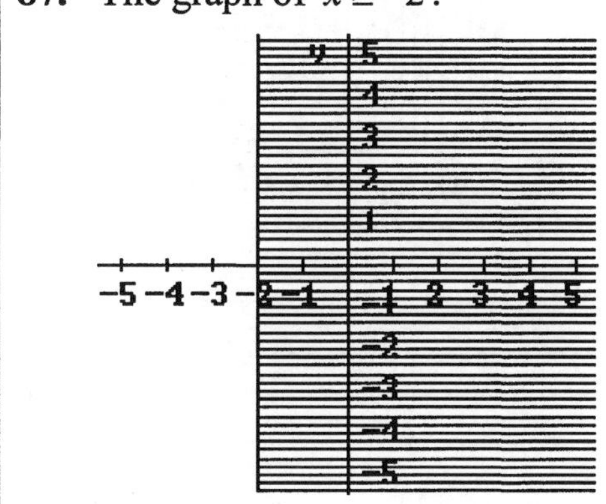

89. The graph of $x^2 + y^2 < 4$:

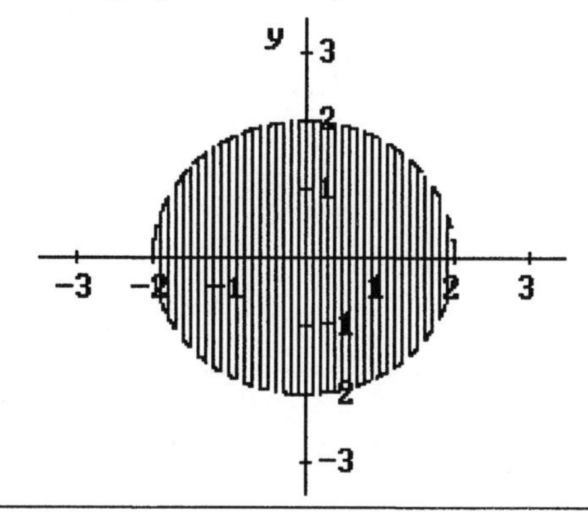

91. The graph of the system of inequalities

$x - y \leq 6$

$x + y \leq 4$

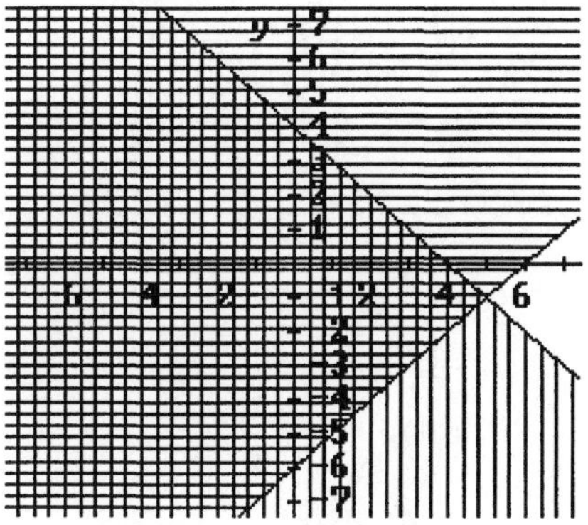

93. The graph of the system of inequalities

$4x - 7y \geq 14$

$3y \leq x + 6$

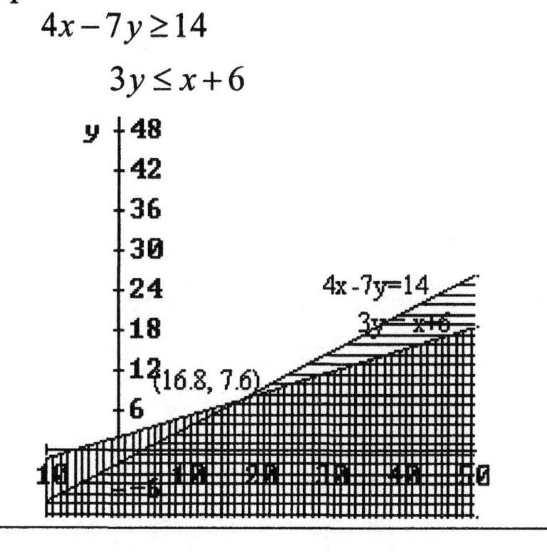

95. The graph of the system of inequalities:

$$-x+4y>8$$
$$-x+y>0$$
$$-x+2y>4$$

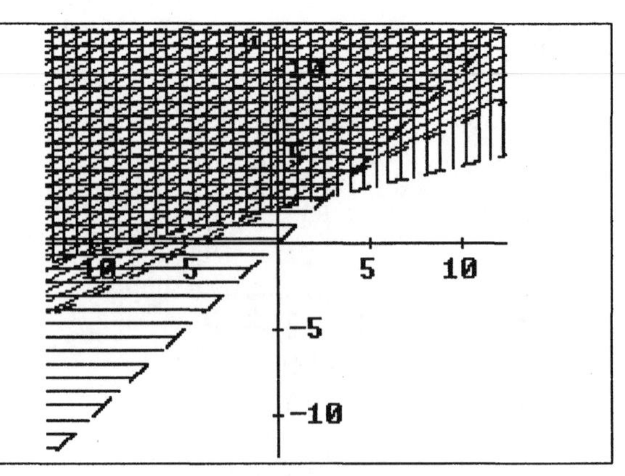

97. The graph of the system of inequalities:

$$y \geq x^2-2$$
$$y \leq x$$

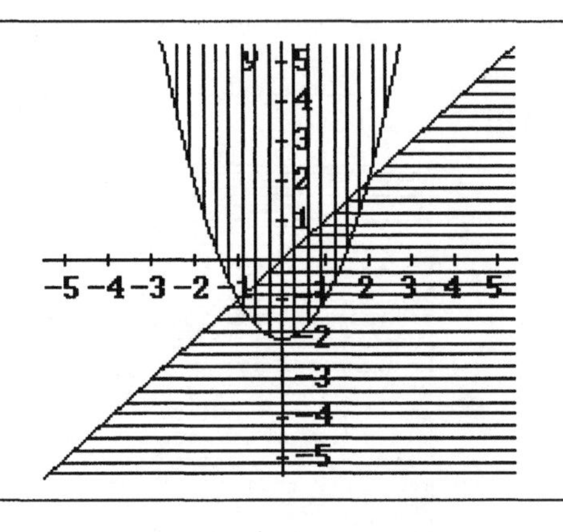

99. The graph of the system of inequalities:

$$x^2+y^2 \leq 25$$
$$x \geq 3$$

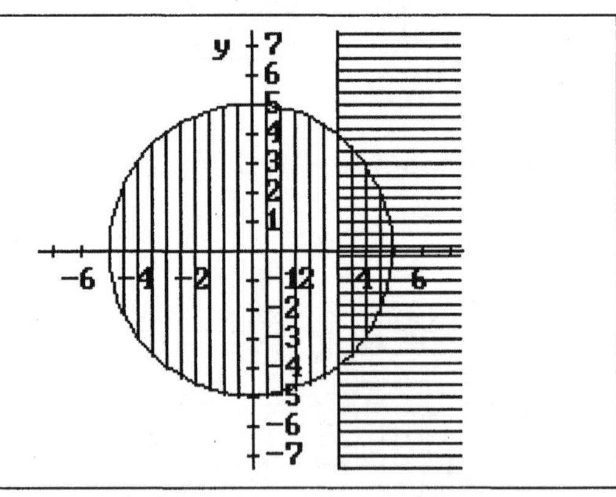

103. To maximize $P=12x+10y$, subject to : $x+y \leq 6$

$$2x+y \leq 10$$
$$x \geq 0$$
$$y \geq 0$$

The feasible set is the quadrilateral with vertices $(0,0)$, $(0,6)$, $(4,2)$, $(5,0)$. The maximum value is $P(4,2)=68$.

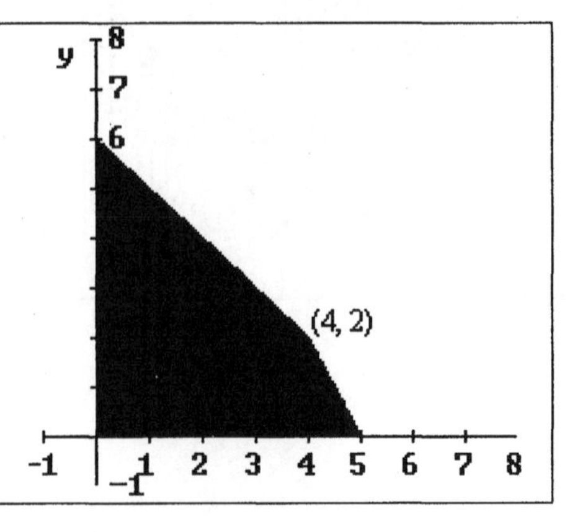

105. To minimize $P = 8x + 10y$, subject to: $2x + y \geq 10$

$$x + y \geq 7$$
$$2x + 3y \geq 16$$
$$x \geq 0$$
$$y \geq 0$$

The feasible set has vertices $(0,10)$, $(3,4)$, $(5,2)$, $(8,0)$. The minimum value is $P(5,2) = 60$.

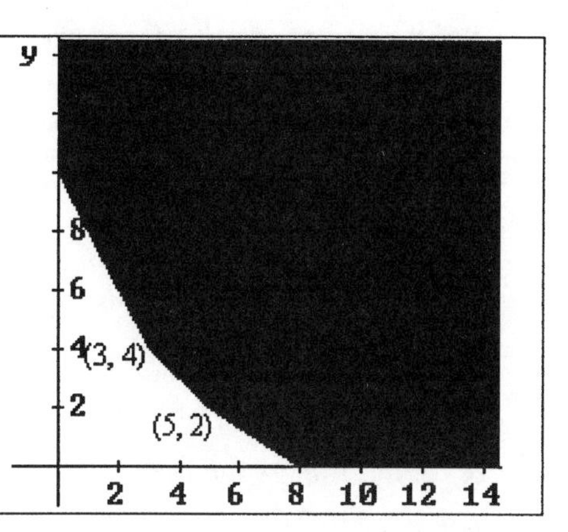

107. If on a right triangle one side is 3 inches less than twice the length of the other side, and the hypotenuse is 51 inches long, then we have the system:

$$y = 2x - 3$$
$$x^2 + y^2 = 51^2$$

The solution of this system is $x = 24,\ y = 48$. The triangle has sides 24, 48, 51.

109. \$100,000 are invested in three funds, a money market fund that pays 4% interest per year, charging an annual maintenance fee of 0.2%, a mutual fund averaging 12% per year with an annual load of 0.5%, and a savings account with no fees that pays 3% per year. In one year, the total interest was \$6550, and the total fees were \$255. Then, if x is the amount invested in the money market fund, y the amount in the mutual fund, and z the amount in the savings account,

$$x + y + z = 100{,}000$$
$$0.04x + 0.12y + 0.03z = 6550$$
$$0.002x + 0.005y = 255$$

The solution of this system is $x = \$40{,}000$ (money market fund), $y = \$35{,}000$ (mutual fund), and $z = \$25{,}000$ (savings account.)

111. The cubic function $f(x) = ax^3 + bx^2 + cx + d$ that passes through the points $(-1,5)$, $(1,5)$, $(2,10)$, and $(3,31)$ is $\frac{19}{2}x^3 - \frac{3}{2}x^2 - \frac{19}{2}x + \frac{13}{2}$. The y-intercept point is $\left(0, \frac{13}{2}\right)$.

113. A farmer intends to grow corn and hay, and has \$50,000 available to invest in the two crops, plus 1200 hours of labor available. The following table gives the amounts of capital and labor required per acre for each of the two crops, as well as the revenue per acre.

	Corn	Hay
Investment (dollars)	125	100
Labor	2	6
Revenue (dollars)	245	210

We want to maximize $R = 245x + 210y$

subject to: $125x + 100y \le 50{,}000$

$$2x + 6y \le 1200$$

$$x \ge 0\,,\, y \ge 0$$

The feasible set has vertices $(0,0)$, $(0{,}200)$, $\left(\frac{3600}{11}, \frac{1000}{11}\right)$, and $(400,0)$. The maximum revenue, that is the maximum value of $R = 245x + 210y$ in the feasible set is \$99,272.73 when $x \approx 327.3$ acres of corn, $y \approx 90.9$ acres of hay.

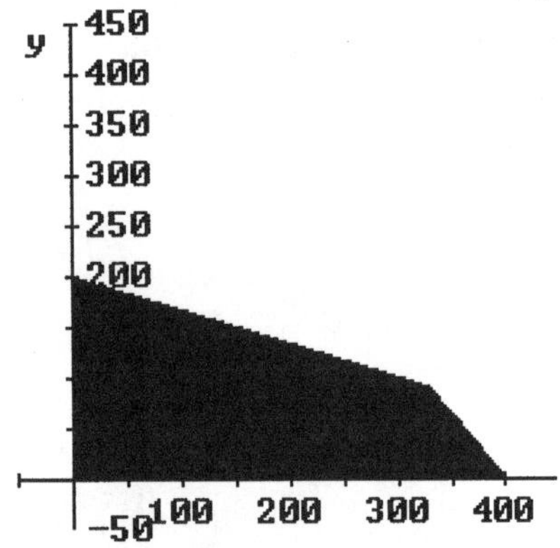

115. \$10,000 are available to invest in selected mutual funds and municipal bonds; at least twice as much should be invested in municipal bonds as it is invested in mutual funds, and at least \$2500 should be invested in mutual funds. The average return of the mutual funds portion is 12%, and the municipal bonds average a 10% return.

To maximize the earnings $P = 0.12x + 0.10y$ subject to $x + y \le 10{,}000$

$$y \ge 2x$$

$$x \ge 2500$$

where x is the amount in mutual funds and y the one in municipal bonds. The feasible set has vertices (2500,5000), (2500,7500), $(3333.33{,}6666.67)$.

The maximum earnings of \$1066.67 occur when $x \approx \$3333.33$, and $y \approx \$6666.67$,

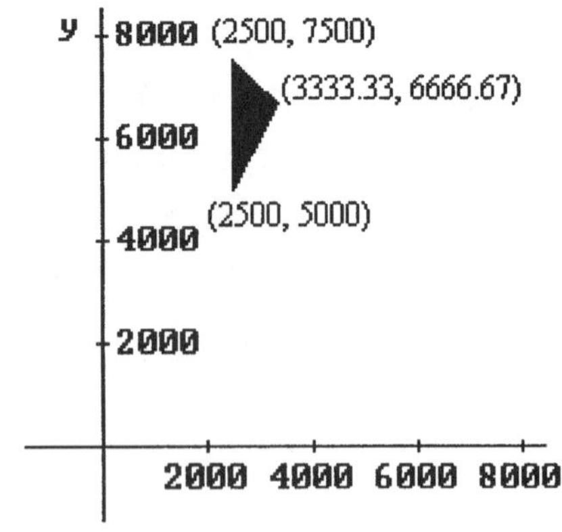

Chapter 9. Test

1. The statement: "Matrices must be of the same order if they are to be added" is true.

3. The statement: "All matrices have inverses" is false.

5. The statement: "All upper triangular square matrices with 1s on the main diagonal are in row-echelon form" is true. (For the matrix to be in reduced row-echelon form, we have the extra condition that the entries above each leading 1 in a row are zeros.)

7. The statement: "The solution set of a linear inequality is a half-plane" is true.

9. Answers may vary. An example of a matrix of order 2×3 is $\begin{bmatrix} 3 & -2 & 0 \\ 1 & 3 & 6 \end{bmatrix}$.

11. Answers may vary. An example of an augmented matrix in row-echelon form but not reduced row-echelon form is $\begin{bmatrix} 1 & -2 & 0 \\ 0 & 1 & 3 \end{bmatrix}$.

13. Answers may vary. An example of an elementary row operation is the interchanging of two rows.

15. Given the matrices

$$A = \begin{bmatrix} 2 & -1 & 0 \\ 3 & 4 & -2 \\ 0 & -3 & 1 \end{bmatrix} \text{ and } B = \begin{bmatrix} 0 & -2 & 5 \\ 6 & 3 & 1 \\ 4 & 0 & -1 \end{bmatrix}, \text{ then } \textbf{(a) } AB = \begin{bmatrix} -6 & -7 & 9 \\ 16 & 6 & 21 \\ -14 & -9 & 1 \end{bmatrix}$$

$$\textbf{(b) } 2A+B = \begin{bmatrix} 4 & -4 & 5 \\ 12 & 11 & -3 \\ 4 & -6 & 1 \end{bmatrix}, \text{ and } \textbf{(c) } A-B = \begin{bmatrix} 2 & 1 & -5 \\ -3 & 1 & -3 \\ -4 & -3 & 2 \end{bmatrix}$$

17. (a) The augmented matrix

$$\left[\begin{array}{ccc|c} 2 & -2 & 2 & 0 \\ 0 & 1 & -3 & -2 \\ 0 & 0 & 4 & 8 \end{array}\right]$$

is equivalent to the system of equations :

$$\begin{aligned} 2x-2y+2z &= 0 \\ y-3z &= -2 \\ 4z &= 8 \end{aligned}$$

(b) The reduced row-echelon form of this matrix is:

$$\left[\begin{array}{ccc|c} 1 & 0 & 0 & 2 \\ 0 & 1 & 0 & 4 \\ 0 & 0 & 1 & 2 \end{array}\right]$$

Chapter 10. INTEGER FUNCTIONS AND PROBABILITY

Section 10.1 Sequences

In Exercises 1-15, the n-th term of a sequence is given, and based on it the first four terms of the sequence are determined.

***n*-th term**	**First Four Terms of the Sequence**
1. $a_n = 3n+1$	$a_1 = 4,\ a_2 = 7,\ a_3 = 10,\ a_4 = 13$
3. $z_n = (-4)^n$	$z_1 = -4,\ z_2 = 16,\ z_3 = -64,\ z_4 = 256$
5. $r_n = \dfrac{1}{n^2}$	$r_1 = 1,\ r_2 = \dfrac{1}{4},\ r_3 = \dfrac{1}{9},\ r_4 = \dfrac{1}{16}$
7. $c_n = 1+(-1)^n$	$c_1 = 0,\ c_2 = 2,\ c_3 = 0,\ c_4 = 2$
9. $v_n = \dfrac{(-1)^{n-1}}{n}$	$v_1 = 1,\ v_2 = -\dfrac{1}{2},\ v_3 = \dfrac{1}{3},\ v_4 = -\dfrac{1}{4}$
11. $a_n = \sqrt{n(n+1)}$	$a_1 = \sqrt{2},\ a_2 = \sqrt{6},\ a_3 = 2\sqrt{3},\ a_4 = 2\sqrt{5}$
13. $b_n = \left(1+\dfrac{1}{n}\right)^n$	$b_1 = 2,\ b_2 = \dfrac{9}{4},\ b_3 = \dfrac{64}{27},\ b_4 = \dfrac{625}{256}$
15. $s_n = 1+2+\ldots+n$	$s_1 = 1,\ s_2 = 3,\ s_3 = 6,\ s_4 = 10$

In Exercises 17-23, a few terms of a sequence are given, and based on them an expression for the n-th term is determined. Answers may vary.

First Four Terms	***n*-th term**
17. $2,\ 4,\ 6,\ 8,\ldots$	$a_n = 2n$
19. $-\dfrac{1}{2},\ \dfrac{1}{4},\ -\dfrac{1}{8},\ \dfrac{1}{16},\ldots$	$a_n = \left(-\dfrac{1}{2}\right)^n$
21. $\dfrac{1}{2},\ \dfrac{2}{3},\ \dfrac{3}{4},\ \dfrac{4}{5},\ldots$	$a_n = \dfrac{n}{n+1}$
23. $1-\frac{1}{1},\ 1-\frac{1}{2},\ 1-\frac{1}{3},\ 1-\frac{1}{4},\ldots$	$a_n = 1-\frac{1}{n}$

25. The sequence $5,\ 8,\ 11,\ 14,\ldots$ is an arithmetic sequence, with common difference 3.

27. The sequence $3,\ 12,\ 60,\ 360,\ldots$ is neither arithmetic nor geometric.

29. The sequence $5,\ 1,\ -3,\ -7,\ldots$ is an arithmetic sequence, with common difference -4.

31. The sequence $\frac{1}{2}, \frac{3}{4}, \frac{7}{8}, \frac{15}{16}, \ldots$ is neither arithmetic nor geometric.

33. The sequence 0.2, 0.02, 0.002, 0.0002,… is a geometric sequence, with a common difference of 0.1.

In Exercises 35-39, a few terms of an arithmetic sequence are given, and based on them an expression for the n-th term is determined.

First Four Terms	***n*-th term**
35. 7, 13, 19, 25,…	$a_n = 7 + 6(n-1)$, or $a_n = 1 + 6n$
37. 100, 85, 70, 55,…	$a_n = 100 - 15(n-1)$, or $a_n = 115 - 15n$
39. $-2, -\frac{3}{4}, \frac{1}{2}, \frac{7}{4}, \ldots$	$a_n = -2 + \frac{5}{4}(n-1)$, or $a_n = -\frac{13}{4} + \frac{5}{4}n$

In Exercises 41-45, some information about an arithmetic sequence is provided, and based on it an indicated term is determined.

41. If $a_1 = 3,\ d = 8$, then $a_{10} = 3 + 9(8)$, that is $a_{10} = 75$.

43. If $a_1 = 10,\ a_5 = -10$, then $-10 = 10 + 4d$, therefore $d = -5$, and $a_{15} = 10 + 14(-5)$, that is $a_{15} = -60$.

45. If $a_8 = 38,\ a_{17} = 92$, then $a_1 + 7d = 38$, and $a_1 + 16d = 92$. Subtracting the first equation from the second, we obtain $9d = 54$, therefore $d = 6$, and $a_1 = 92 - 16(6)$, that is $a_1 = -4$. It follows that $a_{31} = -4 + 30(6)$, that is $a_{31} = 176$.

In Exercises 47-51, the first four terms of a geometric sequence are given, and based on them an expression for the n-th term is determined.

First Four Terms	***n*-th term**
47. 1, 3, 9, 27,…	$a_n = 3^{n-1}$
49. 32, −16, 8, −4,…	$a_n = 32\left(-\frac{1}{2}\right)^{n-1}$
51. $4, 5, \frac{25}{4}, \frac{125}{16}, \ldots$	$a_n = 4\left(\frac{5}{4}\right)^{n-1}$

In Exercises 53-55, some information about a geometric sequence is provided, and based on it an indicated term is determined.

53. If $a_1 = 2,\ r = 3,$ then $a_8 = 2 \cdot 3^7$, that is $a_8 = 4374.$

54. If $a_1 = 6,\ r = -\frac{1}{2},$ then $a_{10} = 6\left(-\frac{1}{2}\right)^9$, that is $a_{10} = -\frac{3}{256}.$

55. If $a_1 = 9,\ a_3 = 1,$ then $1 = 9r^2$, so $r^2 = \frac{1}{9}$. There are two possibilities, $r = \frac{1}{3}$ and $r = -\frac{1}{3}$.

In both cases, $a_7 = 9r^6$, so $a_7 = 9\left(\pm\frac{1}{3}\right)^6$, that is $a_7 = \frac{1}{81}$.

In Exercises 57-64, a recursively defined sequence is provided. The first five terms of the sequence are determined using the recursive definition.

57. If $a_1 = 2$ and $a_n = 1 - 2a_{n-1}$ for $n \geq 2$, then $a_1 = 2$, $a_2 = -3$, $a_3 = 7$, $a_4 = -13$, $a_5 = 27$.

59. If $a_1 = 1$ and $a_n = na_{n-1}$ for $n \geq 2$, then $a_1 = 1$, $a_2 = 2$, $a_3 = 6$, $a_4 = 24$, $a_5 = 120$.

61. If $a_1 = 2$ and $a_n = \sqrt{a_{n-1}}$ for $n \geq 2$, then $a_1 = 2$, $a_2 = \sqrt{2}$, $a_3 = \sqrt[4]{2}$, $a_4 = \sqrt[8]{2}$, $a_5 = \sqrt[16]{2}$.

63. If $a_1 = 1,\ a_2 = 2$, and $a_n = a_{n-1}a_{n-2}$ for $n \geq 3$, then $a_1 = 1$, $a_2 = 2$, $a_3 = 2$, $a_4 = 4$, $a_5 = 8$.

Applications

65. If Juanita's starting salary is \$28,500, and she will obtain a guaranteed \$1200 raise each year, plus an annual bonus equal to 3% of the base salary, then Juanita's potential income (salary plus annual bonus) the first three years is given by $a_1 = 28{,}500 + 0.03(28{,}500)$, $a_2 = 29{,}700 + 0.03(29{,}700)$, $a_3 = 30{,}900 + 0.03(30{,}900)$. That is, $a_1 = 29{,}355$, $a_2 = 30{,}591$, $a_3 = 31{,}827$. This forms an arithmetic sequence with first term 29,355 and common difference 1236. Then, the seventh year potential income will be $a_7 = 29{,}355 + 6(1236)$, that is $a_7 = 36{,}771$.

67. If the membership of an organization has increased 20% each year since it was formed with 10 charter members, at the beginning of the 12th year the membership would be modeled by $a_{12} = 10 \cdot 1.2^{12}$, or roughly 89 members.

69. For an account with an initial deposit of \$1000, in which interest is compounded annually, and the balances at the beginning of each of the first four years are \$1000.00, \$1055.00, \$1113.03, and \$1174.25, the ratios between consecutive balances are $\frac{1055}{1000}$, $\frac{1113.03}{1055}$, and $\frac{1174.25}{1113.03}$, or approximately constant, 1.05500 (rounded to five decimal places.)
A formula for the n-th term of this sequence is $a_n = 1000(1.05^{n-1})$. The balance of this account at the beginning of the 10th year is $a_{10} = 1000(1.05^9)$, that is $a_{10} = \$1551.33$.

71. A pair of baby rabbits is placed inside an enclosed area for breeding. During the first month, the pair does not produce any offspring, but each month thereafter it produces a pair (one male and one female) of rabbits. Each new pair follows the same reproductive pattern. The following table shows this reproductive pattern for the first 12 months.

Month	Pairs at the beginning of the month	Pairs born during the month	Pairs at the end of the month
1	1	0	1
2	1	1	2
3	2	1	3
4	3	2	5
5	5	3	8
6	8	5	13

At the end of 12 months, the number of pairs is 233 if we follow this pattern.

Concepts and Critical Thinking

73. The statement: "The sets $\{1,2,3\}$ and $\{3,2,1\}$ are identical" is true.

75. The statement: "The difference between consecutive terms of an arithmetic sequence is a constant" is true.

77. Answers may vary. An example of a recursively defined sequence is:
$a_1 = 1,\ a_2 = 1,\ a_n = a_{n-1} + a_{n-2}$ for $n \ge 3$ (the Fibonacci sequence).

79. Answers may vary. An example of a geometric sequence with common ratio 3 is
$a_n = 5 \cdot 3^{n-1}$ $\left(5,\ \frac{5}{3},\ \frac{5}{9},\ \frac{5}{27},\ldots\right)$

Chapter 10. Section 10.2 Series

In Exercises 1-3, the n-th term of a sequence is given, and the sum of the first five terms of the sequence is determined.

n-th term	Sum of the First Five Terms of the Sequence
1. $a_n = 4n+3$	$7+11+15+19+23=75$
3. $a_n = \frac{n}{n+1}$	$\frac{1}{2}+\frac{2}{3}+\frac{3}{4}+\frac{4}{5}+\frac{5}{6}=\frac{71}{20}$, or 3.55

5. $\sum_{k=1}^{6}(3k-2)=\frac{6}{2}(1+16)=51$

7. $\sum_{n=1}^{5}-3=-15$

9. $\sum_{j=1}^{4}2j^2=2+8+18+32=60$

11. $\sum_{k=3}^{6}\frac{1}{k}=\frac{1}{3}+\frac{1}{4}+\frac{1}{5}+\frac{1}{6}=\frac{19}{20}$

13. $\sum_{i=0}^{4}5\left(\frac{1}{2}\right)^i=\frac{5-5\left(\frac{1}{2}\right)^5}{1-\frac{1}{2}}=\frac{155}{16}$

15. $\frac{1}{1^3}+\frac{1}{2^3}+\frac{1}{3^3}+\frac{1}{4^3}+\frac{1}{5^3}+\frac{1}{6^3}=\sum_{k=1}^{6}\frac{1}{k^3}$

17. $[5(1)+2]+[5(2)+2]+[5(3)+2]+\ldots+[5(12)+2]=\sum_{k=1}^{12}[5(k)+2]$

19. $3+9+27+\ldots+2187=\sum_{k=1}^{7}3^k$

21. $\frac{1}{1}-\frac{1}{4}+\frac{1}{9}-\frac{1}{16}+\ldots-\frac{1}{256}=\sum_{k=1}^{16}\frac{(-1)^{k+1}}{k^2}$

23. $\frac{1}{2}+\frac{3}{4}+\frac{7}{8}+\ldots+\frac{63}{64}=\sum_{k=1}^{6}\frac{2^k-1}{2^k}$

25. The sum: $6+14+22+\ldots+158$ (20 terms) is equal to $\frac{20}{2}(6+158)$, that is 1640.

27. In the sum: $2+2.5+3+\ldots+12.5$, the first term is 2, the last term is 12.5, and the common difference between consecutive terms is $d=0.5$. Therefore, $2+(n-1)(0.5)=12.5$, and the number of terms being added is $n=22$. The sum of these 22 terms is equal to $\frac{22}{2}(2+12.5)$, that is 159.5.

29. If in an arithmetic sequence $a_1 = 12$, and $a_2 = 8$, the common difference is $d = -4$. Therefore, the sum of the first 15 terms of this sequence is:
$a_1 + a_2 + \ldots + a_{15} = 15(12) + \frac{15(14)}{2}(-4)$, that is $a_1 + a_2 + \ldots + a_{15} = -240$.

31. $\sum_{n=1}^{50}(3n-2) = \frac{50}{2}(1+148) = 3725$

33. $\sum_{k=1}^{30}(200-6k) = \frac{30}{2}(194+20) = 3210$

35. $\sum_{i=10}^{25}\frac{5i+3}{2} = \frac{16}{2}\left(\frac{53}{2}+\frac{128}{2}\right) = 724$

37. The sum of the first 10 terms of the geometric sequence 5, 15, 45,... is $\frac{5-5\cdot 3^{10}}{1-3}$, that is 147,620.

39. The sum of the first 15 terms of the geometric sequence whose first two terms are $a_1 = 3$, and $a_2 = 1$, is $a_1 + a_2 + \ldots + a_{15} = \frac{3-3\cdot\left(\frac{1}{3}\right)^{15}}{1-\frac{1}{3}}$, approximately 4.499999686.

41. The sum $4 - 0.4 + 0.04 + \ldots - 0.0000004$ has 8 terms, $a_1 = 4$ and $a_8 = -0.0000004$, that is $4r^7 = -0.0000004$. Therefore, $r^7 = -0.0000001$, from where we can see that $r = -0.1$.
Then, $4 - 0.4 + 0.04 + \ldots - 0.0000004 = \frac{4-4(-0.1)^8}{1-(-0.1)}$, that is
$4 - 0.4 + 0.04 + \ldots - 0.0000004 = 3.6363636$.

43. $\sum_{k=1}^{8} 3\left(4^k\right) = \frac{12-12(4)^8}{1-4} = 262{,}140$

45. $\sum_{n=0}^{9} 2\left(-\frac{1}{6}\right)^n = \frac{2-2\left(-\frac{1}{6}\right)^{10}}{1-\left(-\frac{1}{6}\right)} \approx 1.714286$

47. $\sum_{k=5}^{16} -3(0.2)^k = \frac{-0.00096-(-0.00096)(0.2)^{12}}{1-0.2} \approx -0.0012$

49. The sum of the first 1000 positive integers is $\frac{1000}{2}(1+1000) = 500{,}500$.

51. The sum of the first 100 positive odd integers is $\frac{100}{2}(1+199) = 10{,}000$.

Applications

In some of the following exercises, we use the alternative form of the sum of n terms of an arithmetic series with first term a_1 *and common difference d:* $S = na_1 + \frac{n(n-1)}{2}d$. *This formula holds since* $S = \frac{n}{2}\left(2a_1 + (n-1)d\right) = na_1 + \frac{n(n-1)}{2}d$.

53. For a pile of sewer pipes formed by 12 layers, with the first having 30 pipes, the second having 29, the third 28 and so on, the total number of pipes is equal to $12 \times 30 + \frac{12 \times 11}{2}(-1) = 294.$

55. If an auditorium has 25 rows of seats, with the first row having 12 seats, the second row having 14, the third row having 16, and so on, the total number of seats is $25 \times 12 + \frac{25 \times 24}{2}(2) = 900.$

57. If the expenditure for cell phone charges for the year n ($n = 0$ being 1990) in billions of dollars is given by $a_n = 4.85n + 5.86$, for $n = 0, \ldots, 9$, the total of all cell phone charges for 1990-1999 is $10 \times 5.86 + \frac{10 \times 9}{2}(4.85) = 276.85$ billion dlls.

59. If \$100 dollars are deposited at the beginning of each month in an account that pays 5% interest compounded monthly, then at the end of 4 years the balance in the account is:

$$100\left(1+\tfrac{0.05}{12}\right)^1 + 100\left(1+\tfrac{0.05}{12}\right)^2 + \ldots + 100\left(1+\tfrac{0.05}{12}\right)^{48} = \frac{100\left(1+\frac{0.05}{12}\right)\left(1-\left(1+\frac{0.05}{12}\right)^{48}\right)}{1-\left(1+\frac{0.05}{12}\right)}$$, or about

\$5323.58

61. If the annual internet-based advertising revenue for the years 1995-2000 is approximated by the formula $a_n = 84\left(\frac{8}{3}\right)^n$ million dollars, where n denotes the year, with $n = 0$ corresponding to 1995, then the total internet advertising revenue for the period 1995-2000 is

$$84 + 84\left(\tfrac{8}{3}\right) + \ldots + 84\left(\tfrac{8}{3}\right)^5 = \frac{84\left(1-\left(\frac{8}{3}\right)^6\right)}{1-\frac{8}{3}} \approx 18{,}073.14$$ million dollars (about 18 billion dollars.)

63. If a loan of \$2400 is paid with 24 monthly payments of \$100 plus 0.5% interest per month on the unpaid balance, the total interest paid is
$2400(0.005) + 2300(0.005) + 2200(0.005) + \ldots + 100(0.005)$
$= 0.005(100 + 200 + \ldots + 2400) = \$150.$

65. If the starting salary is \$1.6 million and the salary for each of the following 14 years is 10% more than the preceding year's, then the total contract value is

$$1.6 + 1.6(1.1) + 1.6(1.1)^2 + \ldots + 1.6(1.1)^{14} = \frac{1.6\left(1-1.1^{15}\right)}{1-1.1} \approx \$50.84$$ million.

Concepts and Critical Thinking

67. The statement: "$\sum_{j=1}^{5} j$ is equivalent to $5j$" is false.

69. The statement: "If successive terms in a series have a common ratio, then we say that the series is geometric" is true.

71. Answers may vary. An example of a finite arithmetic series is $2+5+8+\ldots+20$.

73. Answers may vary. An example of an arithmetic series with 5 terms that evaluates to 60 is $6+9+12+15+18$.

75. An example of a geometric series that evaluates to 0 and consists of 10 terms, none of which is zero, is $\sum_{k=1}^{10}(-1)^k$.

Chapter 10. Section 10.3 Permutations and Combinations

1. $6! = 720$

3. $\frac{10!}{7!} = 10 \cdot 9 \cdot 8 = 720$

5. ${}_6P_2 = 30$

7. ${}_5P_5 = 120$

9. ${}_6C_4 = 15$

11. $\binom{8}{1} = 8$

13. $\binom{5}{0} = 1$

15. ${}_8C_6 = 28$

Applications

17. The number of course combinations to satisfy the math and science requirement, if there are two options for a math course (College Algebra or Calculus I), two options for a physics course (Elementary Physics or University Physics I), and two options for a biology course (General Biology or Zoology) is $2 \cdot 2 \cdot 2 = 8$.

19. If the options offered by a computer score include three sizes of monitors (17″, 19″, or 21″), two styles of computer case (mini-tower or tower), and two types of keyboards (standard or cordless), the number of possible configurations is $3 \cdot 2 \cdot 2 = 12$.

21. The number of 3-digit numbers that can be formed using the digits $\{1,2,3,4,5\}$ is:

(a) $5 \cdot 5 \cdot 5 = 125$ if repetitions are allowed, and
(b) $5 \cdot 4 \cdot 3 = 60$ if repetitions are not allowed.

23. For license plates of the form NLLLDDD, where each N denotes one of the digits 1-4, each L denotes a letter, and each D is one of the digits 0-9, the number of possible license plates is $4 \cdot 26 \cdot 26 \cdot 26 \cdot 10 \cdot 10 \cdot 10 = 70{,}304{,}000$.

25. If a survey of TV viewer preferences asks to rank favorite 3 out 8 sitcoms, the number of possible survey responses is ${}_8P_3 = 8 \cdot 7 \cdot 6 = 336$.

27. If 9 starting players are chosen from a group of 13 assigning a batting order, the number of possibilities is:

(a) ${}_{13}P_9 = 259{,}459{,}200$ if there are no restrictions on the order of the selection

(b) $2\left({}_{11}P_7\right) = 2(1{,}663{,}200) = 3{,}326{,}400$ if the two children of the team's sponsor must bat first and second.

29. If a local election ballot allows each voter to select 3 out of 8 school board candidates, then the number of possible choices is ${}_8C_3 = 56$ (also denoted by $\binom{8}{3}$.)

31. If 6 first-string players will be chosen from a team of 12 girls, the number of possible groups of 6 members is ${}_{12}C_6 = 924$ (also denoted by $\binom{12}{6}$.)

33. If a student has four remaining slots to be filled with 10 available classes, each of which has a section open at those times, then the number of ways in which the slots can be filled is $_{10}C_4 = 210$ (also denoted as $\binom{10}{4}$.)

35. If a state lottery is played by selecting 5 different numbers from 1 through 40, the number of possible choices, assuming that the order of selection is not important, is $_{40}C_5 = 658{,}080$ (also denoted by $\binom{40}{5}$.)

37. If the dial on a combination lock is marked with the numbers 1 through 40, the number of different combinations that are possible is:

(a) $_{40}P_3 = 59{,}280$ if repetitions of digits are not allowed, and

(b) $40^3 = 64{,}000$ if repetitions of digits are allowed.

39. If on an exam a student is asked to select 4 out of 6 essay questions, the number in which this can be done is $_6C_4 = 15$. If, in addition, a student is asked to select 10 out of 15 short-answer questions, the number of ways in which the exam may be completed is: $\left({}_6C_4\right)\left({}_{15}C_{10}\right) = 15 \cdot 3003 = 45{,}045$.

41. The number of handshakes that is possible in a classroom of 30 students is the number of pairs that can be formed choosing students from this group, that is $_{30}C_2 = 435$. If each handshake takes 2 seconds and only 2 students can shake hands at a time, the time that it will take for all handshakes to be completed is 870 seconds, or 14.5 minutes.

43. If a 5-card hand is dealt from a standard 52-card deck, the number of possible hands with all 5 cards of the same suit (that is, a *flush*), is $\frac{4\binom{13}{5}}{\binom{52}{5}} = \frac{5148}{2{,}598{,}960} \approx 0.002$.

45. **(a)** The number of possible mosaics formed by 5000 cards, where no duplicates are allowed, is $_{5000}P_{5000} = 5000!$.

(b) The number of possible mosaics formed by 5000 cards, where duplicates are allowed, is 5000^{5000}.

Concepts and Critical Thinking

47. The statement: "$0! = 0$" is false.

49. The statement: "$31! - 13! = 18!$" is false. (A true statement would be $(31 - 13)! = 18!$.)

51. Answers may vary. A distinction between combinations and permutations is that for combinations, the order is immaterial, whereas for permutations, the order is important.

53. The only two numbers n for which $n! = n$ are $n = 1$, and $n = 2$.

Chapter 10. Section 10.4 Probability

1. For the experiment that consists of drawing a single ball from an urn containing red, blue, green, and white balls, the sample space S is: $S=\{R,B,G,W\}$, The number of outcomes of the experiment is $n(S)=4$. (*Note*: these four outcomes may not be equally likely.)

3. For the experiment of tossing a coin three times, the sample space is $S=\{HHH,HHT,HTH,HTT,THH,THT,TTH,TTT\}$, and $n(S)=8$.

In Exercises 5-7, a standard 6-sided die is rolled. The probability of the described event is determined. We are assuming that the die is a fair one (all outcomes are equally likely.)

5. For the event of rolling a 4, the probability is $\frac{1}{6}$.

7. The probability of rolling a number less than 7 is 1.

In Exercises 9-11, a ball is selected at random from an urn containing 3 red balls, 4 blue balls, and 5 white balls. The probability of the described event is determined in each case.

9. The probability of selecting a red ball is $\frac{3}{12}$, or $\frac{1}{4}$.

11. The probability of selecting a ball that is not white is $\frac{7}{12}$.

In Exercises 13-17, a single card is drawn from a standard 52-card deck. The probability of the described event is determined in each case.

13. The probability of drawing the ace of spades is $\frac{1}{52}$.

15. The probability of drawing an ace is $\frac{4}{52}$, or $\frac{1}{13}$.

17. The probability of drawing a red face card is $\frac{6}{52}$, or $\frac{3}{26}$.

In Exercises 19-21, it is assumed that the probability of a male birth is 0.5.

19. For a family with three children, the probability that all of them are of the same sex is $2\times\left(0.5^3\right)=0.25$. (Each of the outcomes *MMM* and *FFF* has probability 0.5^3.)

21. For a family with three children, the probability that exactly two are male is $\frac{3}{8}$. (The outcomes with exactly two male children are *MMF*, *MFM*, and *FMM*. The sample space has 8 possible outcomes, each equally likely with probability $\frac{1}{8}$.)

In Exercises 23-25, balls are selected (without replacement) from an urn containing 5 red balls and 4 blue balls.

23. The probability that when drawing two balls the two of them are red is $\frac{5}{9}\cdot\frac{4}{8}=\frac{5}{18}$.

25. The probability that when drawing two balls they are of different color is $1-P(\text{both balls are of the same color})=1-\frac{5}{9}\cdot\frac{4}{8}+\frac{4}{9}\cdot\frac{3}{8}=1-\frac{4}{9}=\frac{5}{9}$.

In Exercises 27-29, a pair of standard 6-sided dice are rolled.

27. The probability of rolling the same number on each die is $\frac{1}{6}$.

29. The probability of rolling a total of 10 is the probability of rolling (4,6), (6,4), or (5,5) is $\frac{3}{36} = \frac{1}{12}$.

In Exercises 31-33, a 5-card poker hand from a standard 52-card deck is dealt.

31. The probability of getting four aces is $\frac{12}{\binom{52}{5}} = \frac{48}{2,598,960} \approx 0.000018$.

33. The probability of getting all five cards of the same suit is $4\left(\frac{13}{52} \cdot \frac{12}{51} \cdot \frac{11}{50} \cdot \frac{10}{49} \cdot \frac{9}{48}\right) \approx 0.00198$.

In Exercises 35-39, a fair coin is tossed.

35. The probability of tossing 4 heads in a row is $0.5^4 = 0.0625$ (this can be seen in another way, by noting that there is only one outcome *HHHH* among 16 of them in the sample space that corresponds to the desired event.

37. The probability of obtaining the first head on the third toss is $\frac{1}{8}$, since this event has only one outcome, *TTH*, in a sample space of 8 outcomes.

39. The probability of tossing exactly 3 heads out of the first 5 is the probability of obtaining one of the 10 outcomes: *TTHHH, THTHH, THHTH, THHHT, HTHHT, HTHTH, HTTHH, HHHTT, HHTHT, HHTTH*, and the sample space consists of 32 outcomes. Then, the probability of this event is $\frac{10}{32} = \frac{5}{16}$.

In Exercises 41-43, an experiment is described, and two events E and Fare given. The probabilities $P(E)$ and $P(F)$ are determined, and the events $E \cap F, E \cup F, E'$, and F' are described, and their probabilities are computed.

41. A single card is drawn from a 52-card deck, E is the event that a red card is drawn and F is the event that a face card is drawn. $P(E) = \frac{26}{52} = \frac{1}{2}$ and $P(F) = \frac{12}{52} = \frac{3}{13}$; $E \cap F$ is the event that a red face card is drawn, and $P(E \cap F) = \frac{6}{52} = \frac{3}{26}$. $E \cup F$ is the event that a red card or a face card is drawn. The event $E \cup F$ consists of 26 red cards plus 6 face cards that are not red. Then, $P(E \cup F) = \frac{32}{52} = \frac{8}{13}$ (we can also compute this using the formula $P(E \cup F) = P(E) + P(F) - P(E \cap F)$, or $\frac{13}{26} + \frac{6}{26} - \frac{3}{26} = \frac{8}{13}$.) The event E' corresponds to a black card being drawn, and $P(E') = \frac{26}{52} = \frac{1}{2}$. The event F' corresponds to the event that the card drawn is not a face card; the probability of this event is $P(F') = \frac{40}{52} = \frac{10}{13}$.

43. Two 6-sided dice are rolled. E is the event that the sum is even, and F is the event that the sum is less than 8. $E=\left\{\begin{matrix}(1,1),(1,3),(1,5),(2,2),(2,4),(2,6),(3,1),(3,3),(3,5),\\(4,2),(4,4),(4,6),(5,1),(5,3),(5,5),(6,2),(6,4),(6,6)\end{matrix}\right\}$ and

$P(E)=\frac{18}{36}=\frac{1}{2}$; $F=\left\{\begin{matrix}(1,1),\ldots,(1,6),(2,1),\ldots,(2,5),(3,1),\ldots,(3,4),\\(3,1),\ldots,(3,4),(4,1),\ldots,(4,3),(5,1),(5,2),(6,1)\end{matrix}\right\}$ and $P(F)=\frac{21}{36}=\frac{7}{12}$.

$E\cap F=\{(1,1),(1,3),(1,5),(2,2),(2,4),(3,1),(3,3),(4,2),(5,1)\}$ and $P(E\cap F)=\frac{9}{36}=\frac{1}{4}$.

$E\cup F$ is the event that the sum is even or less than 8, and its probability is $P(E\cup F)=P(E)+P(F)-P(E\cap F)=\frac{1}{2}+\frac{7}{12}-\frac{1}{4}=\frac{5}{6}$. E' is the event that the sum is odd, and $P(E')=1-P(E)=\frac{1}{2}$. F' is the event that the sum is larger or equal than 8, and $P(F')=1-P(F)=1-\frac{7}{12}=\frac{5}{12}$.

Applications

45. If at the beginning of the NCAA basketball tournament it has been determined that the probability that Duke will win is 0.2 and the probability that Michigan will win is 0.15, then the probability that one of these two teams will win the championship is 0.35.

47. If a state lottery is designed so that a player selects 5 different numbers from 1 to 40, the probability that a single choice of 5 numbers will win the lottery is $\frac{1}{\binom{40}{5}}=\frac{1}{658{,}008}\approx 0.0000015$.

49. If the distribution of high school grades of the college freshman cohort in the U.S in 2000 was A- to A+ (42.9%), B- To B+ (50.5%) and C+ or lower (6.6%), and a student in this cohort is selected at random,
(a) the probability that the student had a C+ or lower as a high school average is 0.066
(b) the probability that the student did not have an A- to A+ average is 0.571..

51. If a polygraph test is 90% accurate in the sense that 90% of all employees who steal will fail the test, and 90% of the employees who are not stealing will pass the test, plus it is known that 5% of the employees are in fact stealing, the probability that someone who fails the test is stealing is obtained as follows. The probability that an employee fails the test is $0.9(0.05)+0.10(0.95)=0.045+0.095=0.14$ (or 14% of the employees.) This aggregate of 14% is composed of the 4.5% corresponding to employees who steal and fail the test (that is, probability of $0.9(0.05)$ and the 9.5% corresponding to employees who do not steal who fail the test (probability of $0.10(0.95)=0.095$.) The probability of the event that someone who fails the test is stealing is $\frac{0.045}{0.14}=0.32$.)

53. Using the information from the U.S. census of 2000, approximately 12% of the population lives in California. Also, 32.4% of all Californians are Hispanic. Nationwide, 12.5% of the population is Hispanic. If an American is selected at random:

(a) the probability that the person is a Californian or Hispanic is:
$P(C \cup H) = P(C) + P(H) - P(C \cap H) = 0.12 + 0.125 - (0.324)(0.12) \approx 0.21$

(b) the probability that the person is a Californian Hispanic is $= 0.324(0.12) \approx 0.0388$

(c) the probability that the person is a non-Californian is 0.88

(d) the probability that the person is a Californian but is not Hispanic is:
$(1-0.324)(0.12) \approx 0.08$

(e) the probability that the person is Hispanic but is not a Californian is:
$\approx 0.125 - 0.0388 \approx 0.0862$

Concepts and Critical Thinking

55. The statement: "In general, for events E and F, $P(E \cup F) = P(E) + P(F)$" is false.

57. The statement: "If E and F are mutually exclusive events, then $P(E \cup F) = P(E) + P(F)$" is true.

59. Answers may vary. An example of a real-world event for which the probability is $\frac{1}{7}$ is the event of choosing one person at random whose birthday this year falls on a Sunday.

61. Answers may vary. An example of two events E and F such that $P(E \cup F) = P(E) + P(F)$ are: when drawing a single card from a standard 52-card deck, E is the event of drawing an ace, and F the event of drawing a 3.

Chapter 10. Section 10.5 The Binomial Theorem

1. The coefficient of AB^3 in the expansion of $(A+B)^4$ is $\binom{4}{3}=4$.

3. The coefficient of x^3y^2 in the expansion of $(x+y)^5$ is $\binom{5}{3}=10$.

5. The coefficient of z^5 in the expansion of $(z+1)^6$ is $\binom{6}{1}=6$.

7. The coefficient of xy^2 in the expansion of $(2x-3y)^3$ is $\binom{3}{2}(2)(-3)^2=54$.

9. The constant term in the expansion of $\left(x+\frac{1}{x}\right)^4$ is $\binom{4}{2}=6$.

11. The coefficient of w^4 in the expansion of $\left(1+w^2\right)^5$ is $\binom{5}{2}=10$.

(The third term of this expansion is $\binom{5}{2}1^3\left(w^2\right)^2=10w^4$.)

13. $(x+y)^5=x^5+5x^4y+10x^3y^2+10x^2y^3+5xy^4+y^5$

15. $(2x-y)^4=16x^4-32x^3y+24x^2y^2-8xy^3+y^4$

17. $(3x+5y)^3=27x^3+135x^2y+225xy^2+125y^3$

19. $(x+1)^5=x^5+5x^4+10x^3+10x^2+5x+1$

21. $(10w-1)^6=1{,}000{,}000w^6-600{,}000w^5+150{,}000w^4-20{,}000w^3+1500w^2-60w+1$

23. $\left(1+\sqrt{2}\right)^3=7+5\sqrt{2}$

25. $\left(x^2-1\right)^3=x^6-3x^4+3x^2-1$

27. $\left(x+\frac{1}{x}\right)^5=x^5+5x^3+10x+\frac{10}{x}+\frac{5}{x^3}+\frac{1}{x^5}$

29. $(1+i)^3=-2+2i$

31. $(2+2i)^4=-64$

33. $\left(\frac{\sqrt{2}}{2}+\frac{\sqrt{2}}{2}i\right)^2=i$

35. To verify that $n!\cdot n=(n+1)!-n!$, we not that the right hand side can be expressed as $(n+1)!-n!=n!(n+1)-n!$; therefore, $(n+1)!-n!=n!((n+1)-1)=n!\cdot n$.

37. $\frac{(n+1)!}{n!}=\frac{n!(n+1)}{n!}=n+1$, so $\frac{(n+1)!}{n!}=n+1$ is an identity.

39. $\binom{n-1}{k}+\binom{n-1}{k-1}=\frac{(n-1)!}{k!(n-1-k)!}+\frac{(n-1)!}{(k-1)!((n-1)-(k-1))!}=\frac{(n-1)!(n-k)!}{k!(n-1-k)!(n-k)!}+\frac{k(n-1)!}{k(k-1)!((n-1)-(k-1))!}$

$=\frac{(n-1)!(n-k)}{k!(n-k)!}+\frac{k(n-1)!}{k!(n-k)!}=\frac{(n-1)!(n-k)+k(n-1)!}{k!(n-k)!}=\frac{(n-1)!(n-k+k)}{k!(n-k)!}=\frac{n!}{k!(n-k)!}=\binom{n}{k}$

41. To verify that $\binom{n}{r}=\frac{n}{n-r}\binom{n-1}{r}$ is an identity, we examine the right hand side:

$\frac{n}{n-r}\binom{n-1}{r}=\frac{n}{n-r}\frac{(n-1)!}{r!(n-1-r)!}=\frac{n!}{r!(n-r)!}=\binom{n}{r}$.

Concepts and Critical Thinking

43. The statement: "The symbols $\binom{n}{r}$ and ${}_nP_r$ have the same meaning" is false.

45. The statement: "All binomial coefficients can be found within Pascal's triangle" is true.

47. Answers may vary. A context in which binomial coefficients arise (other than in expanding powers of binomials) is when determining the number of subgroups of a predetermined size of a given collection of items (where the order of the chosen elements is not important.)

49. A method for evaluating binomials that does not involve computation of factorials is to determine the binomial coefficients using Pascal's triangle.

51. Answers may vary. A possible reason why the error of simplifying $(x+y)^n$ as x^n+y^n is that the expression $(x+y)^n$ is very similar in appearance to the expression $(xy)^n$, for which it is true that $(xy)^n=x^ny^n$. For $(x+y)^n$ to be equal to x^n+y^n for all *n*, one of the values *x* or *y* has to be 0.

53. In order to compute 11^7, we can proceed as follows:

$$
\begin{aligned}
11^7 &= (10+1)^7 \\
&= \binom{7}{0}10^7 + \binom{7}{1}10^6 + \binom{7}{2}10^5 + \binom{7}{3}10^4 + \binom{7}{4}10^3 + \binom{7}{5}10^2 + \binom{7}{6}10 + \binom{7}{7}1 \\
&= 10{,}000{,}000 + 7{,}000{,}000 + 2{,}100{,}000 + 350{,}000 + 35{,}000 + 2100 + 70 + 1 \\
&= 19{,}487{,}171.
\end{aligned}
$$

This technique can be used somewhat successfully for powers of $11, \ldots 15$. For example:

$$
\begin{aligned}
12^5 &= (10+2)^5 \\
&= \binom{5}{0}10^5 + \binom{5}{1}10^4 \cdot 2 + \binom{5}{2}10^3 \cdot 2^2 + \binom{5}{3}10^2 \cdot 2^3 + \binom{5}{4}10 \cdot 2^4 + \binom{5}{5} \cdot 2^5 \\
&= 100{,}000 + 100{,}000 + 40{,}000 + 8000 + 800 + 32 \\
&= 248{,}832.
\end{aligned}
$$

Also, we can use this technique for powers of 101, 102, 103, 104, 105. The technique would be useful mainly for powers of numbers close to powers of 10, like 11, 12, 101, 102, 1001, 1002,... but not very useful in other cases.

Chapter 10. Section 10.6 Mathematical Induction

1. To prove the statement: $1+3+5+7+9+\ldots+(2n+1)=(n+1)^2$ by induction:
if $n=1$, the statement is $(2(0)+1)^2=(0+1)^2$, which is true; assuming that the statement is true for some natural number $k\geq 1$, that is $1+3+5+7+9+\ldots+(2k+1)=(k+1)^2$, we now show that the statement for $n=k+1$ is true:
$1+3+5+7+9+\ldots+(2(k+1)+1)=(1+3+5+7+9+\ldots+(2k+1))+(2(k+1)+1)$
$=(k+1)^2+(2k+3)=k^2+4k+4=(k+2)^2=((k+1)+1)^2$
By induction, $1+3+5+7+9+\ldots+(2n+1)=(n+1)^2$ is true for all natural numbers n.

3. To prove the statement: $1^2+2^2+3^2+\ldots+n^2=\frac{n(n+1)(2n+1)}{6}$ by induction:
if $n=1$, the statement is $1^2=\frac{1(1+1)(2(1)+1)}{6}$, which is true; assuming that the statement is true for some natural number $k\geq 1$, that is $1^2+2^2+3^2+\ldots+k^2=\frac{k(k+1)(2k+1)}{6}$, we now show that the statement for $n=k+1$ is true:
$1^2+2^2+3^2+\ldots+(k+1)^2=(1^2+2^2+3^2+\ldots+k^2)+(k+1)^2=\frac{k(k+1)(2k+1)}{6}+(k+1)^2$
$=\frac{k(k+1)(2k+1)+6(k+1)^2}{6}=\frac{(k+1)(2k^2+7k+6)}{6}=\frac{(k+1)(k+2)(2k+3)}{6}=\frac{(k+1)((k+1)+1)(2(k+1)+1)}{6}$
By induction, $1^2+2^2+3^2+\ldots+n^2=\frac{n(n+1)(2n+1)}{6}$ is true for all natural numbers n.

5. To prove the statement: $5+10+15+\ldots+(5n)=\frac{5n(n+1)}{2}$ for all natural numbers n by induction:
if $n=1$, the statement is $5(1)=\frac{5(1)(1+1)}{2}$, which is true; assuming that the statement is true for some natural number $k\geq 1$, that is $5+10+15+\ldots+(5k)=\frac{5k(k+1)}{2}$, we now show that the statement for $n=k+1$ is true: $5+10+15+\ldots+(5(k+1))=\underbrace{5+10+15+\ldots+(5k)}_{\frac{5k(k+1)}{2}}+(5(k+1))$
$=\frac{5k(k+1)}{2}+5(k+1)=\frac{5k(k+1)+10(k+1)}{2}=\frac{5(k+1)(k+2)}{2}$.
By induction, the statement $5+10+15+\ldots+(5n)=\frac{5n(n+1)}{2}$ is true for all natural numbers.

7. To prove the statement: $1+2+2^2+2^3\ldots+2^n=2^{n+1}-1$ by induction:
if $n=1$, the statement is $1+2=2^{1+1}-1$, which is true; assuming that the statement is true for some $k\geq 1$, that is $1+2+2^2+2^3\ldots+2^k=2^{k+1}-1$, we show that the statement for $n=k+1$ is true: $1+2+2^2+2^3\ldots+2^{k+1}=\underbrace{1+2+2^2+2^3\ldots+2^k}_{=2^{k+1}-1}+2^{k+1}$
$=(2^{k+1}-1)+2^{k+1}=2(2^{k+1})-1=2^{k+2}-1$. So, $1+2+2^2+2^3\ldots+2^n=2^{n+1}-1$ for all $n\geq 1$.

9. To prove the statement: $1-2+2^2-2^3+\ldots-2^{2n-1}+2^{2n}=\frac{1+2^{n+1}}{3}$ by induction:

if $n=1$, the statement is $1-2+2^2=\frac{1+2^{2(1)+1}}{3}$, which is true since both sides of the equal sign are equal to 3; assuming that the statement is true for some natural number $k\geq 1$, that is $1-2+2^2-2^3+\ldots-2^{2k-1}+2^{2k}=\frac{1+2^{2k+1}}{3}$, we now prove the statement for $n=k+1$:

$$1-2+2^2-2^3+\ldots-2^{2k+1}+2^{2k+2}=\underbrace{1-2+2^2-2^3+\ldots-2^{2k-1}+2^{2k}}_{\frac{1+2^{2k+1}}{3}}-2^{2k+1}+2^{2k+2}$$

$$=\frac{1+2^{2k+1}}{3}-2^{2k+1}+2^{2k+2}=\frac{1+2^{2k+1}-3\cdot 2^{2k+1}+3\cdot 2^{2k+2}}{3}$$

$$=\frac{1+2^{2k+1}-3\cdot 2^{2k+1}+3\cdot 2\cdot 2^{2k+1}}{3}=\frac{1+4\cdot 2^{2k+1}}{3}=\frac{1+2^{2(k+1)+1}}{3}$$

By induction, the $1-2+2^2-2^3+\ldots-2^{2n-1}+2^{2n}=\frac{1+2^{n+1}}{3}$ is true for all natural numbers n.

11. To prove the statement: $\frac{1}{1\cdot 2}+\frac{1}{2\cdot 3}+\frac{1}{3\cdot 4}+\ldots+\frac{1}{n\cdot(n+1)}=1-\frac{1}{n+1}$ for all natural numbers n:

if $n=1$, the statement is $\frac{1}{1\cdot 2}=1-\frac{1}{1\cdot(1+1)}$, which is true; assuming that the statement is true for some natural number $k\geq 1$, that is $\frac{1}{1\cdot 2}+\frac{1}{2\cdot 3}+\frac{1}{3\cdot 4}+\ldots+\frac{1}{k\cdot(k+1)}=1-\frac{1}{k+1}$, we prove the statement for $n=k+1$: $\frac{1}{1\cdot 2}+\frac{1}{2\cdot 3}+\frac{1}{3\cdot 4}+\ldots+\frac{1}{(k+1)\cdot(k+2)}=\left(\frac{1}{1\cdot 2}+\frac{1}{2\cdot 3}+\frac{1}{3\cdot 4}+\ldots+\frac{1}{k\cdot(k+1)}\right)+\frac{1}{(k+1)\cdot(k+2)}$

$$=1-\frac{1}{k+1}+\frac{1}{(k+1)\cdot(k+2)}=1-\left(\frac{1}{k+1}-\frac{1}{(k+1)\cdot(k+2)}\right)=1-\left(\frac{(k+2)-1}{(k+1)\cdot(k+2)}\right)=1-\frac{1}{(k+2)}=1-\frac{1}{(k+1)+1}$$

By induction, $\frac{1}{1\cdot 2}+\frac{1}{2\cdot 3}+\frac{1}{3\cdot 4}+\ldots+\frac{1}{n\cdot(n+1)}=1-\frac{1}{n+1}$ is true for all natural numbers n.

13. To prove the statement: "$1\cdot 2+3\cdot 4+5\cdot 6+\ldots+(2n-1)\cdot 2n=\frac{n(n+1)(4n-1)}{3}$ for all natural n":

if $n=1$, the statement is $(2(1)-1)\cdot 2(1)=\frac{1(1+1)(4(1)-1)}{3}$, which is true; assuming the statement is true for some $k\geq 1$, $1\cdot 2+3\cdot 4+5\cdot 6+\ldots+(2k-1)\cdot 2k=\frac{k(k+1)(4k-1)}{3}$, we have:

$$1\cdot 2+3\cdot 4+5\cdot 6+\ldots+(2(k+1)-1)\cdot 2(k+1)$$

$$=(1\cdot 2+3\cdot 4+5\cdot 6+\ldots+(2k-1)\cdot 2k)+(2(k+1)-1)\cdot 2(k+1)$$

$$=\frac{k(k+1)(4k-1)}{3}+(2k+1)(2k+2)=\frac{(k+1)(4k^2-k)}{3}+\frac{6(2k+1)(k+1)}{3}=\frac{(k+1)(4k^2+11k+6)}{3}$$

$=\frac{(k+1)(k+2)(4k+3)}{3}=\frac{(k+1)((k+1)+1)(4(k+1)-1)}{3}$. By induction, the statement is true for all $n\geq 1$.

15. To prove the statement: "$1\cdot 2^1+2\cdot 2^2+3\cdot 2^3+\ldots+n\cdot 2^n=2^{n+1}(n-1)+2$ for all natural n": if $n=1$, the statement is $1\cdot 2^1=2^{1+1}(1-1)+2$, which is true; assuming that the statement is true for some natural number $k\geq 1$, that is $1\cdot 2^1+2\cdot 2^2+3\cdot 2^3+\ldots+k\cdot 2^k=2^{k+1}(k-1)+2$,
$1\cdot 2^1+2\cdot 2^2+3\cdot 2^3+\ldots+(k+1)\cdot 2^{k+1}=\left(1\cdot 2^1+2\cdot 2^2+3\cdot 2^3+\ldots+k\cdot 2^k\right)+(k+1)\cdot 2^{k+1}$
$=\left(2^{k+1}(k-1)+2\right)+(k+1)\cdot 2^{k+1}=2^{k+1}\cdot 2k+2=2^{(k+1)+1}\cdot\left((k+1)-1\right)+2$
By induction, the statement is true for all natural numbers n.

17. To prove the statement: "n^2+n is even for all natural numbers n": if $n=1$, the statement is "1^2+1 is even", which is of course true; assuming that k^2+k is even for some natural number $k\geq 1$, then $(k+1)^2+(k+1)=\left(k^2+2k+1\right)+(k+1)=\underbrace{\left(k^2+k\right)}_{\text{even}}+\underbrace{(2k+2)}_{\text{even}}$ is, clearly, even. By induction, the statement is true for all natural numbers n.

19. To prove the statement: "n^3-n is divisible by 3 for all natural numbers n": if $n=1$, the statement is "1^3-1 is divisible by 3", which is of course true; assuming that k^3-k is divisible by 3, for some natural number $k\geq 1$, say $k^3-k=3p$, then:
$(k+1)^3-(k+1)=\left(k^3+3k^2+3k+1\right)-k-1=\left(k^3-k\right)+3k^2+3k=3p+3k^2+3k$
$=3(p+k^2+k)$. By induction, the statement is true for all natural numbers n.

21. To prove the statement: "$n\leq 2^{n-1}$ for all natural numbers n": if $n=1$, the statement is "$1\leq 2^{1-1}$", which is, of course, true; assuming that $k\leq 2^{k-1}$ for some natural number $k\geq 1$, then $k+1\leq 2^{k-1}+1<2^{k-1}+2^{k-1}=2\cdot 2^{k-1}=2^k$, that is $k+1\leq 2^{(k+1)-1}$. By induction, the statement is true for all natural numbers n

23. To prove that if $0\leq a<1$, then $a^n<1$, for all natural numbers n, by induction: if $n=1$, the statement is $a<1$, which is true from the initial assumptions on a; assuming that $a^k<1$ for some $k\geq 1$, we can multiply both sides of the inequality $a<1$ by the nonnegative number a^k, to obtain $a^{k+1}<a^k$, and since we are assuming that $a^k<1$, we conclude that $a^{k+1}<1$. The statement: : "if$0\leq a<1$, then $a^n<1$, for all natural numbers" is true.

25. To prove the statement: $2^n<n!$, for $n\geq 4$, by induction: if $n=4$, the statement is $2^4<4!$, which is true, since the left-hand side is equal to 16, and the right-hand side is equal to 24; assuming that the statement is true for some natural number $k\geq 4$, that is $2^k<k!$, we show that the statement for $n=k+1$ is also true: $2^{k+1}=\underbrace{2^k}_{<k!}\cdot\underbrace{2}_{<k+1}$, then $2^{k+1}<k!(k+1)$, that is $2^{k+1}<(k+1)!$. By induction, the statement $2^n<n!$ is true for all natural numbers $n\geq 4$.

27. To prove the statement: $n+12 \le n^2$, for $n \ge 4$, by induction: if $n=4$, the statement is $4+12 \le 4^2$, which is true, since the left-hand side is equal to 16, and the right-hand side is also equal to 16; assuming that the statement is true for some natural number $k \ge 4$, that is $k+12 \le k^2$, we prove the statement for $n=k+1$: $(k+1)+12 = (k+12)+1 \le k^2+1$, and since $k^2+1 < (k+1)^2$, we have $(k+1)+12 \le (k+1)^2$. Then, $n+12 \le n^2$, for $n \ge 4$.

29. To prove the statement: $n^2+4 < (n+1)^2$, for $n \ge 2$, by induction: if $n=2$, the statement is $2^2+4 < (2+1)^2$, which is true, since the left-hand side evaluates to 8, and the right-hand side to 9; assuming that the statement is true for some natural number $k \ge 2$, that is $k^2+4 < (k+1)^2$, we prove the statement for $n=k+1$:

$(k+1)^2+4 = (k^2+4)+2k+1 < (k+1)^2+2k+1 = k^2+4k+2 < k^2+4k+2 = (k+2)^2$

By induction, the statement $n^2+4 < (n+1)^2$ is true for all natural numbers $n \ge 2$.

Exercises 31-33 deal with the Fibonacci sequence, $F_1 = 1$, $F_2 = 1$, and $F_n = F_{n-1} + F_{n-2}$ for $n > 2$.

31. To prove the statement: "$F_1 + F_2 + \ldots + F_n = F_{n+2} - 1$ for all natural numbers n" by induction: if $n=1$, the statement is $F_1 = F_{1+2} - 1$, which is true, since $F_1 = 1$ and $F_3 = 2$; asumming that the statement is true for some natural number $k \ge 1$, $F_1 + F_2 + \ldots + F_k = F_{k+2} - 1$, we have

$F_1 + F_2 + \ldots + F_{k+1} = (F_1 + F_2 + \ldots + F_k) + F_{k+1}$

$= (F_{k+2} - 1) + F_{k+1} = (F_{k+2} + F_{k+1}) - 1 = F_{k+3} - 1 = F_{(k+1)+2} - 1$

By induction, $F_1 + F_2 + \ldots + F_n = F_{n+2} - 1$ for all natural numbers n.

33. To prove the statement: "$F_1 + F_3 + \ldots + F_{2n-1} = F_{2n}$ for all natural numbers n" by induction: if $n=1$, the statement is $F_{2(1)-1} = F_{2(1)}$, which is true, since $F_1 = 1$ and $F_2 = 1$; asumming that the statement is true for some natural number $k \ge 1$, $F_1 + F_3 + \ldots + F_{2k-1} = F_{2k}$,

$F_1 + F_3 + \ldots + F_{2(k+1)-1} = (F_1 + F_3 + \ldots + F_{2k-1}) + F_{2(k+1)-1} = F_{2k} + F_{2k+1} = F_{2k+2} = F_{2(k+1)}$

By induction, $F_1 + F_3 + \ldots + F_{2n-1} = F_{2n}$ for all natural numbers n.

Concepts and Critical Thinking

35. The statement: "When proving statements by mathematical induction, we actually assume that what we are trying to prove is true" is false.

37. The statement: "If we prove that a statement is true for $n=7$ and we prove that if it is true for $n=k$ then it is also true for $n=k+1$, then we have proven that the statement is true for all positive integers greater than or equal to 7" is true.

39. Answers may vary. A statement about the natural numbers that is true for $n = 1$ and $n = 2$ but is false for $n = 3$ is: $n^2 < 5$.

41. An alternative method for establishing that $n^2 + n$ is even for all natural numbers, is to note that $n^2 + n = n(n+1)$. If n is even, that is $n = 2a$ for some natural number a, then $n^2 + n = 2a(2a+1)$, so $n^2 + n$ is even. If n is odd, then $n = 2a + 1$ for some natural number a, and $n^2 + n = (2a+1)(2a+2) = 2(2a+1)(a+1)$, so $n^2 + n$ is even in this case as well.

Chapter 10. Review

1. The first five terms of the sequence with general term $u_n = 1 - 5n$ are: $u_1 = -4$, $u_2 = -9$, $u_3 = -14$, $u_4 = -19$, and $u_5 = -24$. This sequence is arithmetic, with common difference $d = -5$.

3. The first five terms of the sequence with general term $c_n = ne^{-n}$ are: $c_1 = \frac{1}{e}$, $c_2 = \frac{2}{e^2}$, $c_3 = \frac{3}{e^3}$, $c_4 = \frac{4}{e^4}$, and $c_5 = \frac{5}{e^5}$. This sequence is neither arithmetic nor geometric.

5. The first five terms of the sequence with general term $x_n = \frac{1}{2} + (-1)^n$ are: $x_1 = -\frac{1}{2}$, $x_2 = \frac{3}{2}$, $x_3 = -\frac{1}{2}$, $x_4 = \frac{3}{2}$, and $x_5 = -\frac{1}{2}$. This sequence is neither arithmetic nor geometric.

7. If $a_1 = 2$ and $a_n = 3a_{n-1}$ for $n \geq 2$, the first five terms of the sequence are: $a_1 = 2$, $a_2 = 6$, $a_3 = 18$, $a_4 = 54$, and $a_5 = 162$. This sequence is geometric, with common ratio $r = 3$.

9. The sequence 2, 10, 50, 250,... has general term $a_n = 2 \cdot 5^{n-1}$. It is geometric, with common ratio $r = 5$.

11. The sequence 1, $\frac{1}{2}$, $\frac{1}{3}$, $\frac{1}{4}$,... has general term $a_n = \frac{1}{n}$. It is neither arithmetic nor geometric.

13. The sequence 1, $-\frac{3}{2}$, -4, $-\frac{13}{2}$,... has general term $a_n = \frac{7}{2} - \frac{5}{2}n$. It is arithmetic with common difference $d = -\frac{5}{2}$.

15. The sequence 1, -8, 27, -64,... has general term $a_n = (-1)^{n-1} n^3$. It is neither arithmetic nor geometric.

17. If a sequence is arithmetic, with $a_1 = 3$ and common difference $d = 4$, then $a_{12} = 3 + 11 \cdot 4$, that is $a_{12} = 47$.

19. If a sequence is geometric, with $a_1 = 1$, $a_2 = 4$, then $r = \frac{a_2}{a_1}$, that is $r = 4$ and the general term of the sequence is $a_n = 4^{n-1}$. Then $a_6 = 4^5$, that is $a_6 = 1024$.

21. If $a_1 = 1$, $a_{n+1} = (a_n + 1)^2$, then $a_2 = 4$, $a_3 = 25$, and $a_4 = 676$.

23. $\sum_{k=1}^{4} (2k + 5) = \dfrac{4(7 + 13)}{2} = 40$

25. $\sum_{n=2}^{6} \frac{1}{n^2} = \frac{1}{4} + \frac{1}{9} + \frac{1}{16} + \frac{1}{25} + \frac{1}{36} = \frac{1769}{3600}$

27. $[5(1) - 2] + [5(2) - 2] + [5(3) - 2] + \ldots + [5(11) - 2] = \sum_{k=1}^{11} [5(k) - 2]$

29. $4\left(\frac{2}{3}\right)^3+4\left(\frac{2}{3}\right)^4+4\left(\frac{2}{3}\right)^5+4\left(\frac{2}{3}\right)^6+\ldots+4\left(\frac{2}{3}\right)^{13}=\sum_{n=3}^{13}4\left(\frac{2}{3}\right)^n$

31. The sum of the first 10 terms of the arithmetic sequence 2, 8, 14, 20,… is: $\frac{10(2+9(6))}{2}=290$.

33. $10+\frac{10}{5}+\frac{10}{5^2}+\ldots+\frac{10}{5^6}=\frac{10-\frac{10}{5^7}}{1-\frac{1}{5}}=\frac{39,062}{3125}$.

35. $\sum_{n=1}^{9}5(0.3)^n=\frac{5(0.3)-5(0.3)^{10}}{1-0.3}$, that is $\sum_{n=1}^{9}5(0.3)^n=2.142814965$.

37. $\sum_{k=3}^{10}(1000-9k)=\frac{8((1000-27)+(1000-90))}{2}=4(973+910)=7532$

39. ${}_9C_2=36$

41. ${}_8P_5=6720$

43. $\frac{5!}{3!(5-3)!}=10$

45. $\binom{10}{3}=120$

In Exercises 47-49, a single card is drawn from a standard 52-card deck.

47. The probability of drawing a diamond is $\frac{1}{4}$.

49. The probability of drawing a diamond or a face card is $\frac{22}{52}=\frac{11}{26}$.

In Exercises 51-53, a pair of standard 6-sided dice are rolled.

51. The probability of rolling an even number on both dice is $\frac{9}{36}=\frac{1}{4}$.

53. The probability of rolling a total of 5 is $\frac{4}{36}=\frac{1}{9}$.

Exercise 55 refers to a family of 3 children, where the probability of a male birth is 0.5.

55. The probability that all 3 children are boys is $(0.5)^3=0.125$ (or $\frac{1}{8}$.)

Exercises 57-59 use the Defense Personnel distribution of 1990: Air Force (26.2%), Army (35.9%), Marines (9.5%) and Navy (28.4%).

57. The probability that a member of the armed services selected at random is in the Army is 0.359.

59. The probability that a randomly selected member of the armed services is in the Marines or Navy is $0.095+0.284=0.379$.

61. The coefficient of x^5y^3 in the expansion of $(x+y)^8$ is $\binom{8}{3}=56$.

63. The coefficient of ab^3 in the expansion of $(2a-5b)^4$ is $\binom{4}{3}(2)(-5)^3=-1000$.

65. $(x-y)^4=x^4-4x^3y+6x^2y^2-4xy^3+y^4$

67. $\left(r+\frac{1}{s}\right)^3=r^3+\frac{3r^2}{s}+\frac{3r}{s^2}+\frac{1}{s^3}$

69. $\left(\sqrt{z}+2\right)^5=z^{5/2}+10z^2+40z^{3/2}+80z+80\sqrt{z}+32$

71. To prove the statement "$1+4+7+\ldots+(3n-2)=\frac{n(3n-1)}{2}$ for all natural numbers" :
if $n=1$, the statement is $1=\frac{1(3(1)-1)}{2}$, which is clearly true; assuming the statement true for some natural number k, that is $1+4+7+\ldots+(3k-2)=\frac{k(3k-1)}{2}$, we have

$$1+4+7+\ldots+(3(k+1)-2)=\left[1+4+7+\ldots+(3k-2)\right]+\left(3(k+1)-2\right)=\frac{k(3k-1)}{2}+(3k+1)$$

$$=\frac{k(3k-1)+6k+2}{2}=\frac{3k^2+5k+2}{2}=\frac{(k+1)(3k+2)}{2}=\frac{(k+1)\left(3(k+1)-1\right)}{2}.$$

By induction, $1+4+7+\ldots+(3n-2)=\frac{n(3n-1)}{2}$ for all natural numbers.

73. To prove the statement "$1^3+3^3+5^3+\ldots+(2n-1)^3=n^2\left(2n^2-1\right)$ for all natural numbers" :
if $n=1$, the statement is $(2(1)-1)^3=1^2\left(2(1)^2-1\right)$, which is clearly true; assuming the statement true for some natural number k, that is $1^3+3^3+5^3+\ldots+(2k-1)^3=k^2\left(2k^2-1\right)$, we have $1^3+3^3+5^3+\ldots+\left(2(k+1)-1\right)^3=\left[1^3+3^3+5^3+\ldots+(2k-1)^3\right]+\left(2(k+1)-1\right)^3$

$$=k^2\left(2k^2-1\right)+(2k+1)^3=2k^4+8k^3+11k^2+6k+1$$

$$=\left(k^2+2k+1\right)\left(2k^2+4k+1\right)=(k+1)^2\left(2\left(k^2+2k+1\right)-1\right)=(k+1)^2\left(2(k+1)^2-1\right)$$

By induction, $1^3+3^3+5^3+\ldots+(2n-1)^3=n^2\left(2n^2-1\right)$ for all natural numbers.

75. To prove the statement "$2^{2n}-1$ is divisible by 3 for all natural numbers" :
if $n=1$, the statement is $2^{2(1)}-1$ is divisible by 3, which is clearly true; assuming the statement true for some natural number k, that is $2^{2k}-1=3a$, we have

$$2^{2(k+1)}-1=2^{2k+2}-1=4\left(2^{2k}\right)-1=4\left(2^{2k}-1\right)+(4-1)=4(3a)+3=3(4a+1).$$

By induction, $2^{2n}-1$ is divisible by 3 for all natural numbers.

77. The number of possible choices of president, vice president, secretary, and treasurer, selected from a club of 20 members, is ${}_{20}P_4 = 116,280$ (here the order is important.)

79. If Jermaine has enough money for 5 CDs, and he wants to choose them from his wish list of 6 rock groups and 4 rap groups, then:
(a) the number of possible combinations for the 5 CDs, with no restrictions, is ${}_{10}C_5 = 252$,
(b) the number of possible combinations for the 5 CDs, if 3 are to be of rock groups and 2 of rap groups, is $({}_6C_3)({}_4C_2) = 20 \cdot 6 = 120.$

81. If an exam has a true-false portion that consists of 10 questions, and no question is to be left blank, then the number of possible ways to answer this portion of the exam is $2^{10} = 1024$.

83. If the population of a city is increasing at a rate of 5% per year, and it is 200,000 in the year 2000, then the population in the year n (where $n = 1$ corresponds to 2001) is given by $P(n) = 200,000(1.05)^n$. The sequence is geometric with common ratio 1.05 (the ratio between consecutive years is 1.05.) Then, the population in the next 3 years is 210,000 in 2001, 220,500 in 2002, and 231,525 in 2003. In 2012, the population will be 359,171, if the growth rate stays at 5% per year.

85. If a job offer consists of a starting salary of \$26,000, with guaranteed annual raises of 6%, then the salary for the n-th year is $S(n) = 26,000(1.06)^{n-1}$. The total income for the first 10 years is $\frac{26000-26000(1.06)^{10}}{1-1.06} = \$342,700.67$.

87. **(a)** If a final exam will be composed of 20 problems chosen from a total of 80 problems that appeared on exams during the semester, the number of possible final exams is ${}_{80}C_{20}$.
(b) If a final exam will be composed of 20 problems, with 5 problems from each of 4 exams that consisted of 20 problems, the number of possible final exams is $({}_{20}C_5)^4$.

89. The number of ways in which 5 digits can be arranged is ${}_5P_5 = 120.$

Chapter 10. Test

1. The statement: "No sequence is both arithmetic and geometric" is false. For example, the sequence $5,5,5,\ldots$ is arithmetic with common difference $d=0$, and it is also geometric with common ratio $r=1$.

3. The statement: "A permutation takes order into account, whereas a combination does not" is true.

5. The statement: "If *E* is an event in a sample space *S*, then $0<P(E)<1$" is false. The values 0 and 1 are possible for events in a sample space *S*.

7. The statement: "Each term in the expansion of $(x+y)^n$ has degree $n+1$" is false. A true statement would be "Each term in the expansion of $(x+y)^n$ has degree n."

9. Answers may vary. An example of a formula for an arithmetic sequence with common difference -4 is $a_n=10-4n$.

11. Answers may vary. An example of a series that is neither arithmetic nor geometric is $\sum_{n=1}^{20}\frac{1}{n^2}$.

13. Answers may vary. An example of an event that has probability $\frac{1}{3}$ is the event of obtaining an number less than or equal to 2 when rolling a fair die.

15. The first five terms of the sequence $a_n=3n^2+1$ are: $a_1=4,\ a_2=13,\ a_3=28,$ $a_4=49,\ a_5=76.$

17. $\sum_{k=1}^{7}(5k-2)=\frac{7(3+33)}{2}=126.$

19. The number of ways in which the first 8 problems of this exam, which are true-false, if no question is to be left blank, is $2^8=128$. The probability of answering them all correctly, if they are answered at random, is $\frac{1}{128}\approx 0.008$.

21. If a pair of standard 6-sided dice are rolled, the probability that the sum is a multiple of 3 is $\frac{12}{36}=\frac{1}{3}$, since the corresponding event is:

$E=\{(1,2),(2,1),(5,1),(4,2),(3,3),(2,4),(1,5),(6,3),(5,4),(4,5),(3,6),(6,6)\}$, $n(E)=12$.

23. The coefficient of z^4 in the expansion of $(z+5)^9$ is $\binom{9}{5}\cdot 5^5=393{,}750$.

25. To prove that $1+3+5+\ldots+(2n-1)=n^2$ for all natural numbers *n*, by induction: if $n=1$, the statement is $2(1)-1=1^2$, which is true; assuming that $1+3+5+\ldots+(2k-1)=k^2$ for $k\geq 1$, then, $1+3+5+\ldots+(2(k+1)-1)=(1+3+5+\ldots+(2k-1))+(2(k+1)-1)$

$$=k^2+(2k+1)=(k+1)^2$$

By induction, $1+3+5+\ldots+(2n-1)=n^2$ for all natural numbers *n*.